Plants, Genes, and Agriculture

Plants, Genes, and Agriculture

Maarten J. Chrispeels
University of California, San Diego

David E. Sadava
The Claremont Colleges
Claremont, California

With Foreword by Jeff Schell, Director,
Max Planck Institut für Züchtungsforschung,
Köln, Germany

Jones and Bartlett Publishers

Boston   London

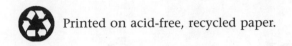 Printed on acid-free, recycled paper.

Editorial, Sales, and Customer Service Offices
Jones and Bartlett Publishers
One Exeter Plaza
Boston, MA 02116
1-800-832-0034
617-859-3900

Jones and Bartlett Publishers International
P.O. Box 1498
London W6 7RS
England

Library of Congress Cataloging-in-Publication Data
Chrispeels, Maarten J., 1938-
    Plants, genes, and agriculture / Maarten J. Chrispeels, David E. Sadava.
      p.   cm.
    Includes bibliographical references and index.
    ISBN 0-86720-871-6
    1. Crops—Genetic engineering.   2. Plant breeding.   3. Crop improvement.   4. Food supply.   I. Sadava, David E.   II. Title.
SB123.57.C48   1994
363.8—dc20
                                                   93-45677
                                                     CIP

Acquisitions Editor: Arthur C. Bartlett
Production Editor: Mary Cervantes
Manufacturing Buyer: Dana Cerrito
Design: The Book Company/Wendy Calmenson
Editorial Production Service: The Book Company
Illustrations: John and Judy Waller
Typesetting: Weimer Graphics, Inc.
Cover Design: Hannus Design Associates
Printing and Binding: Courier Corporation
Cover Printing: John P. Pow Company
Cover Photograph: Jack Dykinga, courtesy of the United States Department of
                           Agriculture

Printed in the United States of America

98 97 96 95 94   10 9 8 7 6 5 4 3 2 1

Brief Contents

Contents

Chapter 11   The Green Revolution and Beyond   298

Foreword

The major challenges facing the world in the next four decades—assuming that we can avoid a global nuclear war—are to feed and provide shelter for a world population that is likely to reach 9–10 billion by the year 2040. In addition, we must protect human health and ensure social and economic conditions that are conducive to peace and the fulfillment of the human potential. Agriculture will play a major role in achieving this agenda. We look to the farmers of our world to provide the food we need, and we expect them to farm in a manner compatible with maintaining the earth's essential natural resources. In the future, agriculture will likely be called on not only to produce food, but to deliver an ever-increasing array of natural products needed by industry for the manufacture of a variety of products. Therefore, the challenge to agriculture in the next few decades is to achieve maximum production of food and other products without further irreversible depletion or destruction of our natural environment. Agriculture must become an integral part of a sustainable global society.

In this book, the problem is presented in a clear, objective, and didactic way. The authors discuss the positive role that can, and must be played by molecular plant breeding. This new branch of crop improvement is made possible by the recent quantum leap in our ability to transfer genes to plants from unrelated organisms, and our understanding of how plant genes function and confer specific desirable traits. The general tone of the book, while realistic, is nevertheless cautiously optimistic. The authors argue that doubling food production will be difficult but not impossible if efficient use is made of scientific knowledge and resulting technologies, and if these technologies do not contribute to further degradation of environmental resources. The authors correctly point out that doubling food production will not by itself satisfy the humanitarian goal of eliminating hunger—the persistence of hunger in the world is the result of poverty and sociopolitical forces.

The authors make a convincing case for the need to use plant biotechnology to address this major challenge to humanity. They provide the reader with the necessary technical and scientific information needed to both understand and evaluate the potential of this new technology. The rapid scientific advances that have taken place in the past 25 years have increased the gap between what scientists know and what most people understand. This book allows the educated layperson to bridge that scientific literacy gap. By explaining what genes are, how they are transferred from plant to plant, both in traditional and molecular plant breeding, and how genes confer important agricultural traits, the book should help its readers evaluate the claims of those who believe that this technology is potentially dangerous, as well as the claims of those who believe that this technology has great beneficial potential. The authors do not claim that plant biotechnology will solve all our agricultural problems, and they point out that the contributions it will make to the sustainability of crop production will depend as much on society at large as on specific technological breakthroughs. From the facts, which are presented in a logical manner, readers can reach their own conclusions.

By discussing agricultural applications of gene transfer technology in the light of fundamental principles of plant growth and development, photosynthesis, plant nutrition, breeding, and molecular biology, the book achieves a special character and should be of great value to all students interested in understanding how fundamental plant biology (formerly called botany) relates to crop production and plant biotechnology. It discusses in relatively simple terms the applications of molecular plant breeding to protecting plants from pests and diseases, to producing more nutritious seeds, to optimizing plant growth in normal as well as stressful environments, to producing valuable chemicals by tissue culture, and to harvesting specialty starches or biodegradable plastics used by industry. A short but important historical context is provided, along with an up-to-date presentation of the major economic developments, using examples from industry and public-sector crop improvement projects. Molecular plant breeding is described as one more step in the long evolution of our crop improvement practices. The authors also discuss some of the major questions that can be raised with regard to biosafety and social factors connected with the widespread use of transgenic plants.

Throughout, this book stresses the notion that to achieve a sustainable "green" agriculture for the benefit of all men and women, whether living in developed, developing, or poor countries, we will need to combine molecular crop improvement techniques with some traditional agricultural management practices, such as crop rotation and biological pest management, that were discarded in our age of high-input crop production.

This book is a solid and clear introduction to a new technology that is developing with great speed, and to the role it will play in achieving food security for our world.

Jeff Schell, Director
Max Planck Institut für Züchtungsforschung
Köln, Germany

Preface

The relationship that humans have with their food sources has been changing in the last 10,000 years, and the pace of this change has recently speeded up. This book is about plants, genes, food, and agriculture, and about the changing relationships among them. Intensive management of wild food plants, coupled with intentional planting and cultivation, led to the beginnings of agriculture between 10,000 and 20,000 years ago. This basic change in culture allowed humans to exist on earth in far greater numbers than would have been otherwise possible. The "invention" of plant husbandry set in motion an ever-accelerating cycle of increased population and increased agricultural production. Each improvement in agriculture made it possible to feed more people, and each increase in human population demanded more food production.

The purposes of this book are to show how agriculture has evolved and how it is changing now, and to discuss the role that genes and genetic engineering play in these changes. This book has three underlying themes: (1) that modern farming has become a sophisticated scientific enterprise, manipulating the relationship between plants and their environments; (2) that scientists who genetically manipulate plants can help solve some of the problems faced by agriculture in seeking to raise crop production; and (3) that agriculture must be carried out in a sustainable manner, ensuring the basis of food production for future human generations.

This book grew out of our desire to teach plant biology in an agricultural context and to bridge the gap between basic and applied science. Basic botany textbooks seldom mention agriculture, let alone human nutrition. Agricultural textbooks describe plants and agricultural practices, but are often short on basic science. We also wanted to introduce our students to the ever-changing relationship between people and their food sources and to the notion that genetic engineering is one more step in that historic process.

This book could be used in existing introductory courses in plant biology, agriculture or economic botany, whether aimed at science majors or non-majors. It is our hope that the text will allow instructors to create new courses that deal with the scientific and societal issues that are so important to all of us: our relation to our food supply, both locally and globally, and the role of biotechnology in this relationship. This is the economic botany of the future.

Humans are by far the most numerous and ubiquitous large animal that has ever lived on earth. By the end of this century, there will be six billion people, and the human population is still expected to double after that, reaching 10 to 12 billion sometime between 2050 and 2100. The purpose of agriculture has always been to produce enough food to feed all the people. That will require a doubling of present crop production capacities in the next 50 years. Doubling food production will be difficult, but not impossible. For example, between 1980 and 1990 agricultural production in China rose 8% yearly, so that it is now the biggest food producer in the world. Wheat production in Mexico increased fourfold between 1950 and 1985, and threefold in India between 1965 and 1983. Such examples show that dramatic increases in food production are possible. They result mostly from applying modern technologies to plant husbandry.

Success in feeding the human species has had unintended and undesirable consequences. First, by using ever more land for food (and fiber) production, people are greatly diminishing the diversity of plant and animal species that inhabit the earth. Second, intensified crop production has degraded the environment in developed as well as in developing countries. Soil erosion, groundwater pollution, desertification, and exhaustion of water reserves are but a few examples of the environmental costs. If people do not improve the efficiency of crop production, they are likely to destroy the very resource base on which this production depends, as they seek to double the amount of food that the earth produces. People need to "invent" a new and sustainable agriculture that relies less, not more, on herbicides, pesticides, inorganic fertilizers, and surface irrigation. Farming practices must be modified so that food can be produced in stable ecosystems that preserve the land resources on which food production depends.

However, increases in food production by themselves will not abolish hunger. Agricultural science cannot achieve this humanitarian goal of feeding everyone. In the past four decades, world population has doubled and so has food production. Yet hunger persists, and the hungry increase their numbers while world markets are glutted with grain and economists struggle with depressed prices. The continued persistence of hunger in the world is a sociopolitical problem that agricultural scientists cannot solve. Many studies have documented that hunger persists because of poverty, not because of insufficient food availability. The social systems that govern so many aspects of human affairs are unable and/or unwilling to eliminate it.

As people increase in numbers and as their expectations rise, there are additional pressures on the land. In many developing countries, newfound wealth demands greater meat production, and the conversion of plant food into animal products (milk, cheese, meat) is an inefficient process that consumes precious resources and creates its own pollution problems. Developed countries are also using more agricultural land to raise products for industry. Vegetable oils and starches are being converted to a variety of industrial products. In the future, crop plants (and by extrapolation, agricultural resources) will be used to produce specialty oils, biodegradable plastics, vaccines, and pharmaceuticals. Therefore, the role of agriculture in the twenty-first century will be to produce enough food to feed all the people, to produce industrial products, and to do so in a manner that does less harm to the environment than is presently the case.

Plant biotechnology and specifically genetic engineering of crops will make major contributions toward reaching these goals, just as plant breeding was of crucial importance to increased food crop production in the twentieth century. Indeed, increasing food production will depend heavily on the continued improvement of the 20 crops that now provide nearly all humanity's food, and genetic engineering as a supplement to traditional plant breeding will play a major role in crop improvement. Genetic engineering will also permit the synthesis by plants and harvesting on a large scale of many new products that will find industrial applications.

The underlying premise of this book is that science can be used to solve some of the problems that exist in the realm of agricultural production. Science has been a powerful agent of change in the world, and some people see science as the solution to all problems. "Every technological problem has a technological fix" is their guiding principle. Others see science and technology as the enemy of true humanity. For example, some gastronomists have gone on record as opposing the use of foods that have been improved by genetic engineering. This could lead to their rejecting foods that have not been sprayed with pesticides in favor of those that have been treated with such chemicals.

Only a small proportion of people understand much about science. Even in countries with a high level of education, probably only one person in 1,000 understands how transgenic plants are created or how a television works. Yet in a democracy everyone participates in the decisions that lead to the use of the technologies. If nonscientists are to make decisions about the application of genetic engineering, then they must know more about it and must understand and accept both potential risks and benefits.

Genetic engineering—the precise transfer of genes from one species to another species—is being used to create new strains of crops that allow farmers to grow more food or do it in a way that taxes the environment less. To improve crops, plant breeders have been transferring genes between plants for more than a hundred years. The new techniques of molecular biology simply allow gene transfer to be more precise than it ever was before. Thus, genetic engineering is just one of the tools that will help create a new agriculture, an agriculture that is finely attuned to all of humanity's food needs. Creating such an agriculture will require more than a revolutionary new technology; it will also require a new ethic that stresses our common responsibility to provide food for all people now and in the future.

In this book, Chapter 1 deals with the problem of population, because bringing population growth under control is the first order of business. If those efforts fail, so will efforts to produce enough food in a sustainable manner. In addition, it discusses the issue of poverty and its relation to hunger and inadequate nutrition in developed and developing countries. Chapter 2 tackles the issues of agricultural production and productivity from a historical perspective. It describes the importance of climate and human-induced climate change and discusses the issue of sustainability. Chapter 3 provides an overview of plant biotechnology, starting with simple technologies such as hydroponics and ending with complex ones such as gene transfer. These first three chapters are meant to familiarize you with the problems and the contribution that biotechnology can make to the solution. A more detailed description comes later.

If you are to raise food production, you must know what food is and how its constituents serve your nutritional needs. That is the subject of Chapter 4. This chapter also discusses diets and the issue of food safety and introduces the idea that plants contain hundreds of natural chemicals whose effect on the human body remains unknown.

An intensive short course in plant growth and development is provided in Chapter 5. We relate the function of specific organs (leaves, tubers, seeds) to their physiology and their food value. How does the function and the physiology of an organ such as a potato relate to the fact that it is a major source of starch? How do the cells of such storage organs differ from other cells?

Chapter 6 discusses the role of photosynthesis in plant growth and the efficiency of this process, as well as the transfer of energy in the ecosystem from one organism to the next. The role of animal agriculture, with its relatively inefficient use of resources, is described in Chapter 6. The source of mineral nutrients that plants need and the importance of the soil ecosystem are the subjects of Chapters 7 and 8. These chapters introduce the ideas that nutrients are recycled, that cereals and legumes have different ways of acquiring nitrogen, and that the soil is an incredibly complex ecosystem about which people know little. Farming practices can upset this normally stable ecosystem, and sustainable agriculture requires sound agronomic practices that keep soil-borne diseases at bay.

Plant breeding and genetic engineering have the same molecular basis. Chapter 9 explains how traits are inherited and genes are transmitted from one generation to the next. We treat the central dogma of molecular biology—"DNA makes RNA makes protein"—as an information transfer problem and explain the molecular basis of recombinant DNA technology. The mechanics of plant transformation are discussed later.

Chapters 10 and 11 deal with the evolution of agriculture as a human enterprise, from the domestication of wheat, rice, and corn thousands of years ago, to the impact of genetics and modern technology in the Green Revolution. The positive and negative impacts of the Green Revolution are discussed. More food is being produced but soil erosion is increasing, landraces are disappearing, and the Green Revolution has had a profound and often negative impact on women in the Third World.

Genetic engineering will have a major effect on the way people protect plants against predators and pests: genetically engineered plants that are resistant to pests will replace the use of chemical pesticides. To explain the perennial war between crops and their pests, Chapter 12 is devoted to the problems caused by weeds, bacteria, fungi, nematodes, and insects. Chapter 13 discusses past, present, and future pest control methods, including those that contribute to a sustainable agriculture. Chapter 13 also discusses genetically engineered herbicide-tolerant plants, virus-resistant potatoes, insect-resistant tomatoes, and crops resistant to bacteria and fungi.

Chapter 14 deals with plant and cell tissue culture and the contributions they could make to producing high-value chemicals. A substantive discussion of the methods, goals, and achievements of genetic engineering is provided in Chapter 15. The methods used to introduce novel genes into plants are discussed first; then we deal with the objectives: improved pest management, agronomic properties, nutritional quality, improvements that affect post-harvest quality, and molecular farming. The chapter ends with a discussion of social issues and the concerns of scientists and citizens.

In the final chapter, we return to the problem of sustainability and some of the nontechnical issues facing agriculture. What is sustainable agriculture, and how does it differ from organic farming? What is the role of the political system in pushing agriculture toward sustainability or away from it? How does the world attain food security? What type of research is needed to make food production truly sustainable? How do people achieve a truly "green" agriculture? These are some of the questions raised in Chapter 16.

We are grateful to our colleagues from around the world, who supplied us not only with data and photographs, but valuable feedback on the text. We especially thank Jeff Schell for his perceptive foreword and Dr. Brian Holl of the University of British Columbia, who provided extensive feedback on the manuscript. A special thanks goes to Lyn Alkan for her valuable assistance in producing the early versions of this book that were tried out in classes at the University of California at San Diego. We also thank the staff at Jones and Bartlett (Art Bartlett, publisher, and Mary Cervantes, production editor) and Wendy Calmenson at The Book Company, who made the sometimes arduous task of turning manuscript into book almost pleasurable.

Plants, Genes,
and Agriculture

Chapter 1

Human Population Growth

Lessons from Demography

People have always been concerned with the balance between their population and food supply. As this century began, this relationship was, as ever, a precarious one, but things seemed to be in reasonable balance. Although crop failures because of uncertain weather had caused periodic famines, the rise of modern science and its application to agriculture, and the farming of new and productive lands, promised to even out the rough edges of food production. Food security seemed assured.

By mid-century, this assurance appeared tenuous at best. The same sciences that had improved agricultural production also improved medical care. People were living longer, and, especially in the less developed (poor) regions of the world, they were having more children, who survived to have children themselves. The population growth rate reached its highest level in human history. Many doubted if there could ever be enough food to feed the teeming millions. Disaster scenarios, using terms such as "population bomb" and "population crisis," dominated discussions of the problem.

Now, as the twentieth century comes to a close, the growth rate of the human population is slowing down. The spectacular gains in food production capacity, which allowed it to just keep up with the population growth, now have the potential to improve the situation of many of the world's poorest people. Yet famines persist, especially in Africa, and grinding poverty and hunger prevail in many of the world's great cities. The food–people dilemma is still with us.

1 | **The past 50 years have been characterized by extremely rapid population growth.**

Human beings have been on earth for a very long time—over 2 million years, according to anthropologists. During this time, their numbers slowly increased. About 2,000 years ago, there were 300 million people, a little more than now live in the United States. In a few years, as we enter the next century, there will be 6.1 billion of us, more than 20 times as many as a

Figure 1.1

The Children of Asia. By the year 2000, the earth's population is projected to surpass 6 billion, and it will continue to rise after that, possibly doubling again in the next century. This rise in the population will create huge cities and put ever-increasing pressures on rural areas to produce sufficient food. Source: United Nations.

short 2,000 years ago (Figure 1.1). We say "short," because 2,000 years is only 0.1% of human history. What happened to population growth? Why this sudden increase? Populations grow when their rate of increase (due to births and, for a region, to immigration) is greater than their rate of decrease (due to deaths and, for a region, to emigration). For humans as a whole, the ancient way of life of hunting game and gathering plant foods probably led to moderately high birth and death rates. This resulted in a slow growth rate, and the population never exceeded 10 million. Then, about 12,000 years ago, agriculture gradually replaced hunting and gathering. With a more reliable food supply and more settled existence, the birth rate increased greatly, and the death rate decreased somewhat. By A.D. 1, the human population had grown to 300 million, and by the year 1650 to about 500 million.

The rise of modern science and technology during the past three centuries, especially during the industrial revolution in Europe, has had a number of effects on population growth, principally by reducing the death rate:

1 Food production has become industrialized and dependent on technology. Improved transportation has given people who are distant from food sources a reliable supply.

2 Rising incomes have allowed more people to afford the available food.

3 Improved housing and public health have reduced the incidence of infectious diseases carried by rats, insects, water, and so on.

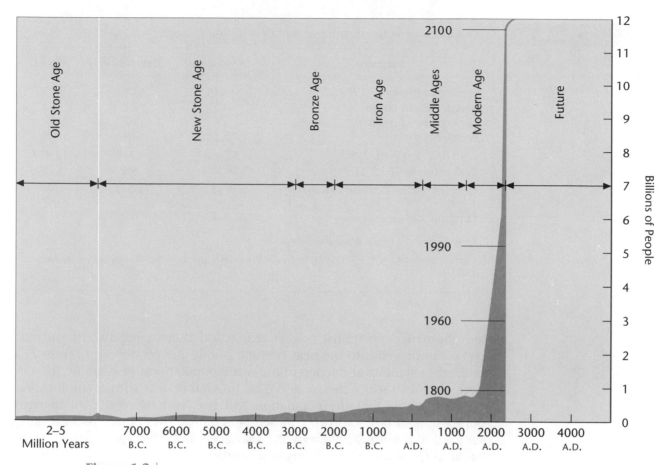

Figure 1.2

World Population Growth. Slow growth changed to rapid growth around 1750, and to very rapid growth around 1950. According to this projection, the population will start leveling off around 2100 and will reach 12 billion. Source: *Population Reference Bureau, Washington, DC.*

4 Medical advances, including the identification of disease agents (such as certain bacteria) and rational treatments (such as antibiotics) have resulted in the control of previously lethal diseases.

In Europe and North America, these changes occurred over several hundred years, and so death rates there declined gradually. But in the poor countries of Asia, Africa, and Latin America, these improvements in science and living conditions were imported from Western countries virtually overnight in demographic terms—in many cases, since 1945. It is not surprising that the populations of these regions have been growing extremely rapidly for the past 50 years. Two thousand years ago, about 500 people were added to the world human population each day. Now, the number is over 200,000!

Another way to look at population growth is to calculate the time it takes for a population to double. This figure can be approximated by dividing 70 by the percent annual growth rate (that is, growth as a percentage of the previous year's absolute population). The doubling time of the human population increased from 1,500 years (between A.D. 0 and 1500), to 150 years, and now stands at 30 years. Although it took over 2 million years to reach our first billion, it took only 12 years to reach our fifth billion (Figure 1.2).

Clearly, this prodigious rate of population growth cannot and will not continue indefinitely. Indeed, there has been a pronounced slowing in the overall growth rate, which peaked at 2.1% per year in 1968 and now is about 1.7%. This implies that birth rates have been falling, for reasons that are discussed later. But even if this trend continues, demographers (scientists who

Table 1.1	Comparison of the developed and developing regions			
Indicator		**Developed**	**Developing**	**World**
Population (millions), 1990		1,210	4,081	5,292
Annual percent growth		0.5	2.0	1.6
Life expectancy (years)		74	62	65
People per room		0.7	2.4	1.9
GNP (U.S. $ billion), 1990		15,998	3,479	19,477
GNP per person (U.S. $)		14,590	840	4,200
Grain production (millions of tons), 1991		838	1,045	1,983
Farmland[a] per person (ha)		1.5	0.7	0.8

Note: [a]Farmland = cropland and pastureland.

Source: Adapted from K. Hill (1992), Fertility and mortality trends in the developing world, *Ambio* 21:83.

study human populations) predict that world human population will not level off until well into the next century. About one person out of three is a child, and a significant fraction of the world population (for example, 45% in Africa, 40% in Latin America, and 37% in Asia) is now within childbearing age. This means that the population will rise further, even if all married couples limited births to two children per couple. The human population is predicted to reach 8.2 billion by the year 2025 and may reach a constant 12 billion in the mid- to late twenty-first century.

The dissemination and use of advances that reduced the death rate are most complete in those industrialized countries collectively called the "developed" world. Europe, North America, and Australia are examples of regions where overall population growth rates are at or near replacement levels. But in those regions where these advances have not yet fully taken hold (collectively—and hopefully—termed the "developing" world), population growth remains high. Although the life expectancy of a person born in the developing world has increased from 42 years in 1950 to 62 years today, it is well below the 74-year life expectancy of people in the developed regions.

A comparison of the developed and developing worlds reveals other profound differences (Table 1.1). The gross national product (GNP) per person of the rich is over 15 times that of the poor, and living conditions, as exemplified by the number of people per habitable room, are much better in the developed regions. Moreover, the environmental cost of living the rich lifestyle is, predictably, much higher than for the poor lifestyle. Rough estimates show a 50- to 100-fold differential in resource consumption between the average person in the United States and in India. As a result, the 15 million people added annually in the rich countries will consume vastly more resources than the 60 million people added each year in the poor countries. Using only energy consumption as a measure, the developed world extracts far more resources from the earth than the developing regions. The U.S. Bureau of the Census calculated energy consumption rates in kilograms of coal equivalent and obtained the comparative data shown in Figure 1.3. A key question in the inevitable attempts of the developing world to "develop" is whether the same rate of resource use that characterizes the currently rich countries can be achieved by the poor countries.

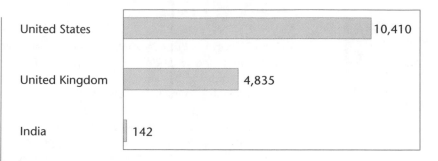

Figure 1.3

Energy Consumption Rates in the United States, the United Kingdom, and India. The rates are expressed in kilograms of coal equivalents per capita per year. Such data clearly illustrate that when a new baby is born in a rich country, it will consume during its lifetime a much greater proportion of the earth's resources than a baby born in a poor country who survives to adulthood. Source: *U.S. Bureau of the Census (1982).*

The rapid population increases that occur in many countries are accompanied by migrations both within countries and between countries. Such migrations fall into several categories: labor migrations, fleeing war (refugees), and ecological migrations (being forced to move because of land degradation). Labor migrations are not new, and in developed countries such as Britain, the nineteenth century was characterized by an enormous migration from the land to the cities. However, because the population was also expanding at the time, there was no diminution of the rural population at that time. Rural depopulation did not occur until the twentieth century. Because of rapid industrialization, jobs were available in the cities. This trend was repeated in countries such as Korea during the 1965–1990 period (see section 5 of this chapter).

Third World cities are growing rapidly because people are moving from rural to urban centers. Researchers project that in the year 2000 Mexico City will have 31 million inhabitants. Worldwide, there will be 25 cities with more than 10 million inhabitants. Of these 25 cities, 20 will be in the Third World (Table 1.2). A major difference between the present growth of cities in the Third World and the growth of cities in nineteenth-century Europe is the availability of jobs for the people who flock to the cities. Industrial jobs were plentiful in those growing nineteenth-century cities, even if working and living conditions were poor by present standards. But many people who move to the rapidly growing cities of the Third World remain unemployed or underemployed.

Table 1.2 | **Ten largest urban areas in the world (populations in millions)**

1980		2000 (estimated)	
1. New York	20.2	1. Mexico City	31.0
2. Tokyo	20.0	2. São Paulo	25.8
3. Mexico City	15.0	3. Shanghai	23.7
4. Shanghai	14.3	4. Tokyo	23.7
5. São Paulo	13.5	5. New York	22.4
6. Los Angeles	11.6	6. Beijing	20.9
7. Beijing	11.4	7. Rio de Janeiro	19.0
8. Rio de Janeiro	10.7	8. Bombay	16.8
9. Buenos Aires	10.1	9. Calcutta	16.4
10. London	10.0	10. Jakarta	15.7

Source: Data from Population Reference Bureau (1992), Washington, DC.

The "Third World," "underdeveloped countries," "developing countries": What's in a name, and what do the names mean? The term "Third World" is left over from the cold war between the West and Soviet communism. It was first used to describe the nonaligned countries that refused to take sides in that confrontation. The capitalist countries saw themselves as the First World and the communist countries as the Second. The remainder therefore constituted the Third World. In time, the emphasis on nonalignment changed to an emphasis on a low level of economic development. Omitting China, the Third World today includes more than 100 countries that cover half of the land surface of the earth and contain about half of all human beings. This single label includes a great diversity of countries, some very poor and others quite rich. Some have enormous inequities in income and property distribution. Third World countries can be subdivided into four groups:

1 *Petroleum exporters.* Iraq, Kuwait, Libya, and Saudi Arabia have small populations, but Mexico, Indonesia, and Nigeria have large populations. These countries' economies depend on the price of oil.

2 *Advanced developing countries.* Brazil, Korea, Taiwan, Hong Kong, Singapore, and Thailand are beginning to close the income gap with the rich countries and are emerging as new centers of manufacturing. Some of these countries, such as Brazil, have very large populations of extremely poor people.

3 *Middle developing countries.* Egypt and Peru are examples, having a per capita income of about U.S. $1,000 and making some progress in development.

4 *Least-developed countries.* These 35 countries are the world's poorest, with per capita incomes of only a few hundred dollars per year. With 30% of the world's population, they only produce 3% of its wealth. In many of these countries, population pressures are causing severe environmental degradation.

The takeover of Kuwait by Iraq in 1990 and the subsequent population dislocation highlighted the extent of the international labor migration. Two million foreign nationals, mostly Palestinians, were in Kuwait at the time of the invasion, along with 80,000 migrant workers, mostly from Asia and Africa. Under the Iraqi occupation and the aftermath of the Gulf War, these foreigners became economic and political refugees and had to be repatriated, under difficult conditions. There are estimated to be 20 million international labor migrants, most of them from developing countries.

2   **Increased national wealth resulting from economic development is not necessarily correlated with a decrease in the rate of population growth.**

Although percent growth rates and doubling times are convenient numbers to summarize the changes occurring in a population, they are not the most sensitive quantitative measures of population change. To predict how populations may be changing, demographers prefer to use crude birth rates or

Table 1.3	Population and fertility rates, 1990			
	Total Population (millions)	Crude Birth Rate (per 1,000)	Total Fertility Rate (TFR)	GNP per Person (U.S.$)
World	5,283	26	3.4	4,200
Low-Income Countries	3,058	30	3.8	350
India	850	30	4.0	350
Middle-Income Countries	1,088	29	3.7	2,270
Mexico	86	27	3.3	2,490
High-Income Countries	1,137	13	1.7	19,590
United States	250	17	1.9	21,790

Source: Data from World Bank (1992), *World Development Report, 1992* (Oxford University Press, UK: World Bank).

total fertility rates (TFR). The **crude birth rate** is the average number of babies born annually per 1,000 people, and this number varies from 12 (northern, western, and central Europe) to 49 (West Africa). The **total fertility rate (TFR)** is the average number of children born to a woman passing through the childbearing years. The TFR varies from 1.7 (northern, western, and central Europe) to 6.8 in Africa. A TFR of 2.1 is the minimal rate for a population to replace itself.

A look at a world map that shows wealth (GNP) and birth rate (or TFR) reveals a general correlation between the two. Regions with a high GNP have a low TFR. This certainly is the case in western Europe, Japan, and Australia. Increased economic development is therefore often suggested as a possible solution to the population problem. However, a close look at the data shows that there are many exceptions to the general rule:

- East Asia and temperate South America have a modest per capita GNP, but a TFR only slightly higher than industrialized countries.
- Oil-rich states in the Middle East have a high GNP per capita and a high TFR.
- Singapore and Hong Kong are newly industrialized states with a low TFR.
- Socialist countries (Cuba, China) have a low per capita GNP and a low TFR.

Religious forces may work against development. Many Middle Eastern countries with strong Islamic traditions have seen little fertility change. The very idea of limiting the number of births is as unacceptable to them as to certain Christian fundamentalists.

China and sub-Saharan Africa are equally poor in terms of GNP per person, but there are dramatic differences in growth rates, fertility rates, infant mortality, and maternal deaths caused by birthing. China has had strong government-supported family planning, whereas in sub-Saharan Africa limiting births is culturally unacceptable and family planning can only be promoted as a means for promoting maternal and child health.

Given these complexities, it is difficult to predict future TFRs in various regions. However, demographers at the United Nations have made an attempt, with the following results:

1   World TFR, which has steadily declined from 5.0 in 1960 to 3.8 in 1990, will continue to decline, to 2.8 by 2010.
2   The TFR in Europe, North America, and the FSU (former USSR) will continue to be below 2.0.
3   In Latin America, where the TFR has declined rapidly in the past 20 years, from 5.5 to 3.5, the pace of decline will slow so that it will approach 2.5 by 2010.
4   The rapid decline in Asian TFR over the past two decades, led by China, will slow down, and the TFR will reach 2.7 by 2010.

3   **Education of women is an important aid in bringing down the population growth rate.**

Although the Third World countries of Africa, Asia, and Latin America contain 77% of the world's population, in 1990 they accounted for 98% of the world's deaths among children under age 5. This statistic merely confirms that developed countries where infant mortality is lower have better health care systems. Infant mortality—the number of children who die in their first year of life per 1,000 live births—is the most widely cited indicator of the development of the health care system of a region. In Japan, this rate is less than 10, while in Europe and North America it ranges from 10 to 20. These rates are far lower than they were in these same countries 100 years ago, before the advent of good nutrition, hygiene, and infection control. Rates then were well above 100 per 1,000 live births.

During the 1980s, infant mortality fell in most poor countries, as conditions gradually improved. In the Near East, it fell from 102 to 77; in Latin America, it fell from 59 to 47; and in Asia, it fell from 53 to 46. A study by the U.N. Food and Agriculture Organization (FAO) determined some characteristics of those countries where infant mortality had fallen most significantly. These included good access to health services, trained personnel attending the birth, and maternal literacy. These factors also come into play within a country. In India, the infant mortality rate is lower among the educated in the cities than it is among the rural population. In the United States, the rate in inner cities (for example, in Washington, DC) and in poor rural areas is over twice that of the rest of the country. As a result, the average infant mortality in the United States is among the highest of industrialized countries.

Many studies have shown that infants are at higher risk of death if they are born to mothers who are adolescents or over the age of 40, or when the interval between births is less than two years. Family planning programs to help space out births and national policies to discourage early marriage and child bearing can substantially reduce infant and child mortality by preventing births to high-risk mothers. Whenever the relationship between child mortality and women's education has been studied, it has been found that

Figure 1.4

The Relationship Between a Mother's Education and the Survival Rate of Her Children (mortality before age 5). The effect, a two- to three-fold difference between the survival rates of children from educated and uneducated mothers, is equally pronounced in all four countries, regardless of the actual death rate. These data are results from selected demographic and health surveys, 1986–1990. Source: Adapted from S. O. Rutstein (1991), "Levels, trends and differentials in infant and child mortality in the less developed countries," paper presented at the seminar on Child Survival Interventions: Effectiveness and Efficiency, at the Johns Hopkins University School of Hygiene and Public Health, Baltimore, MD, June 20–22, 1991.

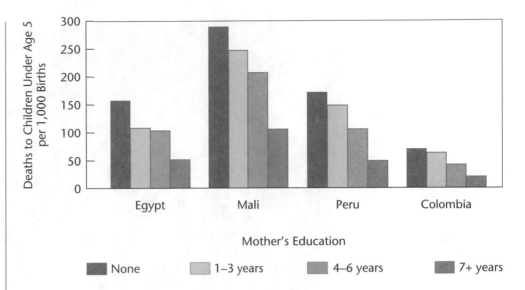

the children's chances of survival beyond the age of 5 improve as the education of their mothers increases (Figure 1.4). Educated mothers have better diets and better hygiene, are more likely to use the available health care services, and probably have more status in the family to take whatever actions are appropriate when a child needs care.

Educated mothers also tend to be more interested in family planning and are able to adopt measures to space the births of their children whether by using contraceptives, through abortion or via more traditional methods, such as long abstinence after childbirth. Strong family planning programs can have a substantial influence on the TFR (Figure 1.5). The history of family planning in Thailand illustrates this relationship. Between 1965 and 1990, the TFR of Thailand declined from 5.7 to 2.8, and the percentage of married women of reproductive age using contraceptive measures increased from 15 to 70%. The easy availability of contraceptives to women in cities and in the countryside, the empowerment of nurses to perform female sterilizations, and the liberal attitude toward abortion, all contributed to the dramatic decline in the TFR. It is unlikely that social and economic development alone, without strong government aid for family planning programs, could have accomplished the same goal.

4 | The United States does not have a population policy.

The United States does not have a national population policy. There are many state and federal laws that encourage or discourage population growth, but these laws are usually designed for other purposes and do not constitute a policy. Laws that affect sex education, the availability of contraceptives, abortion, marriage, child care, income tax deductions, housing, and so forth all help to determine the country's population growth. Although zero population growth, for the world as a whole and for the United States, was much in the news in the 1970s, the urgency seems to have faded in the minds of the people.

However, with a growth rate of 0.8% per year the issue in the United States is as urgent as it was before. If the present growth rate is sustained, in

Figure 1.5

A Family-Planning Class in Jendouba, Tunisia. The United Nations Fund for Population Activities (UNFPA) helps Tunisia and other countries with their family planning programs by providing funds for personnel, training, equipment and materials. The emphasis is on reaching people in rural areas and integrating mother and child health care with family planning and nutrition. Source: United Nations Photo 156350, I. Isaac.

three generations there could be half a billion people in the United States. Current projections indicate that the U.S. population may stabilize between the years 2040 and 2050, at 300 million people. It is good to keep in mind, however, that population projections have often been wrong.

The current growth rate of the U.S. population is caused by both immigration (35%) (legal and illegal) and the excess of births over deaths (65%). The United States has traditionally been a country of immigrants and accepts twice as many legal immigrants as all other countries combined: about 500,000 each year. This number of newcomers is augmented by refugees, about 200,000 each year and an unknown number of illegal immigrants, variously estimated at 100,000 to 300,000 per year. Thus illegal immigrants account for about a quarter of the total immigrants or 10% of the population increase. Illegal immigrants are now a fact of life in most industrialized countries, even in Japan, which is generally inhospitable to immigrants. The breakup of the USSR has resulted in a flood of legal and illegal immigrants to the industrialized countries of western Europe. Immigrants, both legal and illegal, from North Africa, Turkey, Albania, Brazil, Indonesia, and many other countries are found throughout western Europe. In the United States, 80% of the legal immigrants come from Asia and Latin America. This is in contrast to the situation 100 years ago, when 90% of the immigrants came from Europe. The illegal immigrants come primarily from the same areas as do the legal immigrants. One wonders whether the resurgence of nativist sentiment in the United States is related to this change in the racial and ethnic composition of the immigrants rather than to a desire to improve the economic situation of the country. Most economic experts believe that the immigrants, whether legal or illegal, are not responsible for the economic stagnation of the 1990s.

The second cause of population growth in the United States is the excess of births over deaths. Here again there is a strong public perception that the white Caucasian majority will soon become a minority, because minority parents have so many children. At this time the U.S. population is about 78% white Caucasian and 22% minority (African-American, Asian, Hispanic). The

Box 1.2

Landmark Events in International Family Planning

Early Advocates

1860	Malthusian League is founded in England to spread information on birth control.
1873	Comstock Act is enacted in the United States to prohibit advertising or prescription of contraception (later repealed in 1916).
1916	Margaret Sanger opens first free birth control clinic in United States, in Brooklyn, New York.
1921	Marie Stopes opens first birth control clinic in England.
1937	American Medical Association gives qualified endorsement to birth control.

Developments After World War II

1946	U.N. Economic and Social Council establishes a population commission representing member governments and a population division within the secretariat.
1951	India adopts family planning as part of its economic program.
1952	John D. Rockefeller III establishes the Population Council. Constitution for the International Planned Parenthood Federation (IPPF) is drafted at an international conference in Cheltenham, England (ratified in 1953).

Rapid Program Expansion

1960	Oral contraceptives are introduced.
1961	Plastic IUDs become available.
1967	The trust fund for the U.N. Fund for Population Activities (UNFPA) is created.
1968	U.N. International Conference on Human Rights issues the Teheran proclamation, of which article 16 states, "Parents have a basic human right to determine freely and responsibly the number and spacing of their children." Pope Paul VI issues *Humanae vitae,* banning the use of artificial contraception. Paul Ehrlich publishes *The Population Bomb.* U.S. Congress first allots foreign aid funds for family planning.
1969	First noneugenic, nonrestrictive sterilization laws are passed in Singapore and the U.S. state of Virginia.
1973	U.S. Supreme Court upholds *Roe v. Wade* decision, limiting the right of states to interfere in the private decision of a woman and her doctor to terminate a pregnancy.
1974	United Nations holds first world conference on population in Bucharest. The United States urges countries to adopt policies aimed at slowing pitch, primarily through family planning.
1979	The People's Republic of China begins campaign for "one child for one couple."

Political Realignments and New Technologies

1984	The Second U.N. World Conference on Population is held in Mexico City. Most Third World countries favor slower population growth and family planning. The United States shifts its position, stating that population growth is a neutral phenomenon.
1985	A law is enacted that prohibits the U.S. government from supporting any organization that supports or participates in the management of a program of abortion or involuntary sterilization. U.S. support for the UNFPA and IPPF is suspended.
1988	RU-486 is approved for terminating early pregnancy in France.
1993	RU-486 is approved for testing in the United States.

(*Source:* Modified from P. J. Donaldson and A. O. Tsui 1990. The International Family Planning Movement. *Population Bulletin 45:* nr3, p. 8. Population Reference Bureau Inc.)

Table 1.4	The population of North America in 1992	
	Population (millions)	Percent Growth Rate
Canada	27.5	0.8%
United States	258.7	0.8
Mexico	89.5	2.3

minority percentage is likely to increase to 30% by the year 2020. This small change in the racial and ethnic makeup of the population results from a higher TFR among Hispanics (TFR = 2.6) and African-Americans (TFR = 2.3) than among white Caucasians (TFR = 1.8), as well as from the racial composition of the immigrants (see earlier note). It is good to keep in mind that of the 4 million babies born in the United States each year only about 23% are born to minority mothers and 78% are born to white Caucasian mothers. However, there is no doubt that the rapid growth rate of Mexico's population will continue to put pressure on the population of all of North America (Table 1.4). If the economic development of Mexico is inadequate to meet the needs of its population, more Mexicans will seek to migrate north where opportunities are greater. From the Mexican viewpoint, providing jobs for its people in Mexico is a major impetus for the North American Free Trade Agreement.

The principal reason that the United States has no population policy is that political leaders are unwilling to talk about it. The issue of population is potentially as explosive as abortion or the right to die. Imagine what would happen if the U.S. Congress were to discuss a law that permits couples to have only one child or requires couples to have at least two, but no more than three children! Another reason for the absence of a national policy is that many people, and this includes most people in positions of authority or leadership, believe that population growth in the United States has stopped and that the population will soon be leveling off. Nothing could be further from the truth. However, the high rates of population increase in the Third World, and the attention of the western news media given to the poverty there, creates the illusion that "they" have a problem and "we" do not.

Such perceptions produce official expressions of "concern." For example, the Agricultural Trade Development and Assistance Act of 1954 says that the United States should take note of a country's voluntary population control programs before entering agreements with developing countries for the sale of U.S. agricultural products. In 1961, the Foreign Assistance Act stated as an objective, that the United States will offer developing countries help in reducing their rate of population growth. During the 1980s, under the Reagan and Bush administrations, almost no such aid was offered. The United States terminated its financial contributions to the two international funds that are the leaders in population control efforts: the U.N. Population Fund and the International Planned Parenthood Federation. With the election of President Clinton, there has been an abrupt change in philosophy, and the United States is beginning to assert positive leadership in population issues.

Surprisingly, some people still do not view the rapidly rising world population as a source of concern. Rather, they believe that "all will be well in the end." This is an unlikely scenario. On the contrary, a decision to have more than two children is a decision to potentially lower the quality of life for these and future children in the developed, as well as the developing countries.

5 | Different theories of population growth seek to explain why growth rates change.

The causes and consequences of population growth are controversial, and the theories often depend on the ideological leanings of the social scientists involved. The debate was started by Thomas Malthus, who in 1798 published "An Essay on the Principle of Population," in which he noted that human population had the capacity to expand geometrically, while he proposed that food production could only increase linearly. As a result, Malthus predicted, population growth would soon outpace food production, and wars and famines would bring the population back in check. Malthus believed that population increase promoted poverty.

The basic tenet of Malthus's theory has been discredited—we now know that food production can increase geometrically and keep up with, and even rise faster than, population growth. Many of the dire consequences of population increase that he predicted have not materialized. Yet his theory still finds a strong following among neo-Mathusians, who agree with his proposal that population increase has a natural tendency to result in poverty.

The neo-Malthusians believe that rapid population growth dictates that income, land, natural resources, and food must be divided among more and more people and that this distribution reinforces poverty and hunger. In addition, they have formulated some postulates not found in Malthus's original writings:

1 **Ecological Malthusians** stress that population growth undermines the natural resource base. Thus, deforestation, soil erosion, water and air pollution, and other forms of environmental degradation are caused by the growing population and exacerbate poverty. Poverty in turn may lead to further population growth.

2 **Productionist Malthusians** stress a different consequence of population growth: more people require jobs and more services from their government. This severely strains government budgets to provide adequate education, housing, and health care, thereby preventing income growth. Poverty results.

We saw earlier that per capita food production has increased during the last 20 years. Thus both classical Malthusian and neo-Malthusian projections of a declining per capita food production have proven incorrect. In Africa, where per capita food production has stagnated since 1960 largely because of rapid population growth, the increase in population seems to have been caused by poverty rather than being the reason for poverty. The pursuit of economic development by African countries, often with the aid of Western financing, has led to excessive population growth and poverty. In Africa, poverty is not the result of a decrease in the food available per capita, as postulated by the neo-Malthusians.

A different school of demographers, labeled the **non-Malthusians,** believe that although rising population density plays a role in environmental degradation and food shortages, it is not the most important factor in and certainly does not cause poverty and hunger. According to the non-Malthusians, hunger, poverty, and rapid population growth occur together because they have a common cause: the lack of jobs, education, health care, and social security.

Thus fertility control and population control will not eradicate poverty unless the fundamental causes of rising population are also dealt with. These fundamental causes are (1) economic dependency, (2) landlessness and maldistribution of land (a particularly acute problem in Latin America), (3) capital-intensive instead of labor-intensive industrialization, and (4) export-oriented growth. The main evidence advanced by non-Malthusians for their viewpoint is that population growth rates decline in countries where incomes rise or where income redistribution results in more economic security.

The neo-Malthusian versus non-Malthusian debate has profound consequences for policy decisions in the developing countries as they grapple with their enormous problems. If the neo-Malthusians are correct, the greatest assault must be on population growth, and only when it is under control can the other social and environmental problems be dealt with. In contrast, the non-Malthusians propose broad social and economic development first, so that reduced population growth will follow.

Data from demographic studies are on the side of the non-Malthusians. Observations by K. Davis and other demographers showed that industrialization and rising incomes did play a major role in bringing down population growth rates in quite a few countries of Europe, and later in selected countries of Asia and Latin America over the past century. The theory behind this change is called the **demographic transition**, and it has typically occurred in four stages, illustrated in Figure 1.6.

Stage 1   Both the birth rate and death rate are high, essentially canceling each other out so that the growth rate (natural increase) is low. Poor parents have many children because they provide cheap agricultural labor and care for parents in their old age. In addition, social factors such as religion and proof of male virility and female status can encourage couples to have many children.

Stage 2   Improvements in living conditions and health care reduce the incidence of disease and death. These advances are readily accepted by the society. But social mores are much slower to change. Thus the birth rate remains high, while the death rate drops and the overall growth rate increases.

Stage 3   As living conditions and education improve, the birth rate declines to near the level of the death rate. The population growth rate slows down.

Stage 4   The birth rate and death rate are once again near each other, but now at a much lower level than in stage 1. The population growth rate returns to its previous low level.

In country after country, in a diversity of cultures, fertility has declined rapidly as the well-being of the population has improved (stage 3 to stage 4). The data in Figure 1.7 show that there is a temporal correlation between industrialization and the drop in the birth rate. But is there also a causal relationship?

In the 1970s, demographers argued that Third World countries were then undergoing their demographic transitions. Death rates had dropped because of better hygiene and the availability of vaccines, and birth rates would drop as soon as the countries could industrialize. It is now realized that the decline in birth rate in Europe was linked to social change and education, and not caused

Figure 1.6

The Demographic Transition Model. In stage 1, the birth rate and death rate are high, and the rate of population increase is small. In stage 2, better health care and sanitation push the death rate down, but the birth rate stays the same, resulting in a rapid rate of population increase. In stage 3, the birth rate falls, and in stage 4 the rate of population increase is once again low. The population has stabilized once again, but at a much higher level than before.

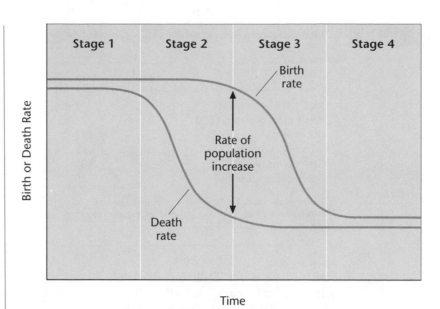

Figure 1.7

The Birth Rate Transition in Different Countries. The decline in the birth rate in different countries occurred at different times and parallels their efforts to industrialize, raise the educational level of the people, and introduce measures that assure more security in old age. Source: Data (from a variety of sources) from W. Murdoch (1990), "World hunger and population," in C. R. Caroll, J. H. Vandermeer, and P. M. Rosset, eds., Agroecology *(New York: McGraw-Hill), p. 6. With permission of the publisher.*

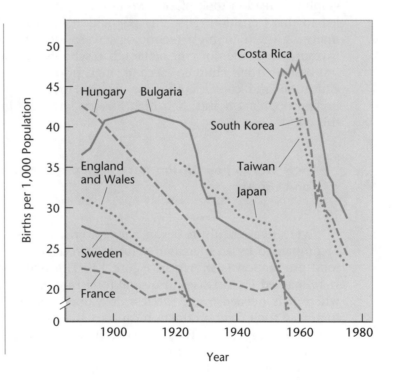

by industrialization. In some countries (oil-rich countries of the Middle East), incomes have risen fast but fertility rates remain high. In other areas (China, Sri Lanka, the state of Kerala in India), population growth rates have steeply declined without industrialization, but because of income redistribution and reduced economic insecurity. Thus, demographic or fertility transitions depend on economic, social, political, and religious factors. In particular, the educational level of women is an important determinant in lowering infant mortality.

A concerted effort by the government can alter fertility rates. A classic recent case that has followed the textbook model is the Republic of Korea

Table 1.5	The demographic transition in Korea: 1965–1990		
		1965	**1990**
Total fertility rate		4.8	1.6
Infant mortality rate/1000		75	24
Life expectancy (years)		55	73
Urban population (%)		28	75
GNP per person (U.S.$)		130	4,500
Contraceptive use (%)		22	80

Source: Data from O. Kim and P. van den Oever (1992), Demographic transition and patterns of natural resources use in Korea, *Ambio* 21:56–62.

(South Korea)—see Table 1.5. In 1960, Korea was a village society, with high birth and death rates, and a life expectancy of about 50 years (stage 1). By 1990, both birth and death rates had fallen precipitously (TFR fell from 6.1 to 1.6), and life expectancy reached 73 years (stage 4). In the meantime, a rapid demographic transition took place. Two factors, both planned by the government, led to this transition. First, the population became urbanized, because jobs, many of them in high technology, were created by encouraging foreign investment. In addition, the better urban schools attracted parents who wanted the best for their children. Second, high priority was given to birth control education and contraceptive use. The relative ethnic homogeneity of the Korean population and their high literacy rate were important factors aiding this effort.

6   Increases in population have been paralleled by increases in food supply.

Ever since agriculture began, the rapid rise in human population has been accompanied by an increased ability to produce food. The recent history of food production is shown in Figure 1.8. The two decades of 1972 to 1992 were marked by increases in overall food production. This was most marked where it is needed most, the less developed countries. Over 20 years, they increased their food production capacity by over 60%, whereas food production in developed countries increased by only 20% over 20 years.

Two questions arise from these data: (1) How has this increase been achieved? and, (2) What foods have been important in the increases? Historically, there have been two ways to increase food production:

1   Increase the amount of land used to produce food
2   Increase the amount of food produced per season on the land already being used for agriculture

The arable land presently being cultivated is about 1.5 billion hectares (ha). Researchers have estimated that cultivated arable land increased by 432 million ha between 1860 and 1920, and again by 419 million ha between 1920 and 1978. Thus, a considerable proportion of the increased food production since 1860 has been caused by putting more land to the plow. The

Figure 1.8

Overall Food Production in Developed (squares) and Developing Countries (circles). Notice that the developing countries, where more people live, have had more dramatic increases than the developed countries. However, the per capita data look quite different (see Figures 1.9 and 1.10). The year 1972 is taken as the basis of 100 for expressing data in subsequent years. Source: Data from Food and Agriculture Organization.

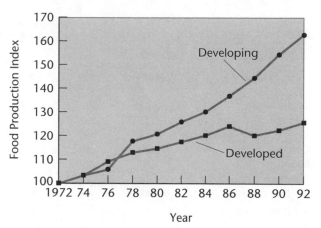

FAO has estimated that about 4 million additional hectares could be brought into cultivation each year. This could put the total arable land at 2 billion ha in 2100, when the population is expected to reach 12 billion. A quick calculation shows that the arable land available per person will fall from 0.28 to 0.17 ha. This means that land productivity must increase substantially to feed everyone, and increasing productivity will require more resources. The present high productivity has been achieved at the cost of an enormous input of energy and other resources (new crop varieties, pesticides, herbicides, fertilizers, and irrigation).

Approximately 300 types of plants are cultivated in the world, but 24 of these supply nearly all our food. More than 85% of our food comes from eight species of plants, and more than 50% comes from only three species: wheat, corn, and rice. These three crops also use most of the arable land. Whereas the area cultivated with these plants increased somewhat over 20 years from 1971 to 1991 (Table 1.6), yield increases (expressed as tons per hectare) were more substantial.

However, average figures mask enormous differences between countries in yield and in yield increases. For example, in 1990 the average yield for wheat in the Netherlands was 6.9 tons/ha, whereas the world average was only 2.4 tons/ha. Similarly, the average yields for rice in Japan and corn in the United States were twice the worldwide averages. In many cases, the reason for these discrepancies is that in some regions—usually the rich ones—the land and resource inputs can be managed to get the most out of the crops' genetic potential.

How did the increase in cultivated area, yield per hectare, and population affect the amount of food available per person? Figures 1.9 and 1.10 show data expressed on a per person basis. In the developed countries, the **food index** (food produced per person) each year rose rapidly from 1972 to 1982, and then plateaued. In the developing countries, there has been a steady rise since 1972, although the trend has slowed since 1986. Separating the data for developing countries by continent (Figure 1.10) shows that food per person in Africa and Latin America has stagnated in the last 10 years, whereas in Asia food per person has continued to rise. However, the increase in Asia in the past 5 years can be accounted for entirely by China.

Opinions differ as to why per capita food production has leveled off since 1985–1988. Some experts blame the worldwide recession and believe that more purchasing power in the developing countries would increase food production there as well as in the developed countries. Others believe that we may have reached the yield potential of the most important crops and that significant scientific advances are needed to break this yield barrier. A third contributing factor may be the environmental degradation that accompanies

Table 1.6	Area and average yield of the main food crops of the world				

| | Area (mha) | | Yield (t/ha) | |
Crops	1971	1991	1971	1991
Total Cereals	667	692	1.6	2.5
Wheat	225	232	1.6	2.5
Rice	133	146	2.4	3.5
Corn	130	140	2.2	3.8
Barley	70	77	1.9	2.3
Sorghum	48	55	1.2	1.7
Millet	48	42	0.6	0.7
Total Roots and Tubers	45	50	10.5	13.2
Potatoes	18	20	13.9	16.1
Cassava	13	16	8.9	9.8
Total Pulses (legumes)	63	65	0.7	0.9
Soybeans	51	55	1.4	1.9
Dry beans	25	23	0.5	0.5

Notes: mha = millions of hectares, t/ha = tons per hectare.

Sources: Data from Food and Agriculture Organization and U.S. Department of Agriculture.

Figure 1.9

Food Production Per Person (1972 = 100) for Developing (circles) and Developed (squares) Regions. Notice that per capita food production has stagnated in the developed countries since 1982, and started leveling off in the developing countries after 1986. Source: *Data from Food and Agriculture Organization.*

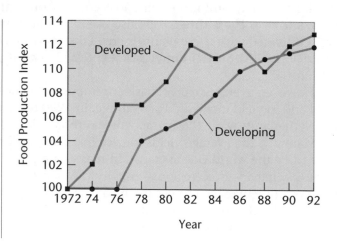

the expansion of agriculture into marginal areas. Finally, a severe drought has affected many important crop-growing areas as a result of a short-term change in the climate.

7 | Widespread hunger persists in a world that produces enough food.

Most developed countries and some developing countries are net food exporters, and government subsidies are often needed to help sell farm surpluses abroad. Economists talk about a "food glut" that severely depresses

Figure 1.10

Per Capita Food Production in the 1980s in Asia (triangles), South America (circles), and Africa (squares). Note that in South America and Africa, food production just barely keeps pace with population whereas in Asia, food production per person is actually rising. Most of the increase in Asia can be accounted for by China. Source: Data from Food and Agriculture Organization.

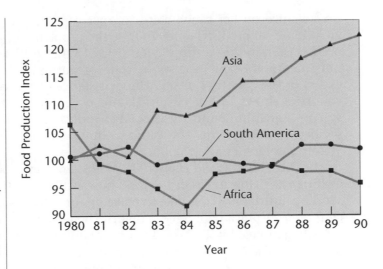

prices on the world markets. That surplus may be the principal reason why food production leveled off in the 1980s and early 1990s. Yet hunger persists and remains widespread in this world of plenty. Hunger, which is best defined as chronic malnutrition resulting from inadequate intake of food, is an emotionally and ideologically charged issue about which misconceptions arise and views widely diverge. The diverging views concern why people have inadequate access to food and what steps should or can be taken to reduce hunger.

One misconception is that hunger is most common in Africa. Because the media pay considerable attention to the drought and war-induced famines in the Sahel, Ethiopia, and Angola, many people confuse famines—which are often human-made—with hunger. Hunger exists in nearly all countries and is most prevalent in Asia. Three-quarters of the world's hungry people are found in the following countries of Asia, South America, and Africa:

- *Asia*: India, Pakistan, Bangladesh, Indonesia, Cambodia, and the Philippines
- *South America*: Brazil
- *Africa*: Zaire and Ethiopia

Many of these countries actually export food, and therein lies the paradox of hunger: it persists and even grows at the same time as the food supply is growing. For example, as many as 200 million people in India are estimated to be malnourished, although India is now a net food exporter and has a hard time disposing of its wheat and rice surpluses. The malnourished Indians do not have enough income to buy the grain that is produced in India. As a result, they remain chronically undernourished.

Experts disagree about how many hungry people live in the world. Estimates range from 200 million to 1,000 million (1 billion) people. Recent estimates by various international agencies put the number at 500 million, or 10% of the world's population. However, no one disagrees that hunger is a growing problem. As the population of the world expands, so does the number of malnourished people. In the 1990s, the number of hungry people is growing annually by 8–10 million people, partially as a result of the worldwide economic recession.

Reliance on per capita statistics (per capita food production, per capita gross national product) tends to hide the problem of hunger. Indeed, hunger persists even in the wealthiest countries. In the United States, malnutrition is not uncommon and persists in spite of an agricultural production far in excess of the country's needs. The persistence of hunger in many developed countries shows that it is caused, not by a failure of the food production system (agriculture and its associated industries), but by the failure of the socioeconomic system. The cause of hunger is the absence of an adequate "safety net" for a nation's poor people.

In their book *World Hunger: 12 Myths*, Frances Moore Lappé and Joseph Collins ask, "How much of the food now available within [each of] the Third World countries would it take to make up for food lacking in the diets of each country's chronically hungry people?" The results of their calculations, based on data gathered by the World Bank, are quite surprising: 5.6% for India, 2% for Indonesia, 7.8% for Tanzania, and 2.5% for Senegal and Sudan. Thus, hunger persists because people are too poor to buy the food that is already available in these countries.

There is general agreement that hunger is caused by poverty and not by a failure of the agricultural system. Nevertheless, most scientists in government, academia, and industry; most international agencies; most philanthropic organizations; and the media all continue to suggest that there is a race between the hunger of a rapidly growing population, on the one hand, and the development of new technologies that will increase productivity, on the other hand. According to this view, if humanity can only win this race hunger will be eliminated, and all will be well. The agricultural biotechnology industry has quickly picked up this theme and has joined "the fight against hunger." The McKnight Foundation, one of the biggest philanthropic organizations in the United States, started a research program in plant biology with the long-term goal of raising food production and eliminating hunger in developing countries. And agricultural production scientists often emphasize that their work is "at the forefront of combating hunger."

Food supplies are, of course, important and will need to grow in the future; yet hunger is primarily a problem of the *distribution* of these food supplies, between countries and within a country, between households, and even within a household. Within a single household, some members may go hungry because they have inadequate access to the small amount of food that can be purchased. A 1986 study by the International Society of Applied Systems Analysis, entitled *Hunger Amidst Abundance,* indicates that increasing the amount of food on world markets would have essentially no impact on the extent of hunger in the world. The real problem is how to increase the purchasing power of the people who need food. Poor people already spend 75% or more of their incomes on food.

Brazil is one of the four or five richest countries in the Third World, with a relatively high per capita GNP. Yet it has some of the highest rates of malnutrition and infant mortality in the Third World. Why is this so? Brazil has one of the most sharply unequal income distributions of all countries, with a small, wealthy upper class and a very large and poor underclass. The manner in which countries such as Brazil have developed in the past 20 years has made the problem of hunger worse, not better (Figure 1.11).

In many Third World countries, there is no trend toward higher and more equal incomes or toward the elimination of poverty among the underclass. And in some developed countries the underclass is growing. As a result, it is

Figure 1.11

Poverty in Honduras.
*Development does not
necessarily abolish pov-
erty, as shown by the
lives of these poor people
in Latin America. Many
countries in Latin
America have undergone
a development boom,
which has largely made
the rich richer, while the
poor have remained poor
and often migrated to
crowded cities. Achieving
unimodal development
requires a strong central
government able and
willing to put in place
laws to ensure that every
segment of the popu-
lation profits. Source:
U.N. Educational,
Social and Cultural
Organization.*

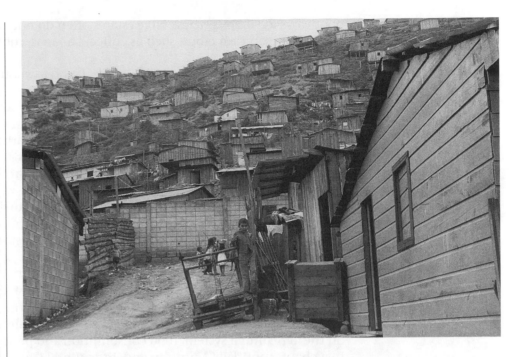

unlikely that hunger, which is the result of economic inequality, can be eliminated any time soon. The extraordinary developments in plant biotechnology described in this book will have little or no impact on hunger in the
world in the foreseeable future.

8 | Economic and agricultural development do not necessarily abolish hunger.

If the root of hunger is poverty and insufficient purchasing power, then
economic development should eliminate or at least significantly reduce hunger. Unfortunately, economic development is not a sufficient condition for
eliminating poverty and hunger. This fact is perhaps best illustrated in Africa
and Latin America where economic development in many countries has gone
hand in hand with ever-increasing poverty and hunger. Earlier we discussed
the Republic of Korea, which has undergone tremendous economic and agricultural development in the past 25 years, resulting in greatly increased food
security for its population. To achieve such a goal, a country must have a
national development strategy in which industrial, social, and agricultural
revolutions are synchronized. The country must have the political will to
achieve a unimodal development in which all segments of the population
share the benefits and contribute to progress. The alternative, a bimodal
development, allows one segment of the population to increase its well-being
at the expense of a large underclass. Countries such as Korea, Taiwan, and
China have pursued unimodal development, whereas much of Latin America,
exemplified by Brazil, have enacted policies that intentionally or uninten-
tionally favored bimodal development. In countries characterized by bimodal
development, there are large numbers of rural and urban poor, who have lost
access to the land. Most land is owned by a small fraction of the population.
Agricultural production is often carried out on large estates where the rural

poor work as laborers. These large estates often do not produce food crops but concentrate on cash crops, such as coffee, bananas, rubber, tea, and coconuts for export.

According to economic theory, the laborers should be able to buy as much food as they were producing under conditions of subsistence agriculture. Unfortunately, theory and practice are not always in accord. The amount of money they earn as laborers is not always enough to buy the necessary food. And once the peasants enter the market economy they may buy cheap staples, rather than the more nutritious traditional foods. Even within a family, all members may not have sufficient food even if money is available. For example, a study in India found that wage income was unrelated to the nutritional status of the children, except when a woman was the head of the household.

When a region makes the transition from subsistence farming to industrial agriculture, food security and nutritional status of the population often decrease. In traditional agricultural systems, nutrition and agriculture are linked: the people eat what they produce and what they gather from wild foods. In general, people engaged in subsistence farming have nutritionally adequate diets. However, when market forces become a dominant influence on the agricultural system, food consumption and production are dissociated. In a chapter entitled "Nutrition and Agricultural Change," Kathryn G. Dewey reviewed a number of case studies in Third World countries and found that agricultural development was not always associated with improved nutritional status for the people (1990, in C. R. Caroll, J. H. Vandermeer, and P. M. Rosset, eds., *Agroecology*, New York: McGraw-Hill). For example, in the municipality of Cunduacan (in the state of Tabasco, Mexico), food production increased sixfold between 1958 and 1971. Yet the percentage of children diagnosed with moderate to severe malnutrition remained essentially constant: 26.1% in 1958 and 22.5% in 1971. Development had apparently benefited the urban middle class, but not the rural poor. This is a typical example of bimodal development, in which there are not enough safeguards to ensure the participation of all segments of society in the material benefits of development.

In many areas, the commercialization of agriculture has resulted in peasant farmers losing access to the land, because land ownership is concentrated in the hands of a few large farming corporations. Commercial agriculture requires investments for equipment, fertilizers, seeds, and pesticides; and peasant farmers do not have access to the necessary credit. Irrigation water is often targeted preferentially for the large landholders because their production is seen as economically more important. They produce the cash crops that earn hard currencies needed for development and for repaying international debt.

Loss of access to the land often means a decline in nutritional status. A study of agricultural development in Peru showed a strong correlation between nutritional adequacy and the proportion of home-grown foods in the family's food budget. Families that had retained access to land and could grow some of their own food were much better off than families that had to rely on cash income to buy food.

These examples should make it clear that it is not easy to achieve unimodal development. First, if it is to lower the birth rate, a country must invest in its infrastructure (buildings, roads, railways, harbors), the production of capital goods and technology, and above all, in educating all its citizens (men and women). Second, it must adopt policies of trade and goods

Box 1.3

A Generous Uncle?

Bilateral nonmilitary aid provided by rich countries to poor countries in 1990 reached a total of $43 billion, amounting to an average of 0.35% of the total GNP of the developed countries. Several international organizations have called for a doubling of this aid, to $80 billion or 0.7% of the GNP. Unfortunately, aid is actually declining, in part as a result of the economic downturn since the late 1980s. In addition, aid is often targeted to specific countries deemed strategically important, and may be tied to the receiving nation's purchase of goods and services in the donor country, constituting in essence a form of export subsidy. The United States provides 39% of its nonmilitary aid to just three countries, Israel, Egypt, and El Salvador, which together have only 1.2% of the world's population.

Development Assistance from Selected Industrial Nations (1989)

Country	Development Aid (billion dollars)	Share of GNP (percent)	Country	Development Aid (billion dollars)	Share of GNP (percent)
Norway	$0.92	1.04%	Germany	$4.95	0.41%
Netherlands	2.09	0.94	Australia	1.02	0.38
France	7.45	0.78	Japan	8.95	0.32
Canada	2.32	0.44	United Kingdom	2.59	0.31
Italy	3.61	0.42	United States	7.66	0.15

It is surprising that a small country like Norway, which has many fewer interactions with Third World countries than the United States, France, or Britain, should be a world leader in development assistance (foreign aid). Norway contributes more than 1% of its GNP—approximately seven times more than the United States on a per capita basis (see the table). In *State of the World, 1991* (L. R. Brown, ed.; New York: Norton, p. 3), S. Postel and C. Flavin describe the contribution of Norway to development assistance in the following terms:

> Norway, in many ways, the world leader in development assistance, might serve as a model. Not only does this small nation provide more aid as a share of its GNP than any other country, it is increasingly focused on sustainable development, as mandated by Parliament in 1987. Agriculture and fisheries receive 19 percent of Norwegian development assistance, and education gets 8 percent. In addition, a special environment fund disbursed more than $10 million to developing countries in 1990. The leading recipients of Norwegian aid are the neediest countries—including, for example, Tanzania, Bangladesh, and India. If the world as a whole had the priorities reflected in Norway's aid budget, Third World environmental reforms would be much further along.

pricing that will not just benefit the mercantile class. Third, it must restructure access to the land and provide inputs for small farmers. It should resist the temptation to import cheap food from abroad, as that will simply remove the incentive from the farmers to increase food production.

Summary

The human population is increasing rapidly and is expected to level off at 10–12 billion sometime between the years 2050 and 2100. This population growth is uneven, with most of it occurring in countries that are in different stages of development. So far, the increase in population has been paralleled by a similar increase in food production. The amount of food available per person has slowly but steadily increased in most areas, except in Africa. In spite of this increase, hunger persists and is actually rising about as fast as the population. The cause is poverty. Poor people do not have access to the available food. In other words, hunger is caused not by a failure of the agricultural system, but by the failure of the socioeconomic system.

Although economic and agricultural development have been paralleled by declines in the death rates and birth rates of a number of countries, this demographic transition is not caused by development. Other developmental paths are possible, depending on the socioeconomic policies of the countries involved. Some have followed a unimodal development policy, while others have followed a bimodal model where a large segment of the population does not share in the benefits of development. In this poverty-stricken underclass, hunger is prevalent and high birth rates persist.

The theory, first proposed by Malthus, that high population growth rates cause poverty has been discredited. Rather, poverty is seen as the root cause of high population growth rates. Embarking on a course of development that will benefit all is no easy task and requires the integration of agricultural and industrial policies.

Further Reading

Abernathy, V. D. 1992. *Population Politics*. New York: Plenum.

Caldwell, J. 1982. *Theory of Fertility Decline*. London: Academic Press.

El-Badry, M. A. 1992. World population change: A long-range perspective. *Ambio* 21:18–23.

Findlay, A., and A. Findlay. 1987. *Population and Development in the Third World*. London: Routledge.

Jones, H. 1990. *Population Geography*. 2d ed. London: Paul Chapman.

Kim, O.-K., and P. van den Oever. 1992. Demographic transition and patterns of natural resource use in the Republic of Korea. *Ambio* 21:56–62.

Lappé, F. M., and J. Collins. 1986. *World Hunger: 12 Myths*. New York: Grove Press.

Meadows, D. 1991. *Beyond the Limits*. London: Chelsea Green Publishers.

Murdoch, W. 1980. *The Poverty of Nations: Population, Hunger and Development*. Baltimore: Johns Hopkins University Press.

Pimentel, D., and C. W. Hall, eds. 1989. *Food and Natural Resources*. San Diego: Academic Press.

Population Reference Bureau. 1992. *World population data sheet*. Washington, DC: Population Reference Bureau.

U.N. Department of Economic and Social Affairs. 1992. *Concise report on the world population situation*. New York: United Nations.

World Resources Institute. 1992. *World resources, 1992–93*. New York: Oxford University Press.

Chapter 2

Farming Systems
Development, Productivity, and Sustainability

Population and food production have been increasing in parallel, and human ingenuity and labor are allowing people to keep up with the ever-increasing demand for food. But how and where is this food produced, and how does the ever-increasing level of production affect the ability to sustain the earth's productivity? Will human-induced climate changes negatively affect the earth's productivity or the earth's ability to feed its human population? Before discussing these complex questions, we need to take stock of the nature of human food, and especially the role played by plant and animal products.

1 | **Directly or indirectly, plants provide all of humanity's food.**

Humans cannot escape the facts that they are part of the earth's ecosystem and that in every ecosystem there are producers and consumers. The producers are the plants; their role in converting solar energy into food energy (chemical energy) is discussed in Chapter 6. The consumers are the animals—from the smallest to the largest—and the micro-organisms that break down dead organic matter and recycle nutrients. To remain in good health, human consumers must drink water, and eat foods that contain minerals, energy-rich molecules such as fats and sugars, and nutrients such as proteins and vitamins that humans cannot make for themselves. Human food needs are explored in greater depth in Chapter 4.

Before agriculture was first practiced, ancient hunter-gatherers had evolved a complex relationship with their environment. They had an intimate knowledge of the plants and animals in their surroundings, and used a wide variety of plant and animal foods. Studies of Native Americans and other aboriginal peoples show that some relied largely on plants, whereas others had diets that included 50% animal products. With the development of agriculture some 12,000 years ago, people narrowed their food selections. Today, a very few cultivated plants account for the overwhelming proportion of the diet.

Subsistence farmers in Africa, Asia, and Latin America still have quite varied diets, consisting of a variety of cultivated plants supplemented by

| Table 2.1 | Comparison of the diets in India and United States |

Food	Source of calories		Source of protein	
	India	United States	India	United States
Cereals, starchy foods	65%	25%	64%	21%
Sugars	6	12	—	—
Beans, lentils	10	4	18	3
Fruits, vegetables	2	6	1	4
Fats, oils	4	19	—	—
Milk, milk products	7	14	11	26
Meat, poultry, eggs, fish	6	20	6	46

Sources: Data from Food and Agriculture Organization and U.S. Department of Agriculture.

plants and animals gathered from the wild. Different cultures rely on different plants as their principal staple, and humanity's three major staples—wheat, rice, and corn—evolved at the same time as the societies that use them. Thus, the Japanese have many different soybean-based foods and sauces, while Westerners eat wheat under many guises.

A look at food sources in various regions of the world reveals another very interesting difference: some societies eat almost entirely plants, while others use a substantial amount of animal-derived foods. This can best be illustrated by comparing the foods consumed in the United States, as an example of a developed country, and India, as an example of a developing country (Table 2.1). Not only is there a difference in the total amount of food available and consumed by the average person in the two countries, but the people of India get a much greater proportion of their calories and protein from plant products directly. They eat only small amounts of eggs, poultry, milk, and milk products. In developed countries, relatively large amounts of meat and animal products are eaten and these animals are themselves consumers—in the ecological sense—because they depend on plants.

The countrywide averages noted in Table 2.1 mask differences between regions, between different social classes, and between people of different religious persuasions. In developing countries, city dwellers typically eat more meat, which has historically been regarded as a sign of affluence, than do rural farmers. Governments often encourage meat consumption by agricultural price supports for feed grains, by tax incentives for feedlot operators, by guaranteed minimum prices, or by government storage of surpluses. In the countries of the Organization for Economic Cooperation and Development (OECD), government subsidies to animal farmers and feed growers totaled $120 billion in 1990. For example, in the United States grazing cattle on public lands is encouraged by charging the ranchers a fee that is much lower than the actual cost. Similarly, in California and other western states, irrigation water to grow alfalfa and other crops is heavily subsidized.

The differences between developed countries and developing countries illustrated in Table 2.1 are quite dramatic. However, we should not forget that there are many more people in the developing countries, and as a result the overall human diet worldwide is skewed toward plants. Indeed, it has been estimated that 85% of the calories and 80% of the protein in the human diet comes directly from plants. But this situation is changing as people become more affluent. Rising affluence therefore puts additional demands on the food system.

2 | The demand for food is growing faster than the population.

Economic development, and the rise in wages and expectations associated with it, lead to increasing demands on the food production system, because when people become more affluent they often want to eat more meat and animal-derived products. Global meat production has increased dramatically in the last 40 years, nearly quadrupling since 1950 (Figure 2.1). Meat consumption per person, calculated as carcass weight, varies from 2 kg per year (India) to 15–20 kg per year (Egypt, Philippines, and Turkey) to 40–50 kg per year (Mexico, Japan, and Brazil) to 50–90 kg per year (CIS, United Kingdom, Italy, Germany, and Argentina) and reaches 90–110 kg per year (France, Australia, Hungary, and the United States). The countries that are developing now and that were in mid development in the period 1960–1990 have seen the most rapid rise in meat consumption. As people get richer, they often want to eat more meat. Many countries, including the United States, have "pro-meat" policies that keep the price of meat low.

Until about 1950 in the industrialized countries, and even today in many parts of the developing world, livestock rearing was integrated with food crop production on ecologically balanced farms that practiced a sustainable mode of farming. Food crops (wheat, potatoes, and sugar beets) were rotated with fodder crops (hay, clover, and alfalfa) that was fed to the livestock. Manures were returned to the soil, and the leguminous fodder crops (clover and alfalfa) added nitrogen to the system (see Chapter 7). Pigs and poultry, which cannot digest grass, were fed crop wastes or grains that were part of the rotation. The modern demand for meat can no longer be sustained by this "antiquated," ecologically sound system, and a large portion of the food crops must now be used as animal feed.

For example, in the United States about 160 million ha of cropland provide food for a population of 250 million people. Considering that at least 20% of the food produced is exported, the potential for domestic consumption is 300 million people. However, 70% of the grain, or 130 million tons, produced in the United States annually is not consumed directly by people, but is fed to livestock (mostly cattle and pigs) on feedlots. If this grain were used directly as human food instead of animal feed, nearly 500 million people could be fed comfortably and with healthy diets (see Chapter 4). Worldwide, about 38% of the grain, especially corn, sorghum, barley, and oats, is fed to livestock.

The feeding of grains to animals is not just a phenomenon found in developed countries. In developing countries, feed grains such as corn (maize) and sorghum are replacing food crops. For example, in Egypt the share of grain fed to livestock has grown from 10 to 36% over the last 25 years as corn replaced traditional food staples such as wheat, rice, sorghum, and millet. In Mexico, where hunger and malnutrition are still widespread, 30% of the grain is now fed to livestock. The area planted to sorghum (a feed crop

Box 2.1

Meat, Manure, and Subsidies

Some choices are difficult. In the Netherlands, a small country in northwestern Europe, the meat and dairy industries produce 85 million tons of manure each year. Feeding animals is an inefficient business, and about 80% of the protein (and larger percentages of other nutrients) that is fed to the animals shows up in the manure in various forms (protein, urea). About 50 million tons of this manure could be disposed of in an environmentally acceptable way, but the other 35 million tons clearly are surplus. Export is not feasible because transport is expensive and the amount of manure is so great. A 25-mile-long manure train would have to leave the country every single day. So the nutrients (100,000 tons of phosphate each year, for example) continue to pollute the ecosystems.

Where does all the animal feed come from? In the past, cows ate grass and chickens ate grain. Until recently, the livestock, poultry, and dairy industries depended on locally produced animal feeds. The most recent rise in consumption of meat and animal products in Europe is driven by feed imports from abroad. Although Europe produces plenty of wheat and other grains (165 million tons annually are produced in the countries of the Economic Community) and actually has a grain surplus of 35 million tons, the animal industries prefer to buy cheaper substitutes. The substitutes are cheaper, in part, because grain production is heavily subsidized by various governments (to the chagrin of other countries—such as Australia—that also have enormous grain surpluses). The grain substitutes that Europe buys—powdered cassava roots, corn gluten, soybean press cakes and citrus pulp—are produced in the United States (40%) and in various countries of the Third World (60%) and enter western Europe primarily through the giant harbor of Rotterdam. Much of the animal feed remains in the Netherlands and is used by the local meat and poultry industries. The rest is dispersed all over Europe by rail and boat. The livelihoods of many Third World farmers and workers now depend on this export of animal feed, which arrives in Rotterdam in giant ships at the rate of 100,000 tons each day.

How can such a problem be solved? The consumers want cheap cheese and steak. The environmentalists want to reduce the manure output. The European farmers want subsidies. The traders and those who promote "development" in the Third World want to continue the trade, which earns hard currency cash for Third World countries. Neither the United States nor the Third World farmers want import taxes that would make European grain more competitive with their products.

Several principles should underlie the search for a solution. First, perhaps too much land is being cultivated in the developed countries. If subsidies are really necessary to give the farmers a decent standard of living, they should be maintained, but their maintenance should be coupled to a lower total output. Land can then be taken out of production and be used for other purposes. Second, the price of steak or eggs should include the cost of disposing of the manure in an environmentally acceptable way. Third, policies in the First World should be examined for their effects on Third World countries. This will involve value judgments. Who benefits if European pork is produced with cheap cassava from Thailand or with soybean press cake from the United States? Do policies in Third World countries that promote the production and export of these new cash crops imperil the ability of these countries to produce food for their own people? Would it be better to take land out of production in the United States, rather than supply the Europeans with grain substitutes so they can have cheaper cheese, eggs, and meats?

Figure 2.1

*Increase in World
Meat Production,
1950–1989. The fastest-
growing meat source in
the United States is poultry, while in eastern
Europe it is pork.* Source:
Data from Food and
Agriculture Organization.

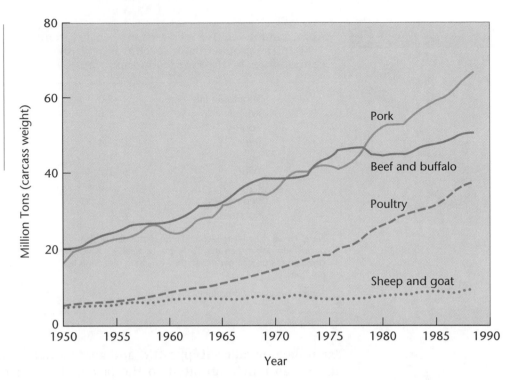

in Mexico), has greatly expanded while the area planted to corn (a food crop in Mexico) has declined from 83 to 69% of the total grain area.

The feed crops produced in the poor countries are often exported to the rich countries to support their meat-eating habits. The countries of western Europe buy groundnuts (such as peanuts) from Nigeria, soybeans from Brazil, and fish meal from Peru to feed their cows, pigs, and chickens. In addition, they raised the productivity of their farmland (Table 2.2). The large increases in oils and grains were not for direct human consumption but made it possible to increase meat production by 50%. The countries of eastern Europe and the FSU (formerly the USSR) aggressively buy grain on the world market to keep feed grain stocks at an acceptable level, even in the face of severe economic problems.

To understand why eating animal products puts additional demands on the food production system, we must also consider the transfer of energy and matter in the ecosystem (see Chapter 6). Remember, plants are the producers and animals are the consumers. If people consume an animal for food, that makes them secondary consumers. Energy and protein are first transferred from the plants, the producers, to the primary consumers, and then to humans. Unfortunately, the transfer step is inefficient. More than 90–95% of the energy and protein are "wasted" (used or excreted) when animals eat plants. The efficiency is higher in feedlots and chicken farms because the animals are confined and do not use energy for getting about. Still, in the United States it takes about 7.0 kg of grain to produce 1 kg of pork, 5 kg of grain to produce 1 kg of beef, and 2 to 3 kg of grain to produce 1 kg of egg, cheese, or poultry.

What matters most to the global food production system is whether the animals eat grains that could also be eaten by humans or whether resources (land, irrigation water, and fertilizers) are used to produce feed grains that could equally well be used to produce food crops. The existence of several successful farming systems based on animal husbandry (for

Table 2.2	Approximate relative increase between 1971 and 1988 in the production of specific agricultural commodities by the European Community

Vegetable oils	2.0
Wheat	1.6
Poultry	1.5
Pork	1.4
Sugar	1.2
Wine	1.2
Eggs	1.0
Vegetables	0.9
Potatoes	0.7

Note: These changes reflect a change in the diet of the population rather than the size of the population, which was constant. As people get richer, they eat more meat and less potatoes.

Source: European Community.

example, nomadic sheep, cattle, and goat herders as well as cattle ranching and sheep ranching) attest to the fact that under certain conditions the inefficient conversion of energy and protein is nevertheless a positive aspect of the food production system. For example, some 30 to 40 million pastoralists living in the world's dryest lands produce animal products (milk, meat, and wool) in areas that are essentially unsuitable for other types of agriculture.

3 | Weather and climate profoundly affect crop production.

The weather is the one factor with which people are familiar that profoundly affects the growth of plants. Everyone knows that to grow, plants need sunlight and rain, and that both are quite variable around the world. **Weather**—defined as short-term (less than a few weeks) variations in the atmosphere—affects the year-to-year yield of crops. The longer-term variations that comprise **climate** determine the geographical conditions where crops can grow at all.

The recent climatic history of the earth is one of change. Over the past 2 million years, there have been repeated ice ages (glacial ages), separated by warmer interglacial periods. About 10,000 years ago, we began the current interglacial period, which started with a rapid rise in global temperatures. This led to more precipitation in many subtropical areas, such as the Indus and Tigris valleys of Asia. As people migrated into these regions, conditions were right for the intensified harvesting of food plants that led to the start of agriculture. Later, about 4,000 years ago, the climate became drier, agricultural productivity declined, and so did these civilizations. A warm phase that affected medieval Europe from 800 to 1200, and the "Little Ice Age" that cooled Europe and North America from 1550 to 1850 also affected crop production during these periods.

Short-term variations caused by the weather that affect crop yields occur even in our modern era of technology-based agriculture:

Table 2.3	Variability in staple food production	

Country	Probability of Production Falling Below 95% Trend
Asia	
Bangladesh	22
India	22
Sri Lanka	29
Africa	
Senegal	39
Tanzania	35
Zaire	15
Latin America	
Brazil	17
Mexico	26
Peru	30

Note: The higher the probability that food production will fall below 95% of the norm, the lower the food security.

Source: Data from J. T. Pierce (1990), *The Food Resource* (London: Longman Scientific), p. 219.

1 *Moisture stress* is caused by insufficient, too much, or ill-timed rainfall. Plants often have precise water requirements, in terms of both amount and time. If these requirements are not met, yields and/or quality suffer.

2 *Temperature stress* occurs commonly when the temperature is either too low or too high for optimal growth. A short period of stress can severely depress growth later. Sudden cold snaps injure plants because they do not have enough time to become acclimatized to the cold weather. High temperature stress is often accompanied by moisture stress.

3 *Natural disasters* occur, such as cyclones, hurricanes, and hailstorms. Although these are more common in such Asian countries as Sri Lanka and Bangladesh, they can also cause severe damage in developed countries, by physically harming crops and the soil that supports them.

Yields of many crops vary considerably year to year in many locations (Table 2.3). The probability of a staple crop yielding the same year after year can be quite low. Over much of North Africa and the Middle East, for example, in half of the crop years yields will fall significantly below expectations.

The question arises, What causes these variations: the weather, or differences in agricultural practice? A hint comes from examining grain yields over time in Canada, where agriculture is technologically advanced (Figure 2.2). Although the yield trend is clearly up, and this rise is obviously caused by improved technology, chemical inputs, and seeds, there is great variability on a year-to-year basis. The reason for this year-to-year

Box 2.2

*Tree Rings
Measure Global
Climate Changes*

Climate has a profound effect on plant growth, and changes in climate (temperature and rainfall) are recorded in the trunks of old trees. Trees grow not only taller as their apical meristems produce more cells, but their trunks grow wider as the cambium, a layer of dividing cells underneath the bark, produces a new layer of xylem cells, or wood, every year (see Chapter 5 for further discussion of plant growth). In conifers, the xylem cells that are produced in the spring are bigger and appear more lightly colored than those that appear in the summer, which are smaller and darker. A cross section of a conifer trunk therefore appears as a series of annual rings, whose widths reflect the growing conditions. Dry years produce narrow rings, whereas the trees produce more xylem cells and wider rings in wet years. Temperature also has an effect on ring width; for example, a long-term cooling trend in the climate causes the rings to become gradually narrower. Scientists called *dendrochronologists* study these tree rings in minute detail, hoping to learn more about climate changes in the past. Ideally suited for such studies are the 5,000-year-old bristlecone pines that grow high up in the White Mountains on the border of California and Nevada. One reason that individual trees have survived for so long is that the dry mountains support little other vegetation, and as a result there are no forest fires.

Initially, dendrochronologists were concerned primarily with dating beams found in archaeological remains. This was done by comparing the ring pattern in such beams with the ring pattern found in cores taken from living trees whose age could be established by counting rings. More recently they have begun to study local as well as global climate changes by studying the ring patterns of various tree species from different areas, and by measuring the ratio of carbon 13 isotope to carbon 12 isotope in individual rings. Carbon dioxide with both isotopes is present in the air and used for photosynthesis, but the ratio of the two forms of carbon dioxide used tells us something about the climate. A short drought lasting a few months will result in wood that has just slightly more carbon 13 in the wood than the usual amount.

The tree ring records also show the existence of a medieval warm period that lasted from about A.D. 1100 to 1375, confirming European writers who described farming by Vikings in the now frozen expanses of Greenland, as well as the existence of the Little Ice Age, a period of cold weather lasting from 1450 to 1850. The tree rings clearly show that in the western United States, the period from 1937 through 1986 was abnormally wet compared with past centuries. The recent (1987–1992) California drought may not be a drought at all, but a return to normality. This could have profound consequences for the future of California's agriculture, which depends heavily on irrigation.

variability is that in Canada wheat is growing at the limit of the climatic zone to which it is adapted. Wheat originated in the Middle East, and that is where its wild relatives still grow. Perturbations in the Canadian weather—an early frost or a dry summer—put stress on the plants, and stress reduces yields.

A similar situation exists in the FSU (former USSR). Complex statistical models have been proposed to explain crop yield variability in other situations, and show that in many cases climate is the major factor. When things go well (the yield trend is a smooth upward or constant curve), it is because the weather is unusually benign.

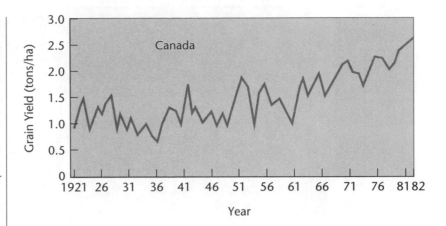

Figure 2.2

Grain Yields in Canada Between 1921 and 1982. Although there is an upward trend since the 1950s, large annual variations remain. These variations are caused primarily by changing weather patterns. The upward trend is caused by the introduction of new genetic strains, the application of more fertilizers and the use of other technologies. New wheat strains must have narrowly defined "bread-baking qualities," and this puts a limit on the yield increase that can be achieved. Source: Data from Agriculture Canada.

Different regions of the world are affected differently by climatic variation. Some of these effects include

1 **Temperate regions.** The north and south temperate regions produce 75% of the world's wheat and corn. The most prominent climatic factors affecting yield are moisture stress and temperature extremes. Wheat and soybeans are more drought resistant than corn.

2 **Humid tropics.** Ranging from fertile valleys to jungles to floodplains, these regions are heterogeneous and show great variability of year-to-year yields. The major determining factor here is rainfall. There are pronounced wet and dry seasons, with the wet season often being in the form of a monsoon, with very heavy, sustained rainfall. But the unpredictability of the intensity of the monsoon, combined with poor water-holding capacity of soils and high rate of evaporation, lead to unstable water supplies for agriculture. In addition, torrential rains on lowland areas, such as river deltas, inundate the crops.

3 **Semiarid tropics.** These regions (for example, the Sahel south of the African Sahara, and much of India) have a long history of periodic crop failures and resulting famines. The major problem here is rainfall or, more precisely, the lack of it. Most of the rain falls in a two- to five-month period (April–October in the Northern Hemisphere; October–April in the Southern Hemisphere), when temperatures are at their yearly peak. This leads to extensive evaporation and less water availability. The sequence of events in growing a crop here are exquisitely sensitive to the annual rainfall cycle. The growth of the crop is timed to coincide with the maximum available water. If this period is very short, so is the growing season.

This relationship of agricultural practices to climate is an ancient one. Every culture has its deities and stories relating the cycle of the seasons to food production. A dramatic way to visualize these cycles is to take "snapshots" of the earth at different times of the year to see which agricultural activities are occurring. Figures 2.3 and 2.4 show two such snapshots, for April and July. The vast range of human activities throughout the globe, many of them tied to climate, is apparent.

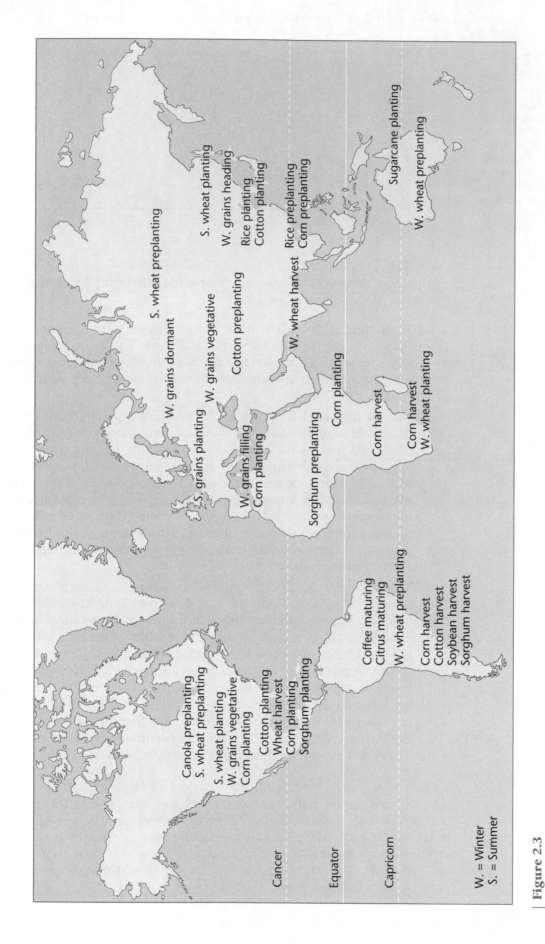

Figure 2.3
Farming Activities Around the World During April.

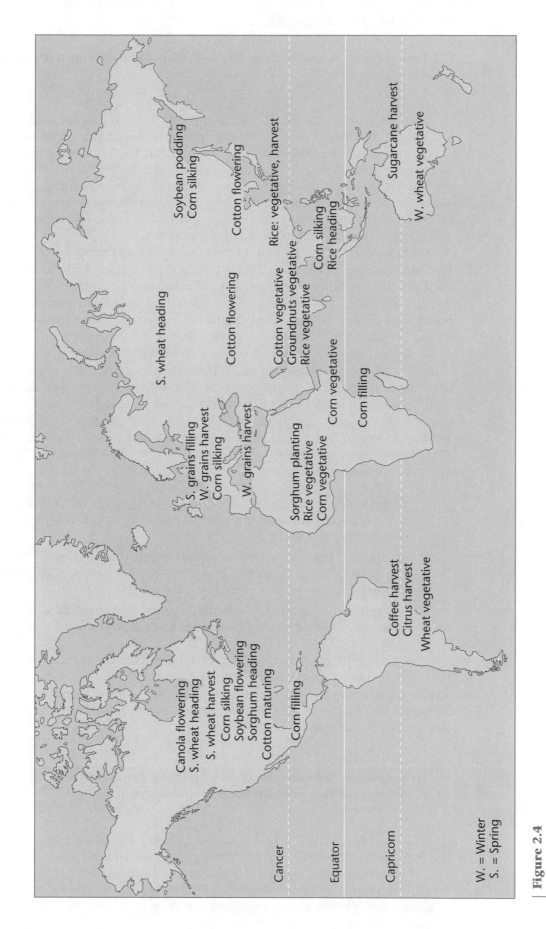

Figure 2.4

Farming Activities Around the World During July.

4 | **Natural events can disrupt normal climatic cycles and affect agriculture.**

Two recent climatic events of natural origin have had great impact on agriculture. The first is the differential **penetration of the monsoon** in its northern region. The semiarid tropics depend on the northward movement of tropical rains (monsoon) for their annual water supply. Above the equatorial tropics, the global air circulation pattern is such that rising, moisture-laden air from the south meets falling, cooler air from the north. This so-called intertropical convergence zone is pushed northward in the summer, and causes the monsoon in the semiarid tropics.

From 1950 to 1965, the high-pressure area pushed the monsoon well into the Sahel and northern India. This led to good rainfall, and adequate agricultural production. Since then, however, the monsoon has not penetrated as far north, and rainfall has been 50–80% of normal. Coupled with population and political pressures, this has had drastically negative consequences for food production and food availability in the Sahel.

A second major climatic event is the **El Niño Southern Oscillation.** Named with the Spanish words for "The (Christ) Child," this phenomenon first was detected early in this century in the South Pacific Ocean. But its "gift" is far from benign. The El Niño condition is caused by a reversal of the normal high-pressure air over Tahiti and the low pressure over Australia. When this occurs, a wave of unusually warm water crosses the ocean in 60 days. When it reached the coasts of South and North America in 1972, and again a decade later, the following occurred:

1 Warm waters caused a failure of the anchovy catch off Peru. Since these fish constitute an important animal feed, this placed extra demand on feed grains.

2 Severe droughts occurred in Central America and Mexico, lowering crop production to the point where formerly self-sufficient countries had to import grain.

3 Droughts also occurred in Australia and southern Africa, in the latter case leading to the deaths of many range animals.

The ability to predict these two disruptive climate events is currently primitive. Climatologists are constructing complex models, using the most advanced computers to try to understand these global symptoms. For El Niño, the models can now pick up its earliest symptoms and warn of anomalies in the coming season.

5 | **Human-induced climatic change may be rapid in the next century, and this may affect crop production.**

The ever-increasing human population and the associated agricultural and industrial activities of humanity are having a profound effect on the climate and therefore on the natural and agricultural ecosystems. Among these human-induced changes are desertification on several continents, the buildup of greenhouse gases and the associated warming of the earth, the increase in

Figure 2.5

Desertification in the Sahel. Desertification of the Sahel, a region in Africa that runs from east to west just south of the Sahara Desert resulted from overpopulation, expansion of herds, and climatic changes. The sequence of events is shown in Figure 2.6. Source: U.N. Educational, Social and Cultural Organization.

acidity of rain over North America and Central Europe, and the increase in ultraviolet radiation resulting from a thinning out of the ozone layer.

The term **desertification** was coined in the 1970s to describe the deterioration of vegetation in the Sahel, a region at the southern edge of the Sahara (Figure 2.5). The northern Sahel is occupied by pastoralists who herd goats and cattle, while the southern Sahel, characterized by dry savanna, is occupied by sedentary farmers. Many observers noted that the Sahara appeared to be marching southward, with scrub giving way to desert, and savanna being replaced by scrub.

This change in the Sahel region coincided with a period of below-average rainfall that started around 1965. It is likely, but by no means proven that the change in rainfall results from both (1) a "natural" climate modification—the failure of the northern penetration of the monsoon—and (2) a human modification of the land itself.

Starting in the 1960s, development assistance was provided, and cash crops for export were introduced (to increase the living standard of the poor subsistence farmers). Thousands of wells were sunk for the nomadic herders (so they could settle down and lead decent lives). Increased sanitation and veterinary care led to dramatic increases in herd sizes and in the human population.

When a period of above-average rainfall (1950–1965) was followed by below-average rainfall (1965–1985), the crops failed, the animals starved, and the people experienced famine. Probably the denudation of the land helped reduce rain. Bare land reflects more heat. Furthermore, less water evaporates because there is less vegetation. These two factors—more heat reflection and less evaporation—cause fewer clouds to form, and as a result less rain falls (Figure 2.6).

This type of human-induced desertification may have happened elsewhere in the past. For example, at one time the Mediterranean basin was covered with forests. Deforestation coupled with overgrazing greatly decreased water evaporation, cloud formation and rainfall, so the whole area is now much drier than before. Such changes tend to be nearly irreversible. Some estimates suggest that 1.5 billion acres—the same amount of land as is now used for crops—were lost

Figure 2.6 |

Chain of Events Contributing to Desertification in the Sahel.
This chain of events was set in motion by the sinking of wells, which changed the lifestyle of the nomadic herders who settled down around the wells. In earlier times, when the herders were nomadic, they moved on before an area was overgrazed. If a drought occurred, they moved hundreds of kilometers to areas where there was food for the animals. This lifestyle prevented any one area from getting overgrazed.

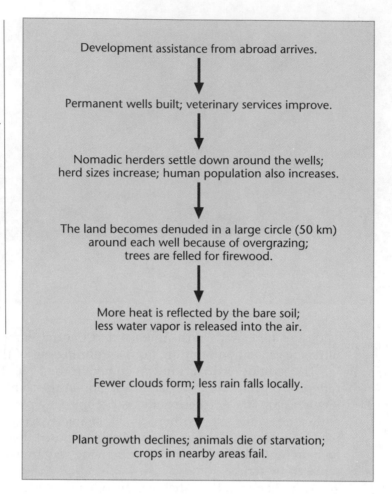

Development assistance from abroad arrives.

↓

Permanent wells built; veterinary services improve.

↓

Nomadic herders settle down around the wells; herd sizes increase; human population also increases.

↓

The land becomes denuded in a large circle (50 km) around each well because of overgrazing; trees are felled for firewood.

↓

More heat is reflected by the bare soil; less water vapor is released into the air.

↓

Fewer clouds form; less rain falls locally.

↓

Plant growth declines; animals die of starvation; crops in nearby areas fail.

to farming as a result of such human activities. Today, desertification caused by overgrazing is also a problem in Patagonia and in the semiarid regions of Iran and Syria, where sheep populations are estimated to be three to four times higher than the carrying capacity of the land.

Bare soil may cause other problems as well. The dust particles that rise from its surface may also prevent rainfall. The region between Lahore and Karachi, on the border between India and Pakistan, was once green and lush, a veritable cradle of civilization. But intense cultivation released tiny soil particles into the air, to the point where now 2 tons of dust or more lie over every square kilometer of what is now a desert. Rainfall may have declined because these particles actually trap moisture, yet are too small to aggregate to form clouds.

The **greenhouse effect**, caused by an increase in the so-called greenhouse gases, is a more global change in the climate, resulting from human activities. When sunlight penetrates the atmosphere and hits the earth, some of it is absorbed by the plants for photosynthesis (see Chapter 6) and some of it is used to heat the planet. But a fair amount of this solar energy is re-emitted to space as heat and light. The principles of physics (thermodynamics) dictate that this re-emitted light is less energetic (has a longer wavelength) than the light that enters the atmosphere.

Carbon dioxide, methane, and chlorofluorocarbons such as the refrigerant freon act along with other gases to allow shorter-wavelength incoming radiation through to the earth, but absorb some of the outgoing, longer-length radiation. This energy is re-reflected back to the earth and significantly warms the planet. In fact, if these gases were completely absent, the average temperature on the surface of the earth would be a chilly $-18°C$. Thus these gases act like the pane of glass in a greenhouse to heat the atmosphere, and so are termed "greenhouse gases."

Clearly, the more greenhouse gases there are, the more the earth will heat up. And greenhouse gases have been increasing in recent years. Carbon dioxide (CO_2), a normal trace constituent of the atmosphere, has been rising rapidly during this century (Figure 2.7), a period that includes both extensive burning of fossil fuels such as coal and oil, and massive deforestation. Such activities produce CO_2. Chlorofluorocarbons are a human invention, and their increase is more recent. Methane, an ingredient of natural gas, comes largely from farm animals (such as cattle), and has increased significantly along with the rising affluence of the human population. The chemical properties of each gas determine how much effect it has on preventing heat loss through the atmosphere. The increase in CO_2 accounts for 50% of the observed greenhouse effect; methane, for 20%; and chlorofluorocarbons, for 14%.

The questions are: Can these gaseous increases, if continued, actually lead to global warming? And if so, how might this warming affect agriculture? Answers to these questions have important policy implications. If an increased greenhouse effect has profound effects on food production around the world, all countries must limit emissions of greenhouse gases. This would mean curtailing the fossil fuels that have driven development in the rich countries and are of increasing importance in the plans of the poor. It might even mean limiting the expansion of animal agriculture.

Already, there is agreement to reduce, and eventually eliminate, the use of chlorofluorocarbons, and since 1988 the world production of chlorofluorocarbons has dropped. The Montreal protocol, signed in 1989, is the first such globally accepted environmental treaty. At the U.N. Environmental Conference in 1992, there was some call to reduce greenhouse gases, but this did not sit well with many developing countries.

The case of CO_2 is difficult and controversial. Although CO_2 has increased, projections for the future are difficult, in part because of recent successful efforts to conserve fossil fuel resources (fewer emissions may be necessary to achieve development), and in part because of the complexities of atmospheric chemistry and possible absorption of the increased CO_2 by the oceans.

In any case, most scientists believe the time for action is now. They are warning the rest of humanity that if things keep going as they are, greenhouse gases will indeed lead to global warming. A consensus has this increase in temperature as being an average increase of about $0.3°C$ per decade, or about $3°C$ by the end of the next century. Climate modeling on this subject has been extensive. Results are uncertain, but the following probable effects emerge:

1 Surface air warms faster over land than water, so the temperature rise in the northern latitudes will be 50–100% greater than the global mean increase.

2 There will be 10% more winter precipitation and 20% less summer precipitation, in central North America and southern Europe.

Figure 2.7

Atmospheric CO$_2$ Concentrations over Time.
(a) *Air bubbles trapped in Antarctic ice show that there has been a dramatic increase in CO$_2$ since the end of the most recent ice age.* (b) *During the last thousand years until quite recently, CO$_2$ has remained constant.* (c) *Monthly averages of CO$_2$ as measured at Mauna Loa Observatory, Hawaii, since 1958. The levels rise during the winter, when respiration and burning exceed plant photosynthesis. During the summer, the reverse occurs and CO$_2$ levels fall.* Source: *F. Salisbury and C. Ross (1992), Plant Physiology (Belmont, CA: Wadsworth), p. 250.*

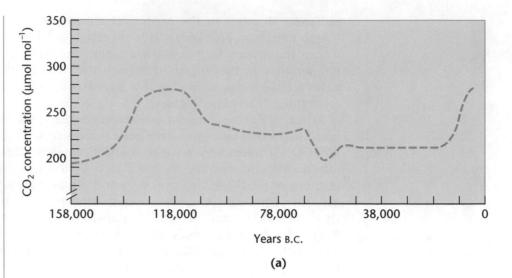

(a)

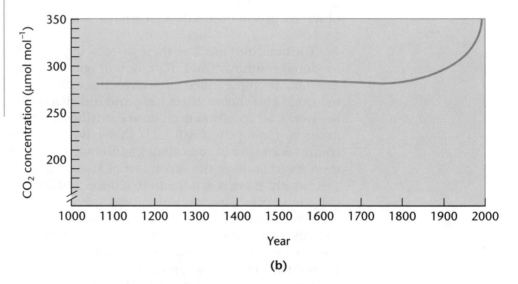

(b)

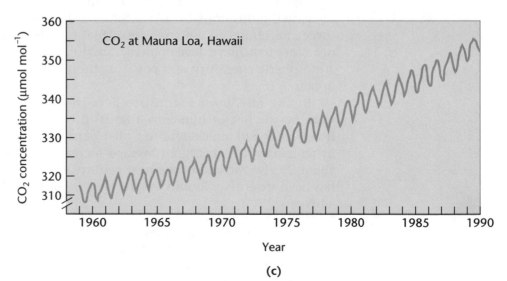

(c)

3 In the southern middle latitudes, there will be warming and increased precipitation.

4 Sea level may rise about 6 cm per decade over the next century because of partial melting of polar ice and glaciers, and thermal expansion of the oceans.

As outlined earlier in this chapter, moisture and temperature stresses can reduce crop yields. Armed with this knowledge, and a map of the probable climate changes due to the greenhouse effect, the agricultural impacts of this climate change can be estimated. Some of these impacts are as follows:

1 The "wheat belt" of the U.S. Great Plains would, because of reduced rainfall, revert to grassland. The same would occur for the major wheat-growing regions of the FSU (former USSR).

2 Corn production in the United States would be reduced significantly.

3 A longer growing season and more rainfall would permit increased wheat production in Canada and more northern latitudes of the FSU.

4 Changing rainfall patterns would play havoc with current irrigation schemes. Land now irrigated might not need it, and new agricultural land would require irrigation.

5 The rise in sea level could inundate the floodplains and deltas on which a great deal of rice grows in Asia. This would severely reduce yields.

Possible climate changes resulting in alterations of food production of this magnitude are hard to ignore. Some governments (such as that of the United States) have called for further studies (few scientists deny the need for this); others have passed laws limiting emission of CO_2, just in case the predictions of the climate model turn out to be right. The dilemma is that by the time additional studies yield definitive data, global warming and related agricultural problems may have already begun—and the process may be irreversible.

6 | How much arable land is there, and where is it located?

With suitable temperature and rainfall, land can be successfully used to grow crops and produce foods. But where is the arable land (cropland) located, and how much is there? Certainly not all the land is used for crops. Of the total land area of the earth, about 11% is used for crops (arable land), another 24% is used for pastures, and 31% is in forests. The arable land is heavily concentrated in the United States, Europe and the FSU, India, China, and Southeast Asia (Figure 2.8). The rest of the land surface, about one-third, is either too cold, too dry, or too steep for plant growth.

The location of arable land is itself a function of climate, soil type, and type of vegetation that grew in the area before the land was cleared. Climate, soil type, and vegetation all interact when soils are formed and the most productive soils were formed in areas that had permanent grassland (such as

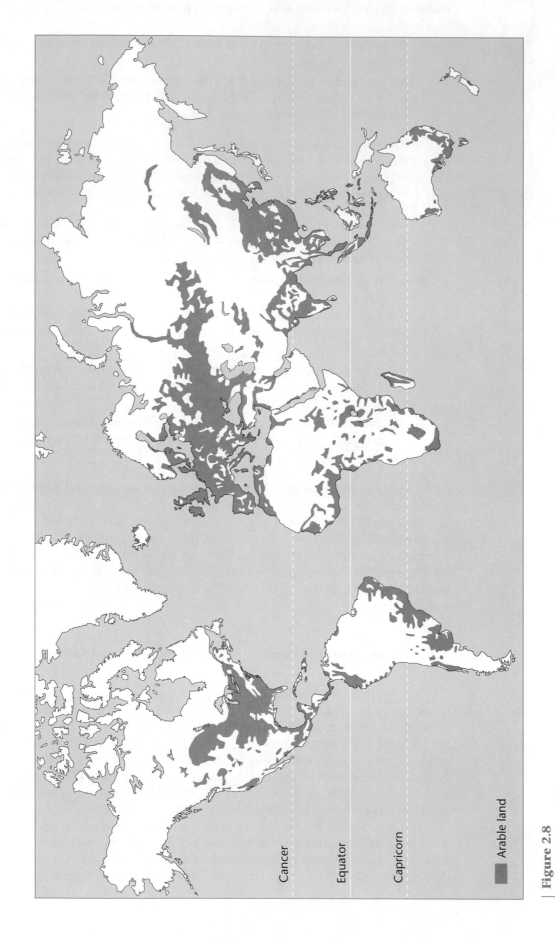

Figure 2.8

The World Distribution of Arable Land. Notice that most of the arable land is in the United States, Europe, Russia, Western Asia, India, and China.

Cancer

Equator

Capricorn

Arable land

the prairies of the midwestern United States) or hardwood forests (most of Europe and India). Grasslands produce rich soils with abundant organic matter that can keep on being productive for many decades when they are converted to cropland. These soils are quite suitable for annual row crops (corn, wheat, beans, and so on).

The process of converting natural ecosystems into pastures or cropland is now being repeated most dramatically in the tropical rainforests of South America. However, converting a tropical rainforest into an agricultural field can be a perilous operation. Most rainforests grow on relatively infertile and highly weathered soil. Because of the high rainfall and the high temperature, the clay minerals that make up a tropical soil (see Chapter 7) are in a much more advanced stage of dissolution, and many mineral nutrients have been leached out of the soil. Whatever nutrients there are, are actually part of the vegetation and the organic matter on the soil.

When such land is cleared and the vegetation is burned, the mineral nutrients are released. That is the principle behind slash-and-burn agriculture (Figure 2.9). However, when the trees are gone, the bare earth is exposed to the sun, soil temperatures rise, and the soil organic matter quickly decays because it is not being replenished by the vegetation. Because of the high rainfall, the soil erodes easily. And when inorganic fertilizers are added, they are often quickly lost (nitrate, for example) or immobilized so they are not available to the plants (phosphate, for example).

Asia, Latin America, and Africa contain many examples of massive soil degradation as a result of clearing away tropical forests and imposing European- or American-type agricultural practices on soils that are totally unsuitable for row cropping. New research shows that many problems can be avoided if the farmers quickly move in with an agroforestry crop system using an upperstory of trees that shades the ground and an understory of crops (annual or perennial). (See Chapter 16 for further discussion of this system.)

Putting more land to the plow may not achieve a greater total of cropland than the 1.5 billion hectares presently used. The FAO estimates that 5 to 7 million hectares are being lost each year as a result of land degradation. One problem is that the areas with the greatest demographic pressures have the least land reserves that could be cleared for crops. This intensifies the use of the existing arable land, and worsens the land degradation problem.

7 | Land use patterns in agriculture show increased intensity of resource use.

Clearing away the natural ecosystem is the way in which all arable land has been created. The development of agriculture in different regions of the world has been accompanied by an increased intensity of land use. Each stage of intensity has its own characteristics and productivity (Table 2.4):

1 **Forest fallow.** Initially, farmers slash and burn parts of the forest. This releases the nutrients contained in the burned plants, which fertilize the land and allow it to support crops. But this fertility lasts only until the crops deplete the soil of its nutrients, and the farmers then move on to another part of the forest, repeating the cycle. In the meantime, the original land is left for up to 20 years, to restore

Figure 2.9

Slash-and-Burn or Forest Fallow Agriculture in Northern Yucatán. This type of agriculture is sustainable as long as the population density is low and the fields are allowed to lie fallow for 15–20 years. **(a)** This field, in which corn and squash are grown together, can only be used for two years because the soil is thin and nutrient poor. **(b)** The ears of corn produced on this field are small because the soil lacks nutrients, especially nitrogen. Source: *A. Chrispeels.*

(a)

(b)

its fertility (fallow period). Parts of Africa and Latin America are still farmed in this way, but it requires a lot of land per person.

2 **Short fallow.** As human population density grows, grassland replaces the forest. The food requirements of the less mobile population reduce the fallow period to one or two years. This is usually not enough to fully restore the soil to allow it to support crops, so added fertilizer—often animal manure—is required. In addition,

Table 2.4	Farming operations in different systems			
Operation	Forest Fallow	Short Fallow	Annual Cultivation	Multiple Cropping
1. Land clearing	Fire	None	None	None
2. Land preparation	None	Plow	Plow; tractor	Plow; tractor
3. Fertilization, compost	Ash	Manure; compost	Manure; compost; chemicals	Manure; chemicals
4. Weeding	Low	Intensive	Intensive	Intensive
5. Animals or machines, transport	None	Plowing; transport	Plowing; transport; irrigation	Plowing; irrigation
6. Percentage of world cropland	2%	28%	45%	25%
7. Grain yield (kg/ha)	250	800	2,000	5,000

Source: Adapted from H. Binswanger and P. Pingali (1988), Technological priorities for farming in sub-Saharan Africa, *World Bank Research Observer* 3:83. Data on yields and areas from P. Buringh (1989), Availability of agricultural land, in D. Pimentel, ed., *Food and Natural Resources* (New York: Academic Press).

increased demands on the land require higher yields, achieved by manual pest control (for example, weeding).

3 **Annual cultivation.** With a still-increasing population, now coupled with the demands of a market at which the crops can be sold (this could be a foreign country, in the case of colonization), the fallow period disappears and a crop is grown every year. Now constant fertilization and pest control are essential to keep up the yield. One way farmers ensure the nutrient content of the soil is by rotating crops that remove different nutrients; for example, planting corn in one year (removes a lot of nitrogen from the soil) and soybeans the next (removes little nitrogen, allowing the soil to replenish for the next corn crop).

4 **Multiple cropping.** In terms of food production, this is the most intensive use of the land. Not only are several crops grown on the same land sequentially in one year, but two or more crops are grown at the same time. This latter pattern, called **intercropping**, is often done for the same reason as crop rotation in that one species removes one type of nutrients, and the intercrop removes other nutrients.

Historically, the transition from stages 1 to 4 has been gradual, as populations grew slowly and the necessary technologies evolved over time. More recently, explosive population growth, coupled with political pressures and available technology, has hastened the transition.

The biological and technological advances that allowed this increasingly intense land use are listed here (many are discussed in detail later):

1 **Introduction of new crops.** This advance began with the exchange of plants between continents after the voyages of European discovery in the fifteenth and sixteenth centuries. Potatoes and corn, domesticated in the Western Hemisphere, had an enormous impact on food production in Europe. The exchange of

Figure 2.10

The Mechanization of Agriculture. Mechanization has been important in raising agricultural productivity. The energy used in the manufacture and operation of machines, such as wheat harvesters shown here, is a major input in modern agriculture. Source: U.S. Department of Agriculture.

plants continued into the eighteenth and nineteenth centuries, with corn and cassava becoming major food staples in Africa.

2   **Mechanization.** The process of mechanization started in the middle of the nineteenth century. New farming techniques depended on scientific advances and the industrial production of inputs. Manual labor and animal power were replaced by steam power and later by the internal combustion engine. Tractors were used to pull plows or the reaper binder, which was later replaced by the combine harvester (Figure 2.10). Milking machines, cotton gins, cotton pickers, sugar beet harvesters, and tomato harvesting machines all have greatly reduced the need for manual labor.

3   **New and improved varieties.** The application of genetics to plant breeding began in the eighteenth and nineteenth centuries, before the heritability of plant characteristics was clearly understood. After the rediscovery of Mendel's work around 1900, crossing different crop varieties led to improved strains. Recently, the principles of improved cultivar selection have been systematically applied to improving rice and wheat, and new varieties have been widely introduced in Mexico, Asia, and Africa, displacing the local varieties in many areas. This process, often called the Green Revolution, has raised crop productivity. Geneticists estimate that roughly 40% of all increases in crop productivity are due to plant breeding and the other 60% to a variety of other inputs (energy, fertilizer, and pesticides).

4   **Inorganic fertilizers.** Originally farmers relied on manure and crop rotation to maintain fertility. Later they found that ground bones and rock phosphate enhanced crop production, as well as nitrates from Chile, in the form of guano. The invention of the Haber process, which combines nitrogen gas with hydrogen gas to form ammonia, allowed the widespread production of nitrogen fertilizers. This led to a tremendous increase in fertilizer use worldwide.

| Table 2.5 | The technological basis of modern agriculture (inputs per hectare of corn in the United States) | | |

Input	Hand Produced	1910	1980
Labor (h)	1,200	120	12
Machinery (kg)	1	15	55
Animal use (h)	0	120	0
Fuel (L)	0	0	125
Manure (kg)	0	4,000	1,000
NPK fertilizer (kg)	0	0	316
Lime (kg)	0	10	426
Seeds (kg)	11	11	21
Insecticides (kg)	0	0	2
Herbicides (kg)	0	0	2
Irrigation (%)	0	0	17
Drying (kg)	0	0	3,200
Electricity (10^3 kcal)	0	0	100
Transport (kg)	0	25	326
Yield (kg)	1,880	1,880	6,500

Source: Data summarized by D. Pimentel, from USDA and estimates, in D. Pimentel, ed. (1989), *Food and Natural Resources* (New York: Academic Press).

5   **Pesticides.** The use of herbicides replaced manual labor for weeding, and insecticides were used to minimize crop losses. These changes in food production had a great impact on commercial grain farming (for example, in the United States, Canada, and Australia), on mixed crop and livestock farming (in Europe) and on intensive wetland rice farming in Asia and Africa.

These technological advances have led to a style of agriculture that is very dependent on technological inputs and is called "high-input" or "conventional" agriculture. A comparison of the production methods for corn in the United States from 1910 to the present clearly shows this dependence (Table 2.5). The technological advances resulted from government-sponsored and industrial research, the main thrust of which has always been to develop technological inputs that let farmers minimize per unit production costs. Many of these advances diminished work opportunities on the farm in favor of jobs in towns and cities where the inputs are manufactured. The benefits of the lower production costs (cheaper food) accrued to the consumers and to agribusiness, while the penalties (pollution, land degradation, pesticides) had to be borne by society as a whole.

8

The application of scientific knowledge to agriculture has resulted in greatly increased yields per unit land area for many of our important crops.

Most of the world's food comes from a small number of plants. Improving their yield has been the main way food production has increased in the past

Figure 2.11

Yields of Wheat in England and Brown Rice in Japan over the Last Several Centuries.
Source: *L. T. Evans (1980), The natural history of crop yield,* American Scientist 68:388–397. *With permission of the publisher.*

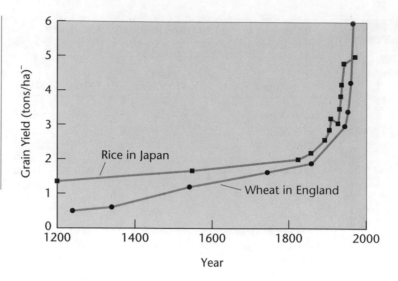

40 years, and is likely to continue so for the next few decades. We noted in the previous chapter that worldwide food production has been rising by 2.3% annually (0.5% on a per capita basis). This increase is the direct result of the development of high-input agriculture, which in turn results from applying scientific knowledge in plant nutrition, plant genetics and breeding, and plant–water relations. This use of scientific knowledge started about 150 years ago in the developed countries and has resulted in yield takeoffs, which can be charted quite accurately for individual crops (Figure 2.11). When Thomas Malthus proposed his theory of population and predicted the world would soon run out of food, for example, he could not foresee that wheat production in Britain (his homeland) was about to take off to keep pace with the population increase.

A classic case of a country turning to yield agriculture is Japan. By 1900, this island nation was growing crops, mostly rice, on all its arable lands. But the population was still growing. The government was reluctant to become dependent on foreign food imports lest a hostile nation cut off food from Japan during an international crisis. Short of lowering the nutritional standard of the people, the only thing to do was to increase the yield of the rice crop. The government mobilized the country's political, social, and scientific resources for the task, and as a result the yield per hectare increased about threefold between 1900 and 1965. Since 1965, rice yields have risen much more slowly in Japan, in spite of additional scientific and technological inputs, leading some to suggest that the yield potential of the present strains of rice may have been reached.

A similar yield takeoff occurred for corn in the United States, starting about 1950. Hybrid corn was introduced in 1935, and subsequent breeding efforts were directed at making the hybrids responsive to nitrogen fertilizer (N-fertilizer). The dramatic increase in the use of N-fertilizer (Figure 2.12) was a major contributing factor to the subsequent fourfold rise in yield per hectare. Some have argued that the introduction of hybrid corn itself was the crucial technology responsible for the yield takeoff for corn, but a similar three- to fourfold increase in yield per hectare occurred for wheat in the United States and in the United Kingdom (Figure 2.11). Wheat is not a hybrid, and such data suggest that a combination of all technologies is responsible for these historic rises in productivity.

Figure 2.12

Corn Yields and the Use of Nitrogen Fertilizers in the United States: 1910–1985. The arrow points to the introduction of hybrid corn. By the time the planting of hybrid corn was complete (1950–1955), yields had increased somewhat. The big yield increase parallels the use of nitrogen fertilizers and is caused by the breeding of strains that respond to nitrogen fertilizers. Source: Data from U.S. Department of Agriculture.

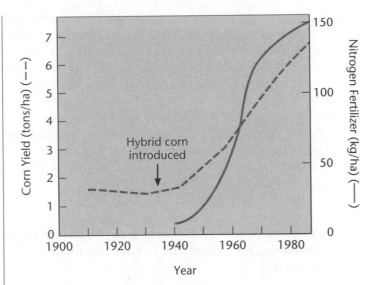

Such dramatic increases in productivity have been repeated more recently (1960–1975) in different regions of the world: wheat in Mexico and southern Asia, rice in the Philippines and Pakistan. These increases are the basis of the "Green Revolution," which is discussed in greater detail in Chapter 11. The increases depended on the same technologies noted earlier: fertilizers, pesticides, mechanization, irrigation, and the breeding of high-yielding strains of wheat and rice that can use these inputs.

This development allowed many Third World countries to raise food production and even become food exporters. Breeding the high-yielding varieties and applying the industrial inputs was seen by international agricultural experts as the quickest way to achieve increased production. As a result, agriculture in developing countries has come to resemble agriculture in developed countries: more heavily dependent on purchased outside inputs and more capital intensive, and less dependent on labor. Earlier, we noted that the money generated from labor is needed for families to buy food. Merely raising production is not enough to eliminate hunger. Because the need for farm laborers goes down as agriculture becomes more technological, the displaced laborers must turn to other sources for income, and too often these are lacking. The inputs are expensive, and even if money or credit are available, there are many regions where people lack the crop varieties and knowledge to effectively use these inputs (for example, much of Africa).

When farmers first apply a new technology—for example, fertilizer or pest control—yield immediately increases as the amount of input rises. A little bit is good, and more is better! During this phase, the money spent on the technology results in a substantial return. Once the plant's inherent capacity to respond to the input is reached, however, the yield increase slows drastically. This is especially true for fertilizers and is discussed in greater detail in Chapter 7. Adding fertilizer and other inputs follows the law of diminishing return. In the United States, 90–100 kg of fertilizer are now applied per ha of cropland, and it is not easy to raise the yield further by adding more. This figure is 65 kg/ha for India, so there is still scope for improvement, but fertilizer use is only 4 kg/ha in Ethiopia and 12 kg/ha for Costa Rica. Gains in crop productivity should be possible in these countries if we can find the proper cropping systems and crop strains that can take advantage of additional fertilizer. When all technologies are maximal, a yield plateau may be reached. The leveling off of rice yields in Japan between 1965 and 1992 could indicate that a plateau has been reached. Some crop physiologists argue that we will always be able to find new crop strains to raise the yield plateau and take advantage of new and existing technologies. Others fear that further increases in yield are not possible with the present agricultural system. Only time can tell whether the yield takeoff curves shown in Figure 2.11 will continue to rise, or whether they are about to take on the familiar S shape associated with so many processes in the world.

9 | **The effect of agricultural techniques on the ecosystems is causing concern about the sustainability of intensive farming.**

Growing populations and increasing affluence demand ever-increased output from agricultural systems. This is true whether we are talking about the high-input agriculture of the United States and Western Europe, about nomadic herds in Africa, or about intensive rice cultivation in Asia. Concerns are now being raised about the **sustainability** of this ever-intensifying productivity.

The World Commission on Environment and Development defined sustainable development as "development which meets the needs of the present without compromising the ability of future generations to meet their own needs." To understand sustainability, we must first take a look at the rate at which productive soils are being lost as a result of present agricultural practices. The FAO of the United Nations has estimated that a quarter of the world's arable land is subject to degradation through salinization, soil erosion, and desertification. Second, there may not be enough energy resources to maintain high-input agriculture. Third, the ever-increasing use of pesticides is arousing concern that we are polluting our ecosystems with synthetic chemicals. Fourth, the trend toward genetically uniform crops increases the potential for serious disasters by eliminating the many strains of a given crop that were previously used by farmers. Fifth, government policies perpetuate conventional agriculture and discourage farming practices that could make agriculture more sustainable.

Land Degradation. The term *land degradation* is used to describe the physical and chemical changes that reduce long-term soil productivity. Such changes are sometimes very obvious, but are often difficult to measure because they occur over long periods of time. They are often compensated for by other improvements, such as more fertilizer or new genetic strains, and as a result the effect of land degradation does not always show up as decreased yield—at least, not in the short run.

An example of declining crop yields as a result of land degradation is found in Malawi, where unfertilized corn yields declined from 1,800 kg/ha in 1960 to 1,000 kg/ha in 1990. As pressure increased to produce more corn, the fallow period was first reduced and then eliminated. Soil organic matter declined, and soil was lost through erosion. The country is now faced with the prospect of a rapidly increasing population and lower yields of its most important staple caused by soil degradation.

Physical degradation takes the form of erosion (Figure 2.13), the carrying off of soil particles by wind and water. In an annual cropping system, the soil is alternatively covered with plants and almost totally bare. When the soil is bare, it is exposed to higher wind velocities and to the force of raindrops, which destroys the structure of the uppermost layer of topsoil. The net result is muddy runoff that carries away soil particles and/or dust storms.

Chemical degradation can take several forms, including (1) acidification from acid rain, (2) alkalinization and salinization (buildup of salts) as a result of irrigation, (3) exhaustion of the mineral nutrients when insufficient fertilizers are used to replace the minerals that are removed with the crop, and (4) leaching of excess mineral fertilizers in the streams, lakes, and groundwater.

Energy. In bygone days, farmers used their own seeds or other planting material for vegetatively propagated crops such as potatoes and employed animal power and human labor. They bought very few things, and the value of the purchased inputs was only 5–10% of the value of the outputs (agricultural products). In high-input farming, this value is now 50% or higher. Farmers buy seeds, fertilizers, pesticides, herbicides, tractors and other machinery, and irrigation equipment, and often must provide and maintain housing for seasonal workers. All these purchased inputs depend heavily on the use and therefore on the price of energy. Most of this energy comes ultimately from coal, natural gas, or oil, our primary energy sources.

Calculations of energy ratios (energy input per kilogram of output) show that modern rice production in the United States requires 375 times as much commercial energy per hectare, as compared to the traditional production in Asia, and uses 80 times as much energy per kilogram of rice produced. This high input of energy results in a 4.5-fold increase in the yield per hectare. Some have argued that we are now "eating fossil energy" (coal, oil, and natural gas), because the energy we eat as food is produced at the expense of energy in fossil fuel. Indeed, this is true. However, compared with food processing and all uses of fossil fuel, modern crop production in high-input, high-output systems is remarkably energy efficient. In the United States, only about 3–5% of the total energy used is required for crop production.

Pesticides. Pesticides are used extensively in the developed countries, and their use in Third World countries is rapidly increasing. The problems that accompany this widespread pesticide usage are extensive and are discussed in Chapter 13. These problems include the emergence of many pesticide-resistant pests, adverse health effects on tens of thousands of farm workers worldwide, pesticide residues on crops, and pollution of lakes, streams, and groundwater with pesticides. Furthermore, the realization that pesticides have not diminished the proportion of the crop lost to pests is prompting many people to try alternate pest control methods.

Genetically Uniform Crops. A main feature of our high-input system of agriculture is genetically uniform crops, because new strains (more resistant to diseases, more responsive to fertilizers, higher-yielding hybrids) have helped raise yields. Yet their use has several drawbacks. First, when farmers start planting seeds furnished by seed companies or state agencies, they stop using and usually discard their own varieties, which were well adapted to the

Figure 2.13

Eroded Soil. Excessive water runoff and an inadequate plant cover have led to this gully erosion. Source: *U.S. Department of Agriculture.*

microenvironment on their farm. Local varieties contain valuable genes that are thus lost to the plant breeders. Second, genetic uniformity means crops are more susceptible to sudden outbreaks of diseases. When a pathogen mutates and overcomes the plant's resistance, the entire crop over a wide geographical area can be ravaged.

Government Policies.   A final area of concern is government policies that actively promote high-input agriculture and prevent sustainable practices and technologies. Such policies are difficult to change because of the vested interests of large and powerful lobbies, including farm organizations and industry advocates.

10 | Can agriculture be made more sustainable?

The high-input agricultural system now used in most developed countries has many undesirable side effects, and the high price of purchased inputs makes it ever more difficult for farmers to earn a fair return on their labor, in spite of dramatic increases in subsidies. To protect their resource base—the soil—and to decrease costs, a number of farmers in the United States and elsewhere have begun to adopt alternative practices. Taken together, these practices constitute "sustainable agriculture," which differs from conventional agriculture not so much by the practices it *rejects* (for example, heavy use of inorganic fertilizers), but by the practices it *incorporates* into the farming system. The more widely used term "organic farming" describes two major aspects of alternative agriculture: (1) the substitution of manures and other organic matter for inorganic fertilizers, and (2) the use of biological pest control instead of chemical pest control. The objective of sustainable agriculture is to sustain and enhance, rather than reduce and simplify the biological interactions on which production agriculture depends. Alternative agriculture is not a single system of farm practices, but encompasses many farming systems also called *organic, biological, low input, regenerative,* or *sustainable* systems. Such systems emphasize management practices as well as biological relationships between organisms; in addition, they take advantage of naturally occurring processes such as nitrogen fixation.

According to a major study by the National Academy of Sciences of the United States, alternative or sustainable agriculture is any food or fiber production system that systematically pursues the following goals:

- More thorough use of natural processes such as nutrient cycles, nitrogen fixation, and pest–predator relationships in agricultural production
- Reduced use of off-farm inputs with the greatest potential to harm the environment or the health of farmers and consumers
- Greater productive use of the biological and genetic potentials of plant and animal species
- Improved match between (1) cropping patterns and (2) the productive potential and physical limitations of agricultural lands, to ensure long-term sustainability of current production levels
- Profitable and efficient production, with emphasis on improving farm management and conserving soil, water, energy, and biochemical resources

A few practices and principles emphasized in sustainable agriculture are

- Rotating crops to mitigate problems with weeds, diseases, and other pests
- Planting legumes to increase soil nitrogen
- Minimum tillage, contour plowing to reduce erosion
- Mulching the soil to conserve water
- Integrated pest management relying heavily on biological rather than chemical control methods
- Using genetically improved crops that are resistant to insect pests and diseases

Studies of alternative agriculture show that the farms derive sustained economic benefits and are not necessarily at a disadvantage. Although yields are often lower, costs are also considerably lower because of eliminating most pesticides. Wider adoption of these techniques would result in significant environmental benefits. Most government policies now strongly favor high-input monoculture agriculture and other agricultural practices that tax the environment. As a whole, government policies in industrialized countries work against environmentally benign practices and the adoption of sustainable cropping systems.

11   Bangladesh illustrates the challenges posed by population, land use, and sustainability.

Surrounding the delta of the Ganges River as it flows into the Bay of Bengal east of India, Bangladesh was formed in 1971 as a primarily Muslim state, formerly East Pakistan. It is the eighth most populous country in the

| Table 2.6 | Effect of submergence on rice yield |

Growth State	Rice Yield (t/ha) at Plant Depth of Water (%)			
	0	25	50	75
Growth of branches (early)	7.9	6.7	5.8	5.7
Initiation of buds	7.9	6.1	5.9	5.5
Flowering	7.9	5.8	5.2	3.2

Note: This study was done in India.

Source: From R. Rajendran, S. Santhanabosu, P. Selvaraj, and V. Shanmu-
gasundaram (1992), Effect of submergence on rice yield, *IRRI News* 17:1.

world, but with an area a third the size of Japan and with about the same
population (120 million), Bangladesh has one of the densest concentrations
of humans, with over 800 people per km². Past high TFR (6.0 in 1980) ensures
a high proportion of young people. In 1990, about half of the total population was under 15 years old. Even though the TFR has gone down significantly (to 4.5 in 1990), the young population means that by 2000 the total
will be 137 million, and by 2010 it will be 155 million. Feeding these numbers will be a major challenge.

The poverty of this increasing population poses another challenge to the
food system: distribution to those in need. Various estimates put the proportion of people growing for themselves, or buying, an adequate quantity of
food, at a mere 5–25%. So over three quarters of the people are not getting
enough food.

Weather, especially the monsoon (called *kharif* in Bengali) and the dry
season (*rabi*), plays a dominant role in the crop-growing cycles in Bangladesh.
During the wet season, crops can usually be grown without irrigation. But
during the dry season, irrigation is often necessary. Although these seasons
are regular, the intensities and lengths may vary. Thus, of about 13 million ha
under cultivation (a remarkable 60% of the total land area of the country),
almost half is prone to flooding and a quarter is susceptible to drought. In
addition, cyclones—which regularly make the news because of people drown-
ing and capsized boats—wreak havoc on the coastal plains, causing loss of
crops and soil.

These stresses on the ecosystem that is supporting agriculture are illustrated by studies of rice yields from river deltas in Asia. In Bangladesh, water
often inundates rice fields to levels far above normal paddy conditions (typically 5 cm standing water). Studies of a single rice variety in the field show
that, depending when the fields flood and to what height water covers the
plant, yields can fall significantly (Table 2.6). Especially when rice is flowering, water control in these regions is essential—but the people often cannot
afford it.

Specific rice crops are adapted to the wet–dry cycle:

1 *Aus* is a rice planted before the monsoon and harvested during it.
2 *Aman* is a rice planted during the monsoon and harvested after it.
3 *Boro* is a rice planted and harvested during the dry season.

Figure 2.14

Transplanting Paddy Rice Seedlings in Bangladesh. Lowland regions such as this are prone to salinization and weather-related disasters. Source: United Nations.

Wetland rice (Figure 2.14) is the principal crop of Bangladesh. Food production has grown at an annual rate of 2% over the past 15 years, with much of the increase due to irrigated dryland rice and wheat. This trend has considerable potential for continuing, and even increasing (Table 2.7). Farmers are using improved strains of rice, wheat, and legumes, specifically adapted to conditions in Bangladesh. More land is being irrigated, so that water supply during the dry season is more reliable. Soil analyses have shown specific deficiencies (for example, much of the land is low in sulfate) that can be simply remedied to improve yields. The land used for cultivating legumes can be increased.

But this growth in food production must occur without the extensive degradation of agricultural resources now occurring. A third of the cultivated land, because of combined climate, drainage, and fertilization factors, is becoming excessively loaded with salts (salinized), to the point of lowering crop yields. Faulty cultivation practices and the weather have combined to bring about the loss of topsoil, with 10% of the cultivated area at risk. Finally, the high intensity of farming—up to three crops a year on the same piece of land—has resulted in the loss of soil nutrients, especially organic matter. The poor farmers generally do not have enough capital to replace the lost nutrients with inorganic fertilizers.

Thus, Bangladesh faces a triple challenge over the next two decades:

1 It must continue to bring its population growth under control.
2 It must increase its food production.
3 It must reverse the decline in its agricultural resource base.

Not only must it do all those things, but it must create rural employment so that poverty and hunger can be abolished. Its industrial policy will have to be integrated with its agricultural policies.

Table 2.7	Food needs and production potential in Bangladesh (millions of tons)

Item	Production 1990	Demand 2010	Potential 2010
Grain	18,450	29,720	42,120
Legumes	0.54	0.91	1.05
Meat	0.29	0.84	1.33
Fish	0.84	1.80	3.83
Fruits, vegetables	3,679	15,338	12,910

Source: Adapted from F. Mahtab and Z. Karim (1992), Population and agricultural land use: Towards a sustainable food production system in Bangladesh, *Ambio* 21:53.

Summary

Plants—especially the cereal grains wheat, rice, and corn—provide well over 80% of the food humans consume directly. Thus increasing agricultural production to keep pace with the growing population requires managing plant growth and development. Rising affluence worldwide puts additional stress on the plant-growing capacity of the earth. People want to eat more animal products, and because conversion of plant to edible animal products is inefficient, the land needed to produce a kilogram of animal food could be used to grow 7–15 kg of plant food for direct human consumption.

The major natural environmental variable affecting crop production is climate. Historically, both long-term (for example, ice ages) and annual (seasons of rainfall and temperature) weather patterns have affected plant growth. Societies have adapted their agricultural practices to these cycles. Both natural and human-induced climatic abnormalities affect crop production. Two examples of the natural type are the failure of the annual monsoon rains to arrive and the El Niño current in the Southern Hemisphere. Two examples of climate changes caused by people are (1) the contribution of overgrazing, with its lessening plant cover, to desertification in the Sahel; and (2) the continuing emission of "greenhouse" gases such as CO_2, which could lead to global warming, with profound effects on agriculture.

Only one-fifth of the earth's land surface is appropriate for growing crops. Converting these lands from their natural ecosystems to artificial, agricultural ones was traditionally done with minimal disturbance to the environment, letting the soil regenerate itself when necessary (allowing a fallow period). But as the need to grow more food, for local or distant human consumption, increases, the intensity of land use increases.

Technological inputs such as mechanization, fertilizers, irrigation, and pesticides are needed to increase the land's productivity. In addition, strains of crop plants have been bred to take full advantage of these inputs. These methods have resulted in "yield takeoffs" for some crops in some regions.

But the increasing technological management has a negative effect on the long-term sustainability of the ecosystem. For example, up to 3% of the world's valuable topsoil is lost every year due to erosion, and salt buildup (salinization) is common in arid regions. Bangladesh is an example of a country facing the problems of population growth, poverty, and an unreliable agricultural ecosystem whose sustainability is a challenge.

Further Reading

Barrow, C. J. 1991. *Land Degradation.* Cambridge, UK: Cambridge University Press.

Binswanger, H. 1988. Technological priorities for farming in sub-Saharan Africa. *World Bank Research Observer* 3:81–92.

Chleq, J. -L., and H. Dupriez. 1988. *Vanishing Land and Water.* New York: Macmillan.

Food and Agriculture Organization. 1988. *World Agriculture: Toward 2000.* Rome: FAO.

Gnaegy, S., and J. Anderson, eds. 1991. *Agricultural Technology in Sub-Saharan Africa.* Washington, DC: World Bank.

Griffiths, J. F., and D. M. Driscoll. 1982. *Survey of Climatology.* Columbus, OH: Merrill Press.

Grigg, D. B. 1974. *The Agricultural Systems of the World: An Evolutionary Approach.* London: Cambridge University Press.

International Rice Research Institute. 1990. *Climate and Food Security.* Philippines: Los Baños. IRRI.

Lugo, A. E., ed. 1987. *Ecological Development in the Humid Tropics.* Arlington, VA: Winrock.

Mahtab, F. U., and Z. Karim. 1992. Population and agricultural land use: Towards a sustainable food production system in Bangladesh. *Ambio* 21:50–55.

Omara-Ojungu, P. 1992. *Resource Management in Developing Countries.* New York: Wiley.

Organization for Economic Cooperation and Development (OECD). 1992. *Future of Agriculture.* Paris: OECD.

Parry, M. 1990. *Climate Change and World Agriculture.* London: Earthscan.

Pierce, J. T. 1990. *The Food Resource.* New York: Longman Scientific.

Ramakrishnon, P. S. 1992. *Shifting Agriculture and Sustainable Development.* Paris: UNESCO (United Nations Educational, Social and Cultural Organization).

U.S. Department of Agriculture. 1986. Crop yields and climate change to the year 2000. Washington, DC: USDA.

U.S. Department of Agriculture. 1987. Major world crop areas and climatic profiles. Washington, DC: USDA.

U.S. National Research Council. 1990. *Saline Agriculture.* Washington, DC: U.S. Government Printing Office.

Wrigley, G. 1982. *Tropical Agriculture: The Development of Production.* 4th ed. London: Longman Scientific.

Chapter 3

Plant Biotechnology
An Overview

We noted in the two previous chapters that the earth's population has increased rapidly and is likely to reach 12 billion by the year 2100. Global food production has also increased rapidly in the past 60 years and has kept up with the ever-expanding population. Raising the yields of food crops will continue to be the main goal of agricultural research. At the same time, people must shift from a high-input agriculture to a sustainable agriculture to minimize the impact of agriculture on the environment. Recent developments in plant biotechnology are changing the ways in which food crops, pharmaceuticals, and some chemicals needed by industry will be produced. For example, plant genetic engineering is revolutionizing the way in which new crop varieties are made. In this chapter, we provide an overview of the changes in agricultural systems brought about by plant biotechnology.

1 | **Biotechnology is the use and manipulation of living organisms, or substances obtained from these organisms, to make products of value to humanity.**

Although *biotechnology* is a relatively new term, its origins go back to the dawn of civilization, when women first began to keep domestic animals and grow crops for food instead of depending on what they could gather in the wild. The domestication of plants and animals gradually changed their characteristics. Selection for certain characteristics may have occurred on purpose, as well as unwittingly. When subsistence farmers set aside a portion of their harvest for planting in the next season, they are selecting seeds with specific characteristics. It is likely that this also happened 12,000 years ago when our first crops were being domesticated by neolithic farmers. When harvesting grain, for example, they unknowingly selected in favor of plants that produce more seeds per plant, have a uniform ripening of the seeds, and do not scatter their seeds before harvest. Thus plants as well as animals were modified by people for their own use. The discoveries that fermenting fruit juice and honey produce alcoholic beverages, and that fungi improve the taste and preservation of clotted milk solids that have been separated from whey, led to the culture, selection, and gradual improvement of micro-organisms for food preparation.

Within the past hundred years, the pace of modifying useful organisms accelerated. As farming and industry became more interdependent and as the food industry developed, plants as well as microbes were often modified to suit industrial needs. Breeders selected strains of wheat, corn, and rice that responded to chemical fertilizers. The architecture of cotton plants was altered to make mechanical harvesting easier, and tomato breeders selected fruits with tough skins that could withstand the bruising treatment of harvesting machines. Plants that produce valuable chemicals needed for food preparation or other human uses, such as the peppermint plant (for mint oil) were carefully selected and propagated. Industrial microbiologists selected strains of micro-organisms that produce drugs such as penicillin or enzymes that are added to laundry detergents or used in the food industry.

Some 30 or 40 years after laboratory scientists began improving micro-organisms, they also entered the field of crop improvement and started working together with plant breeders. Mutagenesis of seeds by X rays, the rescue of embryos arising from previously impossible interspecific crosses, and the sterile culture of plant organs and cells were the first technologies to be used for creating improved crops. Now, with the advent of gene transfer technology, an even more dramatic modification of organisms has become possible. Before, breeders were largely limited to genetic exchanges within and between species. Now we can transfer genes and therefore inherited traits between very different organisms. This ability to transfer genes among humans, plants, and bacteria has revolutionized biotechnology. However, although gene transfer (genetic engineering) is an important aspect of biotechnology, agricultural biotechnology involves other important new techniques, such as plant tissue culture and mass production of biological control agents.

2 | Hydroponics is the simplest form of plant biotechnology.

In 1860, two German plant physiologists discovered that many plants could be grown, not in soil, but in aqueous solutions of four salts: calcium nitrate, potassium dihydrogen phosphate, magnesium sulfate, and a trace of iron sulfate. Given only these, as well as oxygen, carbon dioxide, and sunlight, most plants flourish. Some 50 years later, researchers realized that the salts used by the German plant physiologists were not completely pure and contained impurities that were essential for growth. It then became evident that plants require eight other nutrients in trace amounts, in addition to the six macronutrients (calcium, magnesium, potassium, sulfate, nitrate, and phosphate) originally identified. These discoveries are at the basis of all research in plant mineral nutrition and the formulation of inorganic fertilizers for agriculture (see Chapter 7 for a detailed discussion of plant nutrition). Hydroponics, the culture of plants without soil, became the biotechnological spinoff of this research (Figure 3.1).

At first plants were grown in large containers of aerated nutrient solution or in gravel beds that were flushed regularly with a nutrient solution. In many commercial enterprises, these methods have now been replaced by

Figure 3.1

Hydroponics Can Be Used to Grow Plants on Both Small and Large Scales. (a) *A simple system for the home, called the "Baby Bloomer," uses a lower tray, in which nutrient-rich water is kept and then pumped at intervals into an upper tray, where the plants are in pots in crushed rock.* (b) *A more complex water system in a farm where tomatoes are grown hydroponically.* Sources: (a) *Photo courtesy of A. Douvas;* (b) *Photo courtesy of K. Holt, Holt Farms, Guymon, Oklahoma.*

(a)

(b)

the nutrient film technique, in which a thin film of nutrients dissolved in water continuously flows past a fairly small root system (small compared to plants grown in soil). The continuous flow ensures that nutrients at the right concentration are always available to the plant.

This new type of hydroponics is now widely practiced in several countries for the production of high-value vegetable crops. The extensive experience of the Dutch with growing crops and flowers in greenhouses led naturally to their combining these two technologies. In the United States, hydroponics is rapidly becoming a major industry. Because of the high requirement for capital (hydroponics requires an expensive physical plant) and energy, only high-value crops can be grown in this way. Hydroponics is well suited for growing fresh vegetables (tomatoes, peppers, lettuce, and so on), but is unlikely to be used for growing important staples, such as corn, wheat, or potatoes.

3 | Plant cell and tissue culture form the basis of several new industries.

The knowledge that whole plants could be grown in the laboratory in correctly formulated solutions led people to experiment with plant organs. Would it be possible to grow roots in continuous culture by supplying them only with mineral nutrients and sucrose, the product of photosynthesis? The answer, to everyone's surprise, turned out to be no. Roots also need tiny amounts of three B vitamins (thiamin, nicotinic acid, and pyridoxine) for continuous vigorous growth. In the whole plant, these vitamins are normally made by the shoots and transported to the roots. Conversely, roots also produce some special products that are needed for growth by the shoots.

Micropropagation. Scientists also experimented with culturing very small (embryonic) leaves and the tiny masses of dividing cells that are present in every shoot apex. Such a region of cell division is referred to as a **meristem** (see Chapter 5), and the technique is called *meristem culture*. When placed in culture, an embryonic leaf will grow into a mature leaf and a tiny shoot apex will produce a mature shoot (a stem with leaves). Such shoot development does not require any additives such as vitamins to the culture medium. Occasionally roots appear at the base of such cultured shoots, and a whole new plant is formed. Thus, by starting with a small shoot apex consisting of a meristem (a mass of dividing cells) with a few embryonic leaves, it is possible to grow a whole new plant in culture. If trace amounts of two hormones, auxin and cytokinin, are added in the right concentration to the culture medium, not one shoot, but multiple shoots appear from a single-shoot apex culture (Figure 3.2). A few weeks' growth may produce 20 tiny shoots that can be separated from one another and grown into full-sized plants. This method of multiplying plants is called **clonal propagation**, and all the plants form a **clone**, a group of genetically identical individuals. This does not mean that there may not be some minor variations between different members of a clone, as is also the case with identical twins. The term *cloning* can be applied to organisms, cells, and molecules (especially molecules of deoxyribonucleic acid, or DNA) and refers to the process of producing identical copies. The cloning of plants in tissue culture, as just described, forms the basis of a whole new worldwide micropropagation industry that multiplies plants by this

Figure 3.2

Multiple Shoots Arising in a Sterile Tissue Culture. Source: *Courtesy of S. Satoh.*

clonal method. Within a year, a single plant can produce millions of cloned plants that are essentially identical.

An example of a slightly different type of meristem culture is shown in Figure 3.3. Here the flower meristem of millet was transferred to a sterile culture dish, and the cells multiplied to form a callus, a mass of undifferentiated cells. At the surface of the callus embryo-like structures called *somatic embryos* were formed (**1**). These somatic embryos developed into shoots (**2**) that later formed roots and developed into plants (**3**).

More than a thousand different plant species have been propagated in tissue culture. The method has been used in hundreds of small and large nurseries and laboratories throughout the world. Micropropagation has been especially successful with ornamental plants, and many ornamentals sold in supermarkets are thus propagated. The method has been extended to agricultural plantation crops, such as potato, strawberry, oil palm, and banana, as well as to medicinal and aromatic plants and trees.

One obvious advantage of micropropagation is that elite specimens—especially in the case of trees—can quickly be multiplied without fear that the genetic makeup will change. When plants reproduce sexually, change in genetic makeup usually occurs. Improving tree crops via traditional plant-breeding approaches is very slow and uncertain, and micropropagation has considerably reduced the time needed.

The oil palm was one of the first crops improved by applying tissue culture research. As early as the 1960s, the multinational Unilever Corporation, owner of large oil palm plantations in Malaysia and other tropical countries, understood the potential of clonal propagation for improving the oil palm. Productivity among different oil palm trees varies greatly and generating high-yielding strains has been impossible with classical plant breeding, because an individual tree's capacity for oil production can only be measured after many years of tree growth. The Unilever laboratories started to propagate clones of elite trees that had oil production 30 percent higher than average. The oil palm industry needs 30 million trees annually to replace aging

Figure 3.3

Regeneration of Millet from a Cultured Floral Meristem. The first panel shows a callus in sterile culture; the second panel shows the regenerated millet plant; in the third it is growing in soil. Source: *T. S. Rangan and I. K. Vasil (1982), Somatic embryogenesis and plant regeneration in tissue cultures of* Panicum miliaceum *L. and* Panicum miliare *Lamk., Zeitschrift für Pflanzenphysiologie 109:49–53; photo courtesy of Indra Vasil.*

trees, so there is great demand for the laboratory-propagated, high-yielding trees. In addition to being high yielding, the new trees are also shorter, making mechanical harvesting cheaper and easier.

Obtaining disease-free plants is another major advantage of the sterile tissue culture approach. Before meristems are transferred to a tissue culture medium, they are thoroughly surface-sterilized to kill any bacteria or fungi on the outside of the tissue. However, the real advantage of meristem culture is that it produces virus-free planting material. Viruses live within plant cells and are transmitted from one generation to the next in plants, such as potatoes, strawberries, or sugarcane that are vegetatively propagated (without seeds). As viruses cause serious crop losses by weakening the plants, it is important to start with plantlets that are virus free. For reasons that are not clear, there are generally no viruses in plant meristems, and starting from a meristem ensures that all plants derived from it also lack viruses. Commercial strawberry growers and potato producers start every year anew with virus-free planting material produced via micropropagation.

Large-Scale Production of Plant Propagules. Propagules are the unit structures that plants use to propagate themselves either sexually or vegetatively. Seeds are the most common propagules, but there are also many types of vegetative propagules, such as potato tubers, onion bulbs, and ginger rhizomes. In a cholla cactus, the propagules are the short, spiny segments that make up the stem. These segments scatter when the plant is touched by an animal or shaken by the wind.

During normal plant development, a fertilized egg cell in the female reproductive part of the flower develops into an embryo and then into a seed, the basic unit of plant propagation (see Chapter 5). The same properties that seeds acquired during plant evolution also give them their ease of handling and storage for agricultural purposes. Seeds are usually dehydrated and quiescent, have a strong protective coating that contains antibiotics, and contain the food reserves that are needed for initial growth of the embryo. For some agricultural uses, it would be advantageous if the ease of handling seeds could be combined with the benefit of clonal plant propagation using *in vitro* somatic embryogenesis. To use somatic embryos as functional seeds, they

Figure 3.4

*A Carrot Embryo
Encapsulated in a
Hydrated Gel.* Source:
Courtesy of S. Satoh.

would have to be individually packaged in some way to let them be handled, transported, stored, and dispersed on the field, and then grown into plants.

The technology of choice at the moment is to encapsulate each embryo in a protective hydrated gel (Figure 3.4). The kind of packaging needed will depend on the plant species, the specific agricultural application, and the physiological state of the embryo—whether it is dehydrated or not, and quiescent or not. The capsule must provide physical protection and can carry nutrients, growth regulators, antibiotics, and fungicides that all help establish the seedling in the field. Technologists envision both hydrated and dry encapsulation methods, but widespread use of such synthetic seeds with either method has not yet occurred. The technical problems that still need to be solved are mainly in the area of seedling establishment in the field.

Somatic embryos can be produced (1) from a callus growing on a solid medium or (2) in a liquid suspension culture. Large-scale production from a callus on a solid medium is not economically feasible because of the enormous labor costs involved: The embryos or growing plantlets must be manually teased apart and transplanted. The production of somatic embryos in liquid culture can be scaled up in bioreactors, and millions of embryos can be produced at once. If these can be turned into useful propagules, the units of plant propagation, then the economic bottleneck is solved. Successful production of such propagules produced in tissue culture is possible with food crops such as tomato, celery, asparagus (production of plant material), and potatoes (production of microtubers), as well as for many ornamentals such as lilies, Boston fern, begonias, and African violets. The transfer of plantlets from liquid culture to trays containing a sterile soil mix is done by a machine that transplants up to 8,000 plantlets per hour. Plants that have a greenhouse stage before transplantion to fields or pots for resale seem the most suitable for this procedure.

4 | **Plant organ and cell culture are used to produce specialty chemicals.**

Embryonic regions of plants, such as root tips and shoot apexes, can be cultured in the laboratory, but what about more mature tissues where the cells have ceased to divide? In the 1930s to 1950s, scientists in France and the

Figure 3.5

Diagram of Organ or Callus Formation in Tissue Culture. For most plants, the levels of two hormones, auxin and cytokinin, in the medium and the ratio of one to the other determine whether a small piece of tissue only forms callus (when both hormones are at high levels), sends out shoots (high cytokinin, low auxin) or sends out roots (high auxin, low cytokinin). When both hormones are at low levels, cell proliferation is minimal, and a small amount of callus growth ensues. Source: *Courtesy of S. H. Howell.*

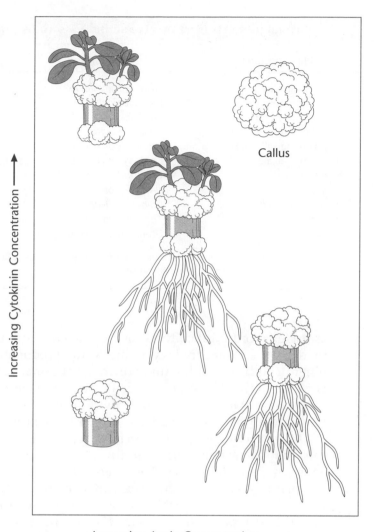

United States discovered that these tissues too could be grown in sterile culture. When small sections of leaves, stems or roots are cut out, sterilized, and put in culture, they form calluses, masses of relatively undifferentiated cells that remind one of the tissues that form around wounds on the stems of trees. The cells in such calluses divide, enlarge, and divide again, provided the medium contains not only minerals and sucrose, but also one or more plant hormones such as auxin and cytokinin. Calluses are normally grown on a nutrient medium that has been solidified with agar, which is a naturally occurring polysaccharide extracted from kelp, and callus cultures have been established for many plant species. For each species, it is necessary to strike the correct balance of auxin and cytokinin in the growth medium.

Vigorously growing calluses can also be cut into small pieces that can be shaken in a liquid nutrient medium (usually the same medium, but without the agar) in a rotary shaker. Single cells or small clumps of cells will separate from the surfaces of the callus pieces and swirl around in the nutrient medium. These cells continue to divide and enlarge, and the small clumps continuously break apart. This process establishes a **cell suspension culture**. Such cultures can be handled in the laboratory or the factory in the same way

Table 3.1	Natural products from plants and their associated industrial uses		
Industry	**Plant Product**	**Plant Species**	**Industrial Uses**
Pharmaceuticals	Codeine (alkaloid)	*Papaver somniferum*	Analgesic
	Diosgenin (steroid)	*Dioscorea deltoidea*	Antifertility agents
	Quinine (alkaloid)	*Cinchona ledgeriana*	Antimalarial
	Digoxin (cardiac glycoside)	*Digitalis lanata*	Cardiatonic
	Scopolamine (alkaloid)	*Datura stramonium*	Antihypertensive
	Vincristine (alkaloid)	*Catharanthus roseus*	Antileukemic
Agrochemicals	Pyrethrin	*Chrysanthemum cinerariaefolium*	Insecticide
Food and drink	Quinine (alkaloid)	*Chinchona ledgeriana*	Bittering agent
Sweetener	Thaumatin (protein)	*Thaumatococcus danielli*	Nonnutritive sweetener
Cosmetics	Shikonin	*Lithosperum erythrorhizon*	Dye

Source: Modified from M. W. Fowler (1983), Commercial applications and economic aspects of mass cell culture, in S. H. Mantell and H. Smith, eds., *Plant Biotechnology* (Cambridge, UK: Cambridge University Press), p. 102.

as bacterial or fungal cultures can be. Bacterial and fungal cultures are widely used in the biotechnology industry for producing industrial feed stocks (chemicals used in manufacturing), pharmaceuticals (penicillin), and enzymes. Over the years, micro-organisms have been modified (domesticated) to maximize their productivity in culture. Similar developments are now under way in the plant cell suspension culture field.

Many species of plants, especially tropical plants, produce unusual chemicals that have found uses as drugs, cosmetics, flavoring compounds, or agrichemicals. Until now, these chemicals have always been extracted from the plants grown in open fields, often in tropical or semitropical regions. However, it is also possible to establish cell suspension cultures from the organs that produce these chemicals, and the plant biotechnology industry aims to grow the cells that produce these chemicals in cell suspension cultures, maximize the production of the chemicals, and eventually replace farms with factories (see Table 3.1). This process imitates the advances in microbial biotechnology that have taken place over the past 50 years.

5   Regeneration of whole plants from single cells is an important new source of genetic variability for plant improvement.

When small pieces of plant tissue are put in sterile culture on a solid medium, the cells proliferate and a callus is formed, and if auxin and cytokinin are present in the correct amounts shoots will be formed. Shoots normally arise from small groups of rapidly dividing cells within the callus. This discovery led scientists to ask a fundamental question: Can a whole plant be grown from a single cell, given the correct nutritional and hormonal environment? The finding by German, Japanese, and U.S. scientists that this is indeed the case let them draw an important conclusion about genes and how they function.

Making an entire organism depends on the correct expression of at least 30,000 genes. When a living plant cell is isolated from a mature tissue, it can

Figure 3.6

Two Somatic Embryos of Corn. Source: *Lu, Indra Vasil, and Ozias-Akins (1982),* Theoretical and Applied Genetics *62:109–112; courtesy of Indra Vasil.*

be induced to start dividing again and all the genes necessary to make an entire organism can be induced to function again in the correct sequence. This ability of a single mature plant cell to give rise to an entire organism is called **totipotency**. Because plant cells are interconnected by their cell walls, in a kind of honeycomb, it is not so easy to isolate single cells. It is much easier to digest the cell walls of a small piece of plant tissue with enzymes and then isolate the protoplasts (naked cells without walls). The protoplasts are quite fragile, but they too can be cultured and regenerated into entire plants.

Protoplasts are usually cultured in a liquid nutrient medium. They first regenerate a new cell wall, and the formation of a complete cell with a wall is followed by the induction of cell division, resulting in small clumps of cells. From such small clumps, there are two routes to regenerating a complete plant. The route of choice usually depends on the species.

When the small clumps are transferred to a solid medium, they will grow into a callus that can form shoots. These shoots can be cut off and transplanted to a medium that induces root formation, and in this way new plants can be regenerated. Alternately, for some plant species, the small clumps can be made to develop into embryos. Such embryos, called *somatic embryos,* resemble the embryos that normally arise as a result of the growth of a fertilized egg cell. In some cases, hundreds of thousands of embryos can be produced in liquid culture. In other cases, embryos or embryo-like structures are formed at the surface of an embryogenic callus growing on solid medium (Figure 3.6). Such embryogenic calluses are formed when tissues obtained from developing seeds are used to start a new tissue culture. By removing the embryos from the surface of the callus, they can then be grown into complete plants. The three different routes for regenerating plants in tissue culture—from meristems, from mature plant parts, and from single cells—are diagramed in Figure 3.7.

Figure 3.7

Outline of Procedures Used in Micropropagating Plants. Sketches not to scale. Source: *Modified from R. M. Klein and D. T. Klein,* Fundamentals of Plant Science *(New York, Harper & Row), p. 379.* © 1988 by Harper and Row, Publishers, Inc. Reprinted by permission of HarperCollins Publishers, Inc.

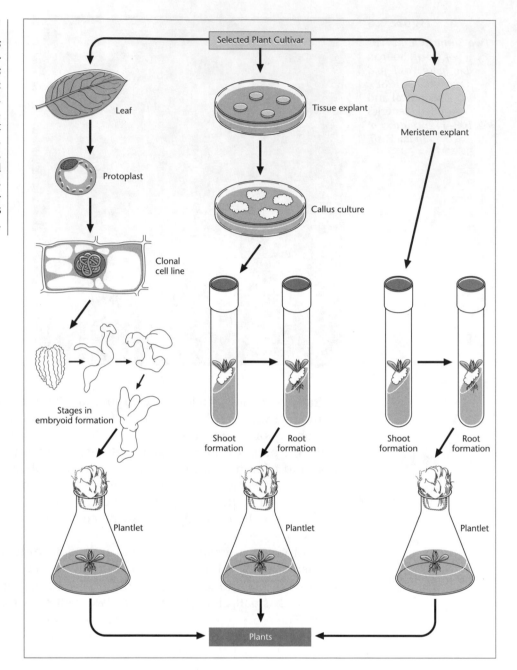

When somatic embryos derived from single cells are grown into mature plants, the plants' characteristics vary somewhat. This phenomenon is called **somaclonal variation**. Scientists were surprised by this variation because all cells in an organism were thought to have the same genetic makeup. So all the plants regenerated from single cells should be exactly alike. This is not the case, because rearrangements of the genetic material and mutations occur as plant cells divide when they give rise to an entire plant. Thus, in a mature plant all the cells do not have *exactly* the same genetic makeup. In addition, genetic material is often rearranged when cells are maintained in tissue culture, and during regeneration of somatic embryos from these cells. Such tissue culture-induced mutations are probably the main source of somaclonal variations. When plant breeders discovered somaclonal variation, they started exploiting it as a new source of genetic variation for crop improvement. Plants regenerated from tissue culture that had desirable properties were further propagated vegetatively (as in the case of potatoes or sugarcane) or used for plant-breeding experiments.

The ability to regenerate plants from single cells (whether a protoplast or a cell that is still part of a piece of tissue) is also an important new technology for producing genetically engineered plants. Genetic engineering involves transferring genes between unrelated organisms. In plants, this is accomplished by a natural gene transfer mechanism (see the next section). Once the transfer has occurred, all the cells that did not receive the new gene must be killed by an antibiotic, and a new plant must be regenerated from the one cell that was transformed by receiving the new gene.

6   The transfer of foreign genes into plants is made possible by a promiscuous bacterium.

The goal of plant genetic engineers is to isolate one or more specific genes and introduce these into plants. The improvement of a crop plant can often be achieved by introducing a single gene, and genes can now be transferred to plants by using the natural gene transfer system of a pathogenic soil bacterium. The soil bacterium *Agrobacterium tumefaciens* causes tumors (called *crown galls*) in many plants. When these bacteria infect a wound site, usually on the stem close to the ground, the infection disturbs the normal wound-healing process. Instead of making a wound callus that covers the wound, cells proliferate into cancerous growth (Figure 3.8). Cells from this tumor can be grown in tissue culture and unlike normal plant cells, they continuously proliferate even when hormones are absent from the culture medium.

Molecular biologists in Belgium, the Netherlands, and the United States found that after the bacteria attach themselves to the wound site, they transfer some of their genes to the plant cells. These genes carry the information necessary to make auxin and cytokinin. The abnormally high levels of these hormones produced as a result of this gene transfer prompt the cancerous growth. This resembles the situation in culture where the continuous availability of the same two hormones in the culture medium causes cell proliferation and callus formation. After the genes have been transferred from a single bacterium to a single plant cell and integrated into a chromosome of the plant cell, the cell—which is now **transformed** with the bacterial genes—begins to proliferate. So the bacterial genes will be present in every cell derived from the transformed cells. This transfer of DNA between the different species could be called "promiscuous," as DNA is usually transferred only between individuals of the same species as part of the process of sexual reproduction. It is important to realize that this transfer of DNA between totally unrelated organisms is not an artificial process, but a natural one.

Soon molecular biologists realized that if they substituted other genes for the genes that the bacterium transferred to the plant cell, they could obtain transformed plant cells that carried any gene that they wished to introduce into the plant. In addition, if they could regenerate a whole new

Figure 3.8

(a) **Crown Gall Disease.** *The soil bacterium* Agrobacterium tumefaciens *infects a wound site on the stem of a tomato plant. The T-DNA, which is part of a circular DNA molecule, is transferred from a bacterial cell to a plant cell and becomes integrated in the DNA of the plant. The T-DNA carries genes that encode enzymes for the synthesis of plant hormones. When the plant cell starts to synthesize these hormones, cell division ensues and a tumor is formed. The plant cells in the tumor are transformed because they carry foreign DNA.*
(b) **Transformation in the Laboratory.** *(1) Pieces of tomato leaves are cultured together with* Agrobacterium tumefaciens *in which the T-DNA genes have been replaced by genes that the experimenter wishes to introduce into the plant. (2) The leaf pieces are moved to a new medium that kills off the bacteria and the plant cells that did not receive the new DNA. (3) The pieces are moved to a medium that will induce the formation of shoots, and eventually a whole new transformed (transgenic) plant is regenerated.*

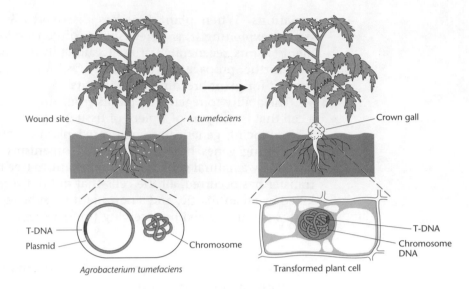

(a) Crown Gall Disease

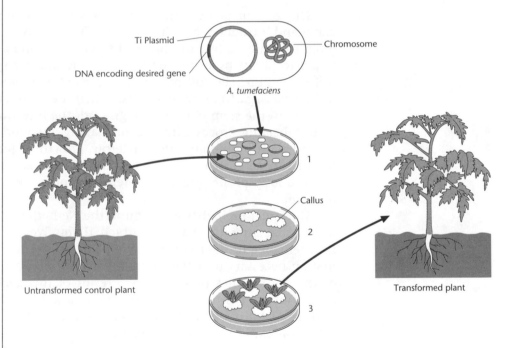

(b) Transformation in the Laboratory

plant from this transformed cell, they would have a transformed plant in which every cell carried the gene. Thus, by combining molecular biology and plant tissue culture and by relying on the natural gene transfer system of this bacterium, it has become possible to transform plants with a variety of interesting and important genes. Transferring genes from one organism to another is the goal of genetic engineering, and plant genetic engineering is an important field within plant biotechnology. Other methods of transferring genes into plant cells, besides the one just described, will be discussed in Chapter 15.

7 | **Improved crop plants that carry one or more foreign genes constitute one of the main goals of plant genetic engineering.**

A principal goal of plant genetic engineering is to improve crop plants by introducing genes from other organisms into them. Many desirable characteristics could be acquired by crop plants by adding just one gene, and the technology to do this is now available. In concept, adding a gene to a plant's genetic makeup is not very different from what plant breeders are doing now. When breeders wish to make wheat resistant to a fungus that infects wheat, they look for a wild wheat variety that is resistant to this fungus. By making the necessary crosses, they transfer the resistance gene from the wild wheat to the cultivated wheat. When this is done by breeding rather than by gene technology, hundreds of other genes are transferred at the same time as the resistance gene (the only one that interests the breeders in this case). The other genes that go along for the ride usually do no harm, but are not needed in the new strain, either.

What are the bottlenecks for improving plants by gene transfer? First, many useful genes have not been precisely identified. In the case of the fungal wheat rust disease, researchers do not know which gene will make the plant resistant, so they cannot even try to isolate it from wild strains of wheat or related grasses that are resistant to the fungus. However, the power of genetic engineering is that no restrictions apply to the organisms between which genes can be transferred. So one can look anywhere for useful genes, and pathogens can often be thwarted by using their own genes. Specific viral genes have been identified that, when introduced into the plant, keep the virus from multiplying or that contain its spread from cell to cell and thus contain the damage and crop losses that viruses cause. The bacterium *Bacillus thuringiensis* carries a gene that makes an insecticidal protein and provides another example of gene transfer between totally unrelated organisms. When this bacterial gene is transferred into plants, the plant cells make the insecticidal protein, and insect larvae are killed when they eat the leaves or roots (Figure 3.9). In these cases, we have identified the genes needed to improve the plants. However, in many cases, while we can identify useful properties of crop plants, such as resistance to fungal or bacterial diseases, greater ability to take up nutrients, greater ability to withstand cold stress, heat stress, drought, or salinity, we have no idea which gene or genes that would have to be isolated and transferred to impart such important characteristics to crops.

The second major bottleneck concerns regenerating plants from the cells into which the new genes have been transferred. There are several methods to transfer genes into cells, but for many species it is difficult to regenerate plants from the transformed cells. Growing complete transformed plants out of single transformed cells is easy for tobacco and tomato; not too difficult for rice, cotton, or canola (oilseed rape); and anything but routine for corn, wheat, or soybean (see Table 3.2).

The major difference between medical biotechnology and agricultural biotechnology is that the former aims to produce high-value products that will be used in small amounts (pharmaceuticals, vaccines, kits to diagnose diseases, and so on) whereas the latter seeks to produce a large amount of a low-value product (seeds of plants transformed with one or more genes from

Figure 3.9

Resistance to Insect Larvae. A tomato plant that has been transformed with a bacterial gene encoding an insecticidal protein is more resistant to insect larvae than are control tomato plants. Source: *Courtesy of Jan Leemans, Plant Genetic Systems, Ghent, Belgium.*

Table 3.2	Examples of crops able to be transformed and regenerated					
Cereals	**Fiber Crops**	**Food Legumes and Oilseeds**	**Horticultural Crops**		**Pastures**	**Trees**
Rice	Cotton	Flax	Carrot	Petunia	Alfalfa	Poplar
Corn		Canola	Cauliflower	Potato	White clover	Apple
Wheat		Soybean	Celery	Sugar beet	Orchard grass	Walnut
Barley		Sunflower	Cucumber	Tobacco		
Rye		Bean	Lettuce	Tomato		
		Pea	Melon			

a different species. Such plants are commonly referred to as *transgenic* plants). Furthermore, one of the goals is, of course, to produce products that will be consumed by humans. This immediately creates regulatory problems: Who will regulate the release of transgenic organisms in the environment? Is it safe to do this? as well as problems of public acceptance (Is this food safe to eat?). These issues are discussed in detail in Chapter 15.

8

Molecular farming is the second major objective of plant genetic engineering.

Some plant genetic engineers see plants as small, cheap factories that need only water, sunlight, minerals, and the right combination of additional genes to produce exactly what industry wants. Given the right genes, plants could produce modified starches, valuable industrial oils, plastics, pharmaceuticals, drugs, or enzymes for the food industry. Since none of these products will actually be eaten, there will be minimal resistance from consumers, and only the release of transgenic organisms in the environment will have to be considered.

One of the first examples of molecular farming was the production of the mammalian neuropeptide Leu-enkephalin in the seeds of canola (*Brassica napus*) by plant biotechnology researchers from Belgium (see Box 3.1). Leu-enkephalin is a short peptide that displays opiate activities and regulates brain functions. The researchers modified the gene of a seed protein in such a way that the mutant protein contained the five amino acids of Leu-enkephalin in the right sequence, and the Leu-enkephalin peptide could be conveniently cleaved (cut apart) from the purified seed protein. This modified gene was introduced into canola, and the seeds of the transformed plants yielded about 1 mg of Leu-enkephalin per 50 g of seed. Although no commercial production is planned, the experiment demonstrates the potential of molecular farming.

In some cases, the capacity to make a product desired by industry can simply be transferred from one plant to another plant that has better agronomic properties or can be grown in a different location. That appears to be the primary reason why Calgene, a California biotechnology company, is genetically engineering canola, a plant closely related to oilseed rape, that grows readily in the northern United States, Canada, and Europe, to produce for industrial purposes an oil that is abundantly found in tropical oils, such as palm oil.

In other cases, introducing one or two new genes will let the plant synthesize one or two more enzymes that can convert a common intermediate of metabolism into a desired product that the plant does not normally make. By transferring two bacterial genes, scientists from Michigan State University generated transformed plants that produce a biodegradable plastic normally produced by bacteria (Figure 3.10). Cheaper production of this plastic could open up new markets for this renewable material, and could help redirect agricultural production from surplus crops, such as corn and soybeans, that require price supports to crops that industry needs.

9 | Plant biotechnology will have a major impact on plant breeding.

Crop improvement through plant breedings has been described as being both an art and a science. It is an art because the trained eye of a plant breeder is needed to distinguish minor, but important variations in hundreds of characteristics of crop varieties, and choose the right parents for future crosses. The number of possible crosses is astronomical, and the plant breeder must select among them *before* he or she knows the outcome. And crop breeding is a science because the principles of inheritance must be applied to crop improvement.

Plant biotechnology will affect plant breeding in at least two important ways. First, plant biotechnology will make producing hybrid seeds much easier. The discovery of hybrid vigor more than 80 years ago by U.S. plant breeder George Shull completely changed corn breeding and established a

Figure 3.10

A Seedling Genetically Engineered to Produce a Biodegradable Plastic. The white spots show the location of polyhydroxybutyrate granules. Source: Y. Poirier, D. E. Dennis, K. Klomparens, and C. R. Somerville (1992), Polyhydroxybutyrate, a biodegradable thermoplastic, produced in transgenic plants, Science 520–523. Photo courtesy of Y. Poirier. Copyright 1992 by the AAAS.

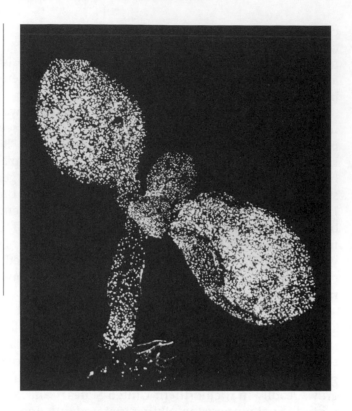

model for improving many other crops. Shull discovered that crossing inbred lines of corn produced plants that showed **hybrid vigor**, meaning that they yielded more corn (about 20% more) than the best lines the farmers had been using up to that time. Producing hybrid plants requires controlled fertilization, which is quite easy for corn, because the male and female parts are in different flowers. For most plants, it is laborious and expensive to produce hybrid seeds, because the male and female parts are in the same flower. For such plants, self-fertilization can only be prevented by removing the male pollen-producing organs (anthers) before they have a chance to pollinate the female organs. Male sterility can now be genetically engineered into any crop species that can be transformed, and male-sterile plants make controlled fertilization much easier. Any seeds such plants produce must be the result of cross-fertilization with pollen from another plant. This recent development, made jointly by scientists from the University of California and Plant Genetic Systems in Belgium, may revolutionize hybrid seed production and plant breeding.

Many important agronomic traits are affected by numerous genes at once, while others are affected by only one or a few genes. Agronomic traits, such as disease resistance, that are encoded by single genes, are easily manipulated by plant breeders, whereas multigene traits, such as yield, are much more difficult to change by breeding. Molecular biologists have come up with a technique in which molecular markers, small pieces of DNA called RAPDs (random amplified polymorphic DNA), can be used to mark a plant's entire **genome** (the total of all its genes). Thus researchers can follow specific traits among the offspring of a cross, between two plants. This technique takes much of the guesswork out of establishing how a trait that is difficult or impossible to visualize in the field is inherited. The technique is rapidly

Box 3.1

*Molecular
Farming:
Synthesis of
Human
Neuropeptides in
the Seeds of
Canola*

Bioactive peptides are increasingly used as pharmaceuticals to treat various disorders. These peptides can be extracted from biological materials, where they are present in extremely small amounts, but this method is very expensive. As the number of conditions treatable with such peptides grows, and as oral delivery methods (much preferred by patients, but which require larger doses due to the smaller amounts that reach the bloodstream) are developed, more economical production methods are necessary. Both chemical synthesis and expression of foreign genes in transgenic organisms are being used. Due to the low cost of growing, harvesting, and processing them, plants are an attractive candidate for the latter approach. Of particular interest is the possibility of using plant seeds, both for economic reasons and because seeds stably store, for long periods of time, large quantities of seed storage proteins. Researchers at Plant Genetic Systems in Belgium have produced several peptides in the seeds of canola (*Brassica napus*), a crop that is widely grown in Canada and Europe. Among the peptides produced so far are the neuropeptide Leu-enkephalin and an antibacterial peptide usually found in the skin of frogs, called magainin.

The research team, under the leadership of Enno Krebbers, studied the genes that encode the small seed proteins in both canola and related plant species. The team found that although the overall structure, in particular the arrangement of intramolecular linkages, was very similar from one protein to the next, one region is more variable in both length and sequence. Furthermore, modeling studies suggested that this region was probably on the outside of the molecule (see figure). The team reasoned that it might be possible

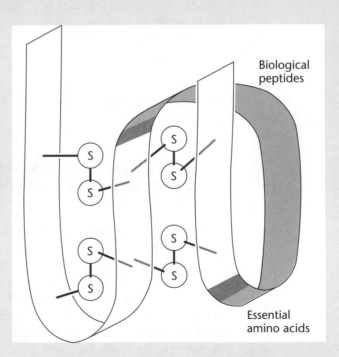

A Model of the Small Seed Storage Protein Present in Canola Seeds. The bridges show how the sulfur (S)-containing amino acid cysteine creates cross-links to stabilize a protein. The cross-hatched loop on the right is the portion of the protein that was altered to contain a neuropeptide or to contain more essential amino acids. Source: Courtesy of Enno Krebbers from Plant Genetic Systems, Ghent, Belgium.

to change some of the amino acids in this variable region without affecting the overall structure of the protein. A series of such modifications, either to produce biological peptides or to change the amino acid composition of the seed (which would be useful to adjust the nutritive value of certain seed crops), has now been made. For example, a storage protein gene was synthesized in which the information encoding a stretch of seven amino acids was replaced by new information encoding the amino acids lysine–tyrosine–glycine–glycine–phenylalanine–leucine–lysine (KYGGFLK, for short). The middle five amino acids form the neuropeptide Leu-enkephalin, which is thus flanked by lysine residues. This modified seed storage protein gene, which already contains sequences that direct seed-specific expression, was used to transform canola, via *Agrobacterium tumefaciens*. The researchers obtained seeds from the transformed plants, purified the storage proteins, and treated them with trypsin, an economical digestive enzyme that cuts proteins next to lysine. This treatment yielded YGGFLK, which was subsequently treated with a second enzyme to remove the terminal lysine to give authentic Leu-enkephalin (YGGFL). Similar procedures were used for other peptides. Calculations show that if the procedure were to be scaled up, yields of peptide per hectare would range from 50 to hundreds of grams, depending on the size and nature of the peptide. Short peptides, like enkephalin, can also be made economically using chemical synthesis, but the cost of the latter procedures rises rapidly with the length of the peptide. For peptides in the 20–50 amino acid range, such as many of the peptides involved in maintaining metabolic balances (for example, peptide hormones) and antibacterial or antifungal peptides isolated from a range of sources, production in plant seeds provides a potentially cheaper alternative.

becoming the major tool for plant breeders who wish to improve so-called quantitative traits.

10 Transgenic plants have the potential to decrease or increase the environmental impact of agricultural practices.

We noted in Chapter 2 that agricultural practices often have a negative impact on the environment. Plants improved by genetic engineering may alleviate some of these negative effects. Creating plants that resist insects, fungi, bacteria, or nematodes would eliminate the need for chemicals to repel these disease organisms and predators. Plants that can use nitrate from the soil more efficiently, or that can dissolve rock phosphate without needing to convert it to superphosphate in a chemical process, would decrease the potential for nitrate contamination of groundwater and could reduce the large amount of energy needed to transport nitrogen and phosphate fertilizers.

A related possibility is that the ability for symbiotic nitrogen fixation could be transferred from legumes to cereal grains. This would indeed save much energy, because manufacturing nitrogen fertilizers is an energy-intensive process. Legumes such as soybeans need no or only small amounts

Figure 3.11

Herbicide Tolerant Beets. *A beet plant that is tolerant of being sprayed with the commonly used herbicide phosphinotricin (Basta®) is shown on the left, and a control plant is shown on the right.* Source: *Courtesy of Jan Leemans, Plant Genetic Systems, Ghent, Belgium.*

of nitrogen fertilizers for maximal crop production, because the bacteria that live within their roots convert atmospheric nitrogen gas into ammonia, which is secreted by the bacteria and thus made available to the plants. Development of this nitrogen-fixing capacity in symbiosis with the bacteria requires the correct expression of a large number of plant genes. Transferring this symbiotic nitrogen-fixing capacity from a legume plant to a cereal such as corn would require transferring a large number of genes and is unlikely to be done in the foreseeable future.

Just as transgenic plants may reduce the use of chemicals, they may also increase the use of chemicals in agriculture. Companies that market herbicides have started to genetically engineer crop plants that tolerate herbicides, usually to the herbicide for which they have the greatest market share. Thus, Monsanto has crop plants resistant to Roundup®, Hoechst has plants resistant to Basta® (Figure 3.11), and Dupont has plants resistant to Glean®. These plants are presently undergoing field testing and will soon be used by farmers. In each case, herbicide tolerance has been obtained by transferring a single bacterial gene into the crop. This approach to weed control could increase the use of chemicals in agriculture.

Industry scientists have argued that plants will be made resistant to biodegradable herbicides such as Monsanto's Roundup® and that this will lessen the impact of the chemicals that are needed. Although this is indeed true, several companies have made plants resistant to herbicides that are not biodegradable and are indeed quite toxic. We later return to the theme that biotechnology by itself will not make agriculture more or less environmentally friendly. The direction of agriculture—toward greater industrialization with more inputs, or toward greater sustainability with fewer external inputs—depends on society at large and not on biotechnologists. The sociopolitical process determines what type of agriculture is practiced. Governments (elected officials as well as regulatory agencies) give substantial incentives that drive agricultural practices in one direction or another. Agricultural biotechnology is just another technology. Given the proper incentives, it can contribute to a reorientation of agricultural practices so that agriculture becomes more sustainable. A more likely scenario, at least in the short run, is that biotechnology will reinforce the trend toward a chemical-industrial agriculture heavily dependent on inputs.

11 | The production of biopesticides is a major goal of the agricultural biotechnology industry.

The adverse impact of plant fungal diseases, insects, nematodes, and weeds on crop production can currently be decreased by breeding resistant plants, by good cultivation practices, or by applying chemical pesticides or introducing natural enemies. We noted earlier that applying ever larger doses of chemicals has been the norm in many countries. The use of pesticides in agriculture has declined in the United States, but this is caused in part by a dramatic drop in the use of pesticides on cotton, a crop that in the past required very large amounts. The decline, usually expressed as the amount of pesticides used per year, is also caused by a switch away from pesticides with low toxicity and used in larger amounts, toward pesticides that are more specific and can be used in smaller amounts.

On a worldwide scale, the use of chemical pesticides is still increasing, and nearly 500 species of arthropods (insects, mites, and so on) are now resistant to commonly applied chemicals. The widespread use of broad-spectrum pesticides often precludes the other major strategy for pest control—the release of natural enemies—if the natural enemies are affected by the same type of chemicals that are routinely used in the fields. This problem is often aggravated by government subsidies for chemical pesticides in the Third World. The release of natural enemies has been successful in a number of well-documented cases. Success has always depended on an in-depth understanding of the biology of the pest and its natural enemies, as well as on the ability to raise the enemies under controlled conditions in the laboratory. Release of natural enemies implies that millions or even billions of individuals must be reared in the laboratory. This amounts to a kind of domestication of these natural enemies.

Biotechnology and gene transfer can be used to select strains of natural enemies or produce transgenic strains that are particularly effective at controlling the plant pests. The following are all attainable goals for this branch of agricultural biotechnology:

1. *Mycoinsecticides.* More than 400 known species of fungi cause diseases of insect pests such as caterpillars, aphids, grasshoppers, and insect larvae. At least two mycoinsecticides are already in wide use in the Third World. One main drawback of mycoinsecticides is that they act very slowly.

2. *Nematode-bacteria complexes (symbioses) that kill insects.* There are 24 known species of nematodes (roundworms) that have symbiotic bacteria within them. The juvenile nematodes penetrate natural openings or muscles of an insect, and the bacteria are subsequently released into the insect's bloodstream. This bacterial infection kills the insect and allows the nematodes to multiply in the insect carcass. These nematodes are highly specific to insects and do not harm plants or mammals.

Effective control of a number of plant pests has been achieved, but the cost-effective rearing of nematodes in the laboratory was until recently a major obstacle to their more widespread use. The problem of growing nematodes efficiently in a factory has recently been solved by the California biotech company Biosys. Previously, the nematodes had to be reared in insects, and the company was only able to produce about 2 billion nematodes a week. By using fermentation technology normally used to grow bacteria and fungi, Biosys was able to grow nematodes in a nutrient broth and now can produce 3.5 trillion a week. At this rate of production, growing nematodes becomes cost effective because about 5 billion nematodes need to be applied to a hectare to control soil-dwelling pests. The nematodes from Biosys are already used in specialty crops, such as cranberries, mint, artichokes, and citrus. The production of bioinsecticides used to be a cottage industry that catered primarily to organic food producers. Recent developments ensure that it will soon be a full-fledged industrial process. Biosys has so far focused all its efforts on a family of nematodes, called Steinernema, that does not harm plants and is unrelated to the nematodes that cause plant diseases.

3. *Bacteria that kill insects.* Few species of bacteria have shown potential as microbial insecticides, but the potential of *Bacillus thuringiensis*, the only one now used commercially, is great. *B. thuringiensis* produces toxic proteins that act as poisons in the gut after the bacteria are eaten by insect larvae that feed on the plants on which the bacteria grow. The several hundred natural strains (isolates) of *B. thuringiensis* have considerable specificity toward various groups of insects such as the lepidoptera (butterflies and moths), coleoptera (beetles), or diptera (mosquitoes). At least six of these strains are being commercially developed, and one that is active against lepidoptera such as cabbage loopers is widely available in gardening stores in the United States. Commercial production of *B. thuringiensis* is done in large fermenters, and after formulation the bacteria can be stored for one to two years. In addition to the use of naturally occurring strains, the gene that encodes the toxin can be transferred to plants (as noted earlier) or to other bacteria, such as *Pseudomonas*, that are more easily grown in mass culture (Figure 3.12).

4. *Mycoherbicides.* Just as fungi can cause crop diseases, they can also cause weed diseases, but controlling weeds with fungal pathogens has not yet found any applications. The problems appear to be mostly technical rather than biological: how to grow the fungi economically in the laboratory, how to store the fungus and the spores, and prolong their shelf life, and how to apply them to the plants (fungi only grow in a moist environment). Development of effective mycoherbicides will require considerable further research on the biotechnology of the production and delivery system. Progress is being made, and the fungal mycoherbicide BioMal—active against smooth-leaved mallow—was registered in Canada in 1992 by a biotechnology company in Saskatoon, called Philom-Bios.

5. *Insect viruses.* Seven major groups of viruses infect insects and could be developed as biological control agents. The known insect viruses are usually highly species-specific and may infect only one or a few species of insects; these viruses are harmless to plants, mammals, and other animals. This specificity stands in stark contrast to the broad-spectrum insecticides that are now used to control insects. It has been estimated that the use of baculoviruses—the best-studied strain of insect viruses—could reduce the use of chemical insecticides by about 60% in California and 80% in Central America. Cost-effective control has been obtained with 30 different baculoviruses against

Figure 3.12

Biopesticides That Kill Insects. **(a)** *Electron micrographs show the round spores and angular Bt toxin crystals that are produced when the* Bacillus thuringiensis *bacteria sporulate. The protein crystal at the center is surrounded by part of the bacterial cell. When the bacteria are collected, they readily break, exposing the Bt toxin crystals. When such preparations are used as a biopesticide, the Bt protein is unstable.* **(b)** *Electron micrograph shows* Pseudomonas *bacteria that have been transformed with the Bt toxin gene. These bacteria can be collected and the crystals remain inside the cells even when the bacteria are heat-killed. Such dead bacteria are an effective biopesticide (see also Chapter 13).* Source: *Courtesy of Mycogen Corporation, San Diego, California.*

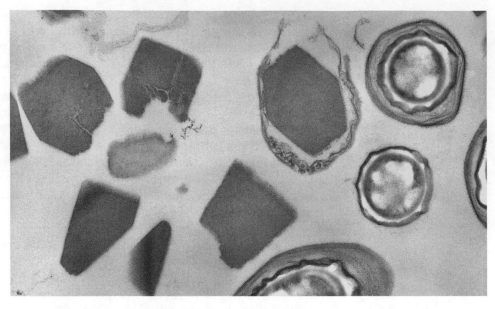

(a)

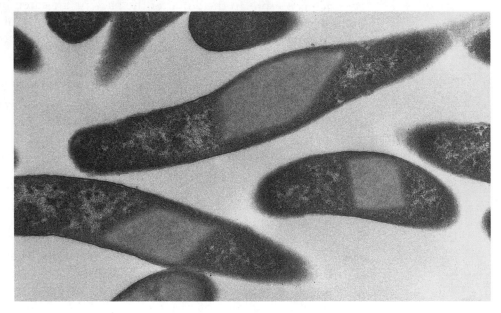

(b)

lepidopteran pests. Although baculoviruses are safe and effective, they are not widely used because of their narrow host range. Agrochemical companies in developed countries generally prefer to develop broad-spectrum chemical or biological control agents, because of the much greater sales potential.

Summary

The present and potential impact of various plant biotechnologies on crop production is great. Hydroponics and clonal propagation of plants, two technologies that do not involve genetic manipulation, are already widely applied, especially in developed countries for the production of high-value crops. The transfer of individual genes between plants and even from other organisms to plants has become possible as a result of two scientific developments: (1) regeneration of whole plants from single cells and (2) understanding the natural gene transfer system employed by *Agrobacterium tumefaciens* to cause crown gall disease. Although technically different from plant breeding, conceptually this type of gene transfer is similar to gene transfer by conventional plant-breeding methods. The aims of gene transfer are crop improvement—breeding crops with improved characteristics—as well as molecular farming, the production of pharmaceuticals, or other specialty chemicals for industry. Plant genetic engineering has the potential to diminish the need for chemical pesticides by making plants more resistant to pests. Tolerance of herbicides, a goal presently pursued by many agrochemical companies, may increase the use of chemicals. Biological control is another proven alternative to the use of pesticides and herbicides in agriculture, and it is likely that biotechnology can help expedite the production and formulation of control organisms.

It should be emphasized however, that plant genetic engineering and agricultural biotechnology are not revolutionary technologies that will magically feed the hungry. Rather, they are the next step in the development of the agricultural food production system. This system produces more than enough food, and the reasons for the persistence of hunger lie elsewhere. Agricultural biotechnology may help produce food more efficiently, with less high-energy, high-value inputs and with less impact on the environment. Combined with traditional agricultural practices, it may make agricultural systems more sustainable. However, it is equally possible that agricultural biotechnology will simply intensify high-input agricultural practices and accelerate the environmental degradation that accompanies these practices.

Further Reading

Cocking, E. C. 1983. Applications of protoplast technology to agriculture. *Experientia* 46:123.

Cocking, E. C. 1986. The tissue culture revolution. In L. Withers and P. G. Anderson, eds., *Plant Tissue Culture and Its Agricultural Applications*. London: Butterworths.

Jensen, M. M., and W. L. Collins. 1985. Hydroponic vegetable production. *Horticultural Reviews* 7:483–558.

Marx, J. E. 1989. *A Revolution in Biotechnology*. Cambridge, UK: Cambridge University Press.

Mason, J. 1990. *Commercial Hydroponics*. Honolulu: International Specialty Books.

Moss, J. P., ed. 1992. *Biotechnology and Crop Improvement in Asia*. Hyderabad, India: International Crops Research Institute for Semiarid Tropics (ICRISAT).

Persley, G., ed. 1991. *Agricultural Biotechnology: Opportunities for International Development*. London: CAB International.

Rogers, S. G. 1991. Free DNA methods for plant transformation. *Current Opinions in Biotechnology* 2:153–157.

Savage, A., ed. 1986. *Hydroponics Worldwide: State of the Art in Soilless Crop Production*. Honolulu: International Center for Special Studies.

Strange, C. 1990. Cereal progress via biotechnology. *BioScience* 40:5–10.

Torrey, J. G. 1985. The development of plant biotechnology. *American Scientist* 73:354–363.

U.S. Office of Technology Assessment. 1982. *Genetic Technology: A New Frontier*. Boulder, CO: Westview Press.

Vasil, I. K. 1991. Plant tissue culture and molecular biology as tools in understanding plant development and plant improvement. *Current Opinions in Biotechnology* 2:158–163.

Withers, L., and P. Alderson. 1986. *Plant Tissue Culture and Its Agricultural Applications*. London: Butterworths.

Chapter 4

Plants and Human Nutrition

1 | What is food?

Food is any substance that provides an organism with energy and nutrients. Humans need **energy** to use as fuel both for conscious actions, such as running or swimming, and for involuntary ones, such as heartbeat or digestion. Energy is also needed for growth and for replacing existing body tissues. The **nutrients** (essential amino acids, a fatty acid, vitamins, and minerals) provide the chemical building blocks for growth and tissue replacement. In addition, they are used to build the molecules that regulate and coordinate all of the body's activities.

The formation of every component of the body's tissues illustrates the need for both energy and nutrients. For example, the synthesis of hemoglobin, the red protein that carries oxygen in the blood, depends not only on the availability of the necessary building blocks (amino acids; carbon skeletons for heme, the pigment that binds oxygen; and iron) but also on the presence of energy to assemble these components. Thyroxin is a hormone, synthesized in the thyroid gland, that regulates the growth of the human body. Its synthesis requires an energy source, various small organic molecules, and the mineral iodine.

Not all living organisms have the same nutritional requirements. An organism can be either autotrophic or heterotrophic, depending on its mode of nutrition. **Autotrophic** ("self-feeding") **organisms** include the plants living on land and in water, and certain bacteria. These organisms have rather simple nutritional requirements, consisting of water, carbon dioxide (CO_2), and a variety of inorganic chemicals derived from the minerals in the soil. The most important feature of green plants, the biosphere's major autotrophs, is that they can use light energy from the sun to provide fuel for all their energy needs. Using CO_2 from the air, water from the environment, and light energy from the sun, green plants can manufacture sugar in the process of photosynthesis (see discussion in Chapter 6). They can then use this sugar, either immediately or after having stored it temporarily, for all their energy-requiring processes (for example, for growth and the uptake of minerals from the soil).

The nutrients required by autotrophs are rather simple. For example, plants must take up such substances as nitrate, phosphate, potassium, and magnesium, usually dissolved in the water of the soil. Therefore, whether or not these nutrients are in adequate supply plays an important role in determining the rate of plant growth and crop production. By using sugar (the product of photosynthesis), water, and the inorganic nutrients from the soil, green plants can synthesize all the complex organic molecules they need to grow and reproduce. Although both plants and animals need many of the same organic molecules to carry out their many functions, only plants can synthesize all these molecules by using only sugar and inorganic nutrients.

Heterotrophic ("other-feeding") **organisms** have more complex nutritional requirements. They cannot use the sun's energy directly, as can autotrophs, but instead must ingest energy-rich organic molecules to carry out their functions. All heterotrophs, whether they eat plants (in which case they are herbivores), animals (carnivores), or dead organic matter (decomposers), ultimately depend on green plants to provide them with these energy-rich organic molecules. Some heterotrophic bacteria can grow only on an energy source (such as sugar), water, and inorganic minerals, and use these to make all the molecules they need. However, higher animals, including humans, cannot synthesize a number of organic molecules and so require them as nutrients. They get these nutrients by eating the plants and bacteria that make them.

To summarize, for its food, an autotrophic plant needs only

1 Water
2 Minerals, including calcium, phosphate, potassium, magnesium, iron, sulfate, nitrate, and traces of many others

In contrast, a heterotrophic human in addition needs specific organic molecules (see Box 4.1 for a definition of the word "organic").

3 Energy is normally provided by carbohydrates and fats, and sometimes by protein
4 Some of the 20 amino acids that make up proteins, including leucine, isoleucine, lysine, methionine, phenylalanine, threonine, tryptophan, valine, and (for infants) arginine and histidine, which must be provided by protein foods
5 The fatty acid linoleic acid, which must be provided by fats in the diet
6 The vitamins A, B_1, B_2, B_6, B_{12}, C, D, E, K, pantothenic acid, biotin, folic acid, and nicotinamide, which must be provided by various foods

When nutritionists estimate the daily needs for food for people, the needs for energy (item 3) and nutrients (4–6) are considered differently (Figure 4.1). When an overall population is considered, energy requirements vary, depending on a number of factors, including body weight and activity (see the discussion later in this chapter). But growth and renewal of body tissues are

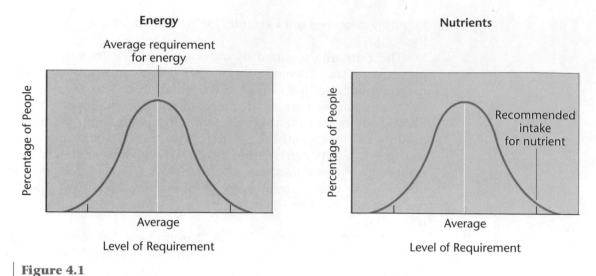

Figure 4.1

The Rationale for Requirements for Energy and Nutrients. Depending on body weight, activity, and genetic background, energy requirements in a population will vary. An average requirement can be set, but this clearly does not apply to all individuals. Overconsumption should be avoided. Nutrient requirements (such as vitamins and proteins) are set at a level that is high enough to meet the needs of all people. (SE=standard error.)

impossible if amounts of even one nutrient are inadequate. Therefore, the RDA (recommended daily allowance) of nutrients is set above the level that everyone needs.

2 | Carbohydrates and fats are our principal sources of energy.

The energy and nutritive values of plants as food depend on their chemical compositions, and foods vary in their ability to supply these dietary components. To understand why dry roasted peanuts are more nutritious than celery stalks, we must examine the chemical nature of food as well as the structure of plant cells and tissues (see Chapter 5).

Carbohydrates are the source of most food energy for people. Not surprisingly, they are abundant in staple foods such as wheat, rice, corn, and potatoes (Table 4.1). They derive their name from the fact that they are composed entirely of the chemical elements carbon (C), hydrogen (H), and oxygen (O), often in the ratio of C:H:O=1:2:1.

There are two major types of carbohydrates:

1 **Simple carbohydrates** consist of one to several sugar molecules.
2 **Complex carbohydrates** (also called *polysaccharides*) are made up of hundreds or even thousands of units of sugar molecules.

One **simple carbohydrate**—and the one most easily broken down by the human body for its chemical energy—is glucose or grape sugar. It has the chemical formula $C_6H_{12}O_6$ and consists of a single sugar molecule (Figure 4.2a). Sucrose (cane sugar or beet sugar) and lactose (milk sugar) are made up of two sugar molecules linked together (Figure 4.2b). These three carbohydrates are readily soluble in water, and are easily taken up from the intestinal tract into the bloodstream. From there, they are quickly transported to tissues

Box 4.1

Cholesterol Is Both Organic and Natural

Throughout this book, the term "organic" is used primarily to describe chemical compounds that have carbon atoms (as well as other atoms such as hydrogen, oxygen, nitrogen, sulfur and phosphorus). Such chemical compounds (such as amino acids and sugars) occur in all living organisms and are therefore described as organic. Organic chemistry is that branch of chemistry that concerns itself with carbon-containing molecules. The term "natural" refers to those molecules that occur in nature. Many of these natural compounds are also organic, such as cholesterol. Other organic molecules, for example many pesticides such as malathion, are synthetic and do not occur in nature. Often they are made in the laboratory by chemists from natural starting materials. The words "organic" and "natural" are therefore value-free in themselves.

Naturally occurring organic compounds can also be synthesized in the laboratory. Thus, vitamin C (ascorbic acid) can be extracted from plants or synthesized in the laboratory. The two chemicals will be exactly alike, although preparations of these chemicals are likely to contain different impurities (nothing is 100% pure). Some advertisers like to distinguish between natural organic chemicals and synthetic organic chemicals, because the public at large equates natural with good and synthetic with bad. There is no scientific basis to suggest that natural vitamin C and natural mint flavoring are in any way different from their synthetic counterparts.

In Chapter 16, the word organic is also used to describe a type of farming in which synthetic pesticides and certain inorganic fertilizers are not used. Inorganic refers to chemicals that do not contain carbon. The mineral nutrients that animals and plants need are all inorganic chemicals. Some, such as rock phosphate or sodium chloride, occur abundantly in nature, whereas others are scarce, but can be made in large quantity by reacting with one or more natural chemicals and can then be used as synthetic fertilizers (e.g., when rock phosphate reacts with sulfuric acid it produces superphosphate, a widely used fertilizer).

Table 4.1 | **Carbohydrate contents of various foods**

Food	Percentage of Carbohydrate Content	Percentage of Carbohydrate in U.S. Diet	Main Carbohydrates
Sugar	99%	39%	Sucrose
Milk	6	6	Lactose
Fruits	10–30	7	Glucose
Vegetables	5–25	10	Starch
Legumes	15	2	Starch
Cereal grains	60–80	36	Starch

Source: Adapted from L. Mahan and M. Arlin (1991), *Food, Nutrition and Diet Therapy*. (Philadelphia: Saunders), pp. 42–43.

all over the body and then broken down for energy. As a result, they provide a source of quick energy.

Plants need energy too, and for this they use the sugars they make autotrophically. Some plant organs are rich in simple sugars, such as sucrose, because the plant uses sucrose as its main way to transport the products of

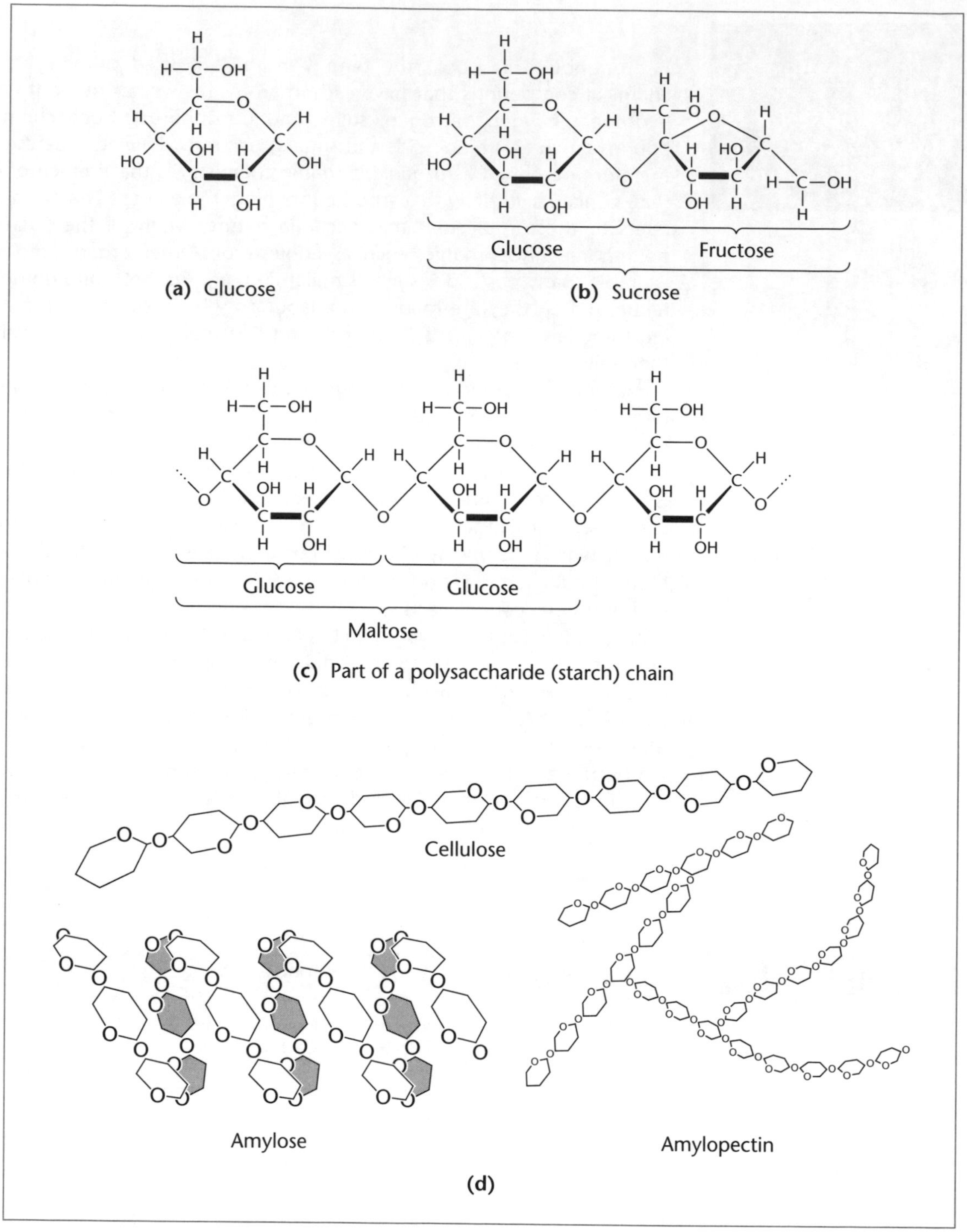

Figure 4.2

Simple and Complex Carbohydrates. *(a) glucose, a simple sugar; (b) sucrose, an easily digestible sugar; (c) part of a polysaccharide, with glucose units bonded together; and (d) three polysaccharides: cellulose, amylose (a starch), and amylopectin (another starch).*

photosynthesis and as one way to store energy. Thus, when we eat plant foods we are often eating foods of plants!

A plant can convert glucose, one of the immediate products of photosynthesis, into more **complex carbohydrates**, such as starch and cellulose. These complex carbohydrates are macromolecules (large molecules), each consisting of hundreds of individual glucose molecules linked together in chains (Figure 4.2c). Starch and cellulose have different chemical and physical properties because the sugar molecules are linked together in different ways. Cellulose forms the fibrous material found in the cell walls of all plant tissues and is an important component of wood. Starch, in contrast, is deposited as stored food by plants in specialized organs such as seeds, roots, and tubers, to be used whenever a source of energy is needed.

The complex carbohydrates shown in Figure 4.2 are linear chains of glucose units. Polysaccharides can also be branched and need not be composed of glucose alone. Plants can convert glucose into about a dozen other sugars, which can also be turned into many different polysaccharides. Some of these compounds are also used to store energy, but most of them become part of the cell wall—the tough, flexible box that surrounds every plant cell.

Complex carbohydrates are usually not soluble in water, and they must be broken down into the simple sugars before they can be taken up and used by tissues in the human body. This breakdown is accomplished by the process of digestion. The enzymes present in saliva and digestive juices break the large molecules in food (such as starch) into smaller components (such as glucose). While people can break down starch, they lack the proper chemistry to digest cellulose, the other polysaccharide in plant foods; so this glucose remains in this complex form and passes through the body largely undigested, and provides no food energy. However, cellulose is not without benefit for human health. Chewing and the acids in the stomach break the cellulose fibers into smaller pieces, which swell up as they pass through the intestinal tract, thereby increasing the bulk of the digesting food. This stimulates intestinal and bowel movements, preventing constipation (see Box 4.2, on dietary fiber).

Some micro-organisms have the enzymes needed to digest cellulose and can therefore use its glucose as food. Ruminant animals, such as cows, goats, and camels, have such micro-organisms living in their digestive tracts, and benefit from this because the micro-organisms digest the cellulose for them. Cows can therefore derive energy from grass, or even newsprint.

Lipids, commonly known as *fats* or *oils*, are present in all organisms, being essential structural components of tissues. All cells (plant, bacterial, and animal) use lipids for the structure of membranes that play a vital role in maintaining the integrity of cells. For example, the plasma membrane separates the cell from other cells and controls which substances are allowed to leave the cell and which are allowed to enter. Lipids occur in large amounts in certain animal tissues (brain, fatty tissues), in milk and eggs, and in certain seeds, such as soybeans and peanuts (Table 4.2).

Like the simple carbohydrates, lipids are small molecules, composed mostly of the elements carbon, hydrogen, and oxygen. But unlike the simple sugars, lipids do not readily dissolve in water. This insolubility is their most important physical characteristic, and allows them to serve a structural role in cells.

Chemically, there are many different types of lipids. Three important lipids are triglycerides (fats and oils), phospholipids, and cholesterol.

1 **Triglycerides** have three fatty acids (Figure 4.3) connected to a molecule of glycerol, a simple three-carbon molecule. Their structure can be likened to a coat hanger (glycerol) to which three long ribbons are attached. Triglycerides are a major storage form of energy in humans as well as in plants.

2 **Phospholipids** consist of a coat hanger and two long ribbons,

Box 4.2

Dietary Fiber

The importance of fiber in the diet has only been recognized fairly recently by the medical profession and the population at large, although people who suffered from constipation have known for a long time that certain foods rich in fiber (bran from cereals, prunes, and small seeds from certain plants) relieve this painful condition. The "price" of relief is increased flatulence. From a dietary point of view, plants have two kinds of fiber, soluble and insoluble, both made up of complex carbohydrates that cannot be digested because humans lack the necessary digestive enzymes. Bran, the outer layer of cereal grains, is a rich source of insoluble fiber. This layer, the pericarp, is composed mostly of dead cells that have thick cell walls containing both cellulose and other polysaccharides made from a variety of sugars. Fruits and beans are excellent sources of soluble fiber, pectins, and mucilages that are also part of the cell wall.

Until the 1900s, when people ate unprocessed foods and especially whole wheat bread, the human diet in industrialized countries contained enough fiber. The replacement of plant food by animal products (meat instead of beans) and the desire to eat white bread (rich people eat white bread, so it must be a good thing), caused a dramatic decrease in dietary fiber. This change came at the same time as the discovery that human feces contained bacteria. The human body was seen to be contaminated by bacteria, whose growth was encouraged by eating fiber-rich foods (more flatulence because of greater bacterial fermentation of the soluble complex carbohydrates in the lower intestine). So they sought to discourage the bacteria.

In 1969, two doctors working in East Africa noticed that Ugandans who ate a diet rich in unrefined plant foods were never constipated and were almost free of diabetes, heart disease, and colon cancer. The publication of these observations stimulated a renewed interest by the medical profession in dietary fiber, leading to further research. Insoluble fiber increases the bulk of the feces and speeds their movement through the large intestine, reducing constipation and the risk of colon cancer. The risk of colon cancer is reduced because chemicals that cause mutations and are carcinogens make a quicker exit from the colon. Soluble fiber increases the viscosity of the half-digested food in the upper intestine, slowing glucose absorption and the reabsorption of bile acids secreted by the gall bladder. The result is a greater excretion of bile acids and a reduced blood cholesterol level.

with a shorter ribbon that has a phosphate group. These molecules are crucial for building all the membranes in the cell.

3 **Cholesterol** is structurally very different from triglycerides and phospholipids. Cholesterol and other molecules like it, such as vitamin D and the steroid hormones, are made out of four linked rings of carbon atoms. People need cholesterol to make bile acids and hormones such as cortisol, testosterone, and estrogen.

Nutritionally, the lipids of greatest importance are the triglycerides (fats and oils). The size of the fatty acids and the number of unsaturated bonds (see Figure 4.3) determine the chemical and physical properties of the fat molecules. The more saturated the fatty acids and the longer the fatty acids, the more solid the fat will be at body temperature. Fats with saturated fatty acids are commonly found in animal products (meat, milk, and cheese), but also in some plant products (cocoa butter). Fats that are liquid at room temperature are commonly called *oils,* and such liquid fats are found in seeds and in fish.

Table 4.2	Fatty acids in some foods			
			Fatty acids (g)	
	Kcalories	Fat (g)	Saturated	Monounsaturated	Polyunsaturated
Olive oil (2 tbsp.)	250	28	3.8	20.6	2.4
Margarine (2 tbsp.)	250	26	6.6	11.6	6.8
Cheddar cheese (2 oz. = 60 g)	230	18	12.0	5.4	0.6
Beef sirloin (3 oz. = 90 g)	240	30	6.4	6.9	0.6
Flounder (3.5 oz. = 100 g)	200	8	6.4	1.5	0.5

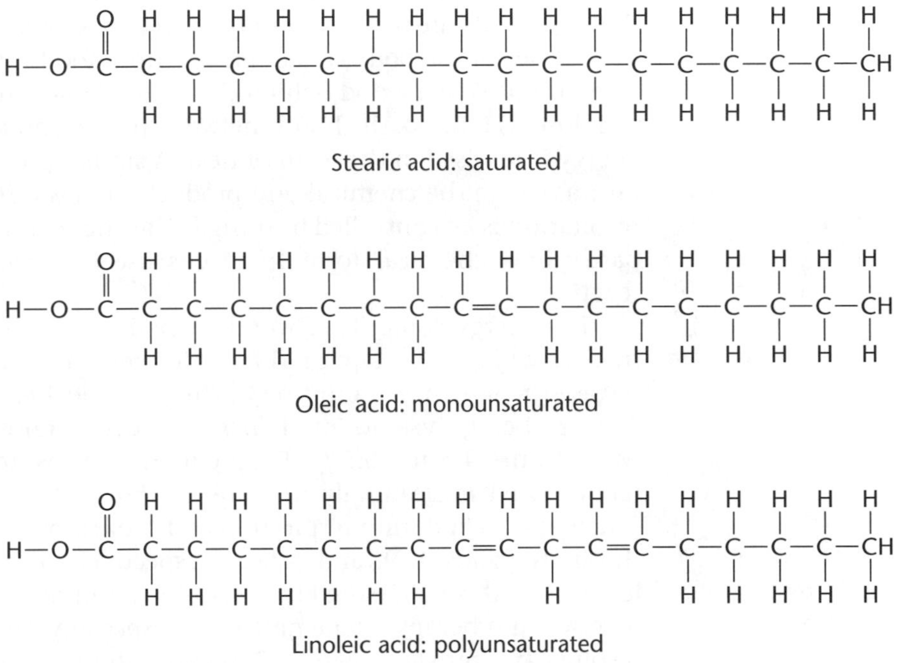

Figure 4.3

Chemical Structures of a Saturated Fatty Acid (Stearic Acid) and Unsaturated Fatty Acids (Oleic and Linoleic Acids). The double bonds between the carbon atoms mean that there are fewer hydrogen atoms attached than if there were a single bond.

Stearic acid: saturated

Oleic acid: monounsaturated

Linoleic acid: polyunsaturated

Fats and oils are so similar chemically that oils can be converted into fats by a chemical process called *hydrogenation*. The manufacture of margarine from various vegetable oils (corn oil, safflower oil) is an example of this interconversion. Triglycerides from which one or two fatty acids have been removed are called diglycerides and monoglycerides, respectively. Strictly speaking they are not fats, and they are the ingredients of fat-free margarine.

Lipids are an important source of energy for humans. As is the case with carbohydrates, fats are broken down into their simplest units—fatty acids and glycerol—during digestion, and are then absorbed into the blood, and transported around the body. Cells have the biochemical machinery for breaking down fats for their chemical energy, and compared on a weight basis they are a better source of energy than carbohydrates (see later discussion).

But lipids also play a nutritive role in the human diet. Different fatty acids are used to manufacture different cellular structures, and plants and bacteria can make all of them by rearranging the atoms of carbohydrates. Humans lack the enzymes to make all the fatty acids. At least one of them (linoleic acid) must be obtained from organisms that make it; that is, this fatty acid is a food nutrient. In addition, two others (linolenic and arachidonic acids) have been found to be beneficial for growth. Lack of linoleic acid in the diet results in poor growth and scaliness of the skin, conditions that can be alleviated by eating fats rich in linoleic acid (most vegetable oils, except olive oil). Dietary fats also help the body absorb certain vitamins, such as A and D, that do not dissolve in water but are soluble in lipids.

3 | Human energy requirements differ depending on age, body weight, and activity.

Carbohydrates and lipids are the principal sources of food energy for people. The body can also derive energy from protein, but only when carbohydrates and fats are in short supply or, at times, when protein is in oversupply. The tissues of the body obtain energy from food molecules by the process of respiration. When wood (which contains a large amount of the carbohydrate cellulose) is burned, the bonds between the carbon atoms are broken, and the energy is released in the form of heat. A similar process happens during respiration, where the chemical end products are also CO_2 and water, except that respiration is a "controlled burning." The energy is released partly as heat and partly in a chemical form that cells use to power their energy-requiring reactions.

The energy contents of wood or food can be measured by burning them in a calorimeter, a device that accurately measures the amount of heat released when they are burned. This released heat is expressed as calories, 1 calorie being the amount of energy needed to raise the temperature of 1 g of water from 14.5 to 15.5°C. Food energy is measured in kilocalories, or kcal (abbreviated misleadingly in some nutrition discussions as C, or Calories). The breakdown during respiration of 1 g of carbohydrate, fat, and protein by the body yields 4, 9, and 4 kcal, respectively. (These values are somewhat lower than those measured by a calorimeter because foods are not completely digested and because some molecules, especially proteins, are not completely respired and broken down to CO_2 and water.)

Human needs for energy depend largely on two factors: the needs of the body to maintain itself and the needs of voluntary activities.

Energy for maintenance includes a wide variety of processes, including growth, tissue repair, the functioning of organs, and compensating for heat losses. The needs of the different organs vary: estimates of the percentage of total basal energy needs are

Liver: 29%

Brain: 19%

Heart: 10%

Kidney: 7%

Resting muscles: 18%

Other: 17%

As the body grows (during its early years), energy needs are most acute. For example, a young growing child has a much higher food energy requirement, per unit of body weight, than a middle-aged adult. On the whole, it

BOX 4.3

*Good Fats and
Bad Fats*

To understand good fats and bad fats, we must first look at the role of cholesterol in a disease called *atherosclerosis* or hardening of the arteries. This disease is life threatening when it affects the coronary arteries that supply the heart with blood. The hardening is caused by the deposition of plaque, a fatty material that consists of cholesterol; small blood particles, called *platelets;* and collagen. The level of cholesterol that circulates in the blood seems to determine the extent of the hardening, and people with high levels of blood cholesterol are at greatest risk of heart disease. Because cholesterol is a fat, it is not soluble in water, and it circulates as a cholesterol-protein complex, a small protein-coated fat globule.

Two types of these complexes circulate in our bloodstream: low-density complexes (a little bit of protein surrounding the fat) and high-density complexes (a lot of protein coating the fat globule). The low-density lipoprotein complexes, called LDL for short, seem to be the major culprits in hardening of the arteries and heart disease. The nutritional advice given out by the American Heart Association and other agencies is predicated on the assumption that lowering LDL (and cholesterol in general) will lower the risk of heart disease. Although the advice is sound, it is good to remember that obesity, smoking, and hypertension are also part of the heart disease picture. However, obesity and hypertension are also linked to poor diets.

Blood cholesterol can be lowered simply by eating less cholesterol, a fat that is only found in animal products. However, the body synthesizes its own cholesterol, so, once again, life is not simple. Foods that are rich in saturated fats will raise cholesterol levels. These include most animal products (meat, cheese, and whole milk), but also certain plant products (coconut oil and palm oil). Foods that are rich in monounsaturated or polyunsaturated fats (many vegetable oils and fish oils) lower blood cholesterol levels. The monounsaturates found abundantly in olive oil are particularly effective at lowering the LDL complexes. Most effective are the unusual unsaturated fatty acids found in fish oil. People whose diets are rich in olive oil or in fish oil have a low incidence of heart disease.

The dietary advice of the American Heart Association is to restrict the intake of cholesterol to 200–300 mg per day (an amount found in one egg or four hamburgers), restricting the calories from saturated fats to 10% or less of the total caloric intake (15% of the calories should come from protein and 55% from carbohydrates, preferably complex carbohydrates).

Source: Saltman P., J. Gurin, and I. Mothner. 1987. *The California Nutrition Book.* Boston: Little, Brown.

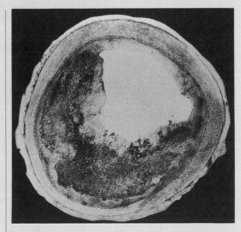

Excess blood cholesterol can lead to atherosclerosis, in which fats and other tissues accumulate on the inner walls of arteries. In this photograph of a cross-section, the white area or lumen, through which blood flows, has been narrowed in the coronary artery that supplies blood to heart. (Source: Drs. M. Brown and J. Goldstein.)

Table 4.3	Energy expenditure for various activity levels	

Activity Examples	Kcalories per Minute
Resting	1–1.2
asleep, reclining	
Very light	1.2–2.5
Seated or standing; painting, cooking, sewing, and so on	
Light	2.5–4.9
Walking 2–4 km/h; carpentry, house cleaning, child care	
Moderate	5.0–7.4
Walking 3–5 km/h; hoeing, cycling, carrying a load	
Heavy	7.5–12.0
Walking with a load uphill; tree felling, manual digging	

Source: Data from Food and Nutrition Board, U.S. National Research Council (1989), *Recommended Daily Allowances,* 10th ed. (Washington, DC: NRC), p. 27.

Table 4.4	Recommended daily intake of energy and protein for people in East Africa		

	Age, Stage, or Lifestyle	Kilocalories per Day	Protein (Grams per Day)
Children	1–2 years	1,000 kcal	40 g
	5–6	1,400	50
Girls	11–12	2,200	65
	13–17	2,500	70
Boys	11–12	2,000	60
	15–18	3,000	80
Women	Sedentary	1,800	55
	Very active	2,500	65
	Pregnant	Add 400	85
	Lactating	Add 900	95
Men	Sedentary	2,200	60
	Very active	3,000	70

Source: M. C. Latham (1965), *Human Nutrition in Tropical Africa* (Rome: Food and Agriculture Organization), p. 243.

takes about 5 kcal to deposit 1 g of new tissue, so every pound of weight gained by a child represents about 2,000 kcal, or about a pound of carbohydrate eaten. This factor is important in pregnant and lactating women, who must take in nutrients for tissue growth not only for themselves but also for the developing fetus and infant. After mature size is reached, energy requirements (all other factors being equal) depend on body size: a larger body has more needs for repair and maintenance than a smaller one. Typically, women, with their smaller frames, have lower energy needs than men.

As mentioned earlier, some heat is released by the digestion and breakdown of foods, and this helps keep body temperature constant. As the environmental temperature goes down, the thermal need from food increases. In tropical regions, there is less need for heat from food, so overall energy requirements fall. A rule of thumb is that for every 10°C rise in temperature above an annual mean of 10°C, food energy needs decline by about 5%. This factor alone reduces the daily need for calories in most of India by 7%, for example, compared to the United States.

Energy for activity depends, of course, on which activities are undertaken (Table 4.3). Clearly, sitting quietly and reading require less energy expenditure than running uphill.

The relationship between energy requirements for maintenance and for activity are seen when recommended daily energy needs are calculated (Table 4.4). For typical adults, maintenance needs are about 1,300–1,600 kcal per day, and activities lead to greater demands for food. Especially needy are the very young and pregnant or lactating women. Not surprisingly, these people tend to be the most affected by food shortages.

4 | To make proteins, people must eat proteins.

Protein is a major constituent of all organisms and accounts for more than half the dry weight of the living substance (protoplasm) of most cells. Like the complex carbohydrates, proteins are built up out of smaller units, in this case called **amino acids**, which are composed of carbon, hydrogen, oxygen, nitrogen, and sometimes sulfur.

Twenty different amino acids occur in proteins, and each of the thousands of different proteins in a plant, animal, or microbe has a specific order and number of amino acids. This order and number together determine the special properties of each protein. Most proteins contain from 100 to 1,000 amino acids, so the possibilities for composition are very large. Unlike the polysaccharides, which are often in an extended configuration, most proteins are globular in shape. The chains of amino acids—which, unlike some complex carbohydrates, are never branched—are first wound helically or pleated. This structure greatly shortens the length of the amino acid chain. The long helices are then folded around each other into a fairly tight globular structure.

Proteins play four important roles in tissues: (1) as enzymes, (2) as structural components of the cells, (3) as regulators of a variety of organism functions, and (4) as sources of energy. In cells, enzymes speed up or catalyze chemical reactions that would otherwise take place at very low rates. An example of an enzyme is amylase, which helps break the chemical bonds linking the different glucose units in a starch molecule. This enzyme is found in human saliva, in the intestines of insect larvae that eat starchy foods such as dry bean seeds or potatoes, and it is also present in plant cells, where it plays a similar role in digesting starch stored within the plant. Most food proteins are not enzymes, however, but are either structural proteins found in muscle cells (meat, fish) or storage proteins such as those found in seeds or potatoes. Both plants and animals use proteins in their defense against pathogens, but the mechanisms are completely different. In addition, plants have unique defense mechanisms that do not rely on proteins. Finally, as noted earlier, proteins can be broken down to provide energy for the organism. But since proteins are needed as a source of amino acids, protein breakdown for energy is often a "last resort" when fats and carbohydrates are not available.

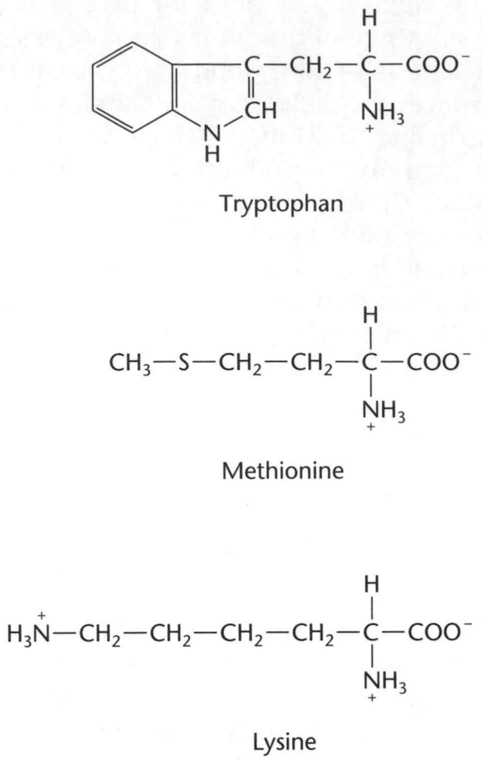

Figure 4.4

Chemical Structures of 3 of the 20 Amino Acids Found in Proteins. These are 3 of the essential amino acids in the human diet.

Tryptophan

Methionine

Lysine

As autotrophs, plants can synthesize all 20 amino acids from the carbohydrates they make in photosynthesis (from atmospheric CO_2), and nitrate (NO_3^-) and sulfate (SO_4^{2-}) taken up by the roots in soil water. The amino acids are then linked together to form the proteins that the plant needs.

The situation is different in people. As heterotrophs, we must take in the carbohydrates from other organisms, usually plants. Also, humans do not use nitrate and sulfate, but derive these as part of larger organic molecules in food. What is more, we cannot take the carbohydrates, fats, nitrogen, and sulfur and fashion them into the 20 amino acids. In fact, we can only make 12 of them, and must eat the other 8 in food.

Nutritionists identified the 8 so-called **essential amino acids** in the human diet by substituting purified amino acids for foods, and feeding people diets lacking one or another of the 20 amino acids. The 8 "essential" amino acids must be obtained from food, and in sufficient quantities to satisfy the requirements of the body. The structures of 3 essential amino acids are shown in Figure 4.4. If even one essential amino acid is missing, or if there is not enough of it, the human body cannot make the proteins it needs for proper functioning. Thus we must eat not only enough protein, but also the right kind of protein, which provides us with the necessary amino acids.

The nutritional value of protein-rich foods depends not only on their digestibility (not all proteins can be digested equally well), but also, and especially, on the relative abundance of the 8 essential amino acids. These criteria can be used to evaluate proteins for their human dietary value. It must be emphasized that these scores are only for the human diet. All plant and animal proteins have immeasurable value to the organism har-

Table 4.5		Essential amino acids in selected foods*							
Food	Isoleucine	Leucine	Lysine	Methionine	Phenylalanine	Threonine	Tryptophan	Valine	Protein Score
Hen's egg	393	551	436	210	358	320	93	428	100
Beef	301	507	556	169	275	287	70	313	80
Cow's milk	295	596	487	157	336	278	88	362	79
Chicken	334	460	497	157	250	248	64	318	72
Fish	299	480	569	179	245	286	70	382	70
Corn	230	783	167	120	305	225	44	303	49
Wheat	204	417	179	94	282	183	68	276	62
Rice	238	514	237	145	322	244	78	344	69
Beans	262	476	450	66	326	248	63	287	44
Soybeans	284	486	399	79	309	241	80	300	67
Potatoes	236	377	299	81	251	235	103	292	34

Source: Food and Agriculture Organization (1970), *Nutritional Study No. 24* (Rome: FAO).

*Amounts are expressed as milligrams of amino acid per gram of protein nitrogen. Egg is considered to have a perfect protein, and others are rated in comparison with it to give a protein score.

boring them, because the proteins are uniquely structured to perform their functions.

Human milk protein and egg protein are defined to have protein scores of 100, which means that they contain the essential amino acids in the exact proportions required by the human body. The amino acid compositions and protein scores for various foods are given in Table 4.5. Animal proteins generally have a higher protein score than the major dietary plant proteins, because the latter tend to be low in the essential amino acids tryptophan, methionine, or lysine.

A marked deficiency in even one essential amino acid drastically lowers the protein score of a protein. Cornmeal, which is low in tryptophan and lysine, has a protein score of 49, and navy beans, which are low in methionine, have a protein score of 44. Foods with protein scores lower than 70 are generally considered unsatisfactory for human growth and maintenance. This does not mean, however, that corn protein and bean protein have no nutritional value for humans. Different foods that are deficient in different amino acids can complement each other when eaten at the same meal (see next section). A meal of corn tortillas with refried beans has a much higher protein score than either food alone.

5 | Protein plays an important role in the human diet.

Studies of human nutrition have yielded fairly accurate figures for the body's daily requirements for calories, although this varies considerably for people in different situations (compare Figure 4.1 and Table 4.4). Estimates of the daily protein requirement are more complicated. The sum total of all the amino acids in the human body at any time can be thought of as a "pool" (Figure 4.5). This pool is like a bank into which deposits are made and out of which funds are withdrawn:

Figure 4.5

The Amino Acid Pool.
The concept of a pool of
amino acids in the body
is a useful way to relate
nutrition, breakdown of
the body's proteins, and
the body's ability to make
its own (nonessential)
amino acids.

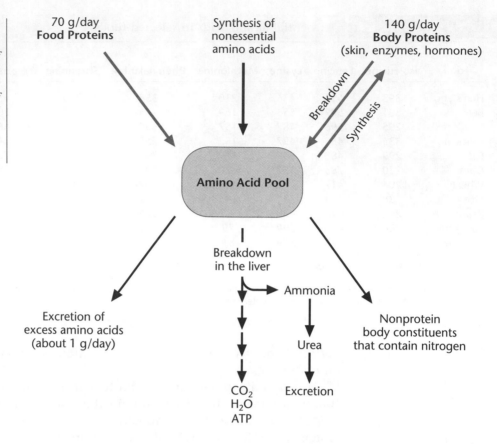

1 Deposits into the pool from the body itself include (a) the synthesis of nonessential amino acids from other molecules and (b) the amino acids released when body proteins are broken down.

2 Withdrawals from the pool include the body's synthesis of the proteins it needs, the synthesis of derivatives of amino acids that are important to the body (for example, some hormones are derived from amino acids), and the total breakdown of amino acids for energy.

Unless there is a massive breakdown of the body's proteins, for the essential amino acids it is obvious that withdrawals from the pool exceed deposits. This is also usually the case for the other amino acids, under normal circumstances. The deficit of needs over supply must be made up by dietary intake.

The U.S. government recommends a minimum daily protein intake of 60 g for adult men and 50 g for adult women. Many Americans eat much more protein; a study of U.S. army mess halls and cafeterias revealed a daily protein consumption of 94 g per woman and 126 g per man. To put this in perspective, a typical beef steak (170 g) contains about 50 g protein, a slice of cheese pizza 10 g, an egg has 8 g, and a slice of bread has 3 g. U.S. estimates of human protein needs tend to be higher than other national agencies. The U.N. Food and Agricultural Organization (FAO) has recently concluded that most adults can easily get by on 40 g of protein a day. This figure agrees closely with the finding that the average person excretes about 30 g of protein (or products derived from protein) each day.

As with calories, the amount of protein required each day depends on many variables (see Table 4.4). Rapidly growing young people, physically active people, and pregnant and lactating women need more dietary protein. People on a high-energy diet can usually get by with less protein; indeed, the body's amino acid pool can be maintained by a protein intake of 20 grams per day if the overall energy intake is 4,000 kcal per day.

If the daily protein intake is very high (over 100 g per day), excess amino acids are excreted (as urea), because the body has no means of storing amino acids, either in the pool or as proteins. This is in contrast to carbohydrates, where there is a limited storage of the polysaccharide glycogen in liver and muscle, and lipids, which are easily stored in fatty deposits. Plants can store proteins in their seeds, leaves, and bark.

In low-calorie diets, body proteins are broken down to replenish the amino acid pool. In this case, the body always satisfies its energy needs first. If the calorie intake is low, amino acids are withdrawn from the pool to be broken down for energy. This leaves the pool low in amino acids for synthesizing the proteins that the body needs.

The amount of dietary protein the body needs is also influenced by the distribution of the protein among the three daily meals. Proteins or protein mixtures containing all the essential amino acids in the right proportions should be eaten at all three meals, because essential amino acids cannot be stored very long. Experiments show that growth may be retarded when not all the essential amino acids are eaten simultaneously.

High muscular activity greatly increases the body's need for energy. Many people think such activity also increases the need for protein—that the muscle tissues become worn out and must be replaced. This is not true. Experiments indicate that when muscular activity doubles the body's caloric requirements, it increases the need for protein by only 5%. College football players and other athletes often habitually eat large amounts of meat. This practice has some psychological value, but nutritional value only insofar as the excess protein may provide some additional energy.

Recent evidence shows that if a woman does not eat enough protein during pregnancy, the intellectual development of the child may be slowed. This finding has focused renewed attention on the importance of proper nutrition during pregnancy and lactation. The growth of the fetus (especially during the final six months of development) and the production of milk require that the normal diet be supplemented with additional calories, proteins, and other nutrients (essential fatty acids, vitamins, and minerals). Women who are pregnant or lactating need to increase their daily protein intake by 10 or 20 g, respectively.

6 Vitamins are small organic molecules that plants can synthesize, but humans cannot.

Vitamins are important biochemical molecules that plants and bacteria can synthesize, but, as with the essential amino acids, the human body cannot. Therefore, vitamins must be obtained in the diet. Whereas daily

Table 4.6	Selected recommended daily allowances for vitamins

| Vitamin | RDA for Adults | | Role |
	Male	Female	
Thiamine (B₁)	1.5 mg	1.1 mg	Breakdown of carbohydrates, synthesis of fats
Riboflavin (B₂)	1.7	1.3	Tissue respiration
Niacin (B₃)	19	15	Breakdown of carbohydrates, synthesis of fats
Vitamin B₆	2.0	1.6	Breakdown of amino acids, synthesis of unsaturated fatty acids
Folic acid	0.20	0.18	Synthesis of DNA; blood cell maturation

Source: RDAs from Food and Nutrition Boards, U.S. National Research Council (1989), *Recommended Daily Allowances,* 10th ed. (Washington, DC: NRC).

protein requirements are measured in grams, vitamin requirements are generally measured in mg, or milligrams—thousandths of a gram (Table 4.6). Vitamins are relatively small molecules, comparable in size to amino acids or sugars. But even in trace amounts they play important roles in the body.

Several vitamins are **coenzymes,** molecules essential for making an enzyme function. For example, vitamin C is needed for the enzymatic formation of collagen, a substance important in wound healing and in the stability of the joints. Vitamin A is required for the synthesis of an eye pigment. Vitamin E is an important anti-oxidant, preventing tissue damage by chemicals. Vitamins are present in foods and are taken up in the body dissolved either in water (vitamins B and C) or in fats (vitamins A, D, E, K).

Although all living organisms contain nearly all the vitamins, some are particularly rich in certain vitamins and so are especially valuable food sources. Vitamin C is abundant in citrus fruits and is found in fresh vegetables. The B vitamins are most abundant in meats, wheat germ, and yeast. Cod liver oil is a good source for the lipid-soluble vitamins A and D. Vitamin A is also found in yellow vegetables (squash, carrots, sweet potatoes), but vitamin D is not abundant in plants.

The concept of minimum daily allowance as a level of intake, including a slight excess for safety, that will satisfy the needs of all people, applies to vitamins (see Figure 4.1 and Table 4.6). These minimum levels are set high enough to prevent specific **deficiency diseases.** For example,

- Deficiency of vitamin A causes night blindness and increases childhood mortality.
- Deficiency of vitamin B₁ (thiamine) causes beriberi, characterized by weak muscles and paralysis.
- Deficiency of vitamin B₃ (niacin) causes pellagra, characterized by skin lesions, diarrhea, and mental apathy.
- Deficiency of vitamin D causes rickets, characterized by weak and misshapen bones.

| Table 4.7 | Selected vitamin contents of cereal grains and products |

Cereal	Thiamine	Riboflavin	Niacin	B$_6$	Folic acid
Wheat					
Grain	0.57	0.12	7.4	0.35	78
Germ	2.01	0.68	4.2	0.92	328
Bran	0.72	0.35	21.0	1.38	223
Flour	0.13	0.04	2.1	0.05	25
Rice					
Brown	0.34	0.05	4.7	0.62	20
Polished	0.07	0.03	1.6	0.04	16

Note: Data are expressed as mg vitamin/100 g food, except folic acid which is expressed as micrograms (μg). Wheat flour is unenriched.

Source: D. K. Salunkhe and S. Deshpande, eds. (1991), *Foods of Plant Origin* (New York: Van Nostrand Reinhold), p. 92.

Unfortunately, the cereal grain staples that make up two-thirds of the human diet are not abundant in the vitamins (Table 4.7). Moreover, the plant parts that are vitamin rich are often not eaten. For example, beriberi became much more common in Asia when mills started polishing rice, a process that removes the thiamine-rich outer layers of the grain. In North America, this happened when bread made from white wheat flour became popular during the late 1800s. Corn contains little tryptophan (Table 4.5), an amino acid that can act as a precursor for the synthesis of vitamin B$_3$. Moreover, the high levels of the amino acid leucine in corn proteins appear to block the conversion of tryptophan to its coenzyme product. These two facts combine so that people who eat corn as a staple often suffer from pellagra.

There are two ways to solve these vitamin deficiencies. First, the people can eat a more varied diet, and thus eat foods rich in the vitamins they lack when they eat the staple. This concept of **complementary foods** is a common one that will be discussed again later in regard to proteins. The second way is to add vitamins artificially to the foods that are eaten. This process, termed **fortification**, has often been very successful. For example, in many developed countries both white and brown wheat flours are fortified at the point of milling with vitamins such as thiamine, riboflavin, niacin, and pyridoxine. In fact, this process makes these flours, and the products made from them, a better source of vitamins than the original whole grain. Another familiar example of fortification is the addition of vitamin D to milk. This has greatly reduced the incidence of rickets.

Fortification is a cheap and effective way of improving the nutritional status of a population. But it works best in places where all the people buy their food at a market (such as in cities). It does not work well in small villages, or where people grow and consume their own food as is often the case in developing countries.

There is some controversy about whether the doses of vitamin intake recommended by nutritionists are too low. Some evidence indicates that larger doses of vitamins may improve health in addition to preventing deficiency diseases. However, diets that are too high in certain vitamins can also result in physiological disorders.

7 | Minerals and water are essential for life.

At least 18 different minerals are essential for human life and must be taken in the diet (Table 4.8). Some of them, such as calcium and phosphorus, are required in large amounts. Others, such as iron or magnesium, are needed in smaller amounts. Still others, such as copper, cobalt, and molybdenum, are required in trace amounts. Because many of these minerals are quite common in the liquids we drink and the foods we eat, nutritionists may not pay enough attention to them. The minerals of special concern are those known to be deficient in certain diets: calcium, phosphate, iron, and iodine.

Calcium, phosphate, and **magnesium** are called the "bone builders" because they are needed in large amounts for bone formation. The body requires vitamin D to incorporate these minerals into bones. Milk and milk products contain abundant amounts of both calcium and phosphate, which are also found in grain products, meat, and a variety of vegetables. In spite of this, many people do not eat enough calcium. A 1985 nutritional survey of the United States revealed that over 40% of the people surveyed had a calcium intake below the recommended allowance. Urban households on low incomes tended to have the most calcium-deficient diets.

The recent upsurge of interest in calcium in the United States and other developed countries illustrates the extent to which human nutrition has been a neglected discipline. Nutritionists recently "discovered" the importance of calcium in the diet to prevent osteoporosis, a weakening of the bones that occurs in women later in life. Food-manufacturing companies were quick to follow, touting their calcium-rich food products, and calcium pills can now be obtained in every health food store, pharmacy, and supermarket. Yet the role of calcium in preventing osteoporosis in cows has been well known to animal nutritionists for 50 years. Recent scientific evidence shows that calcium supplements taken later in life can prevent the onset of osteoporosis even if a person has had a calcium-insufficient diet up to that point.

Sodium, potassium, and **chloride** function primarily as electrolytes because they are present in all our cells and fluids as electrically charged ions (calcium, magnesium, and phosphate also play this role, but to a much lesser extent). These electrolytes maintain blood pressure, play an important role in the acid–base balance of the fluids (also called the pH), and are vital for transmission of nerve impulses and muscle contraction. The main source of sodium and chloride is table salt, while potassium is found in all plant foods. Deficiency of these minerals occurs in cases of chronic diarrhea, where fluid losses are excessive. Such loss can lead to congestive heart failure and death.

Iron is the most important micromineral, and iron deficiency leads to anemia, a disease characterized by a general weakening of the body because there are not enough red blood cells. The red blood cells carry oxygen from the lungs to all the tissues, where the oxygen is used in the biochemical reactions that provide energy. Oxygen is actually carried by the red, iron-containing protein, hemoglobin. When iron is lacking in the diet, the body cannot synthesize hemoglobin. Anemia is more prevalent among women than men, because women must replace the blood lost during menstruation, pregnancy, and childbirth. Therefore, the recommended daily allowance for iron is higher for women (18 mg) than men (10 mg). Normal diets provide between 10 and 15 mg of iron per day, which is not enough for women who lose a substantial amount of blood during menstruation.

Table 4.8	Some mineral requirements in the human diet

Mineral	RDA (mg)	Functions
Calcium	800–1,200	Bone and tooth formation
Phosphorus	800–1,200	Bones and teeth; some metabolic
Magnesium	280–350	Bone, enzyme activator
Sodium	500–3,000	Bone, electrolyte
Chloride	750–3,000	Electrolyte
Potassium	2,000	Electrolyte
Iron	10–15	Hemoglobin, enzymes
Zinc	12–15	Enzymes, insulin
Copper	1.5–3	Enzymes
Iodine	0.15	Thyroxine
Manganese	2.5–5	Enzymes
Selenium	0.07	Fat metabolism
Chromium	0.05	Glucose metabolism

Source: RDAs from Food and Nutrition Board, U.S. National Research Council (1989), Recommended Daily Allowances, 10th ed. (Washington, DC: NRC).

The iron requirements of pregnant women are very high, and anemia is especially prevalent among them. One-third of all mothers' deaths at childbirth result from iron deficiency anemia. A newborn child contains about 400 mg of iron; and the growth of the placenta, and the loss of blood during delivery, use up about another 300 mg. The total iron requirement for a pregnancy is therefore about 700 mg, or 2.5 mg per day, in addition to the basic metabolic requirement for iron. Since only 10 percent of the dietary iron is normally absorbed by the body, the diet should contain an extra 25 mg per day, for a total of 40 to 45 mg per day. This high need for iron can usually be satisfied by iron supplements.

Although the body needs only small amounts of **iodine** per day, the World Health Organization (WHO) estimates that many millions of people suffer from goiter, a disease caused by iodine deficiency. Goiter is characterized by an enlarged thyroid gland, a sluggish metabolism, a tendency to obesity, and an enlargement of the face and neck. The iodine content of many foods is variable, and in many areas of the world it is so low that a normal diet does not provide the body with enough iodine. Iodine deficiency can be most easily prevented by the fortification of table salt with iodide.

We take **water** so much for granted that we often do not realize how crucial it is for life. The water content of the human body is very high, varying from 70% in a lean person to 50% in an obese person. The importance of water as a food was demonstrated when it was shown that animals on a starvation diet still survived after losing all their stored fats and carbohydrates and about half their protein. However, a loss of 10% of their water was very serious, and a loss of 20% usually resulted in death. Water must be taken in continuously to prevent the body from dehydrating and to maintain the proper balance of salts in body fluids.

Water plays several important roles in the body. First, it is the medium in which all biochemical transformation of the other nutrients takes place. Water helps carry nutrients from the gastrointestinal tract into the bloodstream, because the nutrients are dissolved in water. Water is also important in excretion; waste products are dissolved in it and excreted as urine. Water helps regulate body temperature by absorbing the heat released by the respiratory activity of all the tissues; much of the heat is used to transform liquid water into water vapor during the process of perspiration.

An average person (60 to 70 kg) needs to take in 1,800 to 2,500 g of water each day. About half of this is excreted as urine, and the other half leaves the body as perspiration or in the expired air. Half a person's water intake comes from beverages; the rest is contained in food (some foods, such as fruits, contain up to 80% water).

8 | Undernutrition and malnutrition are widespread in the Third World.

Nutritionists have specified recommended daily intakes for people for energy and proteins (Table 4.4), vitamins (Table 4.6), and minerals (Table 4.8). Unfortunately, many people do not get these required amounts. Either they do not get enough food and thus are **undernourished**, or they do not get adequate amounts of all the nutrients and thus are **malnourished.**

One clear indicator of undernutrition is the unavailability of food in a country or region. The daily energy requirement for a group of people can be calculated by knowing the age structure, pregnancy status, body weights, activity, and climate, all of which influence requirements (Table 4.4). Finding how much food is available is a relatively simple calculation in developed countries, with their centralized markets. But it is more difficult in less developed regions, where many people grow their own food.

Nevertheless, the United Nations makes these estimates annually, making it possible to calculate the daily supply of food as a percentage of requirements. For 1992, the world food supply exceeded needs by a small margin, at 105%. But this surplus clearly masked profound regional differences. For developed countries, such as the United States (138%), Canada (135%), France (132%), and Britain (124%), the supply exceeded needs to a considerable extent. But in Africa (Burkina Faso, 91%; Nigeria, 93%; Sierra Leone, 85%) and Asia (Bangladesh, 90%; India, 99%; Pakistan, 95%), the situation was more precarious.

Within countries, there are often great disparities. The well-known specter of undernutrition in the United States, with its oversupply of food, is an example. A striking example of food supply imbalance was found in a study made by the FAO in 1970 of nutritional maldistribution in Indonesia. In one region, 2,087 kcal per capita per day were available; in another region, only 1,076 were available. Similarly, the available protein varied from 46.7 g per person per day down to 22.6 g. The area where food was scarcest (Djokjakarta in Central Java) also had the lowest population growth rate for all the regions studied. The figures suggest that gross malnutrition can result in a decline in the birth rate. However, such severe malnutrition is not characteristic of most underdeveloped countries that have high birth rates.

When these disparities are investigated, the fact almost always emerges that the people who are undernourished are poor. The interrelationships between poverty and nutrition in developing countries are shown in Figure 4.6. When people have been displaced from the farm as a demographic transition is attempted (see Chapter 1), they face limited opportunities for employment in the cities or the village. They take low-paying jobs, and the pressure

Figure 4.6

The Complex Relationships Among Poverty, Malnutrition, and Economic Status. Source: *From M. Pimentel (1989), Food as a resource, in D. Pimentel, ed.,* Food and Natural Resources *(New York: Academic Press), p. 423.*

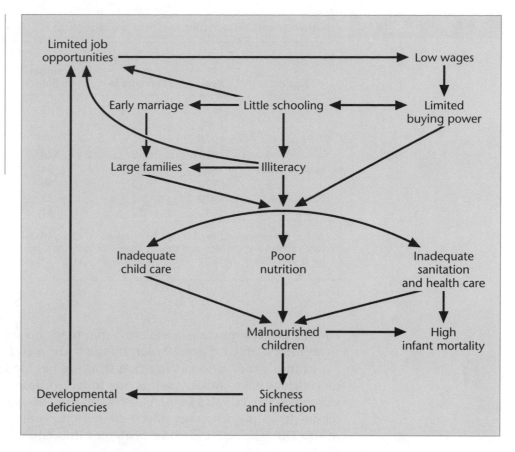

of providing income to the family often results in the children having to work. Demographers have clearly shown that low levels of education often accompany high fertility. Having more mouths to feed places more stress on the food that is available, and with low incomes, families simply cannot afford to buy what is available.

An example of the poverty–undernutrition correlation is shown by an FAO study of people in northeast Brazil. Over a six-year period, two parameters were measured: (1) the percentage of low-birth-weight newborn infants, which is a good indicator of maternal nutrition, and (2) the number of employment hours needed for a person to earn the money to buy food to feed the family for a month. Not surprisingly, these two parameters were strongly related (Figure 4.7).

Estimates of how many people are chronically undernourished are hard to make with any certainty, but the number is certainly large. The United Nations has used similar methods over time, and so comparisons can be made using U.N. data (Table 4.9). Although the total number of undernourished people in the world has gone down (from close to 1 billion in 1970 to about 800 million in 1990), the gains have come mostly in the Far East, especially China. A glance at the world "hunger map" (Figure 4.8) shows that the most serious problems remain in Asia and Africa.

Malnutrition—not obtaining adequate amounts of nutrients—is as widespread as undernutrition. As noted earlier, the cereal grain staples, even if eaten in amounts that supply enough calories, often do not provide the minimum daily allowances of important vitamins and minerals. When calorie intake is too low, these nutritional deficits are exacerbated. Moreover, wheat

Table 4.9	Chronic undernutrition in developing regions

Region	Year	Undernourished Percentage	Number (millions)
Africa	1970	35%	101
	1990	33	168
Far East	1970	40	751
	1990	19	528
Latin America	1970	19	54
	1990	13	59
Near East	1970	22	35
	1990	12	31

Source: Data from Food and Agriculture Organization (1992), World food supplies and prevalence of chronic undernutrition in developing regions (Rome: FAO).

and rice contain phytate, a molecule that binds up some minerals and makes them unavailable for assimilation by the body even if they are eaten.

In India, the National Nutrition Institute has made several recent studies on nutrient deficiencies. Anemia due to lack of iron is common not only in women (65%), but also in men (45%) and preschool-age children (77%). Goiter due to iodine deficiency affects 40 million people, and up to one newborn in six. These minerals are now being put into table salt in an effort to fortify diets.

Mortality, night blindness, and other problems such as increased susceptibility to measles and growth retardation are all caused by lack of vitamin A. They are so common that a major effort has been undertaken in several countries to give at least two doses of the vitamin to all children every year. (This fat-soluble vitamin can be stored in the liver.) This program has dramatically reduced childhood mortality due to vitamin A deficiency (Figure 4.9).

Figure 4.7

Poverty and Malnutrition. The correlation between the rising number of hours of work needed to feed a family and the incidence of low-birthweight infants in Brazil. Source: *Data from Food and Agriculture Organization.*

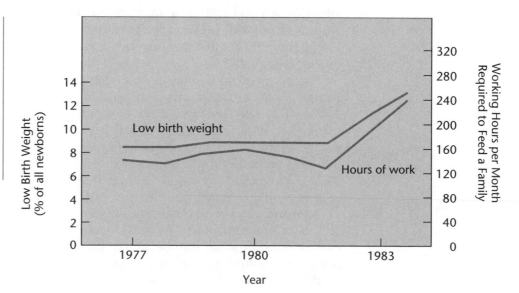

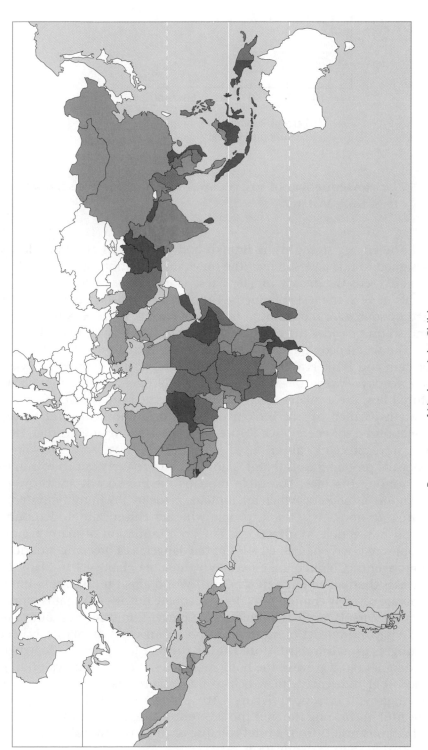

Percentage of Underweight Children
(under 5 years)

0–9   10–19   20–29   30–39   40–49   50–59   60–69

Figure 4.8

The Geography of Hunger. Underweight (<80% of normal) children around the world.

Figure 4.9

Vitamin A Supplements Reduce Childhood Mortality in Children. Source: *Data from Food and Agriculture Organization.*

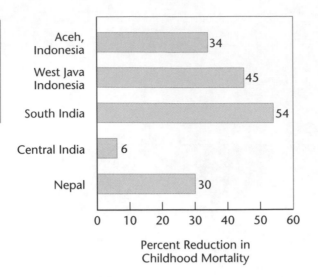

Percent Reduction in Childhood Mortality

9 | The consequences of nutritional deficiencies are serious and often long-lasting.

The energy stores in the human body—glycogen and lipid deposits—are designed to tide a person over the inevitable situations where the demand for food exceeds the dietary intake. These stores are used, for example, between meals to keep the level of glucose in the blood constant (glucose is the major energy source for the brain). A myriad of complex feedback loops, hormones, and so on provides the mechanism for regulation and use of the stores.

But when people are forced to live on diets that provide only 25 g of plant protein and 1,100 kcal per day, the stores are soon exhausted. Only about one to two days' worth of energy stored in glycogen, and at most a month in stored fats, are available for use. After this, the only source of calories left is proteins, which can be broken down into amino acids and these in turn metabolized for energy (see Figure 4.5). But as mentioned earlier, humans do not store proteins—all are used in some vital way. So a breakdown of an essential protein means that it is not available to carry out its role in the body.

Among the most accessible proteins for breakdown are those in blood, including antibodies, which are proteins made by the immune system made to fight infections. As a result, people who are chronically undernourished are highly susceptible to **infections**. The poor sanitation of many regions where people are poorly nourished adds to the danger, as a breeding ground for infectious agents. Children are especially vulnerable (Table 4.10). Many infectious diseases that kill children in the Third World affect those whose immune systems are severely compromised by inadequate nutrition. What is more, poverty puts treatments (such as antibiotics) or prevention (vaccines) out of reach.

A second impact of undernutrition, again most dramatically seen in the young, is **growth retardation**. During the first few years of life, when a child undergoes rapid growth, the body must have the necessary calories, proteins, and other nutrients with which to build more tissues. If these are not supplied, growth is reduced (Figure 4.10). Moderate reductions are generally not harmful in the long run. But the severe stunting too often seen in poor tropical regions results from a total collapse of the biochemical adaptations of the body, due to food deprivation.

| Table 4.10 | Infections that kill poorly nourished children |

Diseases	Estimated deaths, 1990 (millions)
Acute respiratory infections	3.6
Diarrhea	3.0
Malaria (Africa)	0.8
Sexually transmitted (via mother)	0.8
Neonatal sepsis and meningitis	0.3
Vaccine-preventable diseases[a]	2.1

[a]These are diseases for which vaccines exist, but are not sufficiently available in the Third World: polio, tetanus, measles, diphtheria, whooping cough, tuberculosis.

Source: Data from Food and Agriculture Organization.

Figure 4.10

Typical Growth Curves for Children. These curves show the effects of malnutrition on growth and development. Source: *Data from Food and Agriculture Organization.*

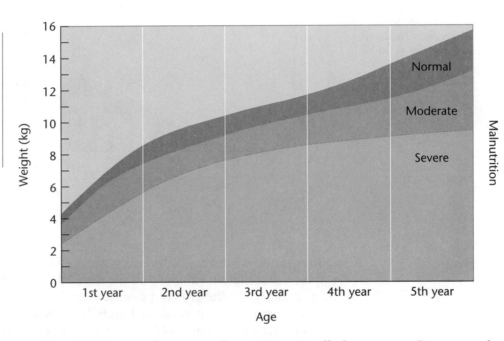

The initial stage of severe undernutrition is called *marasmus*. It commonly results from weaning an infant from energy- and protein-rich mother's milk, to a gruel made from a grain staple (such as corn or millets) that is relatively poor in calories but usually adequate in protein. For example, studies in India of children suffering from undernutrition showed that their daily energy intake (75 kcal per kg body weight versus the recommended 100) was low, but their protein intake (about 2 g per kg) was at the internationally recommended level.

Children suffering from marasmus show poor growth and some wasting of muscles, which begin to be broken down for energy. If undernutrition continues, the child may progress to the more serious disease *kwashiorkor* (a West African word meaning "the disease the child gets when another baby is born"). Children with kwashiorkor are not only severely stunted, but have severe muscle wasting and edema (swelling due to fluid accumulation). These are the children with bloated bellies seen in regions where hunger is prevalent.

If treated in time, children suffering from undernutrition can recover remarkably. The initial treatment is with oral rehydration, in which a

Figure 4.11

Estimated Weight Gain During Pregnancy for a (a) Healthy Woman in Western Europe and a (b) Poor, Undernourished Woman in India. Note that the weight gain by the fetus and supporting organs is the same in both cases. What is sacrificed in the underfed pregnant woman is her own body stores. This means that she is at risk, either during development, delivery, or after the infant is born. Source: Data from Food and Agriculture Organization.

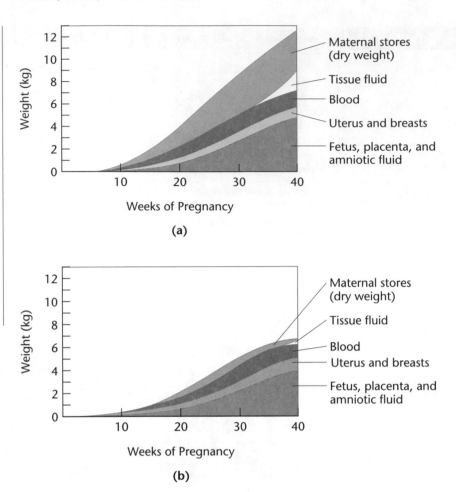

concentrated solution of salts and sugar is given by mouth. Then the child is put on a diet high in calories and protein. Although children so treated never quite make up all the growth potential they lose, the positive results are often seen in a short time, and normal growth follows.

But often a hidden problem remains: nutrient deprivation harms **brain development**. The human brain grows most rapidly during later stages of prenatal development and the first year of life, and is most sensitive to undernutrition at these times. Before birth, the fetus obtains its food from the mother. As she must eat more to gain the necessary nutrition for both herself and the fetus, dietary deprivations at this time can harm both (Figure 4.11). Although the mother's development will be sacrificed in favor of the fetus, both animal studies and human observations show that severe caloric restriction results in lower birth weight and later mental retardation.

Undernutrition during the first year of life, even later compensated for, often results in a physically smaller brain, as reflected by a reduced head circumference. An examination of the brains of children who died from undernutrition during the first year of life showed they had 15 to 20% fewer brain cells than the normal child. These observations confirmed earlier experiments showing that rats and other animals fed protein-deficient diets in early life had physically smaller brains containing fewer cells than normal.

Other experiments show that such animals are also deficient in their learning capacity. Because it is unethical to do controlled feeding experiments with children, it has been difficult to show conclusively that this is also the case in humans. Nevertheless, children who suffered from marasmus or kwashiorkor when they were younger often score lower on intelligence and adaptive behavior tests than their counterparts who are adequately nourished and live in a similar environment. These studies suggest that poor nutrition of infants (especially during the first year) may permanently restrict their mental abilities. This has frightening implications in the poor countries, where 500 million children are now growing up with physical and mental characteristics that may be inadequate to meet the challenges that will face them as adults.

10 | Diets in technologically advanced countries are linked to major diseases.

A consideration of the nutrients that humans need cannot be separated from a consideration of the food they eat. Very few people eat just to nourish themselves in the most efficient way. Eating is a social activity, and many factors other than nutrition play a role, including childhood food habits, taste preferences, and a desire for relaxation (eating can be a relaxing activity), as well as a preference for prepared or cooked foods, and so on. Our diet is the sum total of the foods we eat on a regular basis.

Human diets have changed dramatically over the course of history. The food sources used by hunter-gatherers were extremely varied. Most collected many different plants and supplemented a largely vegetarian diet with meat when it was available to them. Some had abundant supplies of fish and shellfish. (See Box 4.4 on food sources of early native Californians.) The mobile societies followed their food sources, and probably there were few nutritional deficiencies. The major cause of illness or death in these societies was infection, primarily due to a lack of sanitation.

As societies made the transition to agriculture (see Chapter 10), the human diet changed somewhat (Table 4.11). Although there was still a largely vegetarian diet supplemented with some meat, the diversity of the diet decreased, and its composition began to reflect a very few plants. Many people in the Third World still eat this diet. Although infectious diseases and undernutrition are too prevalent in the Third World, the diseases that cause the most deaths in the developed countries—heart disease, stroke, and diabetes—are not as common.

The next transition, to the modern, affluent societies that characterize the developed world, led to a more radical change in diet. Plants—especially fruits, vegetables, and whole grain products—now play a minor role. Processed foods and animal products play a more important role. Particularly, consumption of sugar, salt, fat, and alcohol has increased dramatically (Table 4.12).

However, people's energy and nutrient requirements have remained unchanged or have even diminished, because they lead less active lives. As a result, obesity and high blood pressure are common, and heart disease, cancer, stroke, and diabetes are the leading causes of death. The main challenge for nutritionists is to reduce these diet-related illnesses, which affect the quality of life and result in premature death (Table 4.13). To achieve this goal, they must battle a food industry that is more interested in selling processed foods—processing adds "value," meaning price—than in sound nutrition.

Box 4.4

*Food Sources of
the Early Native
Californians*

At the time of the first Spanish settlement in 1769, there were about 300,000 Native Americans in California. With a population density of one person per 5 km^2, it was probably the most densely populated part of the continent north of Mexico. The area was occupied by many different tribes, who procured food primarily by fishing, hunting, and gathering plants and insects. A few tribes living on the Colorado River also practiced agriculture. Although we cannot be certain, the population of each area was probably near its maximum, as determined by the smallest amount of food available during the leanest year. The population densities in each area certainly reflect the productivity of the ecosystems, with 1.8 Yurok per square kilometer along the fish-rich northwestern coast of California, and about 0.1 Modoc per square kilometer in the high desert of northern California.

The food sources of the native Americans differed markedly depending on the ecosystem they inhabited. Coastal inhabitants collected shellfish and surf fish, hunted mammals (deer, elk, and bear), caught ocean fish (salmon, tuna, halibut, and flounder) and collected seeds (acorns and grass seeds), tubers, and greens. Desert hunter-gatherers collected seeds (pine nuts, mesquite, grasses), tubers, and greens; hunted mammals (rabbit, wood rat, deer, mountain sheep and antelope) and reptiles; and collected insects. Because they lived in such a rich ecological area, the native Californians used about 500 kinds of plants and animals for food. David Prescott Barrows in 1900 studied the Cahuilla—a group that lived in the mountains and desert of southern California, and he recorded the use of 60 different plant species for food and 28 species for narcotics, stimulants, and medicines. A later study of the same tribe describes the use of 174 different plants, and the authors suggest that their list is incomplete. Hunter-gatherers in California, as well as elsewhere in the world, exploited the ecosystem in which they lived to the maximum, using an enormous variety of plants and animals. The coast-dwelling Yurok may have obtained 70% of their food from fish, shellfish, and game and only 30% from plants, whereas this ratio was probably reversed for the inhabitants of the mountains and the deserts.

Acorns, gathered from six different species of oak, were the principal staple of the native Californians. Acorns are an excellent food source, containing about 40% starch, 3–5% protein, 4–6% fat, and 8% fiber, with the remainder being water. Large amounts of acorns were easily gathered—a single tree produces 200–500 kg—and could readily be stored. Early European travelers in California often commented on the substantial stores of acorns that the native Californians gathered in the fall. It was common practice to collect enough acorns in one season to last for two years. The Sierra Miwok had granaries that contained about 2,500 kg of acorns each, enough to feed a family of six for a year. Preparing acorns was laborious because of the presence of tannin, a plant defense chemical that has a bitter taste. After removal of the dark seed coat, the acorns had to be ground to a flour, and the tannins leached out of the flour by repeatedly pouring hot water on the flour and allowing it to drain away.

It is of particular interest that hunter-gatherers and agriculturalists lived side by side in California. The agriculturalists annually planted crops in the floodplains of the Colorado River after the flood water receded. They planted five varieties of corn, three varieties of beans, pumpkins, and gourds. In addition, they sowed the seeds of wild grasses in small plots, presumably because a dense stand would make their harvest easier. These agriculturalists lived on the western boundary of a region (southern Arizona) where the agricultural way of life had found widespread acceptance. The Colorado River Indians produced about half their food by agriculture and the other half in the more traditional hunter-gatherer manner. Like other desert dwellers, they relied on the seeds of several species of mesquite, a desert legume so deep-rooted that its seed production is independent of rainfall. It is a good food producer in both good years and bad.

Table 4.11	Changes in dietary pattern with changes in lifestyle

| | | Percentages of Energy from | | | Fiber |
Lifestyle	Fat	Sugar	Starch	Protein	(g/day)
Hunter-gatherers	15–20%	0%	50–70%	15–20%	40 g/day
Peasant farmers	10–15	5	60–75	10–15	90
Affluent societies	40+	20	25–30	12	20

Source: Estimates from Food and Agriculture Organization.

Table 4.12	Per capita consumption of various foods over the past 200 years in Britain

| | Grams per Day, per Person | | |
	1780	1880	1980
Fat	25	75	145
Sugar	10	80	150
Potatoes	120	400	240
Wheat flour	500	375	200
Cereal (crude fiber)	5	1	0.2

Source: Australian Academy of Science (1990), *Biology: The Common Threads,* Part 1. (Canberra: Australian Academy of Science), p. 204.

Paul Saltman, a nutritional biochemist from the University of California, maintains that "there is no junk food, there are only junk diets." Young, healthy, active people can easily deal with the high-sugar, high-fat, high-salt foods popularly classified as junk food. However, once their food preferences have been established with sugar-rich breakfast cereals, sweetened carbonated drinks, and fat-rich hamburgers, people are hooked on a diet that is less and less appropriate as they get older and their lifestyles become more sedentary. It is very difficult for most people to make even modest changes in the foods they eat each day unless they are on a complete dietary overhaul program. People who go on such programs usually lose weight, then gain it right back again because they have not modified their food habits and their exercise habits. The main culprits in "Western" diets are sugar and saturated fat. Both can be avoided by a high-fruit, high-vegetable diet that also limits store-bought desserts and animal products.

11 | Millions of healthy vegetarians attest to the fact that animal products are an unnecessary component of the human diet.

As was pointed out in Chapter 2, a small number of plants—principally cereal grains, legumes, and root crops—supply most of the energy and nutrients in the human diet. Most grains contain at least 70% starch, whereas root crops contain only 25% starch or other carbohydrates. Thus it takes 3 tons of root crops to supply the same amount of energy as 1 ton of cereal grain.

Table 4.13	Diet-related diseases

Disease	Nutrients Strongly Linked	Nutrients Under Suspicion
Heart disease	High fat (particularly saturated fats)	Low fiber
High blood pressure and related conditions	High salt	Excess alcohol
Cancer of mouth, pharynx, larynx, and esophagus	Alcohol (especially spirits and wine)	
Cancer of stomach		Alcohol (especially spirits), abrasive or irritant foods, low vitamin A or vitamin C
Cancer of colon		Low fiber, high meat intake, low green and yellow vegetable intake, high fat intake
Cancer of rectum		Same as for colon cancer, high beer intake
Cancer of lung		Low vitamin A
Hormone-dependent cancers (such as breasts, uterus)		High fat, low vitamin A, low fiber
Cancer of pancreas		High alcohol, high fat, cholesterol, refined carbohydrate, low fiber
Gallstones		High fat, cholesterol, alcohol, refined carbohydrate, low fiber
Kidney stones		Low fluid intake, high purine, high animal protein, high alcohol; high vitamin C, low magnesium, high vitamin D, high oxalate, high calcium
Diverticulitis and other vein disorders (such as hemorrhoids, varicose veins)		Low fiber
Non-insulin-dependent diabetes		High fat, refined sugar, low fiber, alcohol
Obesity	High-energy diet	Low fiber, alcohol
Liver cirrhosis	Alcohol (especially spirits and wine)	
Dental decay	Sugars	
Joint diseases (gout and osteoarthritis)		High alcohol
Multiple sclerosis		Low iodine, low polyunsaturated and saturated fats, low vitamin D, low calcium

Source: Australian Academy of Science (1990), *Biology, the Common Threads*, vol. 1. (Canberra: Australian Academy of Science), p. 207.

However, root crops have a much greater caloric productivity per unit land area than cereals; a hectare of potatoes yields 2 to 2.5 times as many calories as a hectare of wheat or rice.

In general, seed staples have a favorable protein content (8 to 15% of the dry weight), whereas roots and tubers have a much lower protein content (1 to 3%). The ratio between protein and carbohydrates is often expressed as grams of protein per 100 kcal (see Figure 4.12). Because an average adult should have a daily food intake of 50 g of protein and 2,500 kcal, or about 2 g

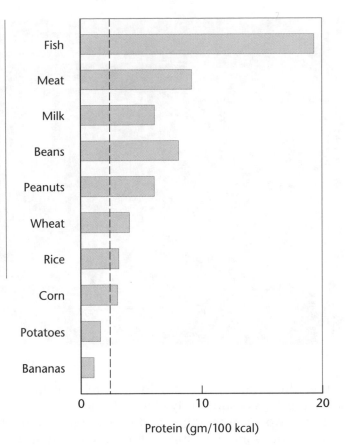

Figure 4.12

Protein–Calorie Ratios of Several Foods. The dashed line shows the approximate adult daily dietary requirement for protein. Note that the three major staples—wheat, rice, and corn—provide sufficient protein if calorie needs are met. However, a single staple does not provide a proper ratio of essential amino acids (see Table 4.14). Source: Data from Food and Agriculture Organization.

of protein per 100 kcal, staples that contain 2 or more g or protein per 100 kcal are therefore also good sources of protein.

It is no accident that societies have adopted as staples foods that meet or exceed the protein–calorie ratio standard. More than half of the human diet (55%) comes from the seed staples wheat, rice, and corn. Another 13% is provided by the protein-rich seeds of legumes (peas, chickpeas, beans, lentils, soybeans, cowpeas). Animal products supply 20% of human beings' dietary protein, but most of this is consumed in the industrialized countries of the Western world.

The lower protein–energy ratio of plant foods in comparison with animal foods (Figure 4.12) and the low protein scores of plant proteins when compared with animal proteins (Table 4.5) also help explain to some extent the correlation between nutritional deficiencies and dependence on plant protein. Plant proteins are usually low in certain essential amino acids, and heavy reliance on a single staple can result in an inadequate supply of these. This deficiency can be corrected by supplementing the diet with plant proteins from different sources (for example, by mixing cereal grains and legumes) or by including small amounts of animal proteins in the diet. Unfortunately, these alternatives are not available to poor people who cannot buy the available foods.

If animal proteins have a higher nutritional value for human beings than plant proteins because they have a higher protein score, why not convert more plant protein into animal protein by raising more livestock and producing more milk and cheese? In the past, producing animal products depended largely on food sources that human beings could not eat. Herbivores such as

Figure 4.13

The Conversion of Plant Protein to Animal Protein. Shown here is a pig-feeding operation in Arkansas. Pigs that are allowed to be in the open air, as shown here, are healthier than pigs that are confined indoors. Source: Courtesy of U.S. Department of Agriculture.

cattle, sheep, and goats can live entirely on grass and other plants that human beings cannot easily digest. Pigs and poultry can live in the farmyard or in the woods, where they find insects, seeds, and kitchen scraps. In many areas of the world, pigs still perform the essential service of converting garbage into meat. The fish and shellfish that human beings eat depend on the plants that live in the sea (either directly or indirectly) as their source of food. When animal protein is produced in these ways, it is said to come "cheaply," because food sources of little value to human beings are converted into foods valuable to humans.

In recent years, the production of animal protein has come to depend increasingly on animals eating food sources that human beings could eat. Human beings and domesticated animals are competing for the same food. Modern animal husbandry uses feedlots and factory farms, where composition of animal feeds is constantly monitored to achieve spectacular results. There are now chickens that lay an egg nearly every day of the year, dairy cows that produce 1,500 L of milk a year, and sows that each produce litters of 15 piglets, which six months later together weigh 2,000 kg (Figure 4.13).

This intensive production depends heavily on wheat, corn, skim milk, soybeans, fish meal, and other foods that human beings could eat directly. Indeed, in the United States and Europe, over two-thirds of all the grain that is grown is eaten by livestock. The feeding of animals has also become extremely sophisticated. Computers are used to calculate the cheapest and best mixtures of foodstuffs, taking into account the composition, protein score (contents of essential amino acids), and price of each, and the requirements of the animals. Synthetic amino acids (especially methionine) are now widely used to adjust protein scores when necessary. Millions of human infants are suffering from undernutrition while the cattle in modern feedlots munch their scientifically controlled rations.

Eating meat every day is so culturally ingrained in technologically devel-

Table 4.14	Protein score and protein content of various mixtures of staples

Cereal	Supplement[a]	Percentage of Protein	Protein Score
Wheat	None	11.2%	62
	Groundnut	14.2	67
	Soybean	13.8	76
Corn	None	9.5	49
	Groundnut	12.6	58
	Soybean	12.2	67
Rice	None	6.7	69
	Groundnut	10.0	73
	Soybean	9.6	77

[a]Adding 8 g of protein concentrate to 100 g of staple.

Source: G. R. Jansen (1974), Amino-acid fortification of cereals, in A. Attschul, ed., New Protein Foods (New York: Academic Press), pp. 94–99.

oped countries that it is easy to forget that many vegetarians never eat any meat at all. Some do so out of choice, while others are vegetarians because of poverty. Some vegetarians do not exclude all animal products from their diets: many drink milk and eat milk products, and others eat eggs.

Can plants and plant products be an exclusive food source for humans? And can people remain in good health if they only eat plants? The answer to this question depends on the meaning of the word *plants*. The answer is no if we only eat the familiar higher plants, but yes if we include microbes such as yeast or bacteria. Neither plants nor animals can synthesize vitamin B_{12}; only microbes can. Meat is a good source of vitamin B_{12} because of the microbes living among the meat fibers. In India, some vegetarians eat seaweed, the slime of which may be rich in B_{12}; some African tribes eat soil, in which there are abundant B_{12}-rich microbes; still others, members of religious sects in Asia, eat neither of these but still seem to survive, possibly because they have microbes that produce B_{12} living in their digestive systems. In most regions of the world, however, people on a strict vegetarian diet should supplement it with yeast or vitamin B_{12}. A deficiency of this vitamin results in a variety of diseases, usually culminating with degeneration of the spinal cord.

Plant proteins have low protein scores because they are low in certain essential amino acids. By carefully selecting a variety of plant proteins, vegetarians can obtain a diet with a high protein score by eating complementary foods. For example, most societies that rely on a cereal grain staple have also come to rely on a legume that complements that grain (Table 4.14). Combinations of proteins should be eaten at the same meal, because the body cannot store amino acids, and indeed this usually occurs (for example, a burrito combines corn and beans).

Careful planning is essential if one wants to derive an adequate supply of protein from a vegetarian diet. It is far easier to ensure a proper diet by eating a small amount of animal protein to supplement a larger amount of plant protein. The large amounts of animal protein now consumed by most people in technologically advanced countries are nutritionally superfluous and are a drain on the global supplies of staple foods.

Table 4.15	Food-related diseases	
Disease Agent (and example)	Where Found	Food Source
Viruses (Hepatitis A)	Wide range include shellfish, raw fruit, and vegetables	Poor hygiene, growing plants in area of raw sewage or organic wastes
Bacteria (*Salmonella*)	Raw and unprocessed foods, such as cereals and fish	Poor hygiene, carried by rodents, birds, humans
Molds (*Aspergillus*)	Nuts and cereals	Storage at high temperature, humidity
Protozoa (*Amoeba*)	Vegetables, fruits, raw milk	Contaminated water
Helminth worms (*Ascaris*)	Vegetables and undercooked meat and fish	Contaminated soil and water
Agricultural agents (pesticides)	All foods treated	Use of banned chemicals or excesses

Source: Adapted from Food and Agriculture Organization (1992), *Nutrition: The Global Challenge* (Rome: FAO), p. 24.

12 | Food safety is an important issue in both developing and developed countries.

Growing crops and making the food available to the people are the major aims of agriculture. But events that happen between the field and the consumer can make the food unsafe to eat. In the developing regions, millions die because food or drinking water is contaminated with pathogens, while in the developed regions food-borne diseases lead to increased production costs and health problems.

These **food-related diseases** (as opposed to the nutritional diseases just considered) usually are caused by micro-organisms (Table 4.15). Some, such as the virus-caused hepatitis A, occur worldwide. But others, such as the bacterial diseases cholera and typhoid, the protozoan disease amoebiasis, and the *Ascaris* worm, occur today only in the less developed regions, where food processing and general sanitation are poor. The increase in urbanization, with its crowding, unsanitary conditions, and proliferation of street vendors, has widened the spread of food-carried infections. Educating consumers about food storage and preparation, as well as importing methods to eliminate pathogens, are the major approaches taken to stem the tide of pathogens.

In the developed countries, concern about food safety has focused on contamination of foods with chemicals used in food production. **Pesticides**, used to kill pests that would otherwise damage crop plants before or after harvest, are an important component of agriculture. An almost unavoidable consequence of their use is that a tiny amount of the pesticide may end up as a contaminant of the resulting food or food product. Some of these pesticides have been shown to be harmful to people in large doses, and to animals in small ones. The question is, what concentration of each chemical constitutes an acceptable risk to the consumer? Of greater importance, but unfortunately of less concern to the public at large, is the exposure of agricultural workers to pesticides and the effect that pesticides have on soil and water ecosystems.

Elaborate and costly systems have been developed to ensure safe levels of pesticides in foods. In the United States, the Environmental Protection

Agency (EPA) establishes a tolerance level based on the **ADI**, or **acceptable daily intake**: this is the level of daily exposure to a pesticide residue that over a 70-year life span will have no negative effect. A tolerance level for a pesticide is usually calculated by a manufacturer, who treats a crop with a pesticide under the most severe conditions (dosage and number of applications) and who then measures the residue on the crop. The EPA then calculates human exposure to the pesticide from all crops and approves a tolerance level for a specific chemical on a specific crop if the calculations show that exposure will be below the ADI.

The ADI itself usually has a safety factor of 100; it is set 100 times lower than the level known not to have a negative effect. The Food and Drug Administration (FDA) tests food in U.S. supermarkets using a method called the Total Diet Study. A large number (234) of food items are bought and prepared for home consumption, and the levels of pesticides are measured in the prepared foods. Such studies show that exposure to pesticides ranges from 0.1 to 1% of the ADI, or 10,000 to 100,000 times lower than doses that have no long-range toxic effects.

Slightly different rules apply to carcinogens, chemicals that cause cancers in laboratory animals. Generally, the regulatory limit for a carcinogen is the dose that poses a negligible risk—causing one additional cancer in 1 million people. (About one person in three will get cancer in his or her lifetime—this risk, therefore, represents an increase in cancer incidence from 330,000 to 330,001.) This dose is estimated by carrying out experiments with rodents or cells with a much higher dose of the chemical, generating a dose–cancer curve, and then working back to a small dose to see what level would cause one cancer in a million people.

Tests in rodents are often done at the maximum tolerated dose, and under these conditions about half the chemicals tested turn out to be carcinogens. Scientists disagree about whether (1) this is an acceptable way of testing for carcinogenic effects of chemicals and whether (2) the extrapolations needed to calculate the dose that poses a negligible risk are justified. In addition to these uncertainties about validity of the test data, considerable evidence shows that our normal diet contains many carcinogens, all perfectly "natural"—either synthesized by the plants, or produced during storage or cooking. The natural carcinogens in our food are part of the plant's defense mechanisms against fungal and bacterial pathogens and against predators (see Chapter 12).

Biologist Bruce Ames, who has studied this issue for over a decade, has calculated that 99.99% of the carcinogens in food are natural organic chemicals, synthesized by plants or derived from such compounds during food preparation. Only 0.01% of the carcinogens in food are of industrial or agricultural origin. Americans each eat about 1.5 g of these natural pesticides each day, which is about 10,000 times more than the synthetic pesticides they eat. Although the public at large is very much aware of the presence of synthetic pesticides in food, it is unaware of the presence of natural pesticides.

The issue of natural pesticides is of particular interest for biotechnology. The use of synthetic pesticides is presently declining for two reasons. First, the chemicals coming into use are much more specific, so much smaller amounts can be used per hectare. Second, the expense of pesticide usage to the farmer encourages a greater reliance on natural pesticides. One goal of genetic engineering is to make plants resistant to pests and predators by transferring genes that will cause crop plants to use natural defense systems that are now present in noncrop plants. Such defense systems rely on organic, natural pesticides, many of which can be shown to be carcinogenic in the same types of tests as are used for synthetic chemicals. Clearly, the policy pendulum will continue to swing back and forth in this very complex issue.

13 | **The plants people eat contain many natural chemicals whose effects on the human body are not known.**

To defend themselves against pathogens and pests, plants use chemical defenses, and these are discussed in detail in Chapter 12. These defense chemicals of plants include more than 10,000 different compounds. The effects of some of them are known, especially the potent poisons, but most of these chemicals have never been studied. Although the overwhelming majority of defense chemicals are present in nonfood plants, crops still contain hundreds of them (see Box 4.5).

Defense chemicals generally do not poison the cells of the plants in which they occur because they are stored in a separate compartment. Only when the plant tissues are disrupted by chewing (by people or insects) are the chemicals released. Some of these defense chemicals are inactivated by cooking, and using fire for food preparation expanded the range of useful foods for the early hunter-gatherers. Although toxic proteins in many seeds are inactivated by cooking, some (such as beans) require long cooking times. Some of these seed protein toxins are so harmful that a single molecule can kill a human cell. Others, such as the lectins present in many legume seeds, interact in a rather complex way with the proteins in the intestinal tract and with the bacteria that live there.

It is well known that raw kidney beans are toxic to humans and other monogastric animals such as pigs or chickens. This toxicity is due primarily to the presence of phytohemagglutinin (PHA), a protein that is classified as a lectin. Lectins are proteins that bind molecules that contain sugars, such as glycoproteins (proteins that have sugars attached to some of their amino acids) and glycolipids (lipids that have attached sugars). Many plants contain abundant supplies of one or more lectins in their seeds. When animals eat these seeds, the lectins bind to the glycoproteins in the membranes of the intestinal epithelial cells.

One antinutritional effect of phytohemagglutinin is a general reduction in food conversion efficiency. Experiments show that PHA is not digested after an animal eats raw beans, but ends up coating the lining of the entire gut. This coating may well be the reason why food absorption is decreased in the presence of the lectin. A second unusual effect of lectins is that although they are toxic, they indirectly act as potent growth factors for the small intestine. When rats were fed a high-quality diet, supplemented by a small amount of phytohemagglutinin (42 mg per rat per day), their growth was severely retarded, but there was a remarkable increase in the size of the small intestine. This may well be a response of the animal to the reduced efficiency of food absorption.

Defense chemicals can also have unexpected beneficial effects. The mustard family (for example, cabbage, kale, and broccoli) defends itself by producing glucosinolates (mustard oils), which are partially broken down by an

BOX 4.5

*Cyanogenic
Glucosides:
Chemicals That
Produce Hydrogen
Cyanide in Our
Food*

Cyanogenesis—the ability of plants to release the toxic gas hydrogen cyanide (HCN) under certain circumstances—is extremely widespread in the plant kingdom. It occurs in 2,050 species of higher plants, distributed throughout 110 families. Two major groups of chemicals can be cyanogenic: glucosides and lipids.

HCN is not released continuously by the plants, but only when the tissues are wounded, crushed by eating, or invaded by fungi. The reason is that the cyanogenic chemical and the enzyme needed to break it down and cause the release of HCN are stored in different cellular compartments. Only when the cellular structures are disrupted do enzyme and chemical substance mix to produce HCN. The best-studied cyanogenic chemicals are glucosides, and plants synthesize at least 26 different ones. They are nitrile derivatives of one of five common amino acids that are combined with glucose.

Amygdalin. This cyanogenic glucoside is found in the seeds of cherries, apples, pears, peaches, and apricots.

Many people know that the seeds of apricots and other fruits that have "stones" are toxic, and children are often warned not to eat them. The flesh of these fruits is not cyanogenic, but the seeds contain both amygdalin and the enzyme necessary to break it down so that HCN will be released. Preparations based on peach pit extracts are for sale in Mexico as alleged anticancer drugs. There is, however, no evidence that they are effective, and for this reason they are not for sale in the United States.

Immediately after they germinate, sorghum seedlings synthesize huge amounts of durrhin, another cyanogenic glucoside. In young seedlings, up to 30% of the dry weight (the weight of the plant excluding the water) can be accounted for as durrhin. Although most of the cyanogenic plants or plant organs—often only a portion of the plant is cyanogenic—are not foodstuffs, some important human foods and cattle feed are cyanogenic. These include lima beans and white clover, sorghum, and cassava, an important tropical staple. The function of cyanogenic glucosides is to protect the plants from predation and infection by fungi. Not only is HCN extremely toxic (it inhibits the respiratory activity of mitochondria), but cyanogenic plants also taste bitter. Mammals (including people), insects, and snails all show a feeding preference for species that are not cyanogenic.

Cassava is the one major food crop in which cyanogenesis is a real problem. This tropical root crop is grown extensively in Africa, Asia, and Latin America. It is used as a staple crop, providing up to 60% of dietary calories in some areas. All strains of cassava are cyanogenic, and they range in taste from sweet to bitter depending on their content of linamarin (the chief cyanogenic compound). The dangers of long-term exposure to HCN are not completely understood, but toxic effects involving the central nervous system, the gastrointestinal tract, and the thyroid have been observed. Such conditions are often endemic in areas where people have a high-cassava, low-protein diet. A low-protein diet contributes to cyanide toxicity because amino acids help detoxify cyanide once it has been ingested or released inside the body.

Cyanogenic cassava roots can be largely detoxified by food processing. The roots are usually grated, and the pulp is allowed to stand in water. This causes the enzyme to come in contact with the linamarin and HCN is released. The pulp is then washed and prepared for consumption. Boiling the roots and fermentation are alternate strategies for lowering the linamarin and HCN content of this important food source.

Biotechnological approaches could be used for lowering or eliminating the linamarin from the roots. Unfortunately, we do not know how this would affect the pest resistance and the herbivory by larger animals when the plants are grown under field conditions.

enzyme released from a separate compartment when the cell structure is disrupted. The partial breakdown releases several toxic compounds, including isothiocyanates, thiocyanates, and nitrites. When these are ingested, the body tries to neutralize or detoxify them by converting them to harmless compounds. The enzymes for detoxification are only made when toxic chemicals appear, but not all toxic chemicals are equally effective in inducing such enzymes. It has been suggested, but not yet proven, that the chemicals released by glucosinolates induce these detoxifying enzymes, which may also inactivate carcinogens in our food. This example emphasizes the general ignorance of the complex ways that the hundreds of chemicals in food affect human cells. The most prudent dietary advice therefore remains to eat like the hunter-gatherers: small amounts from many different food sources.

Summary

Food is a source of energy and nutrients. The human body needs energy as well as specific nutrients such as amino acids, fatty acids, vitamins, and minerals. In the course of evolution, humans lost the ability to synthesize certain organic compounds from basic materials (glucose and minerals), and now the compounds must be provided in the food we eat.

Humanity derives most of its food from plants. Although in technologically advanced countries meat and animal products are consumed quite extensively, most of our global supply of food comes from plants. Converting plants to animal products is relatively inefficient as well as unnecessary. Vegetarians all over the world demonstrate that a healthy diet can consist of plant foods only. Unfortunately, many people suffer from either undernutrition (a lack of food), malnutrition (a lack of nutrients), or both. The foods we eat constitute our diet, and in developed countries, diets are linked to major diseases such as heart disease and many forms of cancer.

Finally, plants contain hundreds of chemicals whose effects on the human body are largely unknown. Many of these plant defense chemicals are toxic to humans, while some may have yet-to-be-discovered beneficial effects.

Further Reading

Alamgir, M., and P. Arora. 1991. *Providing Food Security for All*. New York: New York University Press.

Food and Nutrition Board, U.S. National Research Council. 1989. *Recommended Daily Allowances*. 10th ed. Washington, DC: NRC.

Galloway, R. 1991. *Global indicators of nutritional risk.* Washington, DC: World Bank.

Gopalan, C. 1992. The contribution of nutrition research to the control of undernutrition: The Indian experience. *Annual Review of Nutrition* 12:1–17.

Isaacson, R., and K. Jensen. 1992. *The Vulnerable Brain: Malnutrition and Hazard Assessment.* New York: Plenum.

Kolter, N. 1992. *Frontiers of Nutrition and Food Security.* Washington, DC: Smithsonian Press.

Lipton, M. 1983. *Poverty, Undernutrition and Hunger.* Washington, DC: World Bank.

Mahan, L. K., and M. T. Arlin. 1992. *Krause's Food, Nutrition and Diet Therapy.* 8th ed. Philadelphia: Saunders.

Salunkhe, D. K., and S. Deshpande. 1990. *Foods of Plant Origin.* New York: Van Nostrand Reinhold.

Saltman, P., J. Gurin, and I. Mothner. 1987. *The California Nutrition Book.* Boston: Little, Brown.

Van Gelder, N., B. Drujan, and R. Butterworth. 1990. *Malnutrition and Infant Brain.* New York: Wiley.

Chapter 5

Growth and Development of Flowering Plants

To understand our food plants—how and where they grow, how they store food, how seeds are formed, when plants form flowers, why many plants contain toxic chemicals—we must take a closer look at the structure and function of plant organs, at their development and at the ways plants interact with their environment. That is the subject of this chapter.

The unusual diversity of plant sizes and forms is familiar to everyone. Tiny annual flowering plants grow on the desert floor and complete their life cycle in a few weeks, not far from where the giant redwoods grow centuries old.

In spite of this diversity, all plants have major design elements in common. They carry out photosynthesis and are the primary producers in the ecosystem. They are anchored to the soil, from which they derive mineral nutrients. They must continuously take up water to replace water lost by transpiration. And they have stems that are reinforced so that they can stand erect and grow toward the sun. Taxonomists recognize four major groups of plants: mosses, ferns, gymnosperms (the largest group of which are the conifers), and angiosperms or flowering plants. This chapter describes the development of flowering plants. They have been studied most intensely in part because all our crop plants are flowering plants.

1

Our food plants come from many different families belonging to the major groups of flowering plants, the monocots and the dicots.

There are about 250,000 known species of flowering plants, which belong to two taxonomically distinct groups: the **monocots** (monocotyledons) and the **dicots** (dicotyledons). These two groups have some very distinct anatomical features. Their names derive from the number of cotyledons or seed leaves; monocots have one cotyledon, while dicots have two. Both groups of plants contain a large number of different plant families: the monocots are divided into 57 families and the dicots into 250 families. Many plant families contribute in some way to human well-being by providing food, drugs, timber, fibers, dyes, beverages, or spices. In the course of their existence, human beings have used about 3,000 different species as sources of food, but only about 150 are cultivated on a commercial scale.

Monocots are characterized not just by the presence of a single leaf in the seed, but also by fibrous root systems (many small roots coming from the base of the stem) and by long leaves in which the veins are parallel. In contrast, the dicots have branched root systems and leaves with branched veins (Figure 5.1). The major monocot family that has given us food plants is the grass family (Poaceae). Wheat, rice, corn, barley, sorghum, millet, teff, rye, and oats all belong in this family. The other monocot families that have food plants are the palm family (Palmaceae), with coconuts, oil palm, date palm, and sago; the arum lily family (Araceae), with taro; the banana family (Musaceae), with bananas and plantains; and the lily family (Liliaceae), with onions, shallots, leeks, garlic, and asparagus. Two other monocot families have contributed the pineapple and the tropical yams to our list of food plants.

Among the dicots, the legumes (Fabaceae) are the most important food plants. Legumes, with bacteria, fix nitrogen symbiotically. As a result, they can grow on soils that are very poor in nitrogen and nevertheless have a high concentration of protein in their seeds and leaves. The legume family has given us a number of grain legumes (soybean, garden pea, mung bean, chickpea, black-eyed pea, common bean, peanut, Bambara groundnut, and so on) and a number of important pasture plants (clover, alfalfa, and vetch), as well as tropical trees that are important in agroforestry farming systems.

The family of the Solanaceae, whose members produce many toxic compounds, includes tomato, potato, eggplant, and many varieties of peppers. For reasons that are not yet understood, solanaceous plants are easily manipulated in tissue culture (hormone-induced formation of roots and shoots, and regeneration of plants from protoplasts, as discussed in Chapter 3). And the Solanaceae are excellent candidates for genetic engineering experiments.

Three other families are well represented in our food plants. The Brassicaceae give us cabbage, turnips, radishes, mustards, and rapeseed. The Rosaceae (rose family) give us apples, pears, plums, peaches, cherries, apricots, blackberries, raspberries, and strawberries. And the Compositae (daisy family) offer lettuce, Belgian endive, Jerusalem artichoke, and especially sunflower and safflower, whose seeds yield valuable polyunsaturated vegetable oils.

Finally, this list of dicot food plants would be incomplete without mentioning at least three other families. The cucumber family (Cucurbitaceae), the beet family (Chenopodiaceae), and the carrot family (Umbelliferae) have each given us a number of vegetables.

Such botanical classifications are quite useful to scientists because plants that belong to the same family often share many similar characteristics, not only structural (the main basis for classification) but also chemical. Laypeople are more likely to classify food plants by their nutritional function or by the plant organ that is consumed: seeds, tubers, roots, leaves, stalks, flowers, and fruits. The botanical classification shows that we eat the seeds or roots or leaves from many different families.

Everyone who is at all concerned with the nutritional value of food knows that potatoes are more fattening than celery, that wheat germ is more nutritious than lettuce, and that whole wheat bread is better overall than sweet potatoes, although they are rich in vitamin A. The various parts of different plants contain very different amounts of nutrients. Some seeds are good sources of starch; others, of proteins; and still others, of fats. Leaves and

Figure 5.1

Seedlings of (a) a Monocot (Monocotyledon), Wheat, and (b) a Dicot (Dicotyledon) Bean. Monocots have elongated leaves with parallel veins and a fibrous root system. Dicots have broad leaves with a network of veins, and have a taproot.

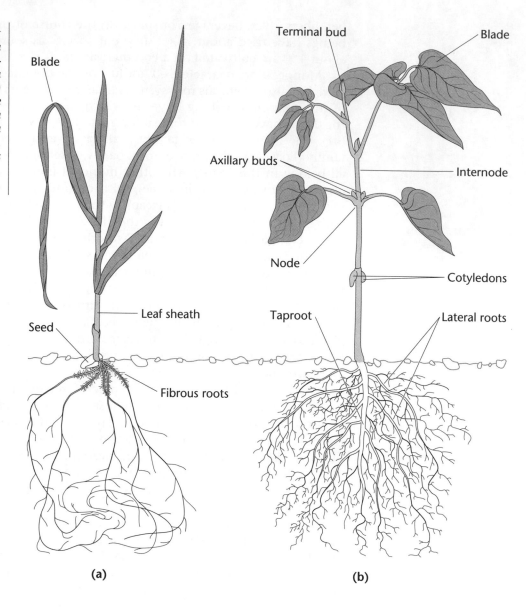

(a)   (b)

stalks are poor sources of these three nutrients, but are rich in some vitamins, minerals, and fiber. Fruits are generally good vitamin and mineral sources, and storage roots and stems are rich in starch. The nutritional values of these plant organs largely depend on the roles the organs play in the plant. A better understanding of this relationship requires a discussion of the structure of the plant and of how it develops from a single cell to a fully mature organism capable of sexual reproduction. An understanding of the role that each organ plays in the plant will lead to a better understanding of its nutritional value.

2 | Plants are made up of cells, tissues, and organs.

The vegetative body of all plants, in spite of their apparent diversity, is composed of three organs: leaf, stem, and root. The leaves attached to the stem together form the shoot system, while the roots together form the root

Figure 5.2

Electron Micrograph Showing the Cellulose Microfibril Orientation in the Plant Cell Wall. Electron micrograph of a cell wall showing cellulose microfibrils magnified about 10,000 times. Each microfibril contains 50–100 cellulose molecules lying parallel. Although cell walls cannot be digested by humans, they serve an important function in our diet by providing fiber (see Box 4.2). Source: D. Sadava (1993), Cell Biology *(Boston: Jones and Bartlett), p. 529; courtesy of T. Itoh.*

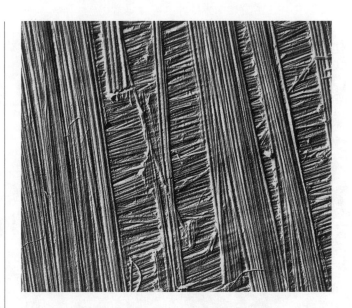

system. The function of the root system is to take up minerals and water, and to store reserves, especially starch or sucrose. The main functions of the shoot system are photosynthesis and the formation of the reproductive organs (flowers). Each organ of the plant is made up of at least three tissues, and each tissue contains one or more types of cells.

The cells are the basic building blocks of the plant. In this respect, plants are very much like animals, whose basic building blocks, the cells, are grouped into tissues, and different tissues together make up organs. In both plants and animals, each cell type is finely adjusted to its function. A cell of the nervous system is very different from a red blood cell or a skin cell, and each has structures and molecules (proteins) that equip it for its function. Similarly, in plants an epidermal cell is very different from a mesophyll cell in the leaf or a vascular cell in the stem, and each has structures and molecules that uniquely equip it for its function.

Plant and animal cells differ in several important ways. First of all, each plant cell is encased in a **cell wall** that is thin and flexible when the cell is young, but can become very thick and rigid in cells that have specialized functions. Perforations, called *plasmodesmata,* through the wall connect the **cytoplasms** (the cell protoplasm outside the nuclear membrane) of adjacent cells to one another and allow molecules to pass freely from cell to cell. These cell walls consist, in part, of cellulose organized in microfibrils (Figure 5.2), and, in part, of other indigestible complex carbohydrates. Another major difference between plant and animal cells is that plant cells have **plastids**, important structures within the cell that fulfill a number of key metabolic roles. In leaf cells, these plastids are called *chloroplasts* and their main function is photosynthesis, but in storage organs (roots, tubers, and seeds) plastids accumulate and store starch until it is needed for growth. A third major difference is that plant cells have a large storage compartment called the **vacuole.**

The vacuole takes up 70–80% of the volume of a cell. Not only does it store minerals, amino acids, sugars, and organic acids (such as vitamin C), but in specialized tissues vacuoles can also store proteins and many other chemicals that help plants defend themselves against predators and pathogens. In dividing cells, vacuoles tend to be very small, but as cells get larger the

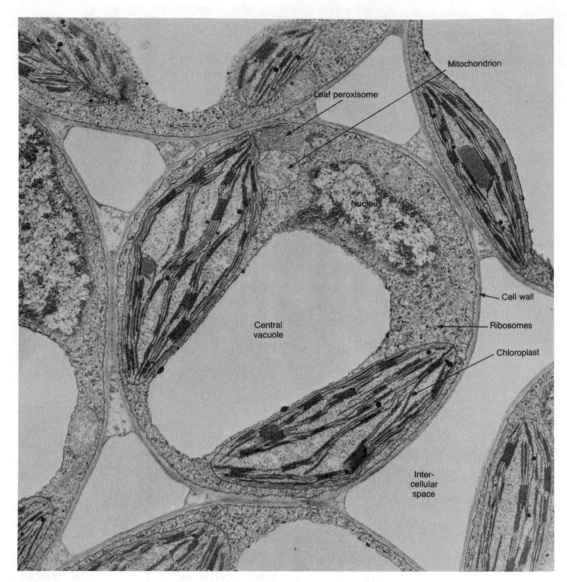

Figure 5.3

Electron Micrograph and Drawing of a Typical Plant Cell Showing a Number of Internal Cellular Structures (organelles). Note the very large central vacuole, the nucleus, the chloroplast, and the mitochondria. Source: D. Sadava (1993), Cell Biology (Boston: Jones and Bartlett), p. 292; micrograph by W. P. Wergin, courtesy of E. H. Newcomb, University of Wisconsin, Madison.

vacuole takes up an ever-greater proportion of the total volume of the cell. An important role of the vacuole is that it provides the turgor pressure that allows plant cells to grow and expand and that keeps the entire plant body turgid, provided there is enough water in the soil. If the turgor pressure in the cells falls, then the plant wilts.

In spite of these important differences, the similarities between plant and animal cells are actually quite striking. Both types of cells have a nucleus that contains the chromosomes in which the genetic material is stored. In both types of cells, a plasma membrane surrounds the cytoplasm and forms the cell boundary; at this boundary, the cell monitors what comes in and what goes out, and where chemical signals such as hormones have their effects. The plasma membrane contains many different types of proteins that regulate the traffic of other molecules, such as sucrose and amino acids, as well as the passage of mineral ions. Both types of cells have mitochondria where energy released from the oxidation of sugars is used for the synthesis of **adenosine triphosphate (ATP)**. Both types of cells have ribosomes for protein synthesis and an extensive internal network of membranes. Both types of cells have a Golgi apparatus for protein secretion, and a cytoskeleton made up of protein filaments that guide the chromosomes during mitosis and the transport of small vesicles within the cell. Plant and animal cells contain many of the same enzymes and carry out many of the same biochemical reactions. In fact some proteins in animal cells are almost identical to plant proteins. Clearly, plant cells resemble animal cells in many different ways. A "typical" plant cell is shown in Figure 5.3.

A characteristic that is shared by all cells of higher organisms (plants or animals) is that each subcellular structure or **organelle** (for example, the chloroplasts, the plasma membrane, and the vacuole) has a unique set of proteins that carries out the function of that organelle. For example, the enzymes necessary for carbon dioxide fixation are only found in the chloroplasts, and those needed for respiration are found in the mitochondria. This means that after proteins are synthesized in the cytoplasm, they must be able to find their way to the correct organelle, where they are to carry out their function. This distribution of proteins within a cell has important implications for biotechnology, and we return to it in Chapters 9 and 15. When a gene from a different plant or another organism is expressed in a crop plant, the protein that is made must be equipped with the right targeting signals so that it will end up in the correct location in the cell.

3 | Cells are grouped in tissues, and several tissues make up an organ.

At the next higher level of organization above the cell, three major tissue systems are found in all organs: dermal tissue, ground tissue, and vascular or conductive tissue. Their location in the plant is shown in Figure 5.4. In addition, the plant contains meristematic tissues or **meristems**, whose primary function is to give rise to new cells. Meristems are very small tissues found near the growing points of plants.

The **epidermis** is the dermal tissue of young plants, or young organs (leaves, flowers) of older plants; it is composed mostly of flattened polygonal cells that block entry of pathogens. In leaves, this barrier is reinforced by a layer of polyester impregnated with wax on the outside of the cell. The main function of this layer is to prevent the loss of water.

Two other important cell types are found in the epidermis: hair cells and guard cells. Individual epidermal cells can grow out into hair cells, and hair cells may be found on roots, leaves, stems, flowers, or seeds. For example, cotton is composed of very long individual hair cells that grow from the epidermis of the cotton seed. These cells have strong cellulosic walls that can be spun into cotton threads. The longer the individual cells, the higher the quality of the cotton. In roots, hair cells are important because they greatly increase the surface area of the root that is in contact with the soil solution, and therefore increase the plant's ability to take up water and minerals. Root hairs have only a short life, and new root hairs are formed constantly as the roots grow through the soil. On leaves, stems, and flowers, the hair cells synthesize chemicals that protect plants against predators.

Pairs of guard cells surrounding microscopic pores, called **stomates** (Figure 5.5), are found in the epidermis of leaves and stems. The hydrostatic or turgor pressure within the guard cells determines the size of the hole between

Figure 5.4

Schematic Representation of the Body of a Plant Showing Three Major Organ Systems: Leaves, Stem, and Roots. Cross sections through each organ show the arrangement of the major tissues in each organ: leaf (a), stem (b), and root (c). Each organ has vascular tissues (xylem and phloem), dermal tissue (epidermis), and ground tissue composed of parenchyma cells (mesophyll, cortex and pith). The leaf primordia are shown greatly enlarged.

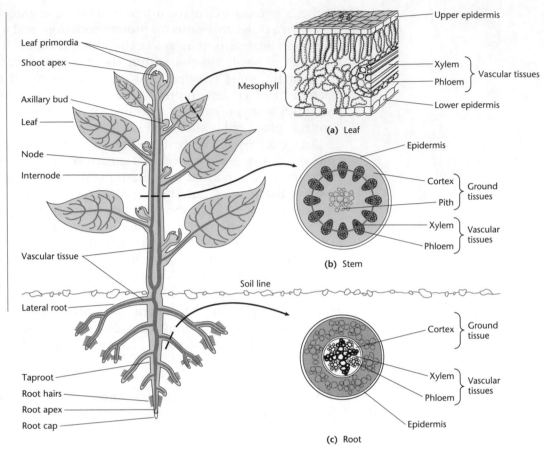

them, and in this way plants regulate their gas exchange (carbon dioxide and water vapor) with the environment. Older plant organs, such as roots and stems, do not have an epidermis, but instead have a dermal tissue that is just below the surface of the organ and gives rise to the **cork.** The cork is the outside layer of the bark, which includes other tissues as well. Cork cells are also impregnated with waxes to protect and provide a barrier against water loss. The cork oak makes a very thick layer of cork that can be stripped regularly and provides the material for making corks.

The basic living tissues of roots, leaves, stems, and flowers are called **ground tissues**, and their most important cell type is the **parenchyma** cell. (See Table 5.1.) In these large cells, the central vacuole occupies most of the volume of the cell, and there is relatively little cytoplasm, and therefore relatively little protein. In leaves, parenchyma cells contain many chloroplasts, but in stems, roots, tubers, and seeds the plastids are rich in starch. In organs that are specialized for food storage, the large parenchyma cells can be completely filled with stored food reserves including starch in plastids, protein in many small storage vacuoles, and oils in tiny oil droplets. These storage parenchyma cells are found predominantly in the storage organs of seeds, but also in potato tubers and in the bark of trees, where nutrients are stored during the winter. In thick, fleshy roots (such as carrots and beets), the storage parenchyma cells store sucrose and minerals in the vacuole. Much of humanity's food comes from these storage parenchyma cells.

Figure 5.5

Scanning Electron Micrograph of a Cross Section Through a Corn Root, Showing Different Tissues.
These are from the outside to the inside: the epidermis (outer ring of cells), the cortex (broad band underneath the epidermis), endodermis (single cell layer), ring of large xylem cells with phloem cells in between and pith (cells at the center of the root). Source: Courtesy of M. C. Drew, Texas A&M University.

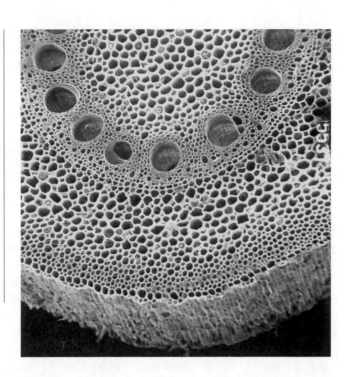

Table 5.1	Organs, tissues, and some cell types of flowering plants

Organs	Tissues	Cell Types
Root	Dermal tissue	Flat epidermal cell
Leaf		Hair cell
Stem		Guard cell
Flower	Ground tissue	Parenchyma cell
Fruit	Vascular tissue	Sieve tube element
		Vessel element
	Meristem	Meristematic cell

The Vascular or Conductive Tissues. Plant organs have two major conductive tissues: the phloem (pronounced *flow-em*) and the xylem (pronounced *zy-lem*). The function of the **phloem** is to conduct organic substances, primarily the products of photosynthesis or the breakdown products of stored reserves that are needed elsewhere for growth. Transport in the phloem goes in both directions, from the leaves to the roots and vice versa. The phloem is a system of live, elongated cells that are organized in vertical files. The cell walls are thick, and the horizontal walls that separate the individual cells in the files are perforated, allowing material to pass freely from cell to cell. This system of cells is referred to as the *sieve tubes*. The cells have no nuclei and are almost devoid of cytoplasm; this emptiness allows a constant stream of organic substances (such as sucrose and amino acids), dissolved in water, to be pumped through the sieve tubes.

The main function of the **xylem** is to transport water and minerals from the roots to the leaves. The most important part of the xylem is a system of large, dead cells that are also organized in vertical files and form tubes called *vessels.* The horizontal crosswalls between the cells are either perforated with large holes or have been completely dissolved. The nucleus and cytoplasm have totally disappeared. Movement of water and minerals through this system of hollow tubes does not require energy from the plant, but is driven primarily by water evaporating from the leaves.

4 | Plant development is characterized by permanent embryogeny.

There is a very fundamental difference in the way plants and animals develop. Plants have small regions, called *meristems*, that are permanently embryogenic; that is, they can continuously give rise to new organs. This gives plants a rather simple way of solving the problem of aging: when an organ is getting old, it is allowed to die and a new one is made. Leaf shedding is a typical example of this phenomenon. When leaves are young, they photosynthesize vigorously, but as they get older their rate of metabolism starts to drop. Eventually they age and are allowed to die. In perennial plants, new leaves are made in each growing season, whereas in annuals new leaves continue to be made at the top of the plant while older leaves near the bottom are dying.

At the tip of each shoot is an **apical meristem** where the cells are continuously dividing (Figure 5.6). Small protrusions of dividing cells, called **primordia**, arise on this dome-shaped mass of dividing cells, and one by one these primordia grow out into small embryonic leaves or become dormant lateral buds. Together, the **apical meristem**, the primordia, and the tiny leaves make up the apical bud that is found at the end of every twig. Apical buds are also present at the ends of other shoots, but are not always as easily identified as in twigs. The activity of meristems is regulated by the environment (temperature and day length) as well as by hormones. When apical buds are actively growing, they produce the growth hormone auxin, and auxin regulates the elongation of the cells just below the meristem. An apical meristem is also present near the tip of every root, just behind the **root cap**, a small structure that protects the meristem as the root grows through the soil.

In addition to these root and shoot apical meristems, cereals and grasses also have growth zones called *intercalary meristems* at specific intervals in the stems. Stems are made up of nodes, where the leaves are attached, and internodes, the piece of stem between two nodes. Corn and grasses, such as sugarcane, have a growth zone at the base of each internode. The existence of these growth zones accounts for the fact that grass can be cut again and again because the stem always can grow again in length after it is topped.

As a root elongates, lateral roots are formed not from the apical meristem at the tip of the root, but from meristems that are newly initiated somewhat back from the tip and below the surface of the root, close to the conductive tissues. Here again, the pattern is one of continued embryogeny and organ formation. The cells of the meristems of the root and the shoot are small, about 1,000 times smaller than the cell types discussed earlier. They have a prominent nucleus, a dense cytoplasm full of ribosomes, and relatively undeveloped mitochondria and plastids. The main function of these cells is to divide and give rise to more cells, and this they do with amazing regularity, once every 36–48 hours. Meristematic cells have a thin, flexible cell wall, and after each nuclear division a new wall separating the two offspring nuclei is formed. At the periphery of the meristem, the cells escape the hormonal

Figure 5.6

Longitudinal Section Through the Shoot Apex Showing the Location of the Meristem and Two Leaf Primordia. Source: *Carolina Biological Supply.*

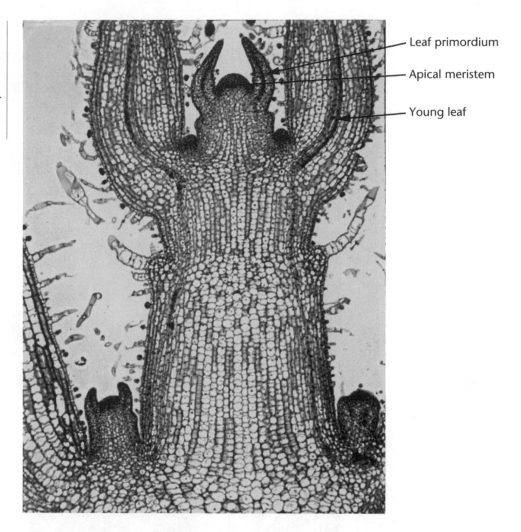

Leaf primordium

Apical meristem

Young leaf

conditions that keep them in a continuously dividing mode. Now the cells begin to enlarge enormously, mostly as a result of the increase in vacuolar volume. By enlarging, they become parenchyma cells. Later, as tissues develop, these parenchyma cells differentiate and give rise to all the major cell types found in the plant.

The function of the shoot apical meristem is to produce the cells that make up the shoot system, leaves and stems. However, given the right environmental and/or hormonal stimuli the activity of this meristem can be redirected toward making flower primordia instead of leaf primordia. These flower primordia then form sepals, petals, stamens, and anthers. At that point, the activity of the meristem ceases, and whatever cells remain, stop dividing. That is the normal process in many annual plants in which the apical bud first gives rise to a number of leaves and finally forms the flower.

5 | Mature seeds develop from fertilized eggs.

Development in organisms that reproduce sexually, whether plant or animal, begins with a single fertilized egg cell. The egg and the sperm that unite in fertilization are normally produced in specialized reproductive organs that, in plants, are located in the flower (Figure 5.7). The female reproductive

Figure 5.7

A Complete Flower Showing Sepals, Petals, Stamens, and Pistil. Source: *Courtesy of D. Ott.*

Style   Filament

Petal   Ovary   Stigma   Anther

Figure 5.8

Scanning Electron Microscopy Image of the Stamens of an Arabidopsis *Flower, Releasing Their Pollen Grains.* Source: *Courtesy of D. Smyth.*

organs, carpels (when the individual carpels are fused, the structure is called a *pistil*), contain a number of ovules, each with a single egg. The male organs, stamens, produce a large number of pollen grains (Figure 5.8), each containing two sperms. Flower structures vary from plant to plant species, but in general the flowers of the important cereal crops occur in clusters. In wheat

Figure 5.9

Scanning Electron Microscopy Images of an Alfalfa Embryo at Three Different Stages of Development.
(a) *Globular embryo,*
(b) *heart-shaped embryo,*
(c) *embryo showing two cotyledons.* Source: *N. Xu and D. Bewley (1992), Contrasting pattern of somatic and zygotic embryo development in alfalfa* (Medicago sativa L.) *as revealed by scanning electron microscopy,* Plant Cell *11:279–284; photo courtesy of N. Xu, University of Guelph, Canada.*

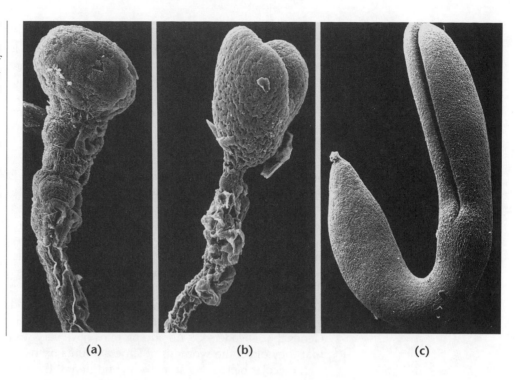

(a) (b) (c)

and rice, small flowers are connected to a central stalk termed a *spike*; here, both male and female reproductive organs are in the same flower. In corn, on the other hand, female flowers are in lateral ears and male flowers in terminal tassels, at different locations on the same plant.

The union of egg and sperm that initiates the development of a new plant occurs within the ovule. Depending on the species, pollen may come from the same plant (**self-pollination**, as in wheat) or from another plant of the same species (**cross-pollination**, as in corn). The type of fertilization that normally occurs for a particular species has important implications for plant breeding.

After fertilization, the cell that results from the union of egg and sperm undergoes many cell divisions and forms a multicellular embryo (Figure 5.9). The growth of the embryo occurs within the ovule in a rich nutrient medium called the *liquid endosperm*. At this stage, the new plant is growing completely heterotrophically, and the liquid endosperm provides the embryo with sugars, amino acids, vitamins, and the proper balance of hormones. These are synthesized by the liquid endosperm from sucrose and amino acids provided by the mother plant. As the embryo grows, it acquires more and more biosynthetic capacities; soon it can make all the complex molecules it needs, but it always continues to depend on sucrose and amino acids from the mother plant.

The most important aspect of embryo formation is the organization of the embryo axis with the two meristems, the root meristem and the shoot meristem, at opposite poles of the embryo. Once the embryo is formed, food reserves (proteins, fats, carbohydrates, and minerals) are synthesized and begin to accumulate in the storage parenchyma cells of the cotyledons and in tissues that are closely associated with the embryo. In the majority of dicots, most of the reserves accumulate in the cotyledons, which themselves become very large and are part of the embryo. In cereals and other grasses, most of the food accumulates in the endosperm, a tissue located just next to the embryo (Figure 5.10). The developed embryo and the food storage tissues then begin

Box 5.1

*Seed Proteins and
Seed-Specific
Genes*

The seeds of our crop plants accumulate very high levels of specific storage proteins. For example, soybeans contain up to 40% protein, and nearly half of this protein can be accounted for by two specific storage proteins: glycinin and conglycinin. Each of these two proteins is encoded by a small family of genes that are very highly expressed during seed development. These proteins are not found in other organs of the soybean plant, because the genes are inactive in these other organs.

The genes for the glycinin and conglycinin proteins of soybean and of many other seed storage proteins have been isolated and studied in considerable detail. Such genes consist of two main parts: the protein-coding region where the sequence of nucleotides in the DNA specifies the sequence of amino acids in the storage protein, and the control region or *promoter*. The promoter of a gene contains the switch that determines in which tissues the gene is "off" and "on," and how much *messenger RNA* will be produced. The messenger RNA is the template that contains the information contained in the gene and that is used by the cell to assemble the protein. Generally speaking, if there is more messenger RNA and if it is stable, then more protein can be produced. (A more detailed discussion of how genes work to produce specific proteins is given in Chapter 9.) Genes that encode seed storage proteins have very strong promoters, meaning that they direct the synthesis of large amounts of messenger RNA.

Molecular biologists can make hybrid genes: the control region of one gene can be linked to the protein-coding region of another gene. If genetic engineers want a specific protein to be produced in a seed, they can fuse one of these strong control regions of a seed protein gene with that part of another gene that encodes the desired protein. When this *gene chimera* is used to transform a plant, the desired protein will be made only in the seeds of the transformed plant and probably at quite high levels.

It is interesting that control regions obtained from one species generally are equally organ or tissue specific in other species. Thus, the glycinin control region that is strong and seed specific in soybean will also be strong and seed specific in oilseed rape, sunflower, or tobacco.

There are obviously many applications of this type of genetic engineering. First, engineers can begin to modify the nutritional quality of seeds. Genes that encode lysine-rich proteins could be expressed in corn or wheat, and genes that encode methionine-rich proteins could be expressed in legumes. This would balance the amino acid profile of these important staples. Second, scientists can modify what is being produced in the seeds at present. Introducing specific enzymes can modify the starches and oils produced by seeds. Starches and oils with very specific physical and chemical properties are highly valued by industry and can be produced in larger quantities in this way. Third, seeds can be used as small factories that produce entirely new products that now they do not synthesize: novel enzymes, new pharmaceuticals, or even biodegradable plastics.

to lose water and dry out. At the same time, the outer layers of the ovule that surrounded the embryo during its entire period of growth, dry down to a hard cover called the *seed coat*. The seed coat plays an important role in protecting the embryo and its stored reserves from being eaten by insects or being attacked by fungi or bacteria.

Seeds are usually formed during the active growing season. Once this season is over and the seeds drop to the ground, conditions for further growth of

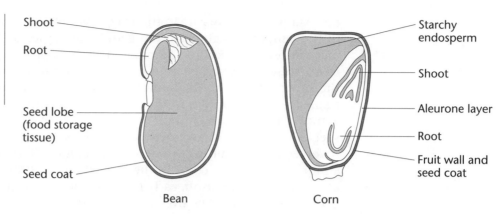

Figure 5.10

Longitudinal Sections Through a Typical Grain (corn) *and a Typical Legume Seed* (bean).

Shoot

Root

Seed lobe (food storage tissue)

Seed coat

Bean

Starchy endosperm

Shoot

Aleurone layer

Root

Fruit wall and seed coat

Corn

the parent plant are often unfavorable. Winter or a dry season may arrive. If the plant is an annual, it dies; if it is a perennial, it often becomes dormant. In either case, the seeds—which are the offspring of the plant—will survive the unfavorable growing period, because they are especially adapted to do so. Seeds have a very low water content (about 15%) and an unusually low metabolic rate. In this dry state, they can survive for weeks, months, years, or even centuries, depending on the species and the environmental conditions.

The seed allows the newly formed plant to survive until environmental conditions are suitable for growth. When conditions are favorable, the seed will germinate: the root and shoot of the embryo start to grow. However, at this point the new plant does not yet have an autotrophic mode of nutrition: it cannot use the sun's energy, because it does not yet have any leaves. It obtains the energy, proteins, and minerals it needs by digesting the food in the storage tissues. This digestion involves the synthesis by the food storage cells of enzymes, called *hydrolases*, that break the starch, proteins, and fats into much smaller molecules that can be transported and used by the growing seedling.

The digestion processes that take place in the growing seedling are very similar to the digestion of food by animals, and the small molecules that the growing seedling needs (amino acids, sucrose, minerals) are the same ones that animals need for their nutrition. As a result, the food reserves stored in seeds (and other storage organs) are eminently suitable for use as human food. The food storage tissues occupy most of the total volume of the seeds that figure prominently in human diets.

The formation of seeds represents an enormous biosynthetic effort for the plant, especially for the annual plants, which include most cereal grains and legumes; these plants die once their seeds are produced. In annuals, photosynthesis occurs primarily in the younger green leaves, while the older leaves are already yellowing. When leaves yellow, not only does their green pigment, chlorophyll, disappear, but also most of their proteins are broken down into amino acids. The amino acids from older leaves, newly synthesized amino acids, and sugars from the photosynthesizing young leaves are transported to the maturing seeds, where they are used to make storage proteins, starch, and fat. Because plant seeds provide more than half the food of the human race, the biosynthesis of food reserves in the seed has been studied in great detail by plant biochemists.

Seed Dormancy.   Once the seed is fully formed, it starts to lose water. This drying-out period is an important part of the maturation process of most seeds. Some plants have seeds that can germinate as soon as they have dried out. Many crop plants are in this category. However, the seeds of many plants in the temperate regions do not germinate immediately after ripening. After maturation, they become dormant, and this dormancy may last from a few weeks to several years. Such seeds will not germinate, no matter how suitable the environment, until dormancy disappears. In some species, the block against germination lies in the seed covering, which is impermeable to oxygen or water and which must be broken for development to occur. In other species, dormancy is governed by hormones. In still others, light seems to be involved in maintaining or breaking dormancy. Dormancy mechanisms were eliminated from our crops in the plant domestication process. By harvesting after one growing season, early farmers quickly selected in favor of seeds that germinate immediately after sowing, thereby eliminating dormancy mechanisms. Sometimes, however, seed dormancy is desirable, especially in regions where there is a great deal of moisture at harvest time. In such a climate dormancy will prevent the seeds of cereals from sprouting on the stalk before they are harvested.

The value of dormancy for the survival of the plant species in natural conditions becomes clear if we look at the role of a cold period in breaking dormancy. Seeds of many plants must spend several months in a moist and cold environment before they will germinate. Normally, they are buried beneath the leaf litter on the forest floor during the winter. This requirement for a cold period ensures that these seeds will not germinate in the fall of the same year that they are formed. Indeed, this would probably be an inappropriate time for germination, because the young seedlings might well be killed by frost. Some seeds may require more than a single winter of chilling, and the germination of seeds produced in a single season is thereby spread over at least two years. This spreading out of germination improves the chance that at least some seedlings will survive to become mature plants.

6 | Cereal grains are humanity's principal source of calories and protein.

The cereal grains (wheat, rice, corn, rye, oats, barley, sorghum, and millet) are humanity's most important source of food (Figure 5.11). Worldwide production of the eight major cereals is around 1,800 million tons per year: 500 million tons of wheat, 500 million tons of rice, 400 million tons of corn, and 400 million tons of other cereals. Enough cereals are produced to provide each inhabitant of the earth with 370–390 kg per year or about 1 kg per day (assuming no postharvest losses, and equal distribution). Human consumption (direct or indirect) of cereals varies from less than 200 kg per person per year in India, Indonesia, Nigeria, Pakistan, and the Philippines, to more than 700 kg per person per year in the United States and Canada. Of the 700 kg consumed per person in the United States, about 100 kg are consumed directly as bread or breakfast foods, while the remaining 600 kg per person are fed to animals, or used for industrial purposes (for example, ethanol production).

Rice. Rice is grown primarily in tropical and subtropical regions of Asia that have abundant rainfall. It is unique among the cereals in being able to germinate and grow in water. Rice can be grown either immersed in water (lowland

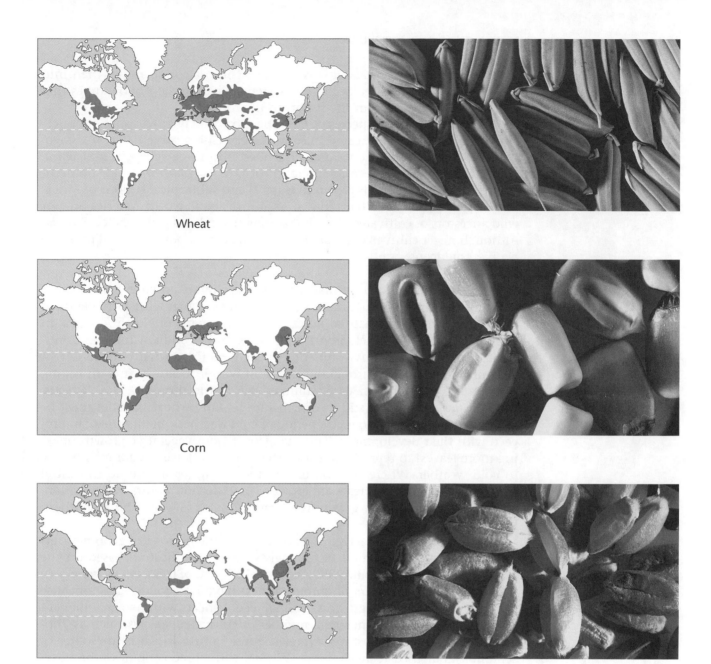

Wheat

Corn

Rice

Figure 5.11

Maps of World Cereal Production and Photographs of the Three Major Cereal Grains.
Note that there is little overlap between the three different cereals suggesting their adaptation to different climates.
Sources: *Maps redrawn from H. Edlin (1967),* Man and Plants *(London: Aldus), pp. 54, 64, 69; photos from the U.S. Department of Agriculture.*

cultivation) or on dry land (upland cultivation), but the yield of lowland rice is greater than that of upland rice. Rice seedlings are usually grown in nurseries and then transplanted to the flooded fields (paddies). Alternatively, rice seeds can be broadcast, but this practice wastes more seed, although it uses much less labor. Rice fields are usually drained a few weeks before harvest, and the cultivation of wetland rice in paddies requires the maintenance of a complex system to irrigate the fields.

Why can rice grow in water, whereas other cereals (and most other plants) cannot? When the root systems of most plants are submerged in water, they die for lack of oxygen. The roots of all plants need to respire, and respiration requires oxygen. Rice plants have interesting adaptations that let them thrive under submerged conditions. First, they have a high level of anaerobic respiration, a type of respiration that does not require oxygen and is found in many yeasts and bacteria; sugar is converted into alcohol, rather than into carbon dioxide and water. Second, rice stems have special hollow tubes that let air be transported from the shoots to the roots.

There are two main species of rice. The most widely grown species of rice, *Oryza sativa*, was domesticated in southeastern China some 6,000 years ago. Its more than 2,000 varieties are subdivided into three subspecies: *japonica*, *indica*, and *javanica*. A second species of rice, *Oryza glaberrima*, was domesticated in West Africa and is grown in tropical and subtropical regions of Africa. The so-called wild rices that grow in Asia, Africa, and North America are not wild ancestors of cultivated rices, but belong to a related genus called *Zizania*. Although never cultivated, *Zizania aquatica* was an important cereal for some Native Americans.

Wheat.   Wheat is grown primarily where the climate provides cool weather and moisture in the spring, followed by a sunny summer with a dry period at harvesting time. This weather pattern is found in parts of North America, in large parts of Europe, including the Ukraine, and in India, China, and Australia. In its original environment, wheat germinated with the onset of autumn rains. It grew throughout the mild winter to flower in early spring and mature its grains in early summer before the drought. Most wheats, called *winter wheats,* are still grown this way. Many wheat varieties must be vernalized—that is, experience a period of cold weather—before they can proceed with their development. The changeover from vegetative growth (making more leaves) to reproductive growth (making a flower) requires exposure to cold weather. Where the winters are too severe (as in the north central United States), spring wheats are grown. These do not require vernalization, and such varieties can be grown in warmer climates as well (China, southern Europe, Africa, India).

Botanists classify wheat species according to the number of genomes (sets of chromosomes), and food technologists add a classification based on the quality and use of the flour. Wild wheats, called *einkorn* (*Triticum monococcum*) with one genome (7 pairs of chromosomes) have no economic significance but are occasionally still grown as animal feed. Wheats with two genomes (14 pairs of chromosomes) include a number of cultivated species, such as *Triticum durum,* that grow best in warmer regions. Durum wheat is the source of semolina flour used to make pasta. Wheats with three genomes—the hexaploid wheats, with 21 pairs of chromosomes—include about six subspecies of *Triticum aestivum*. The main subspecies is *Triticum aestivum vulgare,* common wheat or bread wheat. Grains that yield a strong, coarse flour that is rich in gluten protein (hard wheats) are ideally suited for bread making, while grains that yield a soft, finer flour, with less gluten protein, are good for cookie (biscuit) manufacture.

Corn.   Corn (also called *maize* outside the United States and Canada) or *Zea mays,* is grown in a wide range of environmental conditions, but the yields are highest in areas that have a long growing season and abundant rainfall. Corn is native to tropical Central America and is the only cereal that was domesticated in the Americas. It is now grown from latitude 58°N in Canada to 45°S in the Southern Hemisphere, and its cultivation has been exported to Africa, Asia, and Europe. Corn does not tolerate shade or drought and was originally a short-day plant. Because it grew in a region where wet and dry seasons alternate, its life cycle had to be strictly controlled by day length so that it would come into flower at the right time. As corn moved out of its center of origin, it gave rise to a number of varieties with different properties. The corn grown in the Corn Belt of the United States is called *dent corn* (*Zea mays* var. *americana*), while sweet corn is a variety that cannot convert sucrose into starch in the endosperm. Popcorn, yet another variety, has a hard outer endosperm; when these seeds are heated, the evaporation of water in the inner endosperm causes the seeds to explode.

Corn protein is low in the essential amino acids lysine and tryptophan, so it is not as good a source of protein (if eaten as the sole source of protein) as other cereals. In addition, about 50–80 percent of the niacin (a B vitamin) in corn is unavailable for absorption by humans. In areas of southern Europe and the southern United States, the adoption of corn as the single dietary staple by poor sharecroppers was accompanied by the emergence of a new disease, pellagra. Pellagra was widespread among southern Europeans and North Africans (Egypt) in the eighteenth and nineteenth centuries and appeared in the United States around 1900. The first epidemic occurred at a black mental institution in Alabama where cornmeal mush was the principal staple. Although pellagra was first thought to be an infectious transmitted disease, in the 1930s niacin was discovered and niacin deficiency shown to be the cause. The problem is not with corn itself, but with a diet in which corn is the principal staple that does not include vegetables.

Rye. Rye (*Secale cereale*), one of the most recently domesticated cereals, is grown primarily in Europe. It is believed to have originated in Afghanistan and Turkey. Rye is well adapted to the short summers and colder climates of central Europe (Germany, Russia, Poland, Czech Republic and Slovakia), and the inhabitants there prefer the sour taste and the density of rye bread over the lighter texture and milder taste of wheat bread. Rye can be successfully cultivated on soils that are too poor to grow other cereals.

Oats. Rye and oats probably originated as weed crops in the ancient wheat and barley fields of the Middle East. Oats, *Avena sativa*, are grown primarily in North America and Europe at higher latitudes in cooler, wetter climates. As wheat and barley moved out of their centers of origin, they were accompanied by weeds, some of which, like rye and oats, eventually became crops in the areas to which they were best adapted. Oats are a highly nutritious cereal with a high protein content (15%) and a high content of dietary fiber (12%). The amino acid profile of oats is better balanced than other cereals. In spite of oats' excellent nutritional qualities, little oat breeding has been done, and the production of oats has declined since tractors replaced horses.

Sorghum. Sorghum (*Sorghum vulgare* and other species) is an important food source in Africa and Asia and is widely grown in the southern United States as a cattle feed. Sorghum was probably domesticated in the African savanna about 5,000 years ago. This subtropical plant is well adapted to hot, dry regions, so it can be grown where corn cannot. Sorghum breeding is a relatively recent agricultural development, and no sorghum varieties are known that produce high yields in a cool climate, as are known for corn.

Sorghums tolerate not only periods of hot, dry weather—they can remain dormant if there is insufficient moisture in the soil and resume growth when it rains—but they also tolerate considerable salinity. They also have fewer

Figure 5.12

Millet Cultivation in Niger (Africa). In regions with little rainfall, millet is the only cereal that can be grown. However, if there is a drought, the harvest may still fail. During the 1984 drought in northern Niger, millet yields were only one quarter of their normal level. Source: Courtesy of the Food and Agriculture Organization, Rome.

pests and diseases than other cereals and can be sown quite late in the season after other crops would fail.

Millet. The millets (Figure 5.12) are a group of grain-producing cereals that include several related species. (Sorghum is sometimes mistakenly called *millet.*) They grow in hot dry areas and are used as food crops in Africa and Asia. The yields of millets are usually quite low, but a crop can be produced in a short time on poor soils and with primitive methods of cultivation. Because millets are so well adapted to many regions where it is difficult to produce food, a major effort to improve this crop through plant breeding is warranted.

7   Cereal grains are the cornerstone of human nutrition.

Cereal grains contain 10–15% water, 8–14% protein, 70–75% carbohydrate, and 2–7% fat, as well as a variety of minerals and vitamins. These nutrients are not evenly distributed throughout the grain, but are concentrated in various structures. The outer layers of the grain, which form a protective cover, are composed of dead cells, which have only indigestible cell walls and no protoplasm. The embryo or germ consists of a small root and a small shoot. Its meristematic cells are rich in proteins, fats, several B vitamins, and vitamin E. They also are rich in sugar, which accounts for the sweet taste of wheat germ. The bulk of the seed is taken up by the food storage tissue or **endosperm.** The outer layer of the endosperm, called the **aleurone layer,** consists of cells that contain many protein storage vacuoles and are rich in nicotinic acid (niacin) and several important minerals (calcium, magnesium, phosphate, and potassium). The central portion of the endosperm, called the *starchy endosperm,* consists of large cells tightly packed with starch granules embedded in a matrix of protein. About 70% of the total protein is in the endosperm, but the protein score of the endosperm proteins (the food reserve proteins) is lower than that of the proteins in the embryo.

Throughout history, human beings have devised different ways of preparing cereal grains for eating. If the grains are eaten whole, they simply pass through the body; so humans have learned to pound, grind, parch, cook, roast, or soak the grains so that the body will be able to extract the nutrients more fully. In ancient times, wheat was milled by simply pounding the grain in a mortar with a pestle or by shattering it between millstones. Today, the grain is squeezed first between steel rollers, then separated into its different parts: the bran (outer protective layers, aleurone, and some of the endosperm), the germ, and the endosperm. White flour and many breakfast cereals are made from the starchy endosperm, that is ground up into a powder. The polishing of rice also removes the protective covering structures as well as the aleurone layer.

Is white bread made out of bleached flour nutritious, or is it only a source of "empty" calories? Since ancient times, white bread has had greater social status than brown bread although this trend may now be reversing itself. Claims that brown bread is more nutritious than white have often been made: Greek and Roman writers described brown bread as symbolic of the simple life of country folk and better than the white bread of the rich. When wheat is milled, or rice is polished to remove the bran, the outer layers of fiber-rich dead cells are removed, together with the protein- and vitamin-rich surface layers of the storage tissues (endosperm). Thus the bran fraction is rich not only in fiber but also in important nutrients.

For a long time, brown bread meant not merely whole wheat bread but bread containing flour from other plants (barley and legumes). Whole grain cereal, although not the ideal food, is a complete food, and milling it removes many essential nutrients. To circumvent this problem, white flour can be fortified with the missing nutrients. In the United States, Canada, and Britain, riboflavin, nicotinamide, thiamine, and iron are reintroduced into white wheat flour. The minerals of brown bread—calcium, magnesium, and phosphorus—are present in the aleurone, but in a chemical form unusable by humans; a portion of these minerals is made available to humans when bread dough is allowed to rise for several hours. The protein score of whole wheat flour is higher than that of white flour, but amino acids, such as lysine, can be added to improve the protein score of white flour. Such enrichment increases the protein score to 80%, far above that of whole wheat flour. Lysine-enriched flour is now being used in Japan. Thus, the possibilities of fortification blur the nutritional distinction between brown and white bread. However, brown bread is an important source of fiber.

Apart from bread, the most common form in which grain is consumed in the United States is as breakfast cereal. Most breakfast cereals now consumed in the United States are precooked, rolled into flakes, and dried again so they can be eaten with cold milk. It is odd that an industry that came under fire 15 years ago for giving children little more than empty calories, began as a self-conscious moral movement to rid the United States of its unwholesome turn-of-the-century diet. In recent years, health- and nutrition-conscious Americans have demanded more nutritious breakfast cereals, and the industry has responded accordingly. Still, a number of the available breakfast cereals contain up to 40% sugar. The development of the breakfast cereal industry, involving a uniquely American mix of eccentric health reformers, fringe religion, and commercial interests, is interestingly told in

Harold McGee's fascinating book *On Food and Cooking* (New York: MacMillan, 1984).

An unusual way of processing corn and making it more nutritious is the preparation of masa. The corn kernels are boiled in a 5% lime solution for about an hour, dried, and ground into flour. This ancient method is widespread in modern Mexico and elsewhere in Latin America and goes back to the early civilization of the Aztecs and the Mayas. The alkaline treatment releases niacin from its bound form—thereby avoiding the problem of pellagra—and improves the amino acid balance of corn by *decreasing* the digestibility of the major corn seed storage protein. This protein is particularly poor in lysine and tryptophan, and when it becomes less digestible the *relative* deficiency of these two amino acids is reduced. When corn is made into masa, the relative availability of lysine is increased 2.8 times and of tryptophan, 1.3 times. There is a loss of protein, but the protein that is left has a better amino acid profile.

8 | Leguminous plants produce protein-rich seeds.

The legumes (Fabaceae) produce seeds that contain two to three times more protein than the cereals, and from a human nutrition standpoint the legumes are the second most important plant family. The protein-rich seeds (15–35% protein) are either also rich in starch (peas, kidney beans, lima beans, chickpeas, cowpeas, and mung beans) or in oil (peanuts and soybeans). They contain more iron, five times as much riboflavin, and ten times as much thiamine as the same weight of cereal grains. Legumes are nutritionally similar to meat. In poor countries, legumes are the most important high-protein food, and play the same role in the diet that is occupied by meat and animal products in rich countries. The importance of legumes in human nutrition also derives from the fact that the amino acid profile of proteins complements the amino acid profile of cereals. Legume proteins are rich in lysine and tryptophan, two amino acids that are low in cereals. In the seeds, these proteins are stored within the cells in small protein storage vacuoles.

It is unfortunate that worldwide legume production has not increased at the same pace as cereal production. There are several reasons for this stagnation in production. First, less research has been conducted on legumes than on cereals, with the exception of research on soybeans in the United States, where the cash value of the soybean crop now rivals those of the corn and wheat crops. Second, as incomes gradually rise in the middle-income countries, people eat more meat and less legumes. Eating meat is a sign of affluence, so traditional legume-based dishes are cast aside in favor of animal products. This change decreases the demand for legumes. Third, although legumes are capable of very high yields under ideal growing conditions (for example, a yield of 6.4 tons per hectare of peanuts in Zimbabwe), the yield of legumes is very variable. The yield of legumes seems to be much more susceptible to climatic vagaries than the yield of cereals. Fourth, the push to grow more high-yielding cereals takes land away from other crops, and lower-yielding legumes are the first victim.

Most plants depend entirely on the soil for their supply of nitrogen, the most important element in the manufacture of protein—but legumes do not. They can satisfy some of their nitrogen needs—perhaps up to 50%—from ammonia produced by *Rhizobium* bacteria that live in nodules on the roots of the plants. The bacteria can use the nitrogen gas in the atmosphere. This direct use of atmospheric nitrogen, called "biological nitrogen fixation," is of enormous agricultural importance, and is discussed in Chapter 8. As a result of this process, legumes can grow on nitrogen-poor soils, where wild legumes are often found. The ability to fix nitrogen gives them a competitive advantage over other plants. However, the ability to fix nitrogen does not mean that bumper crops can be produced without synthetic fertilizers, even nitrogen fertilizers. The root nodules take several weeks to develop, and during the first few weeks after planting annual crop legumes depend completely on soil nitrogen.

The Common Bean. The common bean, *Phaseolus vulgaris*, has many other names, including kidney bean, bush bean, snap bean, and pinto bean. Remains of the common bean have been found in archaeological excavations in southwestern Mexico, and have been dated to 5500 B.C. The common bean was domesticated independently in Mexico and the northern Andean region of South America. The common bean remains an important staple in Mexico (with a daily consumption of 40 g per person), as well as in Venezuela and Brazil. A similar pattern of bean consumption is found among the Native Americans, especially in the southwestern United States. The consumption of pinto beans in the U.S. Southwest is supplemented by that of the related tepary bean (*Phaseolus acutifolius*). The cultivation of this drought resistant bean is unfortunately disappearing. The common bean provides an excellent example of the presence in many legumes of antinutritional factors that are part of the plant's defenses against insects. These factors appear to be totally different in every legume species. The common bean contains lectins, proteins that bind to and damage the lining of our intestinal tract, as well as inhibitors of our digestive enzymes. Eating as little as a single uncooked bean can produce an unpleasant intestinal upset. Cooking destroys the effectiveness of these plant defense proteins.

The Soybean. The soybean, *Glycine max*, is native to southeastern China and has been cultivated for thousands of years in China and Japan. The soybean (Figure 5.13) is now the most widely cultivated legume, with 65% of worldwide soybean production taking place in the United States. Soybeans were introduced into North America in the early part of the nineteenth century. They aroused interest because of their high oil content (about 20% of the dry seeds), and the oil was first used for paints and varnishes. Not until the invention of margarine and the hydrogenation process did soybean production really take off. One hundred and thirty thousand tons were grown in the United States in 1925. By 1939, just 14 years later, production increased 18-fold to 2.4 million tons, and by 1968 production reached 29 million tons. The high protein content (40% of dry weight) of soybeans also contributed greatly to the economic success of this crop.

The worldwide demand for soybeans and for the protein-rich residue that is left after the oil has been extracted is caused entirely by the demand for animal products. In Western countries, most of the soybean protein is fed to animals and ends up on dining room tables as animal protein. In Asia, soybeans are consumed directly, either as a curd (tofu) or after fermentation of the cooked beans (as tempeh or miso). Fermentation increases the nutritional value of the beans because the micro-organisms produce vitamin B_{12}, the only

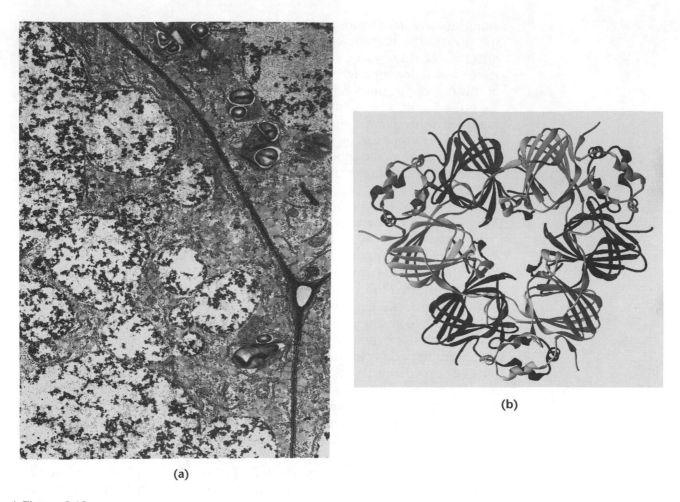

(a)

(b)

Figure 5.13

Legume Seeds Are a Rich Source of Protein Stored in Vacuoles. **(a)** *In this electron micrograph of a soybean cell, the vacuoles are not yet full and protein will continue to accumulate during the 40-day seed maturation period.* **(b)** *The three-dimensional structure of canavalin, a seed storage protein of the Jack bean, similar to the soybean seed protein conglycinin. The chain of amino acids is shown as a long ribbon that is wound around itself to form a globular structure.* Source: **(a)** *courtesy of E. Herman, Agricultural Research Service, Beltsville, MD;* **(b)** *courtesy of A. McPherson, University of California at Riverside.*

vitamin not present in plants. Small amounts of tofu are also eaten in Western countries, and soybean protein is used as a bread flour additive or as a meat extender.

The Cowpea.   The cowpea (*Vigna unguiculata*) was first domesticated in west Africa and is now grown in the entire sub-Saharan region, in India, Brazil, and the United States. Cowpeas are primarily a subsistence crop grown by people for their own use, except in the United States, where they are sold as black-eyed peas. Cowpeas have few antinutritional factors, such as lectins and enzyme inhibitors, and as a result they are more susceptible to attack by storage pests such as weevils that eat the dry mature beans. Some cowpea strains are well adapted to semiarid, subhumid, and humid tropics, and normal plant breeding or genetic engineering could greatly improve the plant's production potential.

9 | **The orderly development of the plant depends on environmental stimuli and a balance of hormones.**

Plant development—the orderly outgrowth of organs and the progression from seed to seedling to vegetative plant and finally to flowering plant—is profoundly affected by the environment. Many plant species use environmental cues, such as changes in day length or a cold period, to proceed from one developmental stage to the next. The amount of light received, the spectral quality of that light, the relative lengths of night and day, and the temperature regime during day and night all affect plant development. There is, first of all, a quantitative relationship both between light intensity and growth, and between temperature and growth. Plants generally grow more vigorously if the temperature is higher and if there is more light. However, because plants have evolved in specific environments to which they are adapted, they may have quite different optimal light intensities and temperatures. Plants adapted to growing in dim light—on the floor of the tropical forest, for example—will suffer from light stress and photo-oxidation when they are planted in full daylight. Björkman and his colleagues at Stanford University compared the growth of two plants: one adapted to the very hot climate of Death Valley in California, and the other adapted to the cool California coast. When both plants were grown at the same temperature in a greenhouse, the researchers found that the cool weather plant died at the higher temperatures while the hot weather plant barely grew in the cooler climate. Apart from these general effects of temperature and light, specific environmental regimes are often needed to let plants progress from one stage of development to the next. We have already discussed seed dormancy and how breaking seed dormancy may require exposure to light or exposure to a period of cold. In addition, the seeds of some plant species only germinate in the dark, thus preventing them from germinating when they are on top of the soil.

As soon as the seedling emerges from the soil, light has a profound effect on its growth. Molecular biologists have found that many genes are activated by light. Thus when light hits the seedling, many genes are activated and many new proteins start being made. This leads to the following changes in growth: the stem elongates less rapidly, and the internodes are much shorter. In dicots, the leaves expand enormously, and the plastids in the leaf parenchyma cells develop into chloroplasts (Figure 5.14). In monocots, the leaves unroll and the chloroplasts develop. The enzymes necessary to carry out photosynthesis are made when the leaf parenchyma cells are exposed to light, but they are not made if the leaves remain in darkness.

Reproductive growth or flowering can also depend on a particular environmental regime. As noted earlier, some biennial plants, including many varieties of winter wheat, absolutely require a period of cold weather before the vegetative apical bud can become a reproductive bud, and give rise to flower primordia. This process is called **vernalization.** Winter wheats are normally planted in the fall and grow 10–20 cm tall before winter arrives. The wheat overwinters under a blanket of snow and growth resumes in the spring. The period of cold is necessary for winter wheat to come into flower and set seed. In many other plants, the relative lengths of the day and the night determine whether flowering will occur. This phenomenon, called **photoperiodism,** is discussed in the next section. The photoperiod has other effects

Box 5.2

*Genes,
Development, and
Genetic
Engineering*

What does development mean in molecular terms, and how is it regulated? Why don't roots make chlorophyll or taste as good as fruits? Why do potato tubers accumulate starch but potato fruits (yes indeed, potatoes have fruits) do not? Development, the process whereby a fertilized egg becomes a flowering plant, requires both cell division and cell differentiation—cells that are initially alike become different from each other. One cell type has chloroplasts, the other one does not. Some cells have thick cell walls, but others do not. These differences are reflected in the enzymes the cells contain. Some cells have the enzymes necessary to synthesize chlorophyll and to transform carbon dioxide into sugar (photosynthesis), but others do not. Other cells may have enzymes to synthesize specific cell wall components.

A cell has at least 30,000 proteins, many of which are enzymes; about half of these proteins are common to all cells, but the other half can be specific to an organ, a tissue, or a cell type. Thus, cell differentiation and development can be thought of as cells acquiring different sets of enzymes and other proteins. These proteins appear in the cells under the influence of stimuli such as hormones, light, or nutrients in the soil. When light first hits a young seedling, it dramatically alters the complement of proteins being synthesized by the cells. Some proteins cease to be synthesized, and many new ones appear. Similar changes occur when hormones exert their action and trigger specific developmental changes, such as the initiation of lateral roots, or fruit ripening. Because the information for making proteins is encoded in genes (see Chapter 9), this means that light or hormones activate genes. Thus, one can think of cell differentiation as differential gene activation.

Although the presence or absence of proteins in a cell is not always controlled at the gene level, gene activation is an important mechanism in the control of development. This finding opens the way to use genetic engineering to alter development. Several scientists have already started with an obvious approach: making genetically engineered (transgenic) plants with genes that encode enzymes for hormone biosynthesis. With such genes, it should be possible to alter the hormone content of specific organs, such as seeds, fruits, or roots, and the hormones will in turn activate other genes that control the accumulation of protein in seeds, the ripening of fruits, or the storage of reserves in roots. The effect would be similar to spraying plants with hormones, but the idea is to make it more specific, by limiting the activity of the genes to certain cells or tissues.

on plant development. For example, the development of fleshy storage roots (such as carrots and beets) is governed by the length of the day.

Many of these environmental effects on plant growth can be mimicked by applying plant hormones (Figure 5.15). In addition, hormone levels within the plant change dramatically during developmental transitions. For example, spinach normally requires long days and short nights to form flowers; this environmental requirement can be circumvented by spraying the plants with the hormone gibberellic acid. Analysis of unsprayed plants showed that the levels of specific gibberellins rose as the plants made the transition from vegetative to reproductive growth. Such observations have led to the hypothesis that the effects of the environment are mediated by changes in the hormone level within the plant.

Plants make at least seven kinds of hormones, and these hormones are produced in ordinary tissues and not in endocrine glands, as is the case in

Figure 5.14

The Effect of Light on the Growth of a Pea Seedling. The plant on the left (a) is grown in the light, the one on the right in the dark (b). Both plants are 9 days old. At the nodes N_1 and N_2, there are no leaves, only scales. At higher nodes, there are leaves (L_1, L_2, L_3) that have expanded in the light, but not in the dark. Note that in the dark the stem remains bent (H = hook).

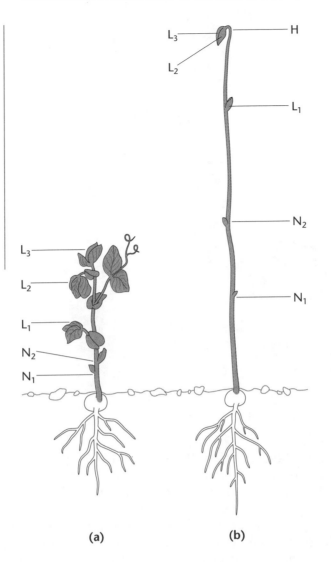

(a)   (b)

animals. The hormones are small molecules (Figure 5.16) that are carried in the transpiration stream of the xylem (cytokinin), are transported from cell to cell (gibberellin, auxin, and abscisic acid) or are released into the air (ethylene, methyljasmonate), like animal pheromones.

Auxin is produced primarily in meristems and young leaves, and it is transported toward the base of the stem and the base of the root, away from the meristems. As auxin is transported, it is broken down so that away from the meristem the auxin concentration gradually declines. The high levels of auxin that occur just below the meristem in the zone of cell elongation are necessary to maintain the high rate of cell elongation. Cell elongation slows down when the source of auxin is removed surgically by cutting off the meristem or chemically by preventing auxin transport. Auxin is also involved in phototropism, the bending of stems toward the light, and gravitropism, the bending of roots toward the earth and of stems away from the earth when plants are placed horizontally.

Gibberellins are a large group of related compounds (more than 80 have been identified) defined not by their biological action, but by their structure. One of their important roles in plant growth is that they are responsible for how tall plants grow. The stem of a tall plant contains more biologically

Figure 5.15

Both Gibberellins and a Cold Period Induce Flowering in Wild Carrot. Wild carrots are biennial plants that flower only the second year after the plant has overwintered in the ground. Stored reserves in the taproot provide food for the initial growth during the second season. **(a)** shows a plant at the end of the first season, while **(c)** shows a plant during the second season. Stem elongation and flowering have been triggered by the cold period. Treatment with gibberellin during the first season **(b)** causes the stem to immediately start elongating and flowering occurs. Source: *From P. Kaufman (1989),* Plants: Their biology and importance *(New York: Harper & Row), p. 590.*

(a)   (b)   (c)

active gibberellin than the stem of a short plant. We noted earlier that gibberellins are involved in the flowering process of certain plant species. In addition, they can break the dormancy of many seeds and obviate the need for a specific environmental regime.

Cytokinins are a smaller group of related compounds that are produced in the roots and move in the transpiration stream to the leaves. They regulate leaf senescence. When an annual plant gets older and its root system stops growing vigorously, it produces less cytokinin. The older leaves receive less cytokinin in the transpiration stream, and this contributes to their loss of chlorophyll, loss of protein, and general senescence. The ratio of auxin to cytokinin in the stem determines the branching pattern of the stem. Beside a terminal bud, every stem also has quiescent buds, which do not grow out into shoots when the ratio of auxin to cytokinin is high. When this ratio falls because auxin levels go down or cytokinin levels go up, the quiescent buds grow into side shoots. The ratio of auxin to cytokinin in a tissue culture medium also governs the formation of roots or shoots from a callus, as discussed in Chapter 3.

The fourth type of plant hormone, **abscisic acid,** is involved in seed and bud dormancy, as well as in the response of the plant to conditions of stress (such as not enough water, too much salt, or too cold). Water and salt stress increase abscisic acid in the roots, and this abscisic acid moves to the leaves in the xylem with the water that forms the transpiration stream. In the leaves, this hormone causes the stomates to close so that further water loss is prevented. At the same time, new proteins are made that help the cells cope with these stresses.

The fifth major plant hormone is **ethylene,** a gas that is released by many different tissues, but especially by ripening fleshy fruits. Ethylene has numer-

Figure 5.16

Structural Formulas of Representatives of the Seven Major Plant Hormones.

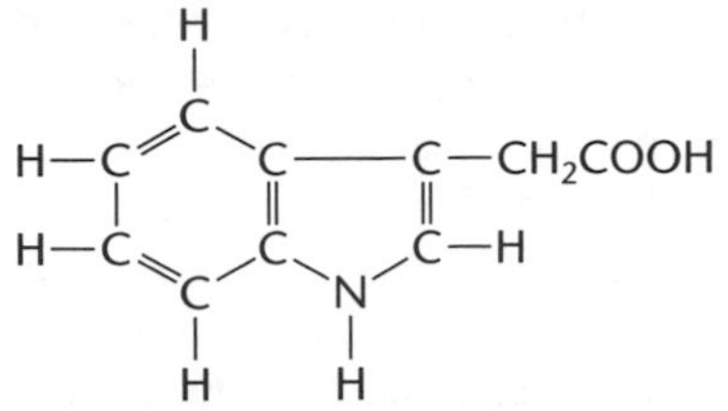

Indole-3-acetic acid (IAA), an auxin

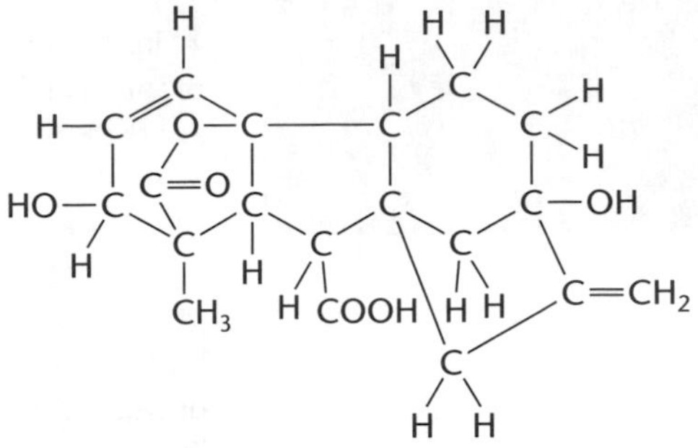

Gibberellic acid (GA$_3$), a gibberellin

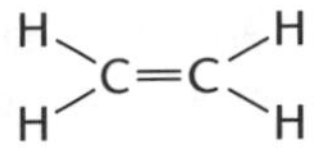

Ethylene

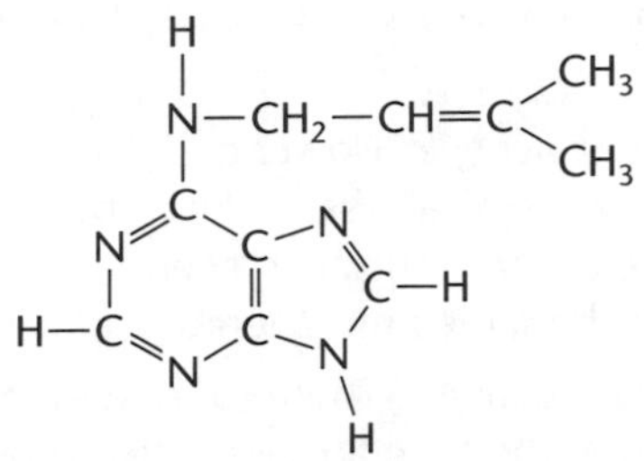

Zeatin, a cytokinin

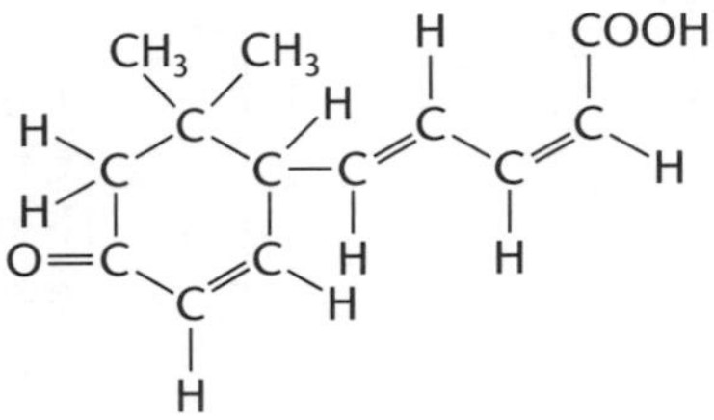

Abscisic acid

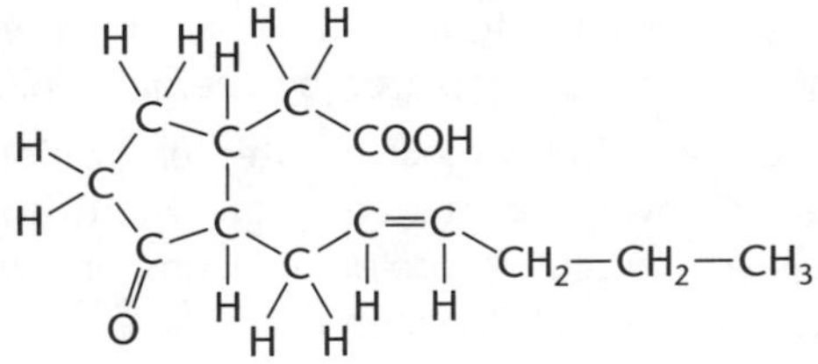

Jasmonic acid

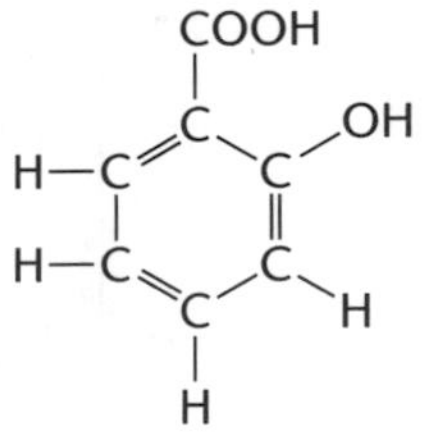

Salicylic acid

ous effects on plants, and these effects vary enormously with the species. It regulates flowering in some plants, and flower aging in others. It promotes fruit ripening in many fleshy fruits and is commercially used to ripen bananas. In rice, ethylene acts to promote stem elongation when the rice is grown in water. In plants that have separate male and female flowers, ethylene promotes the formation of female flowers. However, applying too much ethylene causes flowers and leaves to be shed. Chemicals that cause plants to produce ethylene are therefore used as defoliants.

The five hormones just discussed are primarily responsible for the orderly development of the plant. In the last few years, two new hormones, salicylate and jasmonate, have been discovered that play an important role in turning

Box 5.3

*Plant Growth
Regulators*

Plant growth and development are regulated by hormones synthesized by the plant, usually in specific tissues or groups of cells (such as meristems). Hormones exert their action after being transported elsewhere, and they affect the development of entire organs. It is not necessary to depend entirely on the endogenous hormones synthesized by the plant. The same chemicals can be synthesized in the laboratory and applied to plants—usually as a spray—and they have equally profound effects. In addition, it is possible to synthesize analogs—chemicals that have a somewhat similar structure and produce a similar effect. Finally, there are unrelated chemicals that affect plant development. Chemically they don't look like hormones, but they may exert their action by changing the internal hormone balance in some unknown way.

All these chemicals come under the general heading "plant growth regulators," and many of them are used in agriculture to modify plant development in one way or another. Like herbicides and pesticides, they must be approved for use on specific crops. A few examples of plant growth regulators are chemicals that

- *Promote the formation of roots on cuttings.* This is particularly useful for vegetative propagation of plants.

- *Suppress sprouting of potatoes and onions.* This prolongs the time they can be stored.

- *Induce flowering in pineapple.* Flowering is induced by ethylene and similar gases, as well as by chemicals that induce ethylene synthesis in the plant. It is particularly useful for inducing flowering in pineapples so that all the fruits will ripen at the same time, thereby reducing the costs of harvesting.

- *Inhibit flowering of sugarcane.* These cause greater accumulation of sucrose in the stem because the time for vegetative growth is extended.

- *Cause abscission of fruits such as citrus.* Citrus fruits are very tightly attached to the trees and are difficult to harvest mechanically.

- *Cause fruit drop to occur when the fruits are still small.* Thinning of fruit is an old practice to let the fruits that remain on the trees get bigger; hand thinning has now been replaced by chemical thinning.

- *Induce "fruit set."* Fruit set occurs after fertilization of the flower and indicates that the ovary has started to grow and fruit formation has begun. Chemicals are particularly useful in cucumbers, where under normal conditions only a few fruits form on each plant.

- *Change the ratio of "useful crop" to "useless plant."* Cotton makes very large bushes with a certain number of fruits (cotton bolls). It is possible to grow smaller bushes (easier to harvest) and more bolls by applying certain chemicals.

- *Inhibit water transpiration.* This allows water conservation, while not reducing crop yield.

- *Cause fruits to be red,* by promoting synthesis of carotenoids (orange and red pigments). This helps in marketing apples, cherries, peppers, cranberries.

on the plant's defense systems against pathogens and predators. **Salicylate,** a relative of aspirin, has long been known to be present in plants. It has now been shown to play an important role in the plant's defense against pathogens. When plants are infected with a weak pathogen (fungus, bacterium, or virus), they mount a defense response, and a subsequent infection with a virulent pathogen is now harmless. In this phenomenon, which is very much like the immune response of mammals, salicylate plays the important role of signaling the whole plant that a portion of it has been invaded. **Methyljasmonate,** a volatile derivative of jasmonic acid, also plays a role in turning on defense responses. If one plant is being devoured by caterpillars, it releases methyljasmonate to signal its neighbors that they should turn on their defense systems and prepare for the attack.

10 | Formation of the vegetative body of the plant is the second phase of plant development.

With the appearance of conditions that break dormancy and allow germination, the seed first restores its former water content and swells. This rehydration triggers the two main events of germination: embryo growth and use of the seed's food reserves. Embryo growth depends on the use of the reserves, because not until green leaves have formed can the seedling grow autotrophically. Its initial growth depends on the carbohydrates, fats, proteins, and minerals present in the storage tissues of the seed. The starch, fats, and proteins are first broken down into their simple constituents (sugar, fatty acids, and amino acids), and these are then absorbed and used by the growing embryo.

As the embryo absorbs the digested products of the seed's storage material, it uses them to build more cells and to expand cells once they are made. Within a few days, a recognizable root is growing down through the soil and the shoot (stem and leaves) pushes above ground. Once the shoot is above ground, light causes the stem to elongate less rapidly, and the leaves to expand. In the light, chlorophyll is synthesized and the plant begins to carry out photosynthesis and ceases to depend on seed reserves.

Even in a very young seedling, the three major organs that make up the mature plant—root, stem, and leaf—can be distinguished. The development and growth of these organs are complex, highly integrated processes. When plant growth is examined in terms of what the cells are doing, the development of these organs is seen to involve three basic processes: cell division, cell enlargement, and cell differentiation.

A good example of the sequence of these processes is found at the tip of the growing root. At the very tip, just behind the protective root cap, is a meristem, a region of continuous *cell division* (see Figure 5.17). These new cells then become 100–1,000 times larger; behind the region of cell division there is thus a region of *cell elongation*. Finally, once they are fully enlarged, these cells specialize, depending on location, into the various cell types of the root; behind the enlargement region there is thus a region of *cell differentiation*. This sequence is always maintained as the root grows through the soil.

Lengthening of the root is accompanied by the formation of lateral roots, which grow from the inner tissues of the main root. These lateral roots will in turn form more lateral roots, until the plant has established a widely branched root system capable of taking up water and minerals from a large

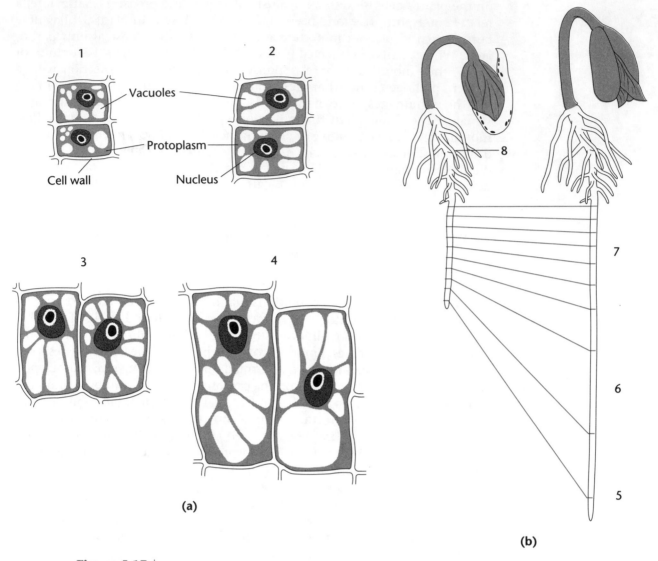

(a)

(b)

Figure 5.17

Cell Enlargement and Root Elongation.
(a) *The cells are drawn enlarged 2,000 times and progressively older stages are shown from 1 to 4. Note that cell enlargement is largely accounted for by the increase in size of the vacuoles.* (b) *The distribution of growth zones is shown for the root: 5 = apical meristem; 6 = zone of rapid cell elongation; 7 = cells are not elongating anymore, but cell differentiation occurs; and 8 = mature portion of the root.*

volume of soil. An average four-month-old rye plant has about 600 km of roots! In addition to having an extensive root system, many plants greatly increase the area exposed to the soil by forming root hairs. These are tiny projections from individual cells on the surface of the root. Most plants have billions of root hairs, which increase the root's absorptive surface area many times over. For an average four-month-old rye plant, this surface area is about 1,000 m² —all for a 1-m-high plant. To provide the plant with water and mineral nutrients, the root system must grow continuously, and new root hairs are always being formed, as the older ones die off. Storage of food reserves is another important function of the root system. Some biennial plants, such as beets and carrots, develop a thick, fleshy taproot at the end of the first growing season. The reserves stored in this taproot are used at the beginning of the second growing season, and a new shoot produces flowers and seeds. Other plants such as cassava develop numerous fleshy, tuberous roots for food storage and vegetative reproduction.

The processes involved in the growth of the shoot are generally similar to those for the root. As noted earlier, a shoot consists of a stem and of leaves that

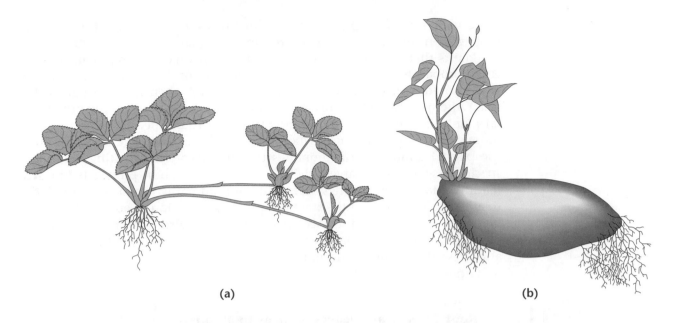

(a)   (b)

Figure 5.18

Vegetative Reproduction. (a) *Strawberry plants send out horizontal stems.* (b) *Tubers develop buds, which sprout.*

are attached to the stem at points called *nodes.* The stem portions between the nodes are the internodes (see Figure 5.1). The growing region of the stem is located at the tip within the terminal bud; each terminal bud contains a meristem, a region of continuously dividing cells, that produces all the cells for the embryonic leaves and the internodes. Below the meristem lies a zone of cell expansion, where these cells enlarge to their full size. While the cells are expanding, they are also differentiating into the different cell types of the shoot.

Vegetative Reproduction and Propagation. In addition to sexual reproduction, marked by flower and seed formation, many plants can also reproduce asexually. In **asexual** or **vegetative reproduction**, new plants are formed, not from seeds, but from specialized structures of the root, stem, or leaf (the vegetative organs of the plant). In nature, such vegetative reproduction is often carried out by means of horizontal stems or roots, which allow the plant to spread over a larger area (see Figure 5.18). A familiar plant that uses this mode of reproduction is the strawberry. It sends out **runners** (horizontal stems), which make new plants by sending down roots and forming leaf clusters at some of the places where they touch the ground. Horizontal stems that grow underground may become thick and fleshy and form **tubers.** Each tuber can give rise to a new plant when the parent plant has died. A potato plant propagates itself thus, by forming tubers (modified, underground stems) that will sprout the next spring. Many of the specialized structures involved in vegetative propagation are at the same time food storage organs and enable the plant to survive adverse conditions. For example, the tuberous roots, that in some plants (sweet potatoes, cassava) are a means of vegetative reproduction, also store food.

Human exploitation of this process of vegetative reproduction is termed **vegetative propagation.** This method is used in propagating potatoes, sweet potatoes, berries, nuts, and a variety of fruit trees and ornamental plants. The advantage of asexual reproduction is that all the offspring have exactly the same characteristics as the parent plant. This is especially important when the parent is a genetic hybrid whose sexually produced offspring would not be the same as the parents. For example, an apple tree grown from a seed does not usually produce apples of the same quality as the parent plant. Vegetative propagation through grafting ensures uniform quality in these plants.

Plants that produce specialized structures for vegetative reproduction are easily propagated. Thus pieces of potato tubers or of the tuberous roots of the sweet potato will sprout and produce new plants when they are put in the soil. Other plants, such as sugarcane, pineapple, cassava, and many ornamental plants, are propagated by stem cuttings. In these species, pieces of stem produce roots spontaneously when placed in moist soil. The discovery that the plant hormone auxin promotes the rooting of stem cuttings in many species has recently enlarged the list of plants that can be propagated in this manner. Many stem cuttings that would not ordinarily produce roots will do so after the lower end of the stem is dipped in a solution of auxin. The use of meristem cultures, as discussed in Chapter 3, is also a form of vegetative propagation. Here the goal is usually not genetic uniformity, although that is also present, but the rapid multiplication of plants free of disease organisms (potatoes) or superior strains (oil palms).

11 | **Roots, stems, and leaves serve as food sources.**

"Root crops," which include a wide variety of plants, are an important source of food for humanity. They provide about 8% of human energy intake. Potatoes, yams, sweet potatoes, cassava, carrots, and beets are all classified as "root crops," because they are all fleshy, underground storage organs (even though not all of them are true roots).

Root crops have a much higher water content than seeds do (70–80% as compared to 10–15% in seeds), and as a result they are more difficult to preserve and more expensive to transport. Their protein contents and protein–calorie ratios are also low; so they are not adequate by themselves for human nutrition. Wherever these crops are used as staples, they must be supplemented with high-protein foods.

The most important root crop is the potato (also called the Irish potato or white potato), which originated in the Andean highlands. It grows best in a climate where cool nights alternate with warm days when the tubers are being formed. Such conditions are found in large parts of Europe, which produces more than 75% of the world's potatoes.

Sweet potatoes and cassava (also called *manioc*) are two important tropical root crops, both of which probably originated in South America. The sweet potato has a higher nutritive value than the white potato, especially in vitamins A, B, and C and in calcium. China and Japan are the major producers and consumers of sweet potatoes.[1] Cassava plants adapt well to poor soils and casual cultivation, and are widely used in some of the shifting agricultural systems of the tropics (Figure 5.19). With little care, a cassava field may yield 15 tons of fresh roots per hectare. The roots contain little protein and are about 30% starch. Sugar beets are another important root crop. They provide 35% of the world's sugar. Like potatoes, beets grow best in areas where warm days alternate with cool nights during the period when the sugar is being deposited in the storage root. As a result, their distribution is similar to that of the potato: Europe accounts for 80% of sugar beet production.

[1]Red-skinned sweet potatoes are mistakenly called "yams" in the United States; the true yams are tubers of an entirely different tropical plant, of East Asian origin.

Figure 5.19

The Roots of the Cassava Plant Are an Important Staple in the Tropics. Source: *Courtesy of C. Lozano, Centro Internacional de Agricultura Tropical, Cali, Colombia.*

The stems and leaves of a wide variety of wild and domesticated plants are eaten as vegetables. They primarily supply minerals and certain vitamins (C and E). Different varieties of the cabbage family (for example, head cabbage, cauliflower, broccoli, kale, collard greens) are humanity's most important leafy vegetables, and are rich in calcium and vitamin C. The amount of vitamin E in leafy vegetables increases with their "greenness"; dark green vegetables, such as collard greens, kale, or spinach, have much more vitamin E than iceberg head lettuce, which is barely green at all. The stem of a plant normally becomes stringier or woodier as it grows older, and only young stems are usually eaten. Asparagus, bamboo shoots, and palm shoots are only a few examples of young stems that are eaten in different parts of the world. The economic importance of leaf and stem crops is difficult to estimate, because most of them are grown in small gardens by the consumers themselves.

Economically the most important stem crop is sugarcane, which supplies 65% of the world's sugar. Sugarcane is a tropical grass that stores sucrose in its stalk. A highly successful plant-breeding program has created strains with a high sugar content, greatly increasing the yield of sugar. Sugarcane is our most efficient crop, because it converts a larger proportion of the sun's energy into food than any other crop. Refined sugar, however, supplies only calories, and contains no other nutrients. The extraction of sugar from sugarcane involves crushing the stalks, squeezing out the juice, and boiling it to concentrate the sugar by removing the water. The remaining sticky liquid contains not only sugar but also minerals and small amounts of proteins as well. If this liquid is dried, it yields a crude form of sugar that is more nutritious than the refined sugar produced in modern sugar mills. This crude form of sugar is a dietary staple in Costa Rica and the Dominican Republic, where it supplies more than half the calories in the diet of many people. Modern sugar refineries first remove the proteins and then allow the sugar to crystallize. The crystallized sugar is then separated from the dark grown, mineral-rich sugary liquid (molasses), and further refined to yield a product free of all "impurities." Molasses is used as an animal feed or in the production of rum.

Sprouts (of mung beans, alfalfa seeds, or sunflower seeds) also fall in this category of stem crops. Sprouts are young seedlings—sometimes grown in the dark—of a variety of nutritious seeds. The commonly available "Chinese" bean sprouts are seedlings of the mung bean (*Vigna radiata*), an Asian legume that was domesticated in India more than 3,000 years ago.

Are bean sprouts more nutritious than the beans themselves? The growth of seedlings is accompanied by the use of the storage proteins in the cotyledons and the resynthesis of new proteins in the shoot. The net result is protein with a better amino acid balance, because the sprouts synthesize those amino acids that are in low supply in the storage proteins. Mung beans, like their cousins the cowpeas, are low in antinutritional factors (lectins and enzyme inhibitors) and can therefore be eaten raw. However, cooking is certainly advised for sprouts of beans (soybeans, kidney beans) that have high levels of antinutritional factors, because these factors do not disappear in the first week of germination. Some vitamins may decline slightly during germination (for example, B vitamins), whereas vitamin C greatly increases. Yet even this dramatic increase does not put them ahead of such vitamin C-rich vegetables as cabbage and green peppers. Finally, germination is accompanied by a decline in the level of those troublesome oligosaccharides that cause flatulence when fermented in the human intestinal tract. So nutritionally, sprouts occupy a middle ground between seeds and vegetables.

12 | Reproductive development involves the formation of flowers, seeds, and fruits.

Sexual reproduction in plants, as in other higher organisms, requires the union of a sperm cell and an egg cell. Gametes (sperm cells and egg cells) are produced in reproductive organs that are contained in flowers (Figures 5.7 and 5.20a). Most dicot flowers consist of four whorls of organs. The sepals, which are often small and green, completely enclose the flower bud before the flower opens. Next are the petals, which are often colored and showy, especially in plants that are pollinated by insects and birds. Inside the ring of petals are the stamens, the male sex organs that carry anthers that produce pollen grains each containing one or two sperm cells. And at the center are the carpels, the female sex organs. Each carpel has at its base an enlarged ovary, usually topped by a slender elongated style and a stigma that receives the pollen grains. On the inner wall of the ovary, one or more knobs or thickenings appear that form the ovule. Each ovule contains only a single egg cell. Several carpels are often fused together in a single structure called the *pistil*.

The elaborate floral organs, each with their own tissues and cell types, originate in the flower bud from **floral primordia**, just as leaves originate as leaf primordia. The development of these floral primordia is itself precisely

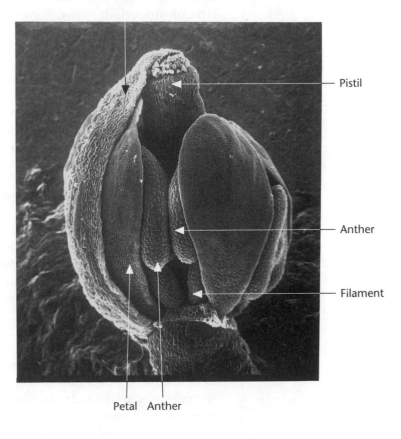

— Pistil

— Anther

— Filament

Petal   Anther

Figure 5.20

(a)

(b)

(a) *Scanning Electron Microscopy Image of a Developing Wheat Floral Meristem. Each protrusion will become a flower and has the potential to develop into a wheat grain.* Source: *U.S. Department of Agriculture.* (b) *Scanning Electron Microscopy Image of a Flower Bud of Arabidopsis. Three sepals have been removed to reveal the petals, pistil, and stamens inside. After the filaments elongate and the anthers mature (see Figure 5.8) pollen grains will fall on the pistil to fertilize the egg cells that are buried within the pistil.* Source: *Courtesy of D. Smyth.*

regulated by genes, and there are many variations to the basic flower structure that result from variations in the developmental pattern. For example, some flowers contain only male or only female sex organs—as in corn—because one set of organs fails to develop in each type of flower. Usually both types of flowers are present on the same plant, but in some species, such as willow trees, each individual is genetically either male or female.

Some of our most important crops—wheat, rice, corn, barley, and millet—are in the grass family (Poaceae), and in this family the flowers have a somewhat different structure. The inner two whorls of floral primordia develop into female and male sex organs, but the correspondence to sepals and petals of the organs that grow from the outer whorls of primordia is less clear. Grass flowers are called *florets,* and each floret has two leaflike structures, the lemma and the palea, that enclose the sex organs (Figure 5.20b). A long, slender awn is attached to the palea in wheat, barley, oats, and other grains. When these flowers are mature, the lemma and palea separate slightly, exposing two stigmas and allowing the anthers to hang free. These wind-pollinated flowers have no need of showy petals to attract pollinators.

In many plants, a terminal vegetative bud that gives rise to leaves and stem tissues can be converted into a flowering bud by a flowering signal (perhaps a hormone) or stimulus. Once the flowering stimulus has been received, the apical meristem starts forming floral primordia (instead of leaf primordia) and these differentiate into floral organs (Figure 5.20). The chemical nature of the flowering stimulus is not yet understood, but we do know that many plant species respond to the relative length of day and night (the **photoperiod**), when it comes to initiating flowers. Plants can be classified as "short day," "long day," or "day neutral" with respect to their flowering response to the photoperiod. Most plants are day neutral, meaning that they are not influenced by the photoperiod.

Short-day plants require that the length of the night exceed a certain minimum value (13, 14, or more hours, depending on the species or ecotype), while long-day plants require that the length of the night falls below a maximum value (11, 10, or 9 hours, depending on the species) before flowering is induced. Response to the photoperiod is an adaptation of the plant to its environment. A well-known short-day plant is the poinsettia. It flowers as the days get shorter in the late fall. Other short-day plants are rice and soybeans. If such plants are kept in a greenhouse under conditions that simulate the long days of summer, they will not come into flower. However, if the light conditions are changed to mimic the normal night length of the winter, they will readily flower, even in the middle of the summer. Wheat, oats, barley, spinach, and lettuce are typical long-day plants that flower during the summer when the days are lengthening and the nights are short.

Because of the plant's response to the photoperiod, certain plants flower only at a particular time of year. This may be a disadvantage for the farmer, because only one crop per year can be grown. Multiple cropping—which is practiced in many tropical areas where there is abundant water and sunlight—is only possible if the plants are insensitive to the photoperiod. The new strains of wheat and rice that formed the basis of much of the increases in grain production in India, China, and Mexico in the 1970s and 1980s, were selected to be photoperiod insensitive, so that several crops could be grown in a 12-month period.

The fertilization of the egg cell by the sperm cell is followed by the growth of the embryo and the formation of the seed. Concomitant with seed formation, the ovary that surrounds the growing embryo(s) is induced to form a **fruit.** Depending on the species, the fruit may also incorporate other tissues that are in close proximity to the developing ovary. A peach is a "typical" fruit. The fleshy part of the peach is the greatly thickened wall of the ovary. In an apple or pear, the fleshy part originally was the receptacle, the end of the stem to which the ovary was attached. The ovary of the apple flower gave rise to the more fibrous "core" of the apple. Thus the "fruit" of the botanist and the "fruit" of the layperson are not necessarily the same thing.

The explanation just given should make it clear that any structure containing seeds is a fruit and not a vegetable. Tomatoes, peppers, eggplants, zucchinis, cucumbers, okra, and green beans are all fruits. Each one has a more or less thick, fleshy wall (the ovary) and contains seeds. So why do we call them *vegetables*? The word *fruit* comes from the Latin *frui*, meaning to enjoy, to delight in, to have the use of something desirable. Sweetness is innately the most preferred taste sensation of mammals, and we call *fruits* those plant parts that are fleshy, sweet, and enjoyable. (So why is a lemon "a fruit"? Those are the mysteries of language and custom!)

The early development of the fruit—also called *fruit set* or *setting*—depends on hormones, especially auxin and gibberellin, produced by the growing embryo within the seed. These hormones stimulate the cells of the ovary to divide and expand. Application of auxin is sometimes used to promote fruit set and cause the development of fruits without fertilized seeds.

The fruit is the vehicle that aids seed distribution. Dandelion fruits (commonly called seeds, but to a botanist they are fruits) are carried away by the wind, berries are eaten by coyotes and other small mammals whose droppings are scattered over the countryside, acorns are carried away and buried by rodents. The later development of the fruit, also called fruit ripening, involves a number of changes that usually cause the fruitwall to soften while acids and starch are converted to sugar. At the same time, aromatic substances are synthesized to attract animals. The ripening process is controlled by the plant hormone ethylene, which is produced by the fruit. Commercial fruit companies can induce ripening at will by picking the fruits when they are green, and then keeping them for a few days in chambers containing ethylene. This procedure has the advantage that the fruits can be handled and transported before they become soft as a result of natural ripening. However, the flavor of such fruits is usually not as good as those that are left to ripen on the plants. Biotechnology companies are now making genetically engineered fruits that ripen much more slowly (see Chapter 15), by suppressing the synthesis of ethylene by the fruits themselves. This allows the fruits to remain on the plant much longer, and tomatoes thus can be left on the vines until they are half red, instead of having to be picked green. That still allows enough time for transport and marketing. Whether such tomatoes will indeed have a vine-ripened taste remains to be seen.

The formation of seeds and fruits often coincides with the advent of environmental conditions not suitable for growth. In annual and biennial plants, aging and death accompany seed and fruit formation. Perennial plants may shed their leaves and become dormant. Leaf aging is often signaled by the yellowing of the leaves, which is caused by loss of the chlorophyll and is accompanied by a rise in abscisic acid. Transport of this abscisic acid to the terminal buds causes them to become dormant. Like flowering, the processes of aging and dormancy are controlled by changes in day length, and so represent a response of the plant to its environment.

13 | Flowers, but especially fruits, provide important essential nutrients.

Most fleshy fruits are eaten raw, and are valued by people because they add variety and flavor to the diet, but they are relatively unimportant as a source of calories or protein. Some, such as the pineapple, are quite rich in sugar, and most are good sources of soluble vitamins and minerals. Citrus fruits and a variety of wild fruits are excellent sources of vitamin C. Other fruits, such as bananas, plantains, dates, coconuts, olives, avocados, mangoes, and breadfruit, supply considerable amounts of calories and protein.

The coconut is a very nutritious fruit, because its white "meat" is actually the endosperm of the seed, and its nutritional value is greater than that of the endosperm in cereal grains. Although coconuts are eaten extensively by the inhabitants of coastal tropical regions, most of the world's coconut production is diverted for the production of oil. The coconut "meat" is first dried in the sun, and oil is then squeezed out of the resulting product. The remaining press cake, which is rich in protein, is usually sold as animal feed. Coconut oil is rich in short-chain fatty acids (such as laurate) that are in great demand in the detergent industry.

Many other fruits and some flowers are also eaten as vegetables. Artichokes, broccoli, and cauliflower are fleshy flower heads, whereas fruits like tomatoes, peppers, okra, eggplant, cucumbers, and squash are nutritionally similar to the leafy vegetables in that they supply human beings with minerals and vitamins.

The most important staple fruit is the banana, which is a native of Southeast Asia, but is now grown all over the tropics. The annual production of bananas is estimated at 40 to 50 million tons. Only a small proportion (15%) of this production enters world trade, and most of that comes from Latin America. Most bananas are grown locally for consumption by the cultivator, often only one or a few trees in each garden. About half the bananas are eaten raw as a fruit, in the way most familiar to Westerners, and the other half is cooked as a vegetable. Cooking bananas, also called *plantains*, are starchy and not as sweet as the yellow-skinned bananas sold in the stores of developed countries.

The ancestor of the banana is *Musa acuminita*, an Asian wild banana that like all other plants has two sets of chromosomes. It produces small fruits, with numerous seeds, that are not particularly prized as a food source. At some stage in its history hybridization occurred with another species (*Musa balbisiana*), resulting in triploid plants (with three sets of chromosomes). Such plants grow more vigorously than plants with two sets of chromosomes, but they are sterile, and cannot be propagated by seeds. However, since bananas are easily propagated from suckers this does not present any problems. More than 300 varieties of bananas are grown worldwide, and most of these are sterile triploids.

The nutritive value of the banana is similar to that of the white potato, with an equally unfavorable protein–carbohydrate ratio, so that it must be augmented with protein-rich food if it is eaten as the sole staple. Bananas are rich in soluble vitamins and minerals, especially potassium. Bananas contain considerable amounts of serotonin, one of the many organic substances found in our foods that can be toxic, and for this reason, dependence on bananas as the sole energy source (24 bananas per day!) would probably be inadvisable. Nevertheless, in many parts of the tropics including East Africa, Brazil, and the Dominican Republic, bananas are the principal food for many people.

Summary

Plants, like animals, are complex organisms developing from a single fertilized egg into a multicellular organism with many different tissues that are grouped together in organs. Cells that in the meristem are originally all alike, differentiate, and different proteins equip each cell type for its own specialized function. The characteristic of permanent embryogeny makes plant development fundamentally different from animal development. In plants, new organs continue to be made from embryonic regions called meristems. As in animals, body growth is integrated by hormones. In addition to hormonal signals, environmental signals also govern plant development. Light—its absence or presence, the length of the day and night—and the temperature regime are examples of environmental signals that affect how a plant develops. The organs humans eat as food have specific roles in development. Nutrient-rich organs such as seeds, storage roots, and tubers store nutrients (starch, protein, and fat) that are needed by the plant later in its development. People obtain food by harvesting these storage organs. Non-storage organs (stems, leaves, flowers, and fruits) provide different nutrients, especially vitamins, minerals, and fiber.

Further Reading

Forbes, J. C., and R. D. Watson. 1992. *Plants in Agriculture*. Cambridge, UK: Cambridge University Press.

Hahn, N., ed. 1989. *In Praise of Cassava*. Ibadan, Nigeria: International Institute of Tropical Agriculture (IITA).

Harlan, J. 1992. *Crops and Man*. 2nd ed. Madison, WI: American Society of Agronomy.

Horton, D. 1987. *Potatoes*. Arlington, VA: Winrock.

Kaufman, P. 1989. *Plants, Their Biology and Importance*. New York: Harper & Row.

Klein, R. M. 1989. *The Green World*. New York: Harper & Row.

Langer, R. H. M., and G. D. Hill. 1991. *Agricultural Plants*. 2nd ed. Cambridge, UK: Cambridge University Press.

Milthorpe, F. L., and J. Moorby. 1980. *Introduction to Crop Physiology*. Cambridge, UK: Cambridge University Press.

Raven, P., R. Evert, and H. Curtis. 1991. *Biology of Plants*. New York: Worth.

Salisbury, F., and C. Ross. 1992. *Plant Physiology*. 4th ed. Belmont, CA: Wadsworth.

Schoonhoven, A., and O. Voysest, eds. 1991. *Common Beans*. Cali, Colombia: Centro Internacional de Agricultura Tropical.

Sprague, G. F., and J. W. Dudley, eds. 1988. *Corn and Corn Improvement*. 2nd ed. Madison, WI: American Society of Agronomy.

Squire, G. R. 1990. *Physiology of Tropical Crop Production*. London: CAB International.

Stern, K. R. 1991. *Introductory Plant Biology*. Dubuque, IA: W. C. Brown.

Taiz, L., and E. Zeiger. 1991. *Plant Physiology*. Menlo Park, CA: Benjamin/Cummings.

Terry, E. R. 1987. *Tropical Root Crops*. Ottawa, Canada: International Development Research Centre.

Tesar, M. B., ed. 1984. *Physiological Basis for Crop Growth and Development*. Madison, WI: American Society of Agronomy.

Vorst, J. J. 1992. *Crop Production*. Stipes.

Wilcox, J. R., ed. 1988. *Soybeans: Improvement, Production and Uses*. Madison, WI: American Society of Agronomy.

Chapter 6

The Role of Energy in Plant Growth and Crop Production

All living organisms require a constant supply of energy. We noted in Chapter 4 that, as heterotrophs, humans need to eat energy-rich organic substances and can use the energy contained in the chemical bonds of these molecules for all their food energy requirements. As autotrophs, however, plants can convert the radiant energy from the sun into chemical bond energy. This process of energy conversion, called *photosynthesis,* is of key importance not only for plant growth, but for maintaining the global ecosystems on which human life depends.

1 | **Solar radiation is the source of heat and light for plant growth.**

The earth receives from the sun a constant stream of radiation in the form of visible light, infrared light (heat rays), radio waves, ultraviolet light, and cosmic rays. Radiation can be considered as particles traveling through space with a wave motion. These waves can be thought of as similar to ocean waves. But while the lengths of ocean waves generally are in the range of a meter or more, those of solar radiation vary considerably over a trillion-fold range (Figure 6.1).

Since all solar radiation travels at about the same speed (300,000 km per second, or 186,000 miles per second), the variation in wavelength has an important consequence. Shorter wavelengths mean that the particle has to move "up and down" much more often per unit distance than longer ones, so shorter wavelength radiation has more energy than radiation with a longer wavelength. For plant growth, the two most important regions of the wavelength spectrum are

1 The **visible** region, which ranges from higher-energy blue light (wavelength about 4,000 Ångstroms, or Å) to lower-energy red light (about 7,000 Å). Radiation in this range is absorbed by plants and converted into chemical energy. Coincidentally, this is also the range that is absorbed by the human eye and provokes sensations of color.

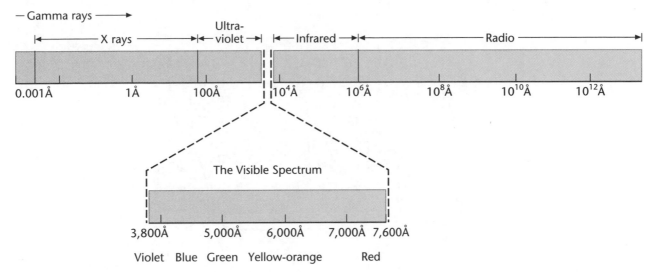

— Gamma rays ——→

X rays

Ultra-violet

Infrared

Radio

0.001Å   1Å   100Å 10^4Å 10^6Å 10^8Å 10^{10}Å 10^{12}Å

The Visible Spectrum

3,800Å   5,000Å   6,000Å   7,000Å 7,600Å

Violet Blue Green Yellow-orange   Red

Figure 6.1

A Portion of the Electromagnetic Spectrum. Plant growth is influenced by infrared and visible light, which constitute only a small part of the entire spectrum. Light of different visible wavelengths is perceived by humans as different colors. Wavelengths are in Ångstroms (= 10^{-10} m).

2 The **infrared** region, which ranges in wavelength from 10,000 Å (1 millionth of a meter) to 1 million Å. This radiation is used to heat the earth (and plants), and temperature profoundly affects the rate of photosynthesis and plant growth (see the following discussion).

The amount of solar radiation (light, heat, and other forms of radiation) that reaches a plant will vary according to the transparency of the atmosphere, the cloud cover, the position of the plant on the earth (latitude), and the time of year. The angle at which sunlight hits the earth has perhaps the greatest effect on the intensity of solar radiation. This angle is determined by the geographical latitude and by the seasons. A plant in the temperate middle latitudes (such as in Europe), will receive more radiation in the summer than during the winter because it faces the sun more directly in the summer. Moreover, the reduced solar energy during the winter leads to a cooler climate, which reduces the rate of plant growth. In contrast, a plant around the equator (such as in central Africa) directly faces the sun all year, and receives a constantly maximal dose of solar energy.

Another and less obvious factor that determines the temperature at Earth's surface is the retention, in the atmosphere, of radiation reflected from the surface. All the reflected heat would be lost into space, except that gases in the atmosphere absorb the infrared radiation. This absorption of the heat rays causes the atmosphere to warm up, a process known as the *greenhouse effect* because the window panes in a greenhouse produce a similar phenomenon. In our greenhouse world, gases such as CO_2 in the atmosphere reradiate heat back to the surface. Atmospheric levels of these gases are increasing, and this may cause the earth's surface to get warmer. These climatic changes may affect where and how much plants grow (see Chapter 2).

2 | **Photosynthesis requires an exchange of gases between plants and the atmosphere.**

In 1771, Joseph Priestley performed experiments that led to the current understanding of the importance of atmospheric gases in photosynthesis. He placed a mouse in a sealed jar and observed that after a while it died. If,

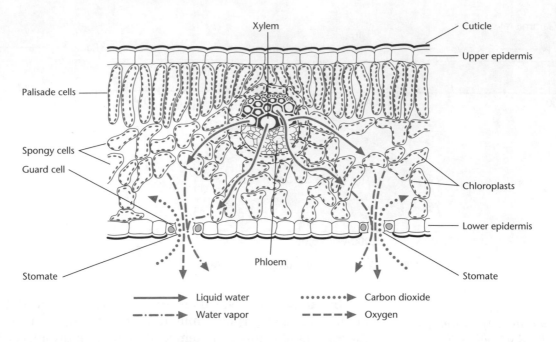

Figure 6.2

Gases Diffuse into and out of the Leaf Through Stomates. A cross section through a leaf, showing pathways along which CO₂, O₂, and H₂O diffuse. Source: Biology: The Common Threads, Part 1 *(1990), (Canberra: Australian Academy of Science), p. 170.*

however, the mouse was put in a jar in which a candle had burnt, it died immediately, because, as Priestly put it, the candle had "exhausted" the air. In a third experiment, he put a sprig of mint plant in with the air "exhausted" by the candle; after a while, he added the mouse. This time, the mouse did not die immediately. "Vegetation," Priestley wrote, "restored the air. From these discoveries we are assured that no vegetable grows in vain but cleanses and purifies our atmosphere."

Chemists now can explain Priestley's experiment in terms of the exchange of gases between organisms and the atmosphere. The burning candle had "exhausted" the air of oxygen, and without a supply of this gas for respiration, the mouse died. The mint sprig used the carbon dioxide remaining in the air to restore oxygen to the air by the process of **photosynthesis**, as shown in the following equation:

$$\text{Sunlight} + 6CO_2 + 6H_2O \rightarrow C_6H_{12}O_6 + 6O_2$$

$$\qquad\qquad\quad \underset{\text{dioxide}}{\text{Carbon}} \quad \text{Water} \quad \text{Sugar} \quad \text{Oxygen}$$

This equation does not show the great complexity of the process of photosynthesis, in which many biochemical reactions and intermediates are involved. But it does indicate that, in general, photosynthesis is the reverse of respiration as carried out by animals, plants, and most micro-organisms. In respiration, oxygen combines with sugar to form carbon dioxide and water:

$$C_6H_{12}O_6 + 6O_2 \rightarrow 6H_2O + 6CO_2 + \textbf{energy}$$

A key difference between the two processes is in the "energy" term. Photosynthesis *uses energy* from the sun to make chemical bonds; respiration *releases energy* when the bonds between these atoms are broken again. This energy is released not as heat or light, forms of energy from the sun, but instead as chemical bond energy in a form that the organism can use to fuel its needs.

Thus, photosynthesis converts one form of energy (solar radiation) into another (chemical bond), whereas respiration converts one form of chemical energy (stored in glucose) into another form of chemical energy (ATP—see later discussion). It is important to note that *both* plants and animals carry out respiration, whereas only green plants are capable of photosynthesis.

Photosynthesis occurs primarily in the leaves of green plants. A cross-section of a typical leaf reveals that it is eminently suited to be a photosynthetic factory (see Figure 6.2). The leaf is thin and flat to maximize the surface area that receives sunlight and to minimize the distance over which CO_2 must diffuse to the area where it is used for photosynthesis. It contains veins of conductive tissue, so that water can be brought in and the sugar formed can be transported to other parts of the plant. The many air spaces between the cells facilitate gas exchange.

In addition, the leaf surfaces, especially the lower epidermis one, are dotted with thousands of tiny pores, the stomates (Figure 6.3a). These pores regulate gas exchange between the leaf and the atmosphere by their number and size. The sizes of these pores are regulated to maximize photosynthetic gas exchange. When the stomates are open, CO_2 can enter the leaf from the atmosphere, but open stomates also allow water to escape. To avoid too much water loss, the stomates open only when CO_2 is needed and photosynthesis is possible. Thus, light is an important regulator of the stomate's opening and closing, and stomates are open during the day and closed at night.

However, in a dry climate the water loss that occurs through the open pores during the day (this process is called **transpiration**) can be excessive. This water must be replaced by the roots taking up water from the soil. If the roots cannot do so adequately because the soil is too dry, then the plant starts to dry out. The response to such drying of the leaves is that the stomates close to prevent further water loss (Figure 6.3b). Thus, there is a tradeoff: the need for photosynthesis necessary for growth must be balanced against the detrimental effect of water loss.

3 | Photochemical and biochemical reactions in the chloroplast result in CO_2 fixation.

As noted earlier, the solar energy of primary concern to plants is in the visible wavelength range (Figure 6.1). When it reaches a leaf, this light energy can

- Be reflected from the leaf back out to the atmosphere
- Pass through the leaf and also end up in the atmosphere
- Be absorbed by the leaf and converted into chemical energy in photosynthesis

The third path is accomplished in green tissues of plants. Cells of these tissues (see Figure 5.2) contain specialized organelles called **chloroplasts**, within which photosynthesis occurs. Chloroplasts are most abundant in the mesophyll cells just below the upper surface of a leaf (Figure 6.2). A single cell may have dozens of chloroplasts (Figure 6.4).

Chloroplasts contain a pigment, chlorophyll, that absorbs light energy in the blue and red portions of the visible spectrum. This absorption of light by

Figure 6.3

Structure and Activity of Stomates. **(a)** *Enface view of a stomate in a hawthorne leaf.* Source: *Carolina Biological Supply.* **(b)** *Fluctuation in the size of the stomatal openings of a eucalyptus leaf during a 24-hour period when the tree is growing in very dry conditions. Note that the pores are fully open for only a short time in the morning.* Source: Biology: The Common Threads, Part 1 *(1990),* (Canberra: Australian Academy of Science), *p. 172.*

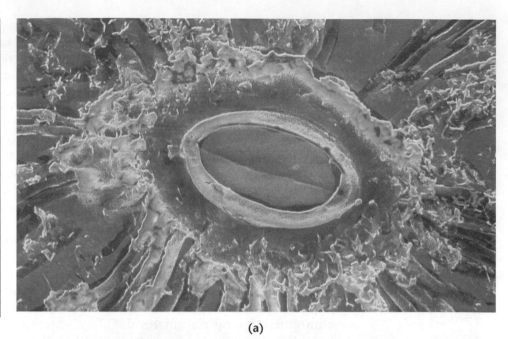

(a)

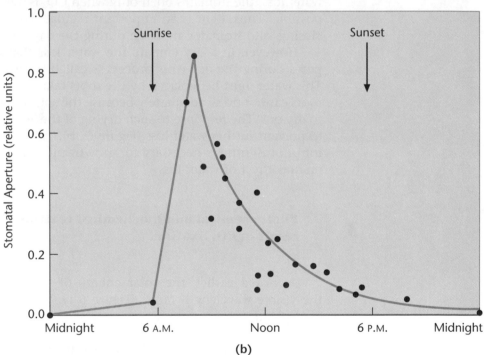

(b)

a molecule is the key event in energy conversion. The unabsorbed wavelengths of sunlight are transmitted through the plant to the observer. Since these comprise the green portion of the spectrum, the leaf appears green.

When a chlorophyll molecule absorbs a photon (energy particle) of light, it makes a transition from its basic ground state to a higher-energy or "excited" state. This excited state, which is very short lived, is characterized by a slightly different and unstable distribution of the electrons in the chlorophyll molecule. The excited chlorophyll rapidly returns to the ground state,

Figure 6.4

A Chloroplast in a Mesophyll Cell of a Leaf. The stacks of membranes are the site of the photosynthetic light reaction. The dark, oblong grains are starch. Source: *Courtesy of E. Newcomb.*

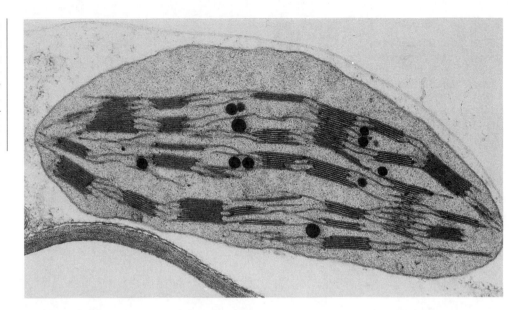

but releases the absorbed energy in the process. In another system, this release could be in the form of heat or light (this is the principle of electric stoves and lights). Instead, a remarkable series of photochemical reactions occurs in which energy of the excited state is transferred to make chemicals that can be used by the plant cell to produce carbohydrate. The photochemical reactions are among the fastest known in nature. An excited chlorophyll molecule can remain in that state for only a few nanoseconds (a nanosecond is one-billionth of a second).

A series of electronic energy transfer reactions in which energy is transferred from one molecule to another is finally followed by the synthesis of two important energy-rich molecules, ATP and NADPH:

- **ATP**, or **adenosine triphosphate**, is an important intermediate in the energy-requiring reactions of all cells. It is synthesized in the chloroplast by coupling a phosphate molecule to adenosine diphosphate (ADP), and this coupling requires energy. When the phosphate bond is broken, the energy is released again and can be used by the cell.

- **NADPH**, or **nicotine adenine dinucleotide phosphate (reduced)**, is another important energy-rich molecule. It is able to impart its energy to other compounds by transferring two electrons and a proton. In the process, it becomes oxidized to NADP, but the actively photosynthesizing chloroplast converts it back again to NADPH so that it can function again.

During the photochemical reaction series (termed the "light reactions" for obvious reasons), water (H_2O) is split into its components: $2H^+$ (these are protons used in part for forming NADPH) and O_2. The latter is not needed for photosynthesis, but is certainly essential for respiration—in all organisms, not just plants. In fact, during evolution photosynthesis probably preceded respiration, because the early atmosphere is thought to have contained CO_2 but not O_2. Humans—indeed, all organisms—depend on green plants not only for food (see Chapter 5) but also for the oxygen in the air we breathe.

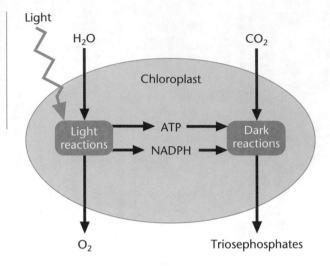

Figure 6.5

The Reactions of Photosynthesis. *The connection between the "light" and "dark" reactions is that the former supplies ATP and NADPH to the latter.*

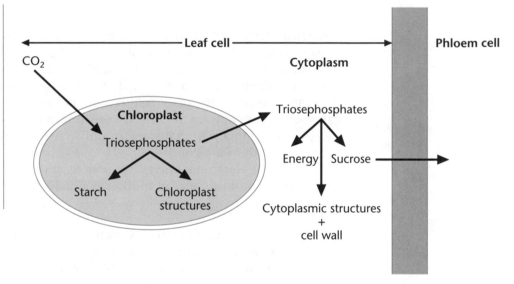

Figure 6.6

The Fates of Triosephosphates, the First Products of CO₂ Fixation, Both Within the Chloroplast and After Export to the Cytoplasm. *Within the leaf cell, the ultimate products of photosynthesis are structural or storage (starch). Sucrose is formed for transport to the rest of the plant.*

The ATP and NADPH formed by the light reaction are used within the chloroplast in a series of reactions to link CO_2 to other chemicals in the chloroplast. Since these reactions do not require light directly, they are often called the "dark reactions"—somewhat of a misnomer, because they do not occur in the dark. However, in the laboratory they *can* occur in the dark if chloroplasts are continuously supplied with the products of the light reactions. The division of light and dark reactions is summarized in Figure 6.5.

The net result of a complex series of biochemical reactions is the formation of **triosephosphates**, sugar molecules that have three carbon atoms instead of the usual six present in sugars such as glucose, as well as a phosphate group. Because triosephosphates have three carbon atoms, this type of photosynthesis is called **C3 photosynthesis**. These molecules can take three paths (see Figure 6.6):

1   Within the chloroplast, triosephosphates can be used to synthesize the organelle's own structural components (membranes, proteins, nucleic acids), or to synthesize starch, which accumulates in the chloroplasts during the daytime.

2 Triosephosphates can be exported from the chloroplasts to the cytoplasm and used there to synthesize the structural components of the cytoplasm (membranes, proteins, nucleic acids) and of the cell wall.

3 Triosephosphates can also be used to synthesize sucrose for export from the leaves. Sucrose travels via the phloem to other organs, especially nonphotosynthetic ones such as roots. There, the sucrose is "digested" to its component sugars, glucose and fructose, and these are broken down in respiration for energy with the release of CO_2.

Some of the triosephosphates themselves are used to drive the energy-requiring reactions of the cells that produce them. The central role of ATP, NADPH, and triosephosphate in all these reactions also illustrates the importance of minerals for cells. All three molecules have phosphate groups, and phosphate is a mineral that is required in large amounts by cells.

4 | **Many herbicides kill plants by blocking photosynthesis or other reactions in the chloroplast.**

So far, we have emphasized the role of the chloroplast in CO_2 fixation and the two sets of reactions that lead to CO_2 fixation: the conversion of radiant energy to chemical energy with the concomitant synthesis of ATP and NADPH, and the subsequent synthesis of triosephosphates. But many other important biochemical processes, all requiring a heavy input of energy, take place in the chloroplasts as well. The synthesis of fatty acids, the synthesis of many amino acids, the assimilation of nitrate as well as sulfate taken up by plant roots, all occur entirely or in part, in chloroplasts. The products of these biochemical processes may be used in the chloroplasts themselves or be exported for use elsewhere in the cell.

It is remarkable, but perhaps not surprising, that many herbicides—chemicals that kill plants, but not animals—affect processes within the chloroplast (Figure 6.7). Many useful herbicides have not been designed by chemists in the laboratory, but discovered by scientists screening tens of thousands of synthetic chemicals for their lethal effects on plants. Having found a compound that kills plants at a reasonably low dose, the scientists conduct toxicology tests to determine its toxicity to humans and other life forms. Organic chemists may alter the structure of a molecule to increase its effectiveness so that it can be used in lower doses.

Chemicals that kill all plants—nonselectively—can be just as useful in agriculture as chemical herbicides that kill certain plants (weeds) without killing others (crops)—selectively. Because chloroplasts and the chloroplast-based biochemical reactions are often unique to plants and absent from animals, many nonselective herbicides have been found to kill plants by blocking processes within the chloroplast. A major type of herbicide interrupts the electron transfer reactions in the chloroplasts: Paraquat, atrazine, and Diuron (dichlorophenyldimethylurea) act in this way.

In areas where a single one of these herbicides has been used extensively for a long time, weed populations have developed resistance as a result of mutation. In some cases, molecular biologists have been able to pinpoint this resistance to a single amino acid change in one of the proteins of the

Figure 6.7

*Herbicides That
Inhibit Biochemical
Reactions Within the
Chloroplast.*

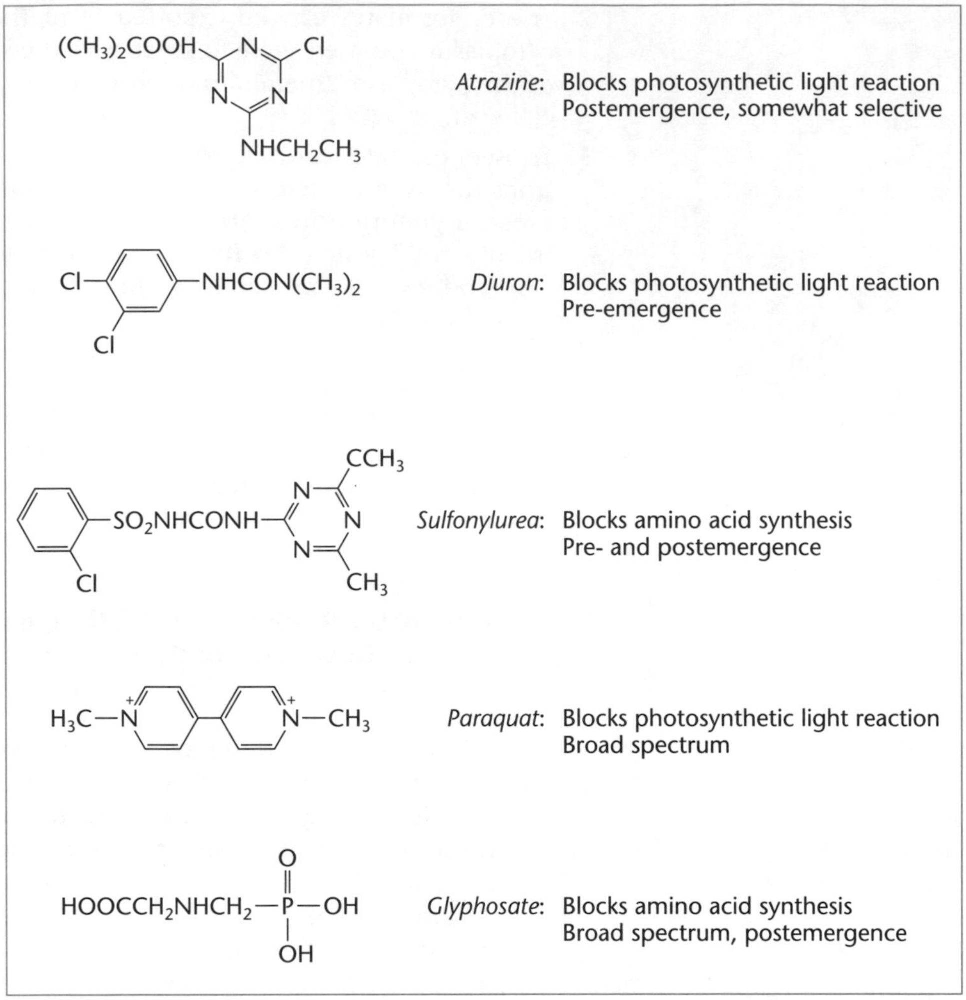

chloroplast electron transport chain of the weed. These discoveries created a great deal of interest in the plant biotechnology industry because the genes of the mutant proteins could be used to make other plants tolerant to the herbicide by transferring the gene for tolerance from the weed to a crop plant. Normally, the crop as well as the weed are negatively affected when the chemical is applied. If a crop plant were truly tolerant, this negative effect might disappear. Only certain crops are usually tolerant to a herbicide, while others are not. Thus, the possibility of making genetically engineered tolerant crops could increase the number of crops on which a particular herbicide could be used and therefore increase the use of that herbicide.

We noted in Chapter 4 that plants synthesize essential amino acids that are not made by people. The syntheses of some of these essential amino acids occur in the chloroplasts and involves enzymes not found in humans. EPSP, or 5-enolpyruvyl-shikimate-3 phosphate, is an important intermediate in the biochemical pathway that leads to the formation of the amino acids tryptophan, phenylalanine, and tyrosine. The enzyme that catalyzes the synthesis of EPSP, not surprisingly called *EPSP synthase*, is the target enzyme for the herbicide Roundup® (glyphosate). Another enzyme, AL synthase (acetolactate synthase), is the target enzyme for sulfonylurea herbicides such as Glean®. This enzyme is part of the pathway that leads to the synthesis of the amino acids isoleucine, leucine, and valine. The genes that encode these two enzymes (EPSP synthase and AL synthase) have been obtained from herbicide-resistant bacteria and have been used in genetic engineering experiments to create herbicide-resistant plants (see Chapter 13).

5 | The products of photosynthesis are partitioned among different sinks.

Photosynthesis in a crop plant is carried on primarily by young and mature leaves; these act as the source of fixed carbon compounds (photosynthate). Other plant organs, such as stems, roots, flowers, fruits, seeds, and very small leaves, are sinks for photosynthate. The **sources** synthesize and export sucrose (or other transport sugars) via the conductive tissue (phloem), and the **sinks** import it for their own development. In the sinks, the transported sugars are used to make storage molecules (starch or sucrose itself), for the synthesis of structural components (cell walls, proteins, and so on), or as a source of energy.

The partitioning of photosynthesis products among different sinks changes in different situations. For example, early in the development of an annual crop plant, the young leaves, stems, and roots are the strongest sinks, while later on, developing storage tissues or organs (seeds, roots, and tubers) may be the most important sinks. In nodulated legumes, the root nodules and their nitrogen-fixing bacteroids are strong sinks because much energy is needed to convert atmospheric N_2 to ammonia.

In vigorously growing plants, from 30 to 40 percent of the photosynthate is transferred to the root system. Sucrose that arrives in the roots is used for the synthesis of more root substance (cellulose, proteins, and so on), for the synthesis of complex carbohydrates that are secreted by roots and used by micro-organisms as a source of energy (this uses half the carbohydrate energy in the root) and as a source of energy in the root itself. In addition to needing energy to fuel normal cell functions, the root requires energy to take up minerals dissolved in soil water.

How a plant allocates photosynthate among these competing sinks is of tremendous importance for crop yields. Although efforts to increase overall plant photosynthesis by breeding have so far not been very successful, significant yield increases have been achieved by increasing the **harvest index**, defined as the ratio of dry matter or protein in the edible portion of the crop (such as corn seeds) to that in the whole plant (stalk + leaves + cob + seeds). (The small roots are usually not taken into account in these calculations.) The idea here is to increase not just the total yield of the entire plant, but the yield only of the grain that people eat.

As Figure 6.8 shows, introducing new strains of cereal crops has resulted in dramatic increases in the harvest index: the percentage of the total weight in the harvestable crop. The new high-yielding strains of wheat and rice, into which dwarfing genes have been introduced, produce much smaller plants (see Chapter 11), and these smaller plants can bear much bigger heads with more and heavier seeds. The net result is a dramatic shift in the harvest index.

Although the harvest indices of some crops in the field have risen, they are probably not yet at their maximum. One problem is that the dwarf plants compete less well with weeds than their normal-sized counterparts, and even

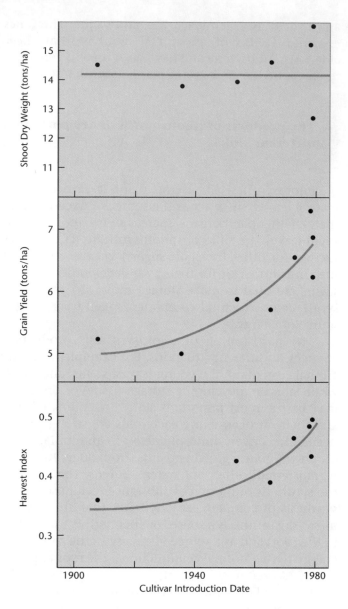

Figure 6.8

Productivity Characteristics of British Winter Wheat Strains Introduced over 80 Years. The strains were all grown under maximal field conditions in 1977–1978. Note that the newer cultivars put more of their fixed energy into grain, thereby raising the harvest index. Source: R. Gifford, J. Thorne, W. Hitz, and G. Giaquinta (1984), Crop productivity and photoassimilate partitioning, Science *225:802–806.*

with careful management this lowers the overall yield in the field. Another is that under the favorable environmental conditions that farmers strive for, both wheat and rice form extra shoots called *tillers,* and these do not bear grain; instead, they act as additional sinks, drawing photosynthate away from the flower-bearing regions. Third, the number of flowers that a typical cereal or legume forms is far less than the number that actually forms seeds. For instance, each flower-bearing spikelet in wheat starts off with about 10 developing flowers, of which only 2 will set grain. The loss of these other flowers appears to be due to limitations in the availability of photosynthate.

Fourth, and finally, there is the poorly understood phenomenon of the influence of the sink on the rate of photosynthesis in the source. If starch accumulates in the leaves, the rate of photosynthesis declines. This can happen if the sinks are full or absent (for example, if storage organs such as the ears of corn are removed). And it may occur when photosynthate is not removed from the leaves fast enough via translocation in the phloem.

Recent genetic engineering experiments indicate that it may be possible to increase the rate of photosynthate removal by introducing new enzymes in the sinks. Invertase is an enzyme that breaks down sucrose into glucose and fructose. In sink tissues, invertase occurs in the cell walls, where it breaks down sucrose as soon as the sucrose exits from the phloem. By breaking sucrose down faster, it may be possible to accelerate the process of unloading sucrose from the phloem. Lothar Willmitzer and his colleagues from the Free University of Berlin tested this hypothesis by making transgenic potato plants in which a gene for invertase is expressed specifically in the developing potato tubers. The result was a big increase in the size of the potatoes, but a decrease in their number. Another approach is to add a gene to potatoes that encodes the enzyme that controls the rate of starch synthesis. This again would increase sink strength, and hopefully reduce the depression of photosynthesis. Such an experiment has led to the production of potatoes with a higher starch content (see Chapter 15).

6 | The photosynthetic apparatuses of different plant species are adapted to different light intensities.

The rate of plant growth is influenced by many environmental variables, such as the intensity of light, the temperature, the CO_2 concentration in the air, the amount of water in the soil, the O_2 concentration in the air, the alternation of night and day temperatures, and the length of the day. Many of these factors have a direct effect on the rate of photosynthesis, while others affect the process of respiration.

The most obvious environmental factor that influences the rate of photosynthesis in a plant is the amount of light that hits the plant. Both the duration of the light period (number of hours of daylight) and the intensity of the light are important variables. The duration of the light period is simply the number of hours the plant can carry out photosynthesis. The intensity of the light affects the rate of the photosynthetic process. A plant with a certain number of chlorophyll molecules in its leaves can use a certain maximum amount of sunlight per second. If the light intensity is less than this maximum for a given plant, that plant will carry out photosynthesis more slowly than it could. For many crop plants, there is a linear relationship between light intensity and photosynthetic rate (measured as the amount of CO_2 "fixed" per unit time into other molecules). The rise in CO_2 fixation with light intensity has its limit at the point where the plant's photosynthetic machinery is working at peak efficiency. Beyond this point, further increase in light intensity has no effect on photosynthesis.

The light intensity required for maximal photosynthesis differs from species to species, and is related to the ecological niche to which the plant is

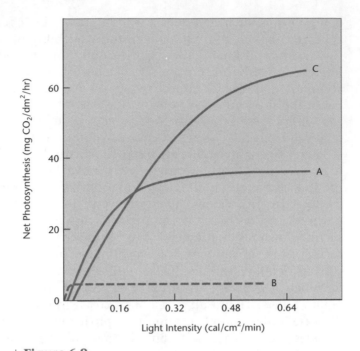

Figure 6.9

***Photosynthetic Light Saturation Curves of* (a) *C3 Sun Leaf,* (b) *C3 Shade Leaf, and*
(c) *C4 Leaf.* *Full sunlight is approximately 0.6 cal/cm²/min. Note that plant C never
reaches its maximum photosynthetic rate, whereas plant A reaches its maximum at half of
full sunlight. Photosynthesis in plant B is saturated at very low light intensity. More light
does not cause its photosynthetic rate to increase.* Source: *P. M. Ray, T. A. Stevens, and
S. A. Fultz (1983),* Botany *(Philadelphia: Saunders College Publishing), p. 141.*

adapted. Figure 6.9 shows a comparison of the responses of sun-loving and
shade-loving plants to increasing sunlight intensity. The sun-loving plant (A)
continues to increase its rate of photosynthesis as light increases up to a
point, and then it levels off, while the shade plant (B) levels off at a much
lower light intensity (note that at very low light intensities the shade plant is
more efficient). Clearly, the sun-adapted plant has the genetic capacity to take
better advantage of the intense sunlight to which it is exposed.

Certain plants are adapted to very high light intensities, and are able to
carry out photosynthesis quite efficiently even with their stomates partially
closed. Under these conditions, for example, at noon on a hot, sunny day
when the plant closes its stomates to reduce water loss, gas exchange with the
atmosphere is reduced. This means that the CO_2 concentration in the leaf
goes down because photosynthesis is proceeding rapidly, and CO_2 is not
replenished from the air. These plants—corn and sugarcane are examples—
have a special mechanism that uses this reduced CO_2 level efficiently and
concentrates it at the site of CO_2 fixation. Because the first product of CO_2
fixation has four carbons rather than the usual three, these plants are called
C4 plants. Their rate of CO_2 fixation may be two to three times greater than
that of other plants, especially under the conditions of high light and high
temperature to which they are adapted (Figure 6.9, curve C). Such C4 plants
also have generally higher production of dry matter and crop yield.

Another important factor in photosynthesis is the proportion of light that
falls on a field and is actually intercepted by the leaves. If the plants are small
(early in the season in a field) or are sparsely placed, they will not intercept all
the light. Photosynthesis in each plant may be maximal, but plant growth

mists therefore select crop plants that are well adapted to dense stands so that the photosynthetic rate per unit area of land can be maximized. Physiologists have shown that corn plants whose leaves stand up straight intercept a greater proportion of sunlight and produce higher yields than corn plants whose leaves droop down. Plant breeders have taken advantage of this by breeding corn varieties that have erect, rather than drooping, leaves. Such corn varieties are now widely used in the United States and elsewhere.

7 | The carbon dioxide concentration within the leaf affects the rate of photosynthesis.

In addition to light, the other important factor that greatly influences the photosynthetic rate is the CO_2 concentration within the leaf. This level, coupled with the affinity of the photosynthetic fixation enzymes for CO_2, together play a major role in determining how much CO_2 fixation will occur. The concentration of CO_2 in the leaf is limited by two factors:

1 The concentration of this gas in the atmosphere (normally about 0.03%). This may be rising (see Chapter 2).

2 The rate of diffusion of CO_2 through the stomates into the leaf cells. This is influenced by whether the pores are open or not, which in turn is determined by physiological mechanisms.

Studies of the effects of elevated CO_2 on photosynthesis have used ingenious methods to raise levels of the gas around plants in the laboratory or field, such as closed chambers and pipes connected to a gas source (Figure 6.10). Such experiments have clearly shown that at high light intensities, increases in CO_2 concentration from the usual 0.03 up to 0.1% (1,000 parts

Figure 6.10

Studying the Effects of CO_2 Enrichment on Cotton Plants. A circular array of pipes vent CO_2 into the air around the plants. Source: U.S. Department of Agriculture.

Figure 6.11

Changes in Photosynthesis as a Function of Ambient Intercellular CO_2 Concentrations in a typical C3 and C4 plant.

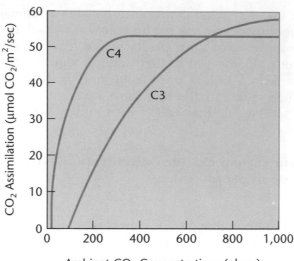

per million, or ppm), increase the rate of photosynthesis because CO_2 can enter the leaf more rapidly. Higher concentrations of CO_2 (beyond 1%) are usually poisonous to plants.

Increases in CO_2 lead to more photosynthesis via increased activity of the CO_2 fixation enzymes in the chloroplasts. Although such findings have considerable relevance to our knowledge of plant physiology, they also have practical applications. Commercial greenhouses often use added CO_2 to raise the CO_2 concentration in the air to 0.08%. This improves yields of the vegetables and ornamental plants thus grown.

As noted, C4 plants such as corn are adapted to high light intensities and high temperatures. They adapt by using a CO_2 concentration mechanism. At low CO_2 concentrations, these plants have an advantage over C3 plants, making photosynthate more rapidly and growing faster (see Figure 6.11). It is not surprising that corn, for example, yields three times as much grain as wheat under similar conditions. Atmospheric CO_2 levels would have to increase to over 700 ppm (0.07%) before C3 plants could do as well as C4 plants.

Of equal interest are the effects of increased photosynthesis on the partitioning of fixed carbon between sources and sinks. The results of experiments with a number of crop species have shown that increased photosynthesis often leads to an increased partitioning of photosynthate to the edible parts of the plant, especially the seeds. A representative experiment is summarized in Table 6.1. Ralph Hardy and his co-workers grew soybeans in air that had different CO_2 concentrations, and showed that a doubled CO_2 concentration led to a slight increase in the total weight of the plant body and its nitrogen content. Remarkably, the weight and nitrogen content of seed pods almost doubled. In this case, nitrogen in the seeds was doubled because the nitrogen-fixing bacteria in the roots of the soybeans converted more atmospheric nitrogen into ammonium, which the plants used to make amino acids and hence proteins. Other experiments have shown increases of about a third in yield when CO_2 levels double to about 0.065%. These experiments indicate that the higher levels of CO_2 in the future could have beneficial effects on crop yield (although the negative impacts of global warming may far outweigh this). But equally important, they show vividly the complex interactions between sources and sinks.

Table 6.1	Effect of CO_2-enriched air on the yield and nitrogen content of field-grown soybeans		
		Air (0.03% CO_2)	Enriched Air (0.06% CO_2)
Dry weight (in grams per plant)			
Plant body		11.8	15.0
Seed pods		8.8	15.7
Nitrogen content (in grams per plant)			
Plant body		0.27	0.30
Seed pods		0.38	0.70
Nitrogen in the plants (in kilograms per hectare)			
Total		295	511
Obtained from soil		219	84
Obtained from nitrogen-fixing bacteria		76	427

Source: R. W. F. Hardy and U. D. Havelka, paper presented at meeting on "Nitrogen Fixation and the Biosphere" held in Edinburgh, Scotland, 1973.

8 | All plant organs carry out respiration.

Many laypeople believe that plants get their energy from photosynthesis, while animals get their energy from respiration. This is only a half-truth. In fact, plants carry out both photosynthesis and respiration. During the day, they photosynthesize and respire simultaneously, and during the night they only respire. Respiration is a process that releases the energy stored in the chemical bonds in carbohydrates, proteins, and fats to synthesize molecules such as ATP, whose energy can be readily used in other cell processes. During respiration, O_2 is consumed and CO_2 is released.

Much CO_2 that is fixed in photosynthesis is released again in respiration, after the photosynthetic products have been used within the leaf or transported out of it to other organs. As noted earlier, plant organs can be divided into sources and sinks with respect to the production or destination of photosynthetic products. The sinks carry out very little photosynthesis, so they respire both during the day and during the night at similar rates. They need a constant supply of energy for growth (the synthesis of new substances). In contrast, photosynthetic sources (that is, leaves) do little true respiration during the day, but do respire at night.

The photosynthate used in respiration amounts to 20–40% of all that is produced by the plant. One factor that influences this rate is temperature. The relative rates of photosynthesis and respiration vary with temperature in different ways. In C3 plants, above 30°C, the rate of photosynthesis begins to decline and that of respiration increases. This means that for optimal production of fixed carbon, cool nights should prevail. This climatic factor has great impact on which C3 crops can be grown where.

However, each plant species has evolved its own optimal rates of photosynthesis and respiration to suit its environment. Because respiration powers all the energy-requiring processes of the cells, we should not think of this as a "loss" of photosynthate. There is no evidence to support the notion that this "loss" is avoidable and that plants would have higher yields if they respired less.

Box 6.1

*The Methanol
Solution for Crop
Plant Productivity*

In the Arizona desert, where all farming requires irrigation, a Japanese-American farmer has recently discovered that spraying plants with a dilute (20%) solution of wood alcohol or methanol, stimulates their growth. Since methanol has long been known to be toxic to plants, its positive effects might appear to be unexpected—but not to Arthur Nonomura. With a doctorate and long experience investigating photosynthesis, Nonomura had left the laboratory for the farm to change his lifestyle. He noticed that his cotton plants wilted during the midday heat. Such wilting is caused by the more rapid loss of water from the leaves than its replacement by water taken up by the roots from the soil. Wilting causes the stomates to close (see Figure 6.2) and reduces photosynthesis because no new CO_2 can enter the leaves. At the same time, the wasteful process called *photorespiration* becomes more pronounced because of the lower CO_2 level inside the leaves.

In 1949, Andrew Benson, a pioneer in photosynthesis research, had shown that methanol could be used by algae for fixed carbon production and indeed Nonomura himself had done similar experiments with Benson in the 1970s. Perhaps methanol at nontoxic concentrations could help out the cotton in this hot, dry climate. Armed with this knowledge and a spray bottle, Nonomura applied some dilute methanol to his wilted cotton. Within hours, the sprayed plants were once again turgid, their stomates were open, photosynthesis resumed at high levels, and photorespiration went down. Not surprisingly, the growth rate increased significantly, and the plants produced cotton bolls two weeks sooner than usual, thereby reducing their overall need for water. The next candidates for methanol spraying were cabbage plants, resulting in heads twice as big as those on untreated plants grown for the same length of time (see figure), then watermelon (yields up by 36%). Methanol also stimulates the growth rates of wheat and barley, and this should result in increased yields.

The Effect of Spraying a Dilute Solution of Methanol on the Growth of Cabbage Plants. *Both cabbage plants have the same age, but the larger one has been sprayed with a dilute solution of methanol.* Source: *Courtesy of A. Benson, University of California at San Diego.*

Not all plants are positively affected by the methanol spray. For example, corn and other C4 plants do not exhibit the midday reduction in photosynthesis and do not have photorespiration, so these crops would not be expected to be stimulated by methanol—and they are not. Nor are plants growing in the shade, an indication that whatever methanol is doing to the plant is mediated by processes involving light. The discovery of this effect of methanol is so recent that almost no research has been done so far to understand the underlying mechanism. It seems unlikely that methanol is being used by these plants as a source of carbon, as it was in the early experiments with algae. Rather, methanol must be regulating one or more essential plant processes. Research into the potential benefits for this simple "fertilizer" treatment is going ahead with full speed on a number of experimental farms in the United States.

Net CO_2 fixation in most C3 plants during the day is strongly inhibited by oxygen gas. This statement might appear odd, given that there is 21% O_2 in the atmosphere. Nevertheless, experiments in which the O_2 level is reduced while CO_2 is held constant clearly show a reduction in this inhibition. The reason for O_2 inhibition is a process called **photorespiration**, in which fixed carbon is lost as CO_2. This is entirely separate from respiration, where glucose is broken down for energy. When photorespiration occurs, the first enzyme of CO_2 fixation combines not with CO_2, but with O_2. A series of reactions ensues, which then results in CO_2 being released. Thus the overall net rate of carbon fixation in these plants becomes:

Net carbon fixation = photosynthesis − respiration − photorespiration

The key to this inhibition by oxygen is the relative concentration of O_2 and of CO_2 fixation in the leaf. Conditions that lower the CO_2 concentration inside the leaf favor photorespiration, because the O_2 concentration is always high and O_2 is therefore always available. CO_2 levels in the leaf will fall if photosynthesis is very rapid: conditions of high light intensity and high temperature accelerate photosynthesis, cause the CO_2 concentration to drop and therefore enhance photorespiration. Water shortage leading to a partial closing of the stomates prevents CO_2 from getting in the leaf, and photorespiration will increase because the CO_2 concentration in the leaf is low. Photorespiration is primarily a problem for C3 plants. C4 plants avoid the problem by concentrating CO_2 at the site of photosynthesis. This means that C4 plants are adapted to conditions of high light, high temperature, and water shortage.

The evolution of the CO_2-concentrating mechanism by the C4 plants is relatively recent (on an evolutionary time scale) and probably was a response to the drop in atmospheric CO_2. When land plants first evolved, CO_2 levels in the atmosphere were much higher than they are now. Plant breeders have tried to eliminate photorespiration, because the process is wasteful from a human point of view, but those efforts were completely unsuccessful. The recent discovery that methanol enhances the growth of C3 but not of C4 plants opens up new possibilities to enhance crop productivity by manipulating photosynthesis (see Box 6.1).

9 | An efficient plant ecosystem converts only 1–2% of the solar energy into dry matter.

There are several ways to determine the efficiency of photosynthesis. One method is to measure the accumulation of dry matter over the growing season of a crop, determine its energy content, and then measure the amount of visible light that fell on the field (incident energy) during the growing season. The ratio of these two energy values is usually between 0.2 and 3.5%, indicating that not much of the incident energy is actually available for people or animals as food. Is photosynthesis really so inefficient?

The losses of light energy that arrives at a field are considerable (Figure 6.12). First, over half of it does not actually hit the leafy canopy, either because the leaves are not always fully formed, or because they do not cover the full surface area of the field. Of the light that remains and actually arrives at the leaf, about half of it cannot be used because it is of the wrong wavelength: chlorophyll and other pigments only absorb certain wavelengths of light. In addition, 5% of the light is reflected by the leaf and another 5% passes through without being absorbed, even though it may be the right wavelength. What remains is only a fraction of the solar light energy that arrived.

The chemical reactions of photosynthesis are actually quite efficient with respect to energy conversion: 2,500 calories of light energy are converted into 700 calories of glucose. Thus the efficiency of energy conversion is about 28%, which is a higher efficiency than automobiles and much higher than household appliances. However, when only about 33% of the light energy hitting the leaf is used with an efficiency of 28%, this means that the energy conversion efficiency is about 9%. A fair proportion of the photosynthate (40–80% depending on the species) is lost in photorespiration and used in respiration; so this reduces the net efficiency of conversion of solar energy to energy permanently stored in chemical bonds in the plant. Only a small percentage of the light energy that is received accumulates as chemical bond energy in plant dry matter (cellulose, starch, protein, and so on).

The amount of energy that actually ends up in plant products (starch, cellulose, protein, and so on) is called the **net productivity** of the plant. It is usually much less than the **gross productivity**, which is the total amount of energy fixed during photosynthesis. These two terms, gross and net productivity, can be applied not only to single plants, but also to entire ecosystems. The annual net productivity of an ecosystem is the amount of dry matter produced by plants each year. The two processes that reduce gross productivity are respiration and photorespiration. Plants that are adapted to high light intensities and high temperatures and carry out C4-type photosynthesis have extremely low levels of photorespiration. Such plants, and ecosystems that are dominated by these plants, have high net productivities.

The net productivities of plants in nature differ greatly, largely because of the climate. Arctic tundra and desert plants have very low net productivity (100 to 200 g of dry matter per square meter per year) whereas tropical forests have net productivity of 5,000 g or more per square meter per year. Productivity appears to be limited primarily by the amount of water available and by the amount of solar energy received. In many areas, plants do not grow year round, because it is either too dry or too cold; so the amount of light received during the growing season is often much less than the total amount of light received during the year.

Figure 6.12

The Fate of Incoming Solar Radiation Energy over a Year on a Field Used for a Grain Crop in a Temperate Climate. Note the small amount of energy that actually ends up stored in fixed CO_2. Source: R. Gifford, J. Thorne, W. Hitz, and R. Giaquinta (1984), Crop productivity and photoassimilate partitioning, Science 225:804.

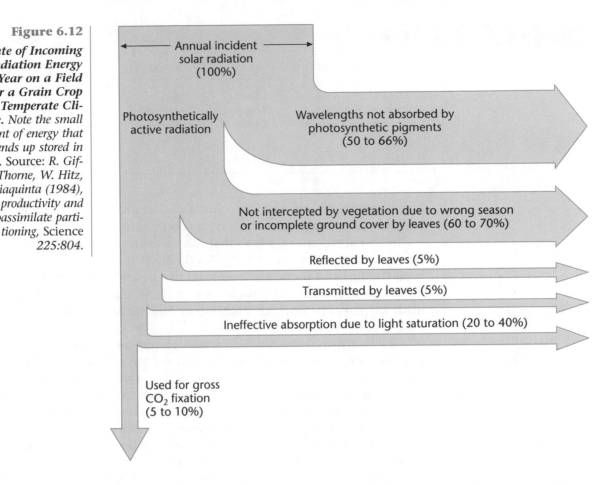

When the efficiency of energy conversion is calculated in terms of light received during the growing season, it is found that natural ecosystems can convert from 0.2 to 3.5% of the incident light energy into net productivity. However, well-managed agricultural systems, especially those consisting of C4 tropical grasses and their cultivated relatives, such as corn and sugarcane, can be even more productive than natural ecosystems. They can convert up to 4% of the sunlight received during the growing period into dry matter.

By using the known figures for the primary productivity of extremely well-managed agricultural ecosystems—wheat, producing 4,000 kg per hectare—it is possible to calculate an **energy budget** for the world. Such calculations show that the earth can support about 5–6 billion people on a mixed diet and at most 15 billion people on a vegetarian diet. An important (but naive) assumption in these calculations is that land degradation will not cause the arable land to decrease significantly in the future.

10 Energy transfer in the ecosystem between successive trophic levels is inefficient.

The living world consists of many kinds of organisms, which do not live independently from one another, but instead interact with each other. How energy and matter are transferred between organisms is an important aspect

Table 6.2	Food chains		
Trophic Level	**Land**	**Ocean**	**Farm**
Primary producers	Grasses	Algae	Corn, soybeans
Primary consumers	Rabbit	Zooplankton	Cow
Secondary consumers	Coyote	Anchovy	Humans
		Herring	
		Tuna	

of the science of ecology. The organisms that are related by consumption of energy and matter are said to form a **food chain**.

In a food chain, organisms that share the same general source of nutrition are said to be at the same trophic level. Green plants, which get their energy from the sun and their nutrition from the air and soil, form the first trophic level: the **producers**. The molecules (such as starch and proteins) that are made by plants become food for organisms that feed on the producers: these feeders are **consumers**.

The first trophic level of consumers comprises the **herbivores**, organisms that feed exclusively on plants. Herbivores are not only familiar vertebrate animals, such as rabbits and deer, but also invertebrates such as insects and worms that eat leaves, roots, or fruits. The next trophic levels contain **carnivores**, predatory animals that eat other animals (see Table 6.2).

In addition to producers and consumers, a food chain also has **decomposers**. These organisms, of which many bacteria and fungi are examples, obtain their energy from dead organic matter. In a system where the producers and consumers are rapidly depleting the environment of nutrients, such as minerals, decomposers play a key role by returning these nutrients to the environment in a form that the producers can use.

Whether herbivores or carnivores, animals use food energy for a variety of purposes. Most is released in respiration to sustain energy-requiring activities, such as the movement of muscles and the complex reactions of biochemistry. Only a small portion is retained within the body, stored as protein, fat, or carbohydrate in a form available as food for the next level of consumers (see Figure 6.13). On the average, about 90% of the energy taken in is lost in respiration or is lost as undigested material; only 10% is stored. The energy flow in a human food chain is diagramed in Figure 6.13b. The energy from the sun that falls on a field of alfalfa is used for photosynthesis and growth of the plants. These plants are used to fatten calves that are fed to humans. The diagram shows that 720,000 kcal of solar energy yield only 1 kcal of energy in new tissue in a growing child.

We have already mentioned that meat is an ecological luxury because the transfer of protein from producer (the corn plant) to consumer (the cattle) is inefficient in terms of the amount of protein available for human consumption (see Figure 6.14). Several (4 to 6) kg of plant protein are needed to produce 1 kg of beef protein, because much of the protein that the cattle eat is used for its biological functions. Only 15–25% is used for making the structural proteins that human beings eat.

One way to graphically demonstrate how "expensive" it is to eat animal protein is to look at how many kilograms of protein (plant or animal) can be produced on a hectare of land (Figure 6.15). Agricultural scientists have attempted to increase the efficiency of these short food chains that convert

Figure 6.13

Energy Flow in a Food Chain. (a) *The fate of food energy ingested by an organism in terms of the energy available for organisms at the next trophic level; and* (b) *energy flow in an alfalfa–cow–child food chain.* Source: *Based on calculations reported in E. P. Odum (1971),* Fundamentals of Ecology, *3d ed. (New York: Holt, Rinehart and Winston).*

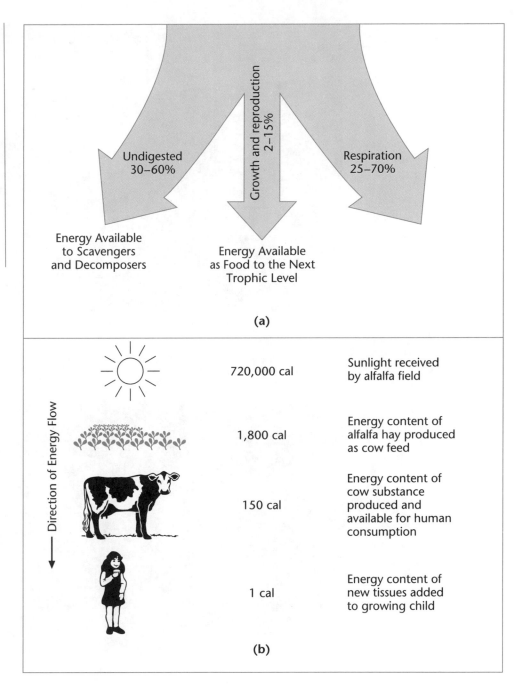

plant protein into animal protein for human consumption by restricting the energy expenditure of the primary consumers and by selecting animals that have a high conversion efficiency. Efficiencies of energy transfer on feedlots now range up to 25% for feed grain to chickens. In natural ecosystems, the undigested material remains in the ecosystem and micro-organisms decompose it and release its minerals. But in feedlots, very large amounts of the undigested materials accumulate as manure and cause serious pollution problems. These nutrient-rich wastes are usually not recycled back to the ecosystems from which they came, and are often not disposed of properly.

Another major difference between an agricultural and a natural ecosystem is that in agriculture a particular part of the food chain (producers) is favored over all others. Human beings modify the environment by irrigation,

(a)

(b)

Figure 6.14

Pastoralism and Extensive Grassland Farming Involve the Conversion of Plant to Animal Protein in Areas that are too Dry, too Cold or too Mountainous for Normal Crop Production. **(a)** *Pastoralism is usually associated with a nomadic way of life and involves herds of sheep, goats and camels. The primary productivity of the land is so low that the herds must keep on moving to go in search of food, as shown in this photograph of Bedouins and their herd in southern Algeria.* **(b)** *Ranching, or extensive grassland farming involves beef cattle, as well as sheep, and emerged in the second half of the nineteenth century as a result of an increased demand for meat and wool in the industrializing countries. Shown here is a herd of beef cattle from the llanos area of Colombia where the introduction of new forage species and careful grassland management have increased productivity substantially.* Sources: *(a) Courtesy United Nations PB/AB; (b) Centro Internacional de Agricultura Tropical, Cali, Colombia.*

Figure 6.15

Protein Yields for Various Crops and Animal Foods. A hectare of land yields 500 kg of soybean protein but only 20 kg of beef protein.

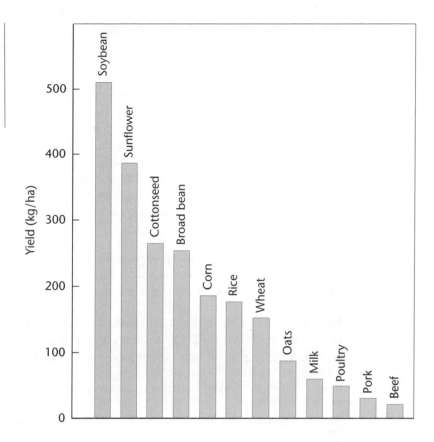

fertilization, and weed control to increase the net productivity of certain plants as much as possible, then keep away competing herbivores and divert the net productivity for their own use. Less energy is available to the decomposers because large amounts of organic matter are removed from the fields.

Summary

Life depends on a continuous input of energy from the sun. This radiant energy is converted to chemical energy in photosynthesis. In this process, CO_2 is fixed into simple carbohydrates and complex substances (cellulose, starch, and proteins) are synthesized and accumulate as plant material. This accumulation constitutes the net productivity of the plant. In an annual cropping system, the efficiency with which energy is trapped this way is only about 1–2% of all the light energy that falls on a plant. Actually, more energy is trapped initially as photosynthetic products, but a considerable amount is "lost" again during respiration and in some cases, photorespiration. Just like animals, plants respire to use energy for many energy-requiring reactions.

Photosynthesis depends on a continuous flux of CO_2 into the leaves and requires that the stomates be open. Two different types of photosynthesis have evolved: C3 and C4 photosynthesis. C4 photosynthesis is an adaptation to high-light and high-temperature conditions, and plants with C4 photosynthesis have a higher net productivity than do plants with C3 photosynthesis.

Once CO_2 has been fixed and converted into triosephosphates and these have been made into sucrose, the products of photosynthesis must be distributed all over the plant so that organs that do not photosynthesize can grow as well. Allocation of photosynthate to competing sinks is a subject that needs much further study, yet it is tremendously important for determining crop much further study, yet it is tremendously important for determining crop yield. People only harvest part of a plant, and they want this part to receive much of the photosynthate.

Plants are autotrophs, and they are the primary producers of the earth's ecosystems. Plants are eaten by herbivores, which in turn are eaten by carnivores. The transfer of energy and protein at each trophic level is only about 10% efficient in terms of the input. Producing a kilogram of animal protein from grain, therefore, requires much more land than producing a kilogram of grain protein. If people in industrialized countries suddenly became vegetarians or even halved their intake of animal products, much land could be taken out of agricultural production and serve other purposes (recreation, habitat protection, future agricultural use, and so on).

Further Reading

American Society of Agronomy. 1990. *Impact of CO₂, trace gases and climate change on global agriculture.* ASA Special Publication no. 53. Madison, WI: ASA.

Amthor, J. S. 1991. Respiration in a future higher CO_2 world. *Plant Cell Environment* 14:13–20.

Baker, N. R. 1992. *Crop Photosynthesis: Spatial and Temporal Determinants.* London: Elsevier.

Bugbee, B., and O. Monje. 1992. The limits of crop productivity. *BioScience* 42:494–502.

Gifford, R., J. Thorne, W. Hitz, and R. Giaquinta. 1984. Crop productivity and photoassimiliate partitioning. *Science* 225:801–807.

Grodzinski, B. 1992. Plant nutrition and growth by regulation of CO_2 enrichment. *Bioscience* 42:517–525.

Hall, D. O., and K. K. Rao. 1992. *Photosynthesis.* Cambridge, UK: Cambridge University Press.

Lawlor, D., and R. Mitchell. 1991. The effects of increasing CO_2 on crop photosynthesis and plant productivity. *Plant Cell Environment* 14: 807–818.

U.S. Department of Energy. 1985. *Direct effects of increasing CO₂ on vegetation.* DOE-ER0238. Washington, DC: U.S. Government Printing Office.

Zelitch, I. 1992. Control of plant productivity by regulation of photorespiration. *BioScience* 42:510–516.

Chapter 7

Nutrition from the Soil

To grow and produce the food on which all humanity depends, plants need more than carbon dioxide and oxygen from the air. They also need water and mineral nutrients from the soil. The soil is truly the "placenta of life," because it supplies essential nutrients to all land plants, and the plants in turn feed all the terrestrial ecosystems. Throughout history, humanity's standard of living has depended on the water content, fertility, and productivity of the soil. The Fertile Crescent of the Middle East, one of the areas in the world where agriculture originated (see Chapter 10), is considerably less fertile now than it was 10,000 years ago. The decrease in fertility is caused, in part, by changes in weather patterns—possibly as a result of deforestation of the Mediterranean basin—and, in part, by the failure of the inhabitants to maintain the productivity of the soils. Soil erosion and salinization are major consequences of poor agronomic practices. Elsewhere in the world, civilizations have declined when the soils that supported them became exhausted. Maintenance of soil fertility should be one of society's important goals.

1 | Water lost from the leaves must be replenished from the soil.

Water is the medium in which soil minerals are dissolved, and therefore it must be present in order for these nutrients to enter the roots of plants. This important role of soil water, however, cannot explain the extremely high water requirements of many crops. For an explanation of these requirements, we must consider the need for CO_2 in the leaf during photosynthesis.

The continuous diffusion of CO_2 from the atmosphere into the leaf occurs through the stomates and photosynthesis normally occurs only if the stomates are open, allowing the CO_2 to enter the leaf. These open stomates also allow water vapor to escape from the interior air spaces of the leaf into the atmosphere (Figure 7.1). The water vapor in the intercellular spaces is quickly replenished by the evaporation of water from the leaf cells. The water evaporates because the leaf absorbs much more solar energy than it can use in photosynthesis. This energy must be dissipated, or it would heat the leaf to a temperature (50°C or more) that would kill the cells. The plant dissipates this

Figure 7.1

*The Transpiration
Stream Through
a Plant.* Arrows indicate
the direction of
movement of water or
water vapor.

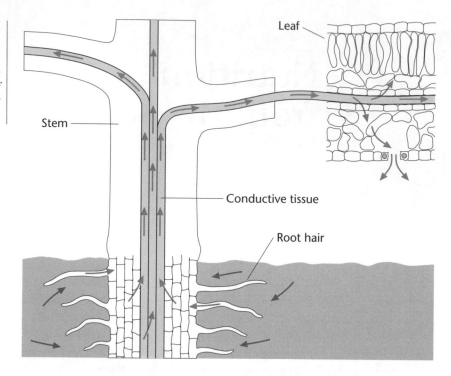

heat partially by allowing the water in the leaf cells to become water vapor in the intercellular spaces. However, the leaf cells must replenish their water supply. Therefore they obtain water from the conductive tissues, which in turn get their water from the soil via the roots. Thus during the daytime, when the stomates are open, there is a continuous stream of water through the plant. This process is called **transpiration** (see Figure 7.1). At night, when the stomates are closed, the transpiration stream stops, although the plant may continue to take up water form the soil until it has become completely rehydrated.

Transpiration dissipates heat in plants, similar to perspiration in humans. It also helps transport throughout the plant minerals taken up from the soil, and organic substances made in the roots. If the rate of water uptake from the soil is inadequate to replenish the water lost from the leaves by transpiration, the leaves will wilt, and this will cause the stomates to close. Closure of stomates prevents further water loss, but it also prevents photosynthesis, since CO_2 cannot enter the leaf. On a hot summer day, plants may experience transient wilting because the water loss is more rapid than the water uptake, even though there is enough water in the soil. This transient wilting can severely depress photosynthesis in the middle of the day. Thus provision of adequate soil water is crucial to plant productivity.

Many crop plants use large amounts of water. It has been calculated that a single corn plant in its growing season transpires about 200 L of water. This means that a hectare of corn uses an amount of water equivalent to a layer measuring 1 hectare in surface and 25 cm in thickness. However, water is lost not only by transpiration but also by evaporation from the soil surface. The sun warms the soil and causes some of the water to escape as water vapor. The term **evapotranspiration** describes the total amount of water that is lost from a plant–soil system. The total productivity of an ecosystem is very closely tied to the amount of water that is lost each year by evapotranspiration (see Figure 7.2). The availability of water in the soil determines plant

Figure 7.2

The Relationship Between the Water Lost from the Soil (by evaporation and transpiration) and Plant Productivity. In general, it is possible to predict productivity from data on temperature and water availability. Source: Adapted from M. Rosenzweig (1968), Net primary productivity of terrestrial communities: Prediction from climatological data, American Naturalist 102:67–74.

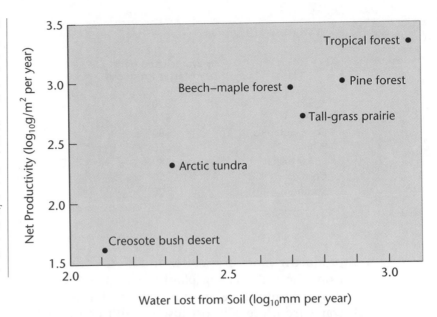

growth more than any other factor. Thus, it is not possible to "make the desert bloom" without irrigation. Although small yields can be obtained with drought-adapted plants, large yields always depend on supplying additional water.

2 | Not all plants have the same water requirements.

Plants differ in their water requirements. Different crops require different amounts of water, and the amount required often determines where these plants can be grown. Some examples of the water use efficiency, expressed as the number of kilograms of water used per kilogram of dry matter produced, are given in Table 7.1. Although a single corn plant uses a lot of water, as noted, corn is not one of the heavy water users, as shown by the data in Table 7.1. Different water use efficiencies reflect different strategies that plants have evolved to cope with conditions of water deficit. We already discussed one such strategy: C4 photosynthesis. Plants that have C4 photosynthesis carry out the first CO_2 fixation reaction with great efficiency even when CO_2 levels in the leaf are low. Water deficit in the leaves results in the partial or complete closing of the stomates, thereby slowing down entry of CO_2 into the leaves and lowering the internal CO_2 concentration because photosynthesis continually uses CO_2. However, C4 plants can cope with these reduced internal levels of CO_2 because the enzyme that binds CO_2 has a greater affinity for CO_2. In Table 7.1, the two plants—corn and sorghum—that use the smallest amount of water to produce dry matter are both C4 plants.

The water requirements of any plant depend on its environment and nutrition. When plants are grown in nutrient-rich soil, they are usually large and have extensive root systems. The total amount of water lost by transpiration increases, but water is used more efficiently. In one experiment with corn plants, the water use efficiency for corn grown on poor soil was 2,000, but it declined to 350 when the soil was well fertilized. Thus the larger, healthier plants used more water, but they used it more efficiently,

Table 7.1	Water use efficiencies

Plant	Kg Water Used per Kg Dry Matter Produced
Alfalfa	850
Soybeans	650
Oats, potatoes	580
Wheat	550
Sugar beets	380
Corn	350
Sorghum	300

producing more dry matter and more corn per kilogram of water extracted from the soil. Placing plants closer together increases the total transpiration, because there are more plants, but it decreases direct evaporation of water from the soil, because the sun does not hit the soil directly. Where the number of corn plants per hectare was increased from 20,000 to 40,000, corn yield increased by 65%, but total water usage increased only 20%. The increase in transpiration was largely compensated by a decrease in evaporation from the soil.

Water use efficiencies as shown in Table 7.1 are usually measured under ideal growth conditions, and do not provide any clues about how plants will perform with a normal (or abnormal) rainfall regime. Crop yield (usually seeds) is the bottom line, not dry matter production. A comparison of millet and cowpea yields in sub-Saharan Africa shows that when the rainfall averages 400–500 mm per year, millet outyields cowpea nearly twofold. However,

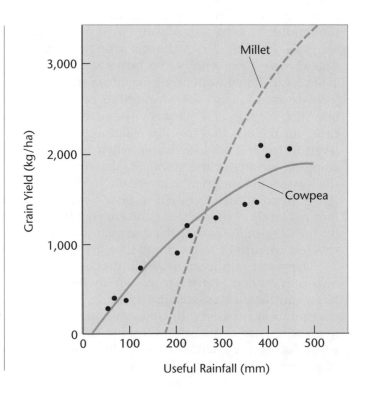

Figure 7.3

Grain Yields of Cowpea (variety Bambey 21) and Millet (variety Souna III) Obtained in Field Experiments at Louga and Bambey in Senegal over Several Years. Source: *From M. Ndoye, C. Dancette, M. Ndiaye, T. Diouf, and N. Cisse (1984), L'amelioration du niebe pour la zone sahelienne: Cas du programme national Senegalais. International Society for Research in Agriculture Publication. World Cowpea Research Conference, International Institute of Tropical Agriculture, Ibadan, Nigeria, November 1984.*

Figure 7.4

Rainfall (lines connecting dots) and Monthly Water Loss by Crop Growth and Evaporation in the Midwestern United States. Note that rainfall during the growing season is not adequate for the needs of the plants, and they must use water stored in the soil. Source: S. Aldrich and E. Leng (1969), Modern Corn Production (Champaign, IL: F and W Publishing), p. 171.

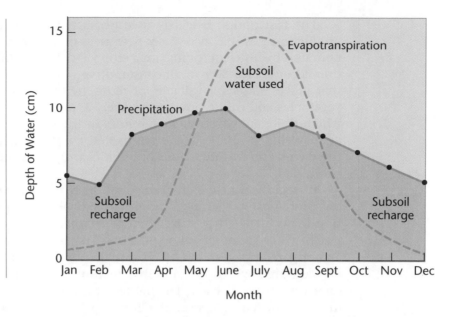

at 150 mm per year, cowpea still produces a meager crop of 600 kg per ha while millet fails completely (see Figure 7.3). Twenty years of drought in the Sahel have resulted in numerous crop failures, but plant breeders have now adapted cowpea varieties that produce acceptable yields even under conditions of low rainfall.

The water lost from the soil to the atmosphere by evapotranspiration must be replenished by precipitation and percolation of the water into the deeper layers of the soil. Crop production can only continue if the annual precipitation and percolation equals the annual evapotranspiration. Water loss from the soil during the growing season often exceeds water recharge, but the reverse is true in winter. Figure 7.4 shows the annual water loss by evapotranspiration, and water gain from precipitation, in the subsoil of a cornfield. Many agronomic practices are aimed at ensuring that the water that falls on the land will percolate into the subsoil rather than be lost as runoff. For example, leaving the stubble and other crop residues on the soil surface increases water percolation. In dry areas, such practices may mean the difference between success and failure in crop production.

3 | **Adaptations to water stress are of particular interest to plant breeders.**

In the course of evolution, plants have adopted many different strategies to cope with water deficits. These are of great interest to plant breeders and crop physiologists because water is the major factor limiting plant growth. We presently have three strategies to cope with water deficits: (1) supply irrigation water to traditional crop varieties, allowing them to be grown in areas to which they are not normally adapted; (2) transfer drought resistance genes from wild relatives to existing crop plants either through plant breeding or through genetic engineering; and (3) improve so-called minor crops that are well adapted to dry areas and were domesticated by people living in arid lands.

The first option is widely used because it is the easiest and makes crop production possible in entirely new areas or enhances yields in areas where rainfall is insufficient for maximal crop yield. In the United States, crop irrigation accounts for 80% of all consumptive water use. Consumptive use is use that makes water unavailable for reuse. Irrigation is not only used to grow vegetables and fruit crops, but is widely applied to corn and wheat, crops for which there are government subsidies in the United States. Irrigation boosts grain yields 40–100% over nonirrigated controls. Increased productivity of these crops on irrigated lands has significantly contributed to the creation of surpluses.

In regions where evapotranspiration is high, irrigation can create serious problems. In such regions rainfall may be low, but rain-fed crop production is nevertheless possible if crops adapted to the area are used and if as much water is conserved as possible. The problems begin when the crops are irrigated. Water cannot be supplied as pure water because it has dissolved salts, which remain in the root zone when evapotranspiration removes the water itself. The net result is a slow buildup of salts in the soil, making it less and less suitable for crop production. Many lands have been taken out of production altogether in the United States, Australia, and other countries because of their increased salinity and alkalinity (high pH) as a result of irrigation.

Existing crop plants are being made drought resistant (see Box 7.1). Plant adaptation to drought appears to be a multigene trait that affects the development of the whole plant. For example, one adaptation to a rainfall regime in which abundant spring rains are followed by summer drought is to limit leaf and shoot growth in the spring. A luxuriant growth of leaves produced in a wet spring cannot be sustained in the summer when the soil dries out. Such a leafy plant is at a disadvantage compared to a plant that has a more moderate leaf growth and uses its soil water more sparingly. Transferring such complex adaptations from one species to another is not possible through genetic engineering, because many genes are involved. However, classical plant breeding can be applied to this problem. In breeding better crops, plant breeders generally aim for plants that grow rapidly at first. Early closure of the canopy (when the plants have grown enough to touch each other) retards water evaporation from the soil and reduces competition from weeds.

When the soil dries out, the concentration in soil water of dissolved minerals increases. When this concentration is greater than that of all the solutes (minerals and organic compounds) in the roots, the plant cannot get any more water from the soil, so the transpiration stream stops. One way to let water uptake continue is for the plant to increase the solute level of its cells. This is exactly how some plants respond to drought or to a buildup of salts in the soil: they start synthesizing specific organic molecules that accumulate in the cytoplasm. This strategy is used by barley, a cereal that does particularly well in semiarid and saline areas. Irrigation of many areas in the Middle East, where wheat originated, resulted in saline soils incapable of supporting wheat production, and barley is now grown there. Relatively few enzymes are involved in the biosynthesis of these organic compounds that accumulate in the cytoplasm of plants under drought stress. Therefore, it may be possible to introduce, by genetic engineering, the genes that encode these enzymes. Experiments to test whether such transgenic plants are indeed drought resistant are now under way.

Box 7.1

*Breeding Drought-
Resistant
Cowpeas*

One way to try to deal with drought is to try to escape it. That means, completing the entire life cycle from seedling to seed when water is available in the soil. Most cowpea varieties grown in the Sahel prior to 1980 required a growing season of 80–120 days. An analysis of rainfall patterns indicated that plants that require only 65 days to produce a crop would be more effective in the drier areas of the Sahel. Legumes such as the cowpea have evolved different strategies to cope with erratic summer rainfall. Some have an **indeterminate growth habit**: they grow vegetatively over a long period and make new flowers and pods sequentially. Such plants often recover after a period of drought, making new flowers and pods when rainfall and growth resume. Others have a **determinate growth habit**: the entire vegetative body of the plant is formed relatively quickly, and the plant then flowers and sets seed. Breeders from the University of California crossed a variety that flowers early under the hot long-day conditions prevalent in the Imperial Valley of California with a day-neutral variety from Senegal, and selected from the progeny a variety that does well under conditions of low rainfall (short rainy season).

Comparison of the yields of different varieties of cowpea under different conditions of rainfall in Senegal

| Cowpea Variety | Grain Yields (kg/ha) | | |
	452 mm Rain	315 mm Rain	135 mm Rain
Local variety	2,260	1,300	50
Senegalese parent	—	1,070	130
U.S. parent	2,350	1,350	195
New variety	2,400	1,800	200

When the rainfall was sufficient (452 mm), the new variety did not perform better than the parents or the local variety. But as the rainfall decreased and yields dropped dramatically, the new variety produced four times more grain than the local variety and 50% more than the parent from Senegal. At intermediate rainfall, the new variety outperformed both parents. It is important to note that when rainfall is low, crop production is also low, even for a drought-adapted variety.

Not surprisingly, these short-cycle, high-yielding varieties also have drawbacks. First of all, the plants are much smaller and so there is less hay for animals. Second, the early flowering varieties do well if the rainfall is evenly spaced, but they recover poorly from a period of drought once growth has begun. Third, in wet years the early-maturing varieties may be devastated by fungi that cause pod rot. As a result of the extreme variability of rainfall in semiarid Africa, farmers are advised to intercrop different cowpea varieties and intercrop cowpeas with millet. Such intercropping of species and varieties adapted to different regimes of rainfall maximizes the changes of producing a crop.

4 | **Slow weathering of rocks produces a layer of soil particles in which plants take root.**

Soil formation is a long, complex process, involving breakdown of the parent rock into small mineral particles, the chemical modification of these particles, and finally the continuous addition and decomposition of organic residues from plants, animals, and micro-organisms. Solid rock is continually being broken down into small particles, a process termed *weathering*. Weathering results from both physical and chemical forces. Even the hardest rocks can be fractured, in time, by alternative heating and cooling of the rock itself or by the freezing and thawing of water, which seeps into cracks. Other physical forces that contribute to soil formation are running water, the scouring action of winds carrying small particles, and glaciers, which creep over the rocks, grinding them into small particles. During the ice ages, this last process began forming what are now rich, agricultural soils in many areas of the world.

Physical forces break up rocks into smaller particles, but chemical forces can change the chemical properties of these particles. The more a rock is fractured by physical weathering, the faster chemical weathering will occur. The earth's crust contains more than 90 different elements, which exist in certain combinations called *minerals*. These minerals usually form small crystalline grains, which are cemented together to form rocks. Many minerals are combinations of silicon, aluminum, iron, and oxygen, because these four elements are by far the most abundant in the earth's crust. A simple granite, for example, may contain three different minerals: feldspar, mica, and quartz (see Table 7.2), which can be recognized as three different types of crystalline grains within it. During physical weathering, the granite breaks up into these small mineral grains. This process is aided by rainwater, because the different mineral grains and the cementing substances are soluble in water to different extents. Rainwater is not pure water, because it contains dissolved carbon dioxide, which forms carbonic acid. Although carbonic acid is a weak acid, it attacks the rocks more vigorously than pure water does.

Once the original minerals have been released, their proportions in the soil are greatly affected by temperature, rainfall, and vegetation. Some of them, such as quartz, are not changed, but are simply broken into smaller and smaller particles, whereas others, such as feldspar or mica, are modified. Feldspar reacts with the carbonic acid and forms soluble potassium carbonate and kaolinite, or clay mineral. Mica is even further degraded. It falls apart into potassium carbonate, oxides of iron and aluminum, and clay. This complete dissolving of certain components of the mineral particles is an important aspect of the entire weathering process. Indeed, the minerals must be dissolved before they can be taken up by plants. Once the minerals have been dissolved, they can either remain in solution, or become bound to the outside surface of the soil particles, or react with other dissolved minerals to form insoluble compounds.

Soil formation depends not only on weathering but also on the simultaneous accumulation of the soil particles. Both water and wind can carry soil particles from their site of formation to other regions. For example, some of the soils that are the most agriculturally important today were formed 10,000 years ago, when winds deposited enormous amounts of clay and silt particles in certain areas. Others were formed at the end of the most recent ice age, when streams flowing from the melting glaciers deposited the particles. When the climate became drier, winds created great dust storms, depositing

Table 7.2	Some primary minerals	
Mineral Group	**Typical Composition**	
Mica	$KH_2Al_3(SiO_4)_3$	
Feldspar	$KAlSi_3O_8$	
Quartz	SiO_2	
Iron oxide	Fe_2O_3	
Carbonate	$CaCO_3$	

the particles over large areas of the world. But just as these factors help to form soils, so they can remove them, in the process of erosion. Plants play an important role in preventing erosion, because their roots hold the soil together, a lesson painfully learned in Oklahoma and Texas during the formation of the 1930s "dust bowls." Whenever the soil is not covered by plants, winds can whip up huge dust storms. These are not "acts of God," but "acts of man" and are caused by inappropriate agronomic practices.

A layer of soil particles may form rapidly or slowly, depending on the nature of the parent rock, the weather, the vegetation, and the topography. Flat lands with a forest vegetation, a parent rock that is easily broken down, and a warm, humid climate favor rapid soil formation. High rainfall and high temperatures promote rapid chemical weathering, because the downward movement of water carries dissolved mineral components to lower layers of the soil and also because chemical reactions go faster at higher temperatures. The percolating water also leaches acids out of the decaying plant residues. The leachates from the decaying needles of conifers are more acid than leachates of leaf litter in deciduous forests, which in turn are more acid than the decay products of prairie grasses. Thus, the type of vegetation affects chemical weathering.

Percolating rainwater causes dissolved minerals, bits of organic matter, and the smallest mineral particles to be slowly redistributed in the soil. This process, called *leaching*, is one of the most important in the formation of a mature soil. Once formed, a mature soil usually shows at least three distinct layers or **horizons** (see Figure 7.5). The A horizon, or topsoil, is the layer richest in the decaying organic matter and dissolved minerals that sustain plant growth. On the average, it is no more than 20 to 30 cm thick, but human survival on Earth is intimately linked with the preservation of this very thin layer (on a 24-inch globe, the topsoil would be three-millionths of an inch thick). Below the topsoil is the B horizon, or subsoil. Minerals leached out of the A horizon accumulate here. Most of a plant's roots are in the topsoil, but good farming practice often involves breaking up the subsoil to allow the plants to find additional water and nutrients. The third layer, or C horizon, consists of parent rock in the process of being broken up.

5 | **The size distribution of the soil particles has important agronomic consequences.**

The soil particles formed by the weathering of the parent material are not uniform in size, and the size distribution of these particles and the way in which they aggregate greatly affect the water- and air-holding capacities of the soil and therefore the soil's ability to support plant life.

Figure 7.5

A Soil Profile, Showing the Different Soil Horizons.

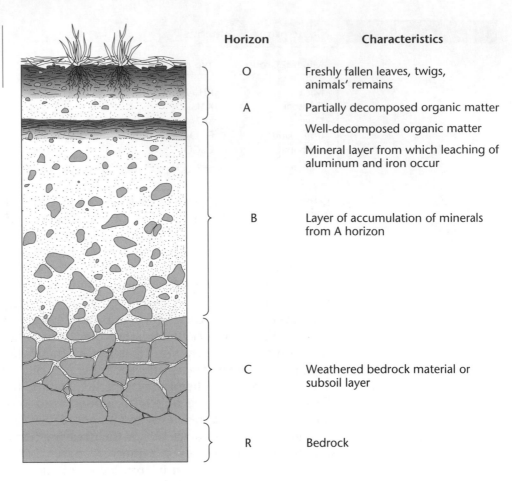

Horizon	Characteristics
O	Freshly fallen leaves, twigs, animals' remains
A	Partially decomposed organic matter
	Well-decomposed organic matter
	Mineral layer from which leaching of aluminum and iron occur
B	Layer of accumulation of minerals from A horizon
C	Weathered bedrock material or subsoil layer
R	Bedrock

Soil Texture and Structure. The soil particles are classified according to their size as sand, silt, and clay. Sand particles range in diameter from 2 to 0.02 mm, and silt particles from 0.02 to 0.002 mm; all particles smaller than 0.002 mm are considered clay particles. Soil scientists classify soils into different types according to the proportion of sand, silt, and clay particles in each (see Table 7.3). These proportions, generally called the **texture** of a soil, are determined by the nature of the parent material and by the extent to which the minerals have been weathered. Sandy soils are often formed on sandstone, limestone gives rise to loam soils, and shale results in clay-rich soils.

Sand and silt particles play an important role in the movement of air and water in soil, whereas clay particles let soil bind and store water and nutrients. Both water and nutrients are **adsorbed** on the surface of the soil particles. Obviously, the more finely divided the particles, the greater their surface area per unit mass, and the greater their capacity to bind water and nutrients. Clay particles have often been weathered so much that they become porous and thus present an even greater binding surface. Thus sandy soils, having relatively more large particles, do not bind much water or plant nutrients in comparison to clay soils, which can bind enormous amounts of water and nutrients.

The individual soil particles generally clump together to form larger aggregates of varying size and shape. This is especially common with the smaller silt and clay particles. The "glue" that sticks the particles together is a combination of lime, hydroxides of iron, and humus and other decaying

| Table 7.3 | Composition of three typical soils according to the size of the mineral particles | | |

| Soil Type | Type of Particle | | |
	Sand	Silt	Clay
Sandy loam	85%	6%	9%
Loam	59	21	20
Clay	10	22	68

organic matter. The degree of aggregation of the particles is generally referred to as the **structure** of the soil. Although soil structure is difficult to define, it is readily recognized from the size and shape of the lumps that are formed when a soil is crumbled.

An important property of an agricultural soil is its **tilth**. When a soil is tilled (plowed, leveled, raked, rolled) the original structure is partially disturbed and new air spaces are created. The new pores may occupy up to half the soil's volume. The tilth of a soil determines how well water percolates into it and whether seedlings can easily push through the surface layer. With time and as a result of gravity and rain, the tilth will change and become less favorable. In soils that are never tilled—as with no-till agriculture—the existence of continuous channels created by earthworms and by roots, that permit water percolation and gas exchange, is very important. Tilling disturbs these natural continuous channels, but creates new channels for water percolation and gas exchange.

Air and Water. The size of the individual pores in the soil is related to the size of the particles or aggregates. Small pores normally occur between small particles, whereas larger pores exist between large particles or aggregates. The small pores, called *capillaries*, are usually filled with water, whereas the larger ones are filled with air. A good, fertile soil should have half the pore space filled with water and the other half with air. Such a distribution provides a good balance between aeration, water percolation, and water storage capacity.

The importance of air for plant roots and for most other soil organisms is commonly underestimated. Plant roots need energy to grow and to take up minerals, and they obtain this energy by respiring the sugars made in the leaves. This respiration requires oxygen and causes CO_2 to be given off. To maintain the proper balance of oxygen and carbon dioxide in the soil, gases must move continuously in and out of the soil. Oxygen from the atmosphere must move into the soil, and carbon dioxide must be allowed to escape, or it will build up to toxic levels. This movement of gases requires that continuous air channels be present in the soil. Such channels are made by small burrowing animals, such as worms, or are formed when dead plant roots decay. Gas exchange between the atmosphere and the soil readily occurs when the soil has an adequate proportion of large pores as well as a system of channels.

To understand how water behaves in the soil, consider what happens to dry soil during a prolonged slow rainfall or irrigation. As soon as the water touches the soil particles and organic matter, it binds to them. The water molecules nearest the particle are held with the greatest force and cannot be dislodged from the particle. The rest of the water fills the smaller pores and percolates downward through the larger pores. This binding process is then

Table 7.4	Water-holding capacities of different soils	

Soil Type	Cm of Available Water[a] per 30 Cm of Soil
Coarse sand	1.25
Very fine sand	3.00
Very fine sandy loam	4.75
Silt loam	5.25
Silty clay	6.50
Clay	7.00

[a] Water that plants can take up from the soil.

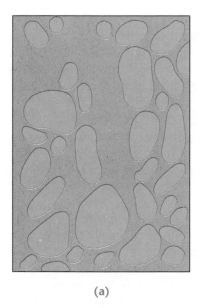

(a)

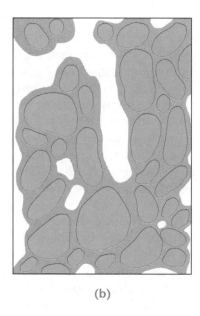

(b)

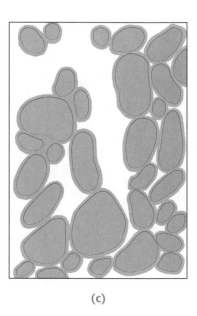

(c)

Figure 7.6

Distribution of Water and Air in a Structured Soil. In (a) the soil is waterlogged and all spaces are filled with water; in (b) the soil is at field capacity. There is air in the air spaces and water in the capillary spaces; in (c) the soil is beginning to dry out; most of the capillary water is gone and only a thin layer of water surrounds each particle. Source: *Adapted from P. M. Ray, T. A. Steeves, and S. A. Fultz (1983), Botany (Philadelphia: Saunders College Publishing), p. 183.*

repeated for the lower soil particles. A soil that becomes saturated with water in this way is said to be at **field capacity**: all the small pores are filled with water, but the large pores are filled with air (Figure 7.6). If enough water falls on the soil and if percolation is adequate, it may eventually become saturated down to the underground water table. If still more water is added, water will fill not only the small pores but also the large ones. This process starts deeper in the soil, and when this filling of air spaces reaches the top layers of the soil, it becomes waterlogged. A waterlogged soil literally suffocates most plants and soil organisms. They die from lack of air.

The texture, the structure, and the tilth of a soil together determine how well it can store water and whether the water will be able to percolate through the soil (see Table 7.4). A clay soil, with its preponderance of small particles, has many small pores and a large water storage capacity. There may be so few large pores, however, that water cannot percolate downward. Rainwater then gathers on the surface of the soil and eventually runs off, often taking the vital topsoil with it. Adding organic matter to such soils causes the formation of aggregates, which create larger pores, causing more rapid percolation. Maintaining plant cover on the land thus helps prevent erosion in three different ways: (1) the roots of the plants stabilize the topsoil, thereby keeping

Table 7.5	Soil organisms

Organism Type	Amount in fertile soil (kg/ha)
Bacteria	800
Fungi	3,300
Protozoa	330
Algae	275
Worms, insects	1,020

Note: How many organisms there are in the soil depends largely on the amount of fresh organic matter that has recently been added. Addition to the soil of manure and crop residues will quickly raise the activities and numbers of bacteria, fungi, and protozoa, if there is enough moisture and the temperature is sufficiently high.

Source: Modified from T. Brock (1966), *Principles of Microbial Ecology* (Englewood Cliffs, NJ: Prentice-Hall), p. 3.

it from being carried away; (2) the decaying plants also supply the organic matter necessary for aggregate formation, causing more rapid percolation; and (3) root penetration of the soil offers ways for water to enter after these roots die and decay.

The important role of organic matter in sustaining the soil ecosystem is discussed in much greater detail in the next chapter. The main physical function of organic matter is to bind together small soil particles into larger aggregates. Its main chemical role is to provide nutrients for plants resulting from its slow decomposition (mineralization) and to provide food for thousands of species of bacteria, fungi, protozoa, nematodes, earthworms, and arthropods that live in the soil (Table 7.5).

6 Plants require six minerals in large amounts and eight others in small amounts.

The idea that plants need to take up minerals from the soil was experimentally documented for the first time in 1699 by the British botanist John Woodward. He grew small sprigs of mint in rainwater, river water, and water to which some soil had been added. When he weighed the mint plants some time later, he concluded that growth was related to the amount of dissolved substances in the water. To understand plant nutrition and the use of fertilizers to increase crop yields, we must look at the chemical elements that are actually *required* for plant growth, and at the ways in which the soil supplies the plants with these essential nutrients.

The earth's crust is made up of over 90 different chemical elements. An analysis of plant ash, the residue that remains after plants are burned, reveals that plants may take up as many as 50 or 60 different elements from the soil. Are all these essential for growth, or does the plant take up whatever it happens to find in the soil? This question was first investigated in a systematic way in 1860 by the German plant physiologist Julius Sachs, who grew plants with their roots immersed in solutions of minerals. He found that many plants could be grown satisfactorily in solutions containing only three mineral salts: calcium nitrate, potassium phosphate, and magnesium sulfate.

Figure 7.7

Corn Plants After Several Weeks' Growth in Solutions Containing All the Essential Elements (complete) or in Media Lacking N, P, or K, Respectively. Source: *A. Glass (1989),* Mineral Nutrition *(Boston: Jones and Bartlett), p. 165.*

These salts provided the plants with six elements: calcium, potassium, magnesium, nitrogen, sulfur, and phosphorus—termed the **major nutrient elements**. If any one of these elements was omitted from the culture solution, the plants did not grow well, so Sachs concluded that these six elements were essential for the plant. Later he also found iron to be essential for growth. These experiments showed that plants did not require any organic substances (such as vitamins), a question that was hotly debated at the time. Using hydroponics it is relatively easy to demonstrate that the major nutrients are required for healthy plant growth (Figure 7.7).

Modern research suggests that the nutrient solutions used by Sachs were contaminated with small amounts of many other minerals. Advances in chemistry made it possible to obtain much purer chemicals, and it was shown later that plants also require trace amounts of seven other minerals. These seven, termed the **minor nutrient elements**, are boron, copper, chlorine, manganese, molybdenum, nickel, and zinc. Iron, already known to be essential by Sachs, is also on the list of minor nutrients. Together, these 14 nutrient elements are known as the **essential plant nutrients**. A mineral nutrient is said to be essential if without it the plant cannot complete its normal life cycle (flower and set seed). (It is not impossible that some additional mineral nutrients will be found to be essential.)

A few other elements are apparently needed by only some plants. Sodium is required by a number of plants that prefer salty environments for growth. Diatoms, a major component of oceanic phytoplankton, and some cereals, such as rice, need silicon. The micro-organisms that live in the roots of leguminous plants and fix nitrogen need cobalt. As a result, many legumes can only be successfully cultivated if the soil contains sufficient cobalt.

The other 40 or 50 elements present in plant ash are taken up by the plants from the soil even though they may not be required for growth. This somewhat indiscriminate uptake of elements may benefit the animals that eat the plants. For example, animals require sodium and iodine, two elements not needed by plants. Plants may also accumulate elements that are toxic to animals. Certain plants in the genus *Astragalus*, called *locoweeds*, accumulate the element selenium. Selenium apparently does the plants no harm, but it kills the sheep, cattle, and horses that eat the locoweeds.

Table 7.6	Plant nutrients in soil water		

Element	Cation Form in Soil Water	Element	Anion Form in Soil Water
Calcium	Ca^{2+}	Phosphorus	$H_2PO_4^-$ (phosphate)
Magnesium	Mg^{2+}	Sulfur	SO_4^{2-} (sulfate)
Potassium	K^+	Chlorine	Cl^- (chloride)
Manganese	Mn^{2+}	Boron	BO_4^- (borate)
Iron	Fe^{2+}	Molybdenum	MoO_4^{2-} (molybdate)
Copper	Cu^+	Nitrogen	NO_3^- (nitrate)
Zinc	Zn^{2+}	Nitrogen is also present as ammonia (NH_4^+)	
Nickel	Ni^{2+}		

7 | To be taken up by the roots, minerals must be dissolved in the soil solution.

Chemical weathering causes mineral particles in the soil to slowly disintegrate. Some minerals dissolve easily, others more slowly, and still others precipitate out after they have been dissolved. The 14 essential elements or nutrients just discussed must be dissolved in the soil water before they can be taken up by the plants. Indeed, essential elements that are locked up in the mineral particles of the soil are unavailable to the plants until they enter the soil solution after the particles have been weathered. Thus, the fertility of a soil depends on the extent to which the minerals have become dissolved and have remained in a usable form in the A and B horizons.

When minerals dissolve in water, they break up into smaller charged particles called **ions**. Ions with a positive electrical charge are termed *cations*; their negative counterparts are *anions*. Table 7.6 shows that the 14 essential elements occur in the soil as cations or anions. Whether an essential element is normally present in the soil as a cation or an anion is of agricultural importance, because some negatively charged ions are more easily leached out of the topsoil. To understand this, we must consider again the surface properties of the soil particles, both mineral and organic, which determine not only the soil's ability to bind water, but also its ability to bind many plant nutrients.

The mineral and organic soil particles have an overall negative charge. Because opposite charges attract, cations bind to these particles. As a result, most of the cations in a soil are more or less firmly bound to the particles, although some remain in the soil solution. The anions, in contrast, remain in the soil water. Because of their negative charge, they do not bind to the soil particles. As rainwater percolates through the soil, it carries anions and cations with it down to the groundwater. As noted earlier, this process is called *leaching*. Since the anions are not bound to the soil particles, they are more easily lost by leaching than the cations, which are replenished as soon as they are removed. Indeed, as soon as cations are removed from the soil solution, either by leaching or because they are taken up by plants, they are replaced by others that are released from the particles. Thus there is an equilibrium

Figure 7.8

Equilibrium Between Cations in the Soil Solution and Those Bound to the Surface of Clay Particles. **(a)** *Removal of nutrients by the roots from the soil solution causes more nutrients to leave the surface of the particles.* **(b)** *Addition of nutrients, in the form of fertilizer or from decay of organic matter, causes more nutrients to be bound to the clay particles.*

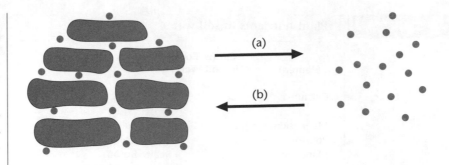

between the cations in the soil water and those bound to the particles, as illustrated in Figure 7.8.

The total amount of available plant nutrients bound to soil particles is called the soil's **exchange capacity**, a parameter that determines soil fertility. Because the surface area available for binding nutrients depends on soil texture, a soil's texture greatly influences its exchange capacity and its fertility. The greater the surface area of the particles, as in clay soils, the more plant nutrients they can bind. As water percolates down through the soil, it often carries nutrients down with it, especially the unbound anions, such as nitrate. In the groundwater, these nitrates may be quite abundant, and are dangerous if the groundwater is tapped for human drinking. As a result of heavy fertilization with nitrogen fertilizers, the groundwater in some agricultural areas is not fit for human consumption, because it is contaminated with nitrate.

8 | **The acidity and alkalinity of a soil help determine its ability to produce crops.**

An important property of soil is its degree of acidity, which influences its physical properties, the availability of certain plant nutrients, and the biological activity in the soil. As a result, acidity strongly influences plant growth. A soil's degree of acidity depends on the concentration of hydrogen ions (which have a positive charge) dissolved in the soil water. In a neutral soil, the hydrogen ion concentration is about 1 part per 10 million parts of water. An acid soil may have a concentration of hydrogen ions that is 100 to 1,000 times higher, while an alkaline soil has a lower hydrogen ion concentration. Scientists express the acidity and alkalinity of a solution by a single number called the pH. A pH of 7 is neutral, while a pH of 5.0 (100 times more hydrogen ions) is acidic, and a pH of 9 (100 times fewer hydrogen ions) is basic. Neither extreme acidity nor extreme alkalinity are suitable for the growth of plants and most other organisms. In addition, these conditions upset soil weathering and the availability of the nutrients. Although some plants can grow in heavily acidic or alkaline soils, most crop plants grow best in neutral or slightly acidic soils.

The acidity of the soil can change as a result of the type of vegetation (some plants produce more acids than others, especially when the organic matter decomposes), as a result of the types of fertilizers that are used (long-term use of superphosphate and especially of soluble nitrogen fertilizers cause acidification), and as a result of rainfall that is loaded with acids. The pH of rain is normally just below neutral, but polluting gases from power plants and

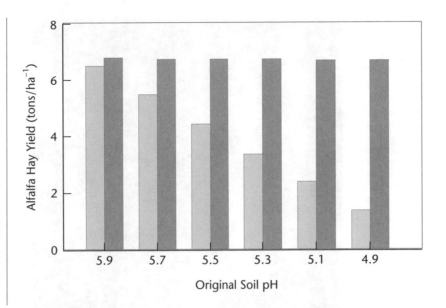

Figure 7.9

Yields of Alfalfa as a Function of Original Soil pH, With or Without Lime. Notice that as the original pH drops from left to right, the yield of alfalfa also drops dramatically (gray bars). When lime is added to the soil to neutralize the acidity (low pH), all the yields are the same (black bars). The final pH after adding lime is not shown. Source: *Agriculture Canada (1974), Publication no. 1521 (Ottawa: Agriculture Canada).*

cars (usually oxides of nitrogen and sulfur) create strong acids (nitrous acid and sulfurous acid) when they dissolve in the rain drops. When this acid rain falls on the land, it lowers the pH of the soil. The increased acidity results in the release of aluminum ions from the soil minerals, which can lead to aluminum toxicity. The direct effects of air pollutants on plants and the indirect effects caused by acid rain and aluminum toxicity are the major cause of decline in forests in large areas of central Europe and North America.

The minerals that are dissolved by chemical weathering sometimes interact with each other to form new, insoluble complexes. A soil's degree of acidity plays an important role in this process, and can greatly reduce the availability of certain nutrients to the plants. For example, phosphate, an essential plant nutrient, can form a variety of insoluble complexes with other ions in the soil solution. If the soil solution is too alkaline, phosphate readily combines with calcium to form insoluble calcium phosphate. When the soil solution is too acidic, phosphate combines with iron, aluminum, and manganese to form insoluble products. Thus phosphate is most readily available to plants when the soil solution is neutral or slightly acidic.

A soil's degree of acidity can be adjusted to make it more nearly neutral. Lime neutralizes excess acid in the soil, and acidic fertilizers, such as ammonium sulfate, neutralize excess alkalinity. Farmers may also use such treatments to optimize the growth of certain crops. For example, potatoes grow best in a somewhat acidic soil, whereas alfalfa thrives in a soil that is very slightly alkaline, but grows very poorly on acidic soil (Figure 7.9).

9   In plants, as in animals, minerals have specific functions.

To take up nutrients from the soil solution, most plants must develop an extensive root system. The root system provides the plant with the large surface area it needs to exploit a large volume of soil. In many plants, this surface area is enormously increased by the presence of root hairs close to the tips of the roots (Figure 7.10). The ions absorbed from the soil by the root come in contact with the root surface, either because they have moved

Figure 7.10

*Cross Section of a Root
Showing that
Roothairs Enormously
Increase the Surface
Area of the Root.*
Source: *U.S. Department
of Agriculture.*

through the soil with the soil water, or because the root has grown into a previously unexploited area of soil. The rapid uptake of nutrients required by most crop plants for maximal yield necessitates continuous growth by the root system. Much evidence indicates that the uptake from any one region of the soil has nearly ceased five or six days after it was first penetrated by a root. When growing normally, different species of plants have very different root systems. Some have shallow roots that can quickly take advantage of a light rain; others are very deep rooted and may tap into the groundwater supplies (Figure 7.11). If the soil is poor in mineral nutrients, the plant will develop a much more extensive root system than when the soil is nutrient rich. In a hydroponic system with the nutrient film technique (see Chapter 3), the root system of the plants remains very small, because a nutrient-rich solution is constantly flowing past the roots. Thus the roots do not need to develop. This is advantageous, because it means that all the products of photosynthesis can be used for growth of the harvested crop (such as lettuce or tomatoes).

Minerals do not pass into the outer root cells by passive diffusion, because each ion is surrounded by water molecules and such a complex cannot pass through the lipid-rich plasma membrane. Special proteins are engaged in ion transport, forming either *channels* through which many ions pass rapidly, or *pumps* that transfer one ion at a time to the other side of the membrane. This transport process requires energy, which is initially supplied as ATP. The ATP maintains the plasma membrane in an "energized" state with a much greater acidity on the outside than the inside of the cell, and a much greater concentration of negative charges on the inside than the outside of the cell. These gradients of charge and protons (acidity) provide the necessary energy for mineral ions to move across the membrane. After the mineral nutrients are inside the root cells, they pass from cell to cell until they reach the conductive xylem tissues in the center of the root. Once in the xylem, dissolved minerals can be transported throughout the plant.

Figure 7.11

Root System of a Soy-bean Plant. The cut-off stem is at the top (left of center). Notice the extensive and highly branched root system. Source: *U.S. Department of Agriculture.*

Movement of nutrients within the plant depends on the rate of water movement and on the plant's supply of nutrients. If the plant is poorly supplied with nutrients, more of the entering nutrients will be retained in the root system. Water movement greatly influences nutrient movement within the plant, and the organs with the greatest rate of transpiration, the fully expanded leaves, receive most of the nutrients because they receive most of the water. However, nutrients are most needed by the growing parts of the plant, and consequently must be redirected from the mature leaves to the growing ones.

In many crop plants, seed maturation occurs at the same time as the aging of the vegetative body of the plant: reproduction and death are linked in time. Root growth and nutrient uptake slow down at the onset of seed maturation. In bean plants, for example, the uptake of phosphate from the soil is not continuous throughout the life cycle of the plant, but levels off once seed maturation begins. The accumulation of phosphate in the developing seeds occurs at the expense of the phosphate already present in the leaves and stem. Senescence and death of these vegetative organs are accompanied by a decline in their phosphate content. There are similar redistribution patterns for many other nutrients. This redistribution also involves a breakdown of the larger organic molecules (such as proteins) and the transport of smaller organic molecules (such as amino acids) to the maturing seeds.

The various nutrients have specific roles in plants that are often similar to their roles in animals. For example, both plants and animals require nitrogen and sulfur to synthesize amino acids, and they need phosphorus to make phospholipids and nucleic acids. Many minor nutrients, needed only in trace amounts, serve to help enzymes carry out biochemical reactions. Nutrients that have unique roles in plants include calcium, which participates in maintaining the integrity and rigidity of the cell wall, and magnesium, which is part of the chlorophyll molecule.

Nutrients also play an important role in maintaining the turgidity of the plant's cells and consequently of the organs as a whole. Plants actively take up ions from the soil, with the result that the total concentration of ions inside the root is greater than that outside. As a result of this difference in concentration, water moves from the soil solution into the roots, and from there to the rest of the plant. This difference in the concentration of minerals between the root cells and the soil drives the uptake of water by the roots. The influx of water causes the cells to swell, and creates pressure within each cell. The water pressure gives rigidity to cells and plant organs that lack strong cell walls. When water loss by evaporation exceeds water uptake, the pressure inside the cells drops. As a result, plant organs that lack strong cell walls (leaves, in many plants) do not retain their shape but collapse, and the plant wilts. Water pressure also drives cell enlargement in the growing zones of the plant, thus playing a key role in plant growth.

10 | Mineral deficiencies are prevented by applying organic or inorganic fertilizers.

If the soil is deficient in even one plant nutrient, plant growth will be retarded and crop yield may be diminished. If the deficiency is severe, the plants will develop the visible symptoms of a **deficiency disease**. Such diseases are in many ways similar to those caused by the lack of vitamins or minerals in humans. Each disease is diagnostic of the element that is deficient. Farm advisers are familiar with the symptoms of the common deficiency diseases, and can recognize them when the plants are still young. Treatment usually involves fertilizing the soil with the deficient nutrient. Unless this is done at an early stage in plant growth, severe crop damage may occur.

The appearance of symptoms such as stunted growth, yellowing of the leaves, "burned" leaf margins, or death of the terminal bud usually indicates a rather severe deficiency in one or more plant nutrients. Plants do not develop these signs of deficiency when the nutrients are only marginally deficient. However, such marginal deficiencies may also cause a loss of crop yield. This "hidden hunger" is difficult to diagnose. It can sometimes be uncovered by measuring the amounts of plant nutrients in the soil or in the plant sap, although neither method is completely satisfactory.

Although the total amounts of particular nutrients present in the soil can be measured, it is not possible to discover by chemical tests what proportion of the nutrients is available to the plants. Farmers, when deciding on a fertilizer program, are primarily interested in available nutrients. Availability can only be measured by experiments with plants. In spite of this problem, much progress has been made in developing soil tests that reflect the fertility of the soil. Agricultural scientists use the results of such tests to make recommendations to farmers about the use of fertilizers. The development of soil tests for

Table 7.7	Nutrients contained in the total above-ground plant material in a hectare of corn yielding 10,000 kg of grain		
Nutrient	**Kg/ha**	**Nutrient**	**Kg/ha**
Nitrogen	200	Iron	2.3
Phosphorus	42	Manganese	0.4
Potassium	205	Copper	0.1
Calcium	41	Zinc	0.42
Magnesium	48	Boron	0.19
Sulfur	24	Molybdenum	0.01
Chlorine	86.0		

Source: S. Barber and R. Olson (1968), *Changing Patterns of Fertilizer Use* (Madison, WI: Soil Science of America), by permission of the American Society of Agronomy.

available nutrients involves first of all, measuring how the yield of crop plants responds to the addition of fertilizers on a specific soil. After the yield responses are known, the plants are analyzed to measure how much of the nutrients they took up from the soil. This lets the soil chemists develop analytical methods that accurately measure the available nutrients.

Soils can be deficient in plant nutrients for a variety of reasons. Harvesting crops removes large amounts of the major nutrients (see Table 7.7). These must be restored if the fertility of the soil is to be maintained. Failure to restore the nutrients leads to a gradual decline in soil fertility and crop productivity. This is adequately demonstrated by the low yields obtained on experimental plots that have not been fertilized for many years. Soils may look deficient in certain nutrients because the nutrients are unavailable to the plants, as can happen when the soil is either too acid or too alkaline. Highly weathered soils in tropical and semitropical areas can often be deficient in certain nutrients because leaching has removed them from the plant root zone. The fact that such soils support a luxuriant vegetation does not necessarily indicate that they contain large amounts of available plant nutrients. Rather, the luxuriant vegetation is maintained by an extremely efficient recycling of nutrients from the dead organic matter (the litter on the forest floor) back to the plants. Sandy soils are often deficient in certain nutrients, because they contain too small a proportion of clay particles and are low in organic matter. As a result, they have a small exchange capacity, and thus a low fertility. Finally, soils may be deficient in nutrients because the parent material from which they developed contained only small amounts of certain nutrients, as sometimes happens for certain trace elements.

There are many ways to increase or restore the fertility of the soil so that it will support maximal crop production. One of the oldest methods is to allow the land to *lie fallow* for a long time. If no crops are grown and no plant nutrients are removed, the natural processes of soil weathering will slowly restore the fertility of the soil. The system of shifting agriculture traditionally practiced in the tropics was based on this principle. A second widely practiced method is to *incorporate organic residues* and wastes into the soil. A survey of agricultural practices around the world reveals that all kinds of organic residues are incorporated into the soil: manure, fish wastes, algae, human excrement, crop residues, sawdust, composted kitchen scraps, and many others. These organic fertilizers are decomposed in the soil, and their nutrients are released. Organic fertilizers have the advantage that the plants can obtain a steady supply of nutrients, but have also the disadvantage that they may not release sufficient nutrients during the period of rapid vegetative growth, when demand is greatest. Slower growth in the early summer often means a reduced crop in the fall. It is a popular misconception that farmers in technologically advanced countries underestimate the value of organic fertilizers. The efforts made by most farmers to reincorporate the crop residues (leaves and stems in all seed crops) into the soil proves the opposite. Still, much more could be done to return animal manures to the field and to compost the organic portion of household garbage, so that the plant nutrients contained in it could be used again by other plants.

A third method to restore the fertility of the soil is to *add inorganic fertilizers* (also called "chemical" or "synthetic" fertilizers). Chemical fertilizers offer the advantage that the nutrients are rapidly released and can be made available when the plant has the greatest need for them. A disadvantage is that some plant nutrients, especially nitrate, may be leached out of the root zone before the plants can use them. This problem is greatest if the fertilizers are applied just before or at the time of planting. The seedlings will take several weeks to develop a root system large enough to take full advantage of the fertilizer.

The promotion of crop production by using inorganic fertilizers is governed by two principles: the law of the minimum and the law of diminishing returns. The **law of the minimum** was first formulated in the nineteenth century by Justus von Liebig, a German chemist who realized that plant growth is limited by the one nutrient that is in shortest supply in the soil. Adding this nutrient will increase plant growth until some other nutrient becomes the limiting factor. However, plants differ in their nutrient requirements. Thus the addition of phosphate to a soil is only beneficial to those crops for which phosphate is the limiting nutrient in that soil. If some other factor, such as water, light, or disease, limits the growth of the plant, then fertilizers will have no effect at all.

The second principle that governs the effect of fertilizers was first formulated by E. A. Mitscherlich, another German scientist, who was concerned with the response of crop plants to added fertilizers. This principle, the **law of diminishing returns**, says that the amount of yield increase is at first proportional to the amount of fertilizer increase, but above a certain threshold the increment in yield is always less for a given increment in fertilizer, until a plateau is reached. As shown in Figure 7.12, with each increment of nitrogen (N) fertilizer (of 25 kg of nitrogen) there is a smaller increment in the yield of corn. The yield keeps going up, but the increase is less, each time. For example, increasing fertilizer from 100 to 125 kg of N per hectare raises the corn yield by 832 kg per ha. However, increasing fertilizer from 200 to 225 kg per ha raises the corn yield only 128 kg per ha.

The law of diminishing returns also applies in economic terms. Does the price received for the last increment in yield compensate for the cost of the last increment in fertilizer? At first glance, this would seem simple to calculate. However, excessive fertilizer usage may have detrimental effects whose costs are difficult to calculate: more contamination of the groundwater and

Figure 7.12

Response of Corn Grown in Central Indiana (United States) to Nitrogen Fertilizer. The numbers in the staircase refer to the increment in yield (kg/ha) for each increment of 25 kg/ha of nitrogen fertilizer (measured as nitrogen). Source: *J. T. Pierce (1990),* The Food Resource *(New York: Halsted Press), p. 271.*

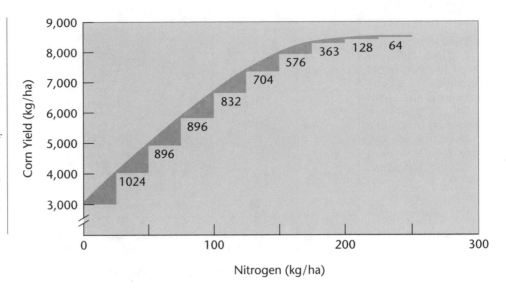

stress, less disease resistance so that more pesticides need to be used, greater energy costs to harvest, transport, and dry the seeds. The price support system for agricultural commodities in place in the United States and other advanced countries promotes farming practices near the plateau of the curve of the law of diminishing returns, without taking into account the environmental costs associated with this type of farming.

Inorganic fertilizer can be broadcast on the land or placed in narrow bands underneath or alongside rows of seeds. Fertilizers must be applied carefully; if the concentration of nutrient in the immediate environs of the root is too high, the plant will be unable to take up water, and it will wilt, dry out, and die. Although inorganic fertilizers are usually applied to the soil to be taken up through the roots, the minor nutrient elements are sometimes applied directly to the leaves as a spray. Weak solutions of iron, zinc, or copper are commonly used on crops such as pineapple, citrus fruit, or avocado. Such sprays can be used as an emergency treatment if the crops develop certain deficiencies on particular soils. For example, in certain parts of Hawaii that have iron deficient soils, pineapple plants are routinely sprayed with a solution of iron sulfate to prevent iron deficiency.

The three plant nutrients used most widely in inorganic fertilizers are nitrogen, phosphorus, and potassium. Large amounts of these elements are removed from the soil when crops are harvested, and they are therefore commonly deficient in agricultural soils. Adding fertilizer to the soil tends to stimulate the overall growth of the plant and hence the production of the crop. However, all crop plants do not respond in the same way to the addition of a given amount of fertilizer to a certain soil. Figure 7.13 shows how the yields of three different crops responded to the addition of phosphorus fertilizer to a particular soil. Even within a given species there can be much variation from strain to strain. Many of the new high-yielding strains of corn, wheat, and rice respond to added fertilizer with increased grain production, whereas the older strains do not. Increased grain production can be accounted for, in part, by more efficient nitrogen use and, in part, by a different plant architecture that tolerates more crowding in the field and can carry a larger seed weight.

Figure 7.13

Yield Responses of Three Different Crops to Phosphorus Fertilizers on a Low-Phosphorus Silt Loam Soil. Different crops respond differently to a particular fertilizer. Source: *U.S. Department of Agriculture.*

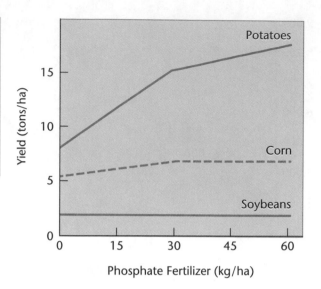

11 | Cereals and legumes have different ways of acquiring nitrogen.

Nitrogen is perhaps the single most important plant nutrient for agriculture. It is, of course, no more "essential" for plant growth than any of the other essential nutrients; but because large amounts of nitrogen are removed from the soil when crops are harvested (see Table 7.7), most crop plants respond to the addition of nitrogen fertilizers by increasing their growth and crop production. Furthermore, nitrogen is an important constituent of protein (each amino acid contains at least one nitrogen atom), so the plants must have a supply of nitrogen to make not only their enzymes, but also the seed proteins that are such an important part of the human diet.

In the soil, most of the nitrogen is present in organic matter, from which it is slowly released as ammonia. Once released, ammonia is rapidly converted to nitrate by soil bacteria, and in this form it is normally taken up by most plants. In the plant, the nitrate must again be transformed into ammonia before it can be used to synthesize amino acids. To synthesize amino acids, the plant uses various molecules made from the glucose derived from photosynthesis, and combines them with the ammonia.

Research by R. Hageman at the University of Illinois has shown that the conversion of nitrate to ammonia is sometimes the limiting process in plant growth and crop production in cereal grains such as corn and wheat. Strains of corn and wheat that are particularly efficient at converting nitrate to ammonia produce large yields and have a high proportion of protein in the grain. The best strains are those that can also reuse the largest proportion of the proteins in the aging leaves and stalks, and transport the amino acids to the developing seeds. These findings led to the selection of strains of corn that are very responsive to large doses of added nitrogen fertilizers, and the high yields of corn and wheat obtained in recent years depend on high doses of chemical nitrogen fertilizers. The selection of these high-yielding strains of corn and wheat has always been carried out in monocultures with high levels of fertilizers and pest control chemicals. Almost no plant-

Figure 7.14

Nodules in the Roots of a Legume. Source: *U.S. Department of Agriculture.*

breeding work has been done to select strains of corn that perform well in crop rotation programs without added chemical fertilizers and with minimal pest control.

As noted earlier, soybeans and other legumes have, in addition to the ability to take up nitrate from the soil, an entirely different method of obtaining nitrogen. Their root systems are characterized by the presence of numerous small nodules (see Figure 7.14), which are filled with bacteria that maintain a symbiotic relationship with the plants. (**Symbiosis** means "living together for mutual benefit.") The plants supply the bacteria with a source of food energy, by providing them with photosynthetic products. The bacteria have the unique ability to convert atmospheric nitrogen (N_2) into ammonia, by the process of nitrogen fixation (see Chapter 8). The products of photosynthesis move from the leaves to the root nodules, and amino acids that are ready to be used by the plants move from the root nodules back to the leaves. Under agricultural conditions, soybeans and peanuts derive about a third of the nitrogen in their protein from the nitrogen-fixing activities of the bacteria in their roots. The rest of the nitrogen is taken up from the soil as nitrate.

In natural ecosystems, legumes derive a much greater proportion of their nitrogen from symbiotic fixation. Many soils are nitrogen poor, and it is precisely because legumes can fix nitrogen symbiotically that they compete well on such nitrogen-poor soils.

One important characteristic of soybeans and other legumes is that they respond less to the addition of nitrogen fertilizers to the soil than other crops, especially cereals. If the addition of nitrogen fertilizers does not result in bigger plants or greater yields, it is because the nitrogen-fixing bacteria fix less and less nitrogen as the nitrate level in the soil rises. In Chapter 6, we

Figure 7.15

Corn Yields in Kilograms per Hectare for "Continuous Corn" or "Corn After Soybeans" with Different Levels of Nitrogen Fertilizer. The rotation of crops is clearly superior to the continuous culture. The effect of soybeans on the yield of corn is equivalent to applying about 150 kg of nitrogen per hectare. However, this effect of crop rotation cannot be attributed solely to the capacity of soybeans to fix nitrogen. Crop rotations have many other beneficial effects, especially on the suppression of pests and pathogens.

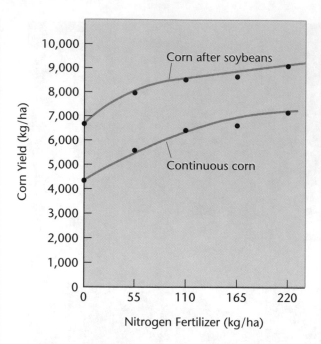

described the research of R. Hardy and his co-workers, showing that increased photosynthesis led to increases in soybean protein (see Table 6.1). This occurred because the nitrogen-fixing bacteria were more active when supplied with large amounts of photosynthate. Thus the key to increased crop production in soybeans may be the breeding of strains that are more efficient in photosynthesis.

From an agronomic point of view, relying on nitrogen fixation may have drawbacks as well as advantages. The nodules that contain the bacteria take several weeks to develop, and the plants still depend on soil nitrate when they are young. This is a problem in the spring when temperatures are low and mineralization of soil organic matter is slow. However, if available soil nitrogen is low, better nodules are formed and the plants quickly catch up once the nodules become active. So the lack of nitrogen fixation by young plants is not really a problem. As the plants get older, the nodules age and the bacteria stop fixing nitrogen when the seeds have set and the main period of seed development is about to begin. During this period, the plant must rely on the proteins already synthesized in the leaves. These proteins are broken down again, and the resulting amino acids are transported to the developing seeds. On the plus side, legumes are particularly well suited for crop rotations and intercropping. Precisely because they can be grown without added fertilizer and actually add nitrogen to the ecosystem, they are extremely valuable in all sustainable agricultural systems. Prior to the manufacture of relatively cheap chemical nitrogen fertilizers, the planting of legumes was a widely recommended procedure to improve soil fertility. Seed legumes such as soybeans or cowpeas, or perennial forages such as alfalfa or clover can be used for this purpose. It is remarkable that maximal yields of specific cereal crops such as corn and sorghum are not as high in continuous culture with the highest level of nitrogen fertilizer, as when the cereal is grown after soybeans *without* nitrogen fertilizer (see Figure 7.15).

12 | **Many different organisms are involved in recycling nitrogen in the biosphere.**

All but one of the inorganic nutrients that plants obtain from the soil are made available to the plants by the weathering of soil particles. The one exception is nitrogen, which is normally taken up by the plants from the soil solution as nitrate (NO_3^-). Soil particles have no nitrogen-containing minerals. The nitrate dissolved in the soil water originally came from nitrogen gas (N_2), which makes up almost 80 percent of the earth's atmosphere. Because green plants cannot use this atmospheric nitrogen directly, they depend on micro-organisms to transform it into a form usable by plants, principally by means of nitrogen fixation. Once nitrogen is in a form where it is combined with hydrogen (as ammonia) or oxygen (as nitrate), it can be used by the plants for synthesizing proteins and other macromolecules. The macromolecules can be eaten by animals, but eventually the proteins return to the soil as dead organic matter that is then degraded by decomposers. In addition, nitrate can be converted to N_2 by yet other micro-organisms, and released again into the atmosphere. Together, these various transformations of nitrogen make up the nitrogen cycle, as illustrated in Figure 7.16.

Nitrogen normally enters the living world through the process of **biological nitrogen fixation**, which converts nitrogen gas to ammonia. Only certain bacteria, blue-green algae, and fungi have the necessary enzymes to fix nitrogen, and they fix an estimated 100 million tons a year. Some of the nitrogen-fixing organisms live in lakes and rivers (algae), and others live in the soil (fungi and bacteria). For agriculture, the most important group of nitrogen fixers are those that live symbiotically in the roots of certain plants, especially the legumes. Natural systems such as forests and grasslands rely heavily on the sustained activity of the nitrogen-fixing organisms in the soil to supply the ecosystem with usable nitrogen.

The nitrogen cycle includes three other significant processes involving soil organisms: ammonification, nitrification, and denitrification. The first two are of special importance, because they make nitrogen available to plants, such as corn or wheat, that have no symbiotic nitrogen-fixing bacteria in their root systems. The organic matter in soil has much nitrogen that is not immediately available to plants, because the large molecules that contain nitrogen cannot enter plant roots. These large molecules are broken down by soil micro-organisms, the decomposers, in a process called **ammonification**. The ammonia released into the soil is usually modified before it enters the plants. During the process of **nitrification**, certain bacteria use the ammonia and secrete nitrite (NO_2^-) into the soil; other bacteria use this nitrite and release nitrate (NO_3^-). Nitrogen is usually taken up by plant roots as nitrate. In a fertile soil, nitrification is much faster than ammonification, and if the organic matter content of the soil is high and the conditions for microbial activity are good, the plants will be supplied with a slow but steady stream of nitrate.

Denitrification completes the nitrogen cycle by releasing nitrogen gas back into the atmosphere. Certain soil bacteria use nitrate and release nitrogen gas into the air. They are actually using nitrate as an oxygen source, and so thrive in oxygen-free environments where plants and the

Figure 7.16

*A **Simplified Version of the Nitrogen Cycle**. The source of all nitrogen in the biosphere is the N_2 in the atmosphere. This N_2 is fixed by free-living N-fixers or symbiotic N-fixers. Plants, such as corn (shown on the right) depend on nitrate that is generated by soil bacteria; animals depend on plants.*

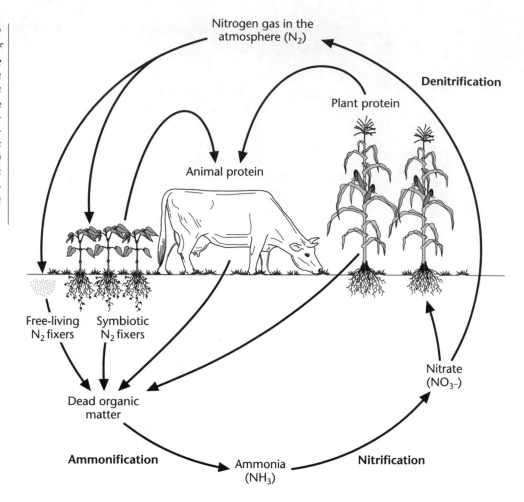

other micro-organisms involved in the nitrogen cycle cannot grow. Denitrification is quite rapid in soils that have few air spaces or that are waterlogged. In heavy rains, much precious nitrogen is lost to the plants by leaching of nitrate into the groundwater and by denitrification if the soil gets waterlogged. On a worldwide scale, nitrogen fixation and denitrification balance each other out, keeping a constant amount of fixed nitrogen in the biosphere.

The nitrogen fertilizers discussed throughout this chapter all contain nitrogen that has been combined with either oxygen or hydrogen by industrial processes. Until 1915, the world's principal source of fixed nitrogen to make nitrogen-rich inorganic fertilizers was the deposits of guano, fossilized bird feces, found along the coast of South America. Then, during World War I, two German chemists discovered a way to make ammonia by combining nitrogen and hydrogen. The ammonia was converted to nitrate, which was used in the manufacture of explosives. This process, called the Haber-Bosch process, is still the basis for the manufacture of most nitrogen fertilizers. Ammonium salts, nitrates, urea, and ammonia are the most widely used forms of nitrogen fertilizers.

13 | Poor agronomic practices result in soil degradation and reduced crop yields.

Poor management of our natural resources—deforestation and misuse of agricultural lands—has led to extensive soil degradation all over the world. **Soil degradation** is the decline in soil quality caused through its use by humans. It includes physical, chemical, and biological deterioration, such as decline in soil fertility, decline in structural condition, erosion, adverse changes in salinity, acidity or alkalinity, and the effects of toxic chemicals, pollutants or excessive inundation. Soil degradation results in a diminution of the soil's current or potential capacity to produce crops. Soil depletion caused by the removal of nutrients by water, or by cropping and removing harvested produce, is a less severe problem, but can be the beginning of soil degradation. Soil degradation takes many forms: the erosion (removal) of the topsoil by wind or water; compacting as a result of traffic; alteration of the soil properties caused by intensive cropping; laterization as a result of deforestation and an increase in the temperature of tropical soils; decline in the organic matter and the biological activity of the soil; increase in soil salinity or alkalinity caused by irrigation.

Although land degradation is widespread, it is extremely difficult to predict either the short-term or long-term impacts on food production. For example, although soil erosion is a problem in the United States, with about 5 billion tons of topsoil being lost annually (about 18 tons per hectare per year for arable land with seasonal crops), productivity in the United States has nevertheless increased steadily. These increases have come about as a result of greater applications of fertilizers, and some have argued that present agronomic practices allow U.S. farmers to ignore the steady decline in soil fertility, resulting from soil erosion. If the present rate of soil erosion continues, when will the United States have to start paying the price for these poor agronomic practices? Good agronomic practices such as contour plowing (Figure 7.17) and reduced tillage can minimize, but not eliminate soil erosion. Experiments in Iowa have shown that a specialized type of cultivation called *ridge tillage* can reduce water runoff to about 40% and soil loss to 10% of that observed with conventional tilling practices.

In tropical regions with abundant seasonal rainfall, erosion rates of 50–100 tons per hectare per year are not uncommon and often affect 30–50% of a country's arable land. Soil erosion generally does not greatly reduce yields on deep soils, especially if fertilizers are used to compensate for the loss of nutrients, but can be a real problem on shallow soils. An example for a shallow tropical soil in southern Nigeria is shown in Table 7.8. The loss of 20 cm of topsoil caused a massive decline in cassava yield, even with supplemental fertilizer.

As noted in Chapter 2, a different type of soil degradation called **desertification** occurs in many arid and semiarid regions. Some 50 years ago, observers noted gradual changes in the vegetation of the Sahel, a vegetation zone in West Africa, bordered to the north by the Sahara and to the south by dry savannas. The changes were described as the advance of the Sahara, because along its northern edge the Sahel was becoming like the Sahara, and along its southern edge the vegetation that characterized the Sahel was moving into the dry

Figure 7.17

Contour Plowing Helps Prevent Soil Erosion.
Source: *U.S. Department of Agriculture.*

savanna. The changes in vegetation were caused by a combination of climatic changes and poor land management resulting from population pressures. Overgrazing brought about irreversible changes in the land because it was denuded of its protective cover.

Overgrazing and the resulting desertification are not confined to the Sahel. In other semiarid regions such as Patagonia, Iran, Syria, India, and Kenya, desertification has also occurred, often because pastoralists kept three to five times more grazing animals than the meager vegetation of the area could support.

Earlier in this chapter, we described the accumulation of salts as a result of irrigation with water that contains dissolved salts (all groundwater and river water contains dissolved salts, but rainwater does not). **Salinization**, another form of land degradation, occurs in areas where the mean annual potential evapotranspiration exceeds the precipitation by a factor of 1.3. The source of the salts that accumulate can be irrigation water, but salts can also come to

Table 7.8	Effect of soil removal depth on cassava yield from a tropical alfisol	
Depth of Soil Removed (cm)	**Cassava Tuber Yield (t/ha)**	
	With Fertilizer	**Without Fertilizer**
Control	36.0	39.5
10	21.4	12.7
20	17.1	7.8

Source: Unpublished data of R. Lal (1989), cited in D. Pimentel and C. W. Hall, eds., *Food and Natural Resources* (New York: Academic Press), p. 103.

the surface from groundwater, or from the weathering of soil particles. They accumulate because there is not enough precipitation to carry them out of the topsoil into the groundwater and/or the rivers. If water is abundantly available, the salts can sometimes be leached out of a salinized soil and be carried away into streams. However, this often creates problems downstream.

Summary

Plants need carbon dioxide and sunlight, as well as water and minerals for growth. The latter two are supplied by the soil. If the plant is to take up CO_2 for photosynthesis, then the stomates must be open. As a result, water will move through the plant in a continuous transpiration stream. During evolution, plants have become adapted to certain rainfall regimes, and not all our crops require the same amounts of water to produce a given amount of plant material. The old solution to the problem of "not enough water" was to irrigate. A new possible solution is to try to understand which genes are involved in drought adaptation and to try to breed drought-resistant plants. Continuous irrigation of land with river water may cause salinization (buildup of salts), which is one aspect of the general soil degradation now occurring all over the world as a result of poor soil management practices.

Minerals become available to plants when the mineral particles derived from rocks slowly weather and dissolve. Some minerals are needed in large amounts, whereas others are needed in very small amounts. Harvesting removes substantial amounts of minerals from the agricultural ecosystem, and these must be replaced by adding either chemical fertilizers or organic material.

Plant growth depends also on the structure of the soil: the clumping of small clay particles into larger aggregates. Structure is important because it determines the availability of air and water to the roots. A good soil has particles of different sizes, with the smallest particles clumped in aggregates. The soil organic matter plays an important role in this clumping. The slow decomposition and mineralization of this organic matter releases minerals that become available to the plant. For this reason, it is important to maintain the level of organic matter in the soil. A slow, but continuous decline in this level seems to be the unavoidable consequence of continuous cultivation.

Nitrogen, the chemical element that is most important to crop growth, is a special case. It is not found in the mineral soil particles, but enters the soil ecosystem from air, via the route of nitrogen fixation. Nitrogen fixation is carried out primarily by bacteria that live symbiotically with legumes. The decay of organic matter supplies nitrogen in the form of ammonia or nitrate to plants that do not have symbiotic nitrogen fixers.

Further Reading

Alexander, M. 1977. *Introduction to Soil Microbiology.* Melbourne, FL: Krieger.

Baker, F. W. G., ed. 1989. *Drought Resistance in Cereals.* London: CAB International.

Bould, O. C., and E. J. Hewitt. 1983. *Diagnosis of Mineral Disorders.* London: Her Majesty's Stationery Office.

Follett, R. F., and B. A. Stewart. 1985. *Soil Erosion and Crop Productivity.* Madison, WI: American Society of Agronomy.

Gordon Press. 1991. *Soil Erosion: A Source Guide.* New York: Gordon Press.

Lai, R., and P. Sanchez, eds. 1992. *Myths and Science of Soils of the Tropics.* Madison, WI: Soil Science Society of America.

Peskin, H. M. 1986. Cropland sources of water pollution. *Environment* 28:30–36.

Sprent, J. I. 1987. *The Ecology of the Nitrogen Cycle.* Cambridge, UK: Cambridge University Press.

Tate, R. 1992. *Soil Organic Matter: Biological and Ecological Effects.* Melbourne, FL: Krieger.

Tisdale, S., W. Nelson, J. Havlin, and J. Beaton. 1992. *Soil Fertility and Fertilizers.* New York: Macmillan.

Troeh, F., J. Hobbs, and R. Donohue. 1991. *Soil and Water Conservation.* Englewood Cliffs, NJ: Prentice Hall.

Chapter 8

Life Together in the Underground

People usually think of roots as having three major functions: they anchor plants to the ground, they take up water to replace what is lost by transpiration, and they extract mineral nutrients from the soil solution. But this short list omits a major newly discovered function of roots: the creation and feeding of an entire and distinct soil ecosystem that can profoundly influence the growth of the plant. This ecosystem inhabits the rhizosphere: the outer layers of the root and the layers of soil particles adjacent to the root. Beneficial, neutral, and harmful bacteria, fungi, algae, protozoa, nematodes, and microarthropods live in the rhizosphere in an intricate food web. This entire underground ecosystem depends for its food on carbohydrates and other molecules, collectively called *exudates,* secreted by the roots into the soil. The health of this underground ecosystem can be profoundly affected by agricultural practices, tipping the balance in favor of beneficial or harmful organisms. The abundance of beneficial organisms can be manipulated by practices such as crop rotations, by chemicals that kill harmful microbes or by inoculation with beneficial ones. Some of the organisms in this underworld benefit the plants directly, others kill harmful pests. The interactions are enormously complex, and we know very little about them.

1 | **Fungi, bacteria, protozoa, and plant roots interact in the rhizosphere.**

It has been known for a long time that plant roots are colonized by a variety of micro-organisms, but the beneficial and detrimental relationships between plants and micro-organisms have only recently been studied. The organisms that interact with the roots may live within the roots themselves, as do the nitrogen-fixing bacteroids of *Rhizobium*, or may live in the soil in immediate proximity of the root. Certain fungi grow within the outside cell layers of the root, but their hyphae, the long cellular strands that make up the fungal body, extend into the soil. This entire space, comprising the outer tissues of the root (cortex and epidermis), the root surface, and the soil

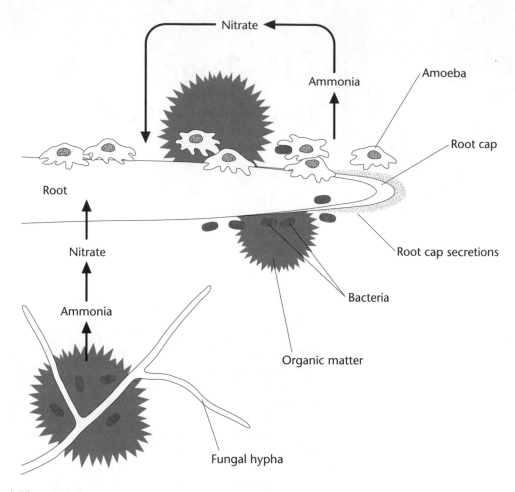

Figure 8.1

Model of Proposed Interactions in the Rhizosphere and in the Bulk Soil. A root is growing in the soil from left to right. Under the influence of root-derived carbohydrates (dots) bacteria using organic matter are supplied with energy to grow and to mineralize nitrogen from the organic matter, which will be immediately immobilized in the bacterial biomass. Bacteria will be consumed by amoebae that are attracted to the site. When digesting the bacteria, the protozoa release part of the bacterial nitrogen or ammonium on the root surface, where it can be taken up by the root. Below the root, in the bulk soil, a fungal hypha is decomposing organic material. Ammonium will be released as a waste product and it can diffuse toward the root as ammonium or, after nitrification, as NO_3^-. Source: From a model proposed by M. Clarholm, Swedish University of Agricultural Science, Uppsala.

around the root, is referred to as the **rhizosphere**. The rhizosphere can be thought of as a microbial continuum that stretches out from the endodermis of the root (a layer of cells within the root) into the soil. The plants stimulate this microbial colonization by secreting complex carbohydrates that can be used by the micro-organisms as a source of food and energy. In return, the plants benefit from some of these associations through a greater availability of mineral nutrients. A special type of interaction between roots and soil particles is found in many grasses and crop plants in the grass family, such as corn. When such plants are gently removed from the soil, a layer of sand and soil particles, embedded in mucilage, covers the roots like a sheath. These sheaths are extensively colonized by bacteria and fungi.

Figure 8.2

Relative Degree of Foot Rot Infection of Wheat Seedlings Caused by Helminthosporium sativum *with Various Soil Saprophytes.* Source: *A. W. Henry (1931), The natural microflora of the soil in relation to the foot rot problem of wheat, Cana-dian Journal of Research 4:69–77.*

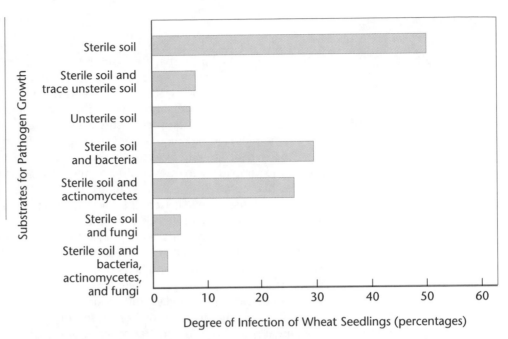

A number of the micro-organisms in the rhizosphere have specific func-tions that benefit plants, such as nitrogen fixation or phosphate uptake, while others simply aid in the decomposition of the organic matter. Organic matter contains nutrients, such as nitrogen or phosphate, that are tied up in larger molecules, and cannot be used directly by the roots. The secretion of muci-lage by the growing roots stimulates the growth of bacteria that break down this organic matter, and use the nutrients for themselves. The growth of the bacteria provides food for protozoa that digest the bacteria and excrete ammonia that can be used by the plants. Fungi growing in the same environ-ment can also break down organic matter and release ammonia; bacteria usu-ally convert this ammonia to nitrate before plants use it. The conversions of different forms of nitrogen are part of the nitrogen cycle discussed in Chap-ter 7. Other nutrients, such as sulfur and phosphate, are also released by these biological processes and made available for plant growth (Figure 8.1). Because mineralization depends entirely on the growth and activities of micro-organisms, it occurs most rapidly in the late spring and summer when the soils are moist and the temperatures are high. Micro-organisms that are bene-ficial to plants by simply breaking down the organic matter and releasing nutrients are called *saprophytes.*

So far, we have emphasized the beneficial organisms that live in the soil. However, many harmful organisms lurk there as well, including bacterial and fungal pathogens and nematodes. The numbers are often held in check by yet other and often uncharacterized micro-organisms. An early example of such interactions between micro-organisms in the soil is provided by the work of A. W. Henry. Sixty years ago, he studied foot rot infection, by the fungus *Helminthosporium sativum,* of wheat seedlings grown in sterile soil. He observed that adding nonsterile soil greatly diminished the intensity of the disease in sterile soil, and was able to show that soil fungi in the nonsterile soil were responsible for this effect (Figure 8.2).

Similar experiments were occasionally reported by others, but progress in this field was extremely slow because the interactions are complex and we are

How little we really know about life underground is underscored by an article in the weekly scientific journal *Nature* in the spring of 1992, in which Canadian researchers reported the existence of a 40-acre fungus that lives beneath a forest floor. This fungus is about 1,500 years old and weighs 100 tons. No one has, of course, seen the entire organism, but its size was confirmed by taking samples over the entire site, isolating fragments from different locations. By using the same gene-mapping technique now widely used in forensic medicine to determine the genetic identity of sperm in rape victims, the researchers established that all the samples taken over the 40-acre site were genetically identical. If these samples had come from another fungus, or from the sexual offspring of the same fungus, the genetic fingerprint would have been different. The researchers deduced the age of this fungus from a measurement that shows the fungus expanding outward from its site of origin at a rate of 20 cm per year. Assuming a constant growth rate, that would mean that it is about 1,500 years old. However, unfavorable growth conditions or diseases may have slowed it down at times. The species of fungus, called *Armillaria bulbesa,* lives underground and produces edible fruiting bodies (mushrooms).

But is such a fungus really one organism, or is it the vegetative offspring of one fungal spore? That is a matter of semantics. Such a fungal "spread" should probably be likened to a stand of bamboo that covers a large area and is the result of vegetative propagation starting from a single plant. But whatever name we attach to these organisms, their discovery tells us that we still know very little about the mysterious life in the underground.

relatively ignorant of the thousands of species of micro-organisms living in the soil. Progress has been more rapid in the past 20 years, and the study of the rhizosphere is an exciting new branch of biology. The opportunities to improve crop productivity by introducing organisms—possibly genetically engineered organisms—into the rhizosphere is inducing more biologists to enter this challenging field. Much remains to be learned about life in the underground (see Box 8.1).

2 | The root cap is the principal source of mucilage for bacterial growth.

Every root tip is covered by a root cap, a specialized tissue that forms a caplike structure at the end of the root tip, and originates from a root cap meristem (Figure 8.3). Cell divisions in this meristem assure that root cap cells are continuously produced. As the cells mature, they pass through the cap and are shed or abraded from the cap surface as the root grows between the soil particles. The cap of a corn root produces up to 20,000 cells per day, which are left behind as the root grows through the soil. These root cap cells are differentiated cells whose main function is the production of mucilage or slime. Corn roots produce so much slime that small droplets form at the end of the roots.

The mucilage, which is left behind as the root grows through the soil, has several functions. It may act as a lubricant when the root is pushing its tip

Figure 8.3

Alfalfa Root Tip Showing Numerous Cells Being Shed From the Root Cap. Source: *Courtesy of M. Hawes.*

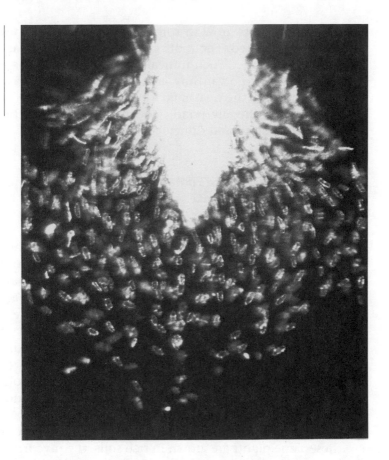

forward between the soil particles. Because the complex carbohydrates that make up the slime bind water, the slime protects the root against drying out and ensures that a continuous water film covers the soil particles and the root. The mucilage binds soil particles into larger aggregates, and serves as a source of food for the bacteria that grow in the rhizosphere.

In addition to the cells and mucilage produced by the root cap, the roots themselves also secrete soluble molecules, especially sugars and amino acids. Secretion of sugars and amino acids is an active process and not simply a passive loss as a result of leakage. As the roots grow through the soil, root hairs die, break off, and are left behind. Thus, as the roots grow, they leave behind root cap cells, broken root hairs, secreted mucilage, and other molecules, all of which provide food for micro-organisms. Soil biologists use the term **rhizodeposition** to describe this transfer of molecules from the roots to the soil. The magnitude of rhizodeposition is difficult to measure, especially in natural ecosystems, and is usually expressed as the percentage of all the carbon fixed in photosynthesis that is released by the roots into the soil. Reliable estimates of photosynthetic products transferred to the soil by rhizodeposition indicate a range from 15% for wheat plants to 40% for a tropical forest ecosystem.

In healthy roots, the microbes live mostly near the surface (epidermis) of the roots, although fungal hyphae can grow into the cortex and between the cortex cells (as discussed later). Broken root hairs provide sites of bacterial infection, and emerging lateral roots that tear through the cortex tissues also provide sites of entry for bacteria. Colonization by pathogenic bacteria elicits a defense response that causes the root cells close to the site of infection to die. When roots grow in soils that contain herbicides, root growth is often adversely affected. The roots are under stress, and bacteria are more likely to invade the cortex. Such unhealthy roots provide less rhizodeposition for the beneficial micro-organisms, and the balance tips in favor of disease organisms that can attack the weakened root system. These effects of herbicides on plant growth are generally poorly documented because the affected plant organs are underground and difficult to sudy.

3 | Mycorrhizas are root–fungus associations that help plants take up phosphate.

The majority of higher plants have fungal hyphae closely associated with their actively growing roots, and such an association is called a **mycorrhiza**. These types of associations were described more than a hundred years ago, but their beneficial effects on plant growth have been discovered relatively recently. We now realize that a mycorrhiza is a true symbiotic relationship. The flow of photosynthetic products from the shoot to the roots benefits the fungus, which uses these products as a source of food; the plant benefits from growing in association with the fungus, especially if the soil is poor in nutrients. When plants are growing in phosphate-poor soils, or under other types of nutrient stress, mycorrhizal plants grow faster and more vigorously than nonmycorrhizal plants because the fungal hyphae are able to take up phosphate from this phosphate-poor soil and transfer it to the plant (Figure 8.4). If these same plants are grown in rich soils, the mycorrhizal associations are not measurably beneficial. Some plants cannot be grown without their fungal symbionts, as is the case for certain species of pine. When such pines are transplanted to a new non-native area, it is necessary to inoculate the soil with the soil fungi from the area in which they were previously grown.

The fungal partner of a mycorrhiza grows on the surface of the root, but in addition its hyphae can extend far into the soil between the soil particles, as well as into the root cortex growing either between the cells or even into the cells. Because fungal hyphae are so thin and can extend into the soil much farther than root hairs, they can draw nutrients from a larger volume of soil particles than a root can by itself.

There are two major types of mycorrhizal associations: ectomycorrhizae and endomycorrhizae. They differ not only in the species of fungi that can form the different associations, but also in the way they invade the root and the extent to which they modify root morphology. Ectomycorrhizal fungi grow as a sheath on the surface of the root, and their hyphae penetrate into the soil and between the cells of the root cortex. The sheath on the root may account for 20–30% of the volume of the fungus–root association. Endomycorrhizal fungi invade the cells of the epidermis and their hyphae form vesiclelike or treelike structures within these cells, allowing for nutrient transport between the two organisms.

Scientists are not sure how the fungus can bring about a more efficient use of nutrients in a nutrient-poor soil. For example, mycorrhizal roots of beech trees take up phosphate five times faster than nonmycorrhizal roots. Micro-organisms generally secrete enzymes that help mineralize organic matter. Secretion of the enzyme phosphatase helps phosphate solubilization and uptake by breaking down larger molecules that contain phosphate. Whether

Figure 8.4

Cultivated Citrus Species Show a Marked Dependency on an Endomycorrhizal Association for Adequate Growth. Shown here are rough lemon seedlings after six months, with and without mycorrhizas, and with weekly applications of no, half, or full-strength nutrient solution without phosphate. The presence of mycorrhizal fungi greatly helps the citrus seedlings to get phosphate from the soil when other nutrients are supplied.

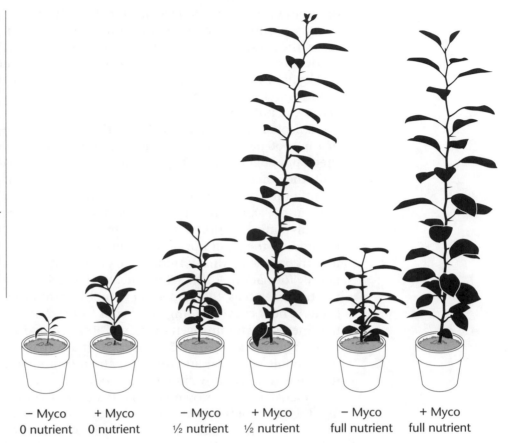

| − Myco | + Myco | − Myco | + Myco | − Myco | + Myco |
| 0 nutrient | 0 nutrient | ½ nutrient | ½ nutrient | full nutrient | full nutrient |

mycorrhizas can also solubilize the insoluble mineral phosphates is not entirely clear. Another way that mycorrhizas may obtain scarce nutrients is by secreting special organic molecules. After such an organic carrier molecule is secreted, it forms a complex with a single nutrient atom or molecule and is then taken up again by the organism that produced it. Although this mechanism has not yet been demonstrated for mycorrhizas, it is reasonable to hypothesize that mycorrhizal fungi acquire scarce minerals in this manner.

4 | ### *Rhizobium* bacteroids live as symbionts within the root nodules of legumes.

Aside from photosynthesis, biological nitrogen fixation is the single most important biological process that supports life on Earth. Its importance in the nitrogen cycle is explained in Chapter 7.

Especially important for nitrogen fixation both in natural ecosystems and for agriculture are the *Rhizobium** bacteria that live symbiotically with leguminous plants. In this symbiosis, the *Rhizobium* bacteria live in root nodules off the photosynthate provided by the plant and they are able to fix (reduce) molecular nitrogen to ammonia, which they export for the plant's benefit. Our understanding of how this symbiosis is established has greatly advanced

---

*The term *Rhizobium* is used here to describe the bacteria belonging to several species in three different genera.

in recent years. Nodulated legume plants can be grown hydroponically in the absence of nitrogen (as ammonia or nitrate) in the nutrient solution, emphasizing the importance of nitrogen fixation for the nutrition of the plant. Symbiotic nitrogen fixers, such as *Rhizobia* in legume nodules, can add from 120–250 kg of nitrogen per hectare per year to the ecosystem. This biological input of nitrogen is of great importance to agriculture because in many crops, nitrogen is the nutrient that most limits yield.

Rhizobia are free-living soil bacteria that attach themselves to the surface of the root hairs of legume plants and cause the root hairs to curl around them and engulf them. By producing enzymes that partially dissolve the cell wall, they manage to penetrate the root hair cell wall. In response, the plant forms a tube or infection thread in which the bacteria multiply and are confined at the same time. The bacteria stimulate the cells of the root cortex to start dividing and form a new meristem. It is not yet clear whether the bacteria produce the plant hormone auxin or whether they stimulate the root cells to start producing this hormone, but an increased auxin level seems to trigger nodule formation. At the same time, the bacteria continue to divide and soon fill the cortex cells, living in small vacuoles within the cytoplasm. Most of the bacteria now cease dividing and begin to make the enzymes and other proteins necessary for nitrogen fixation. Capable of fixing nitrogen, they are now called bacteroids (Figure 8.5). Some bacteria continue to divide and to fill the new nodule cells that are formed as the root nodule grows. Indeed, the root nodule is not just a swelling on the root, but a modified lateral root with its own meristem and its own vascular tissue. When the nodule cells eventually die, the bacteroids die with them. They cannot survive in the soil and join their free-living cousins. However, exudates from the plant ensure that there will always be free-living *Rhizobium* bacteria in the soil.

The interaction between *Rhizobium* and its legume host is very specific. Usually one species of *Rhizobium* can infect only one or a few closely related species of legumes. This specificity resides in the signaling molecules that both organisms use. Each *Rhizobium* species produces small lipid–carbohydrate molecules with a unique structure; a particular legume species is triggered by only one type of molecular structure, but not by a slightly different one. The plants release flavonoids (complex organic molecules related to the red and blue pigments of plants) into the soil, and these also can have many different structures with different plant species producing different flavonoids. A particular *Rhizobium* species will respond best to the flavonoids of one or a few legume species, and not as well to the flavonoids of other plant species. In this way, plants and bacteria signal their identity to each other in the underground.

Not all *Rhizobium*–legume symbioses that are established are equally effective. Even within a single species of bacterium and legume, effectiveness as measured by the ability to fix nitrogen, will vary depending on the strain of the bacterium and the variety of the plant. A really effective strain forms healthy-looking nodules with a distinctive pink color. An ineffective strain forms colorless or greenish nodules. Once a plant has been colonized by one strain, others seem to be excluded, so it is important to start the plants off right. In many farming areas, the *Rhizobia* present in the soil are not particularly effective for fixing nitrogen in symbiosis with the legume crops best suited to those areas. However, effective strains are now available to farmers and can be applied at the time of planting as a seed inoculant. The live bacteria needed to inoculate the seeds are produced as a bacterial culture

Figure 8.5

Electron Micrograph of a Soybean Nodule Cell Showing that Nitrogen-Fixing Bacteroids Fill the Entire Cell. The nucleus is at the center of the cell. Notice the thin cell walls separating different cells. Source: Courtesy of M. A. Webb, Purdue University.

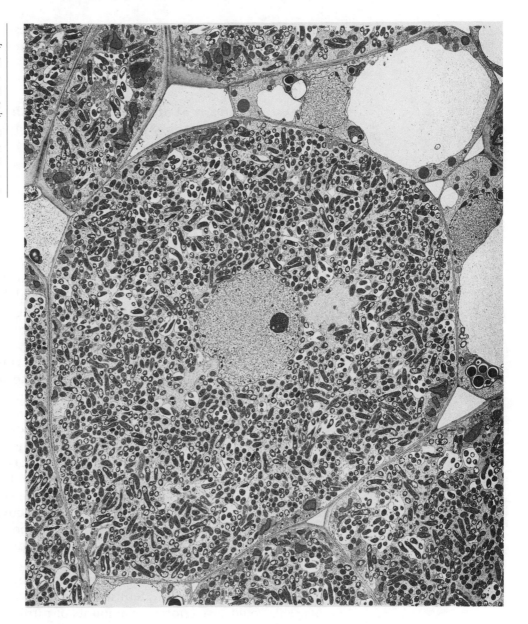

incorporated with a carrier, such as finely ground peat. Seed may also be supplied precoated with *Rhizobium* using one of a number of proprietary processes. Even so, indigenous *Rhizobia* often compete and supplant the introduced ones.

The biotechnology industry is now at work devising stable formulations of effective *Rhizobia* that can be shipped and stored in developed countries as well as in the Third World. An additional problem that limits the effectiveness of *Rhizobia* is that nodulation is inhibited when the soils are nutrient rich, especially with fixed nitrogen. Thus, when crops are grown in a legume–cereal rotation, nitrogen fertilizers added to increase yields of the cereals may inhibit nodulation the next year when legumes are grown. Advocates of sustainable agriculture suggest that we should rely more on nitrogen fixation and less on chemical nitrogen fertilizers. The benefits of symbiotic nitrogen fixation can be realized in several different ways: by crop rotations (soybeans one year, corn the next year), by intercropping (seeding vetch in a rice field that is close to harvest, or planting alternate rows of corn and beans), or green manuring (plowing under a fresh growth of alfalfa).

5 | **The establishment of an effective legume–*Rhizobium* symbiosis requires the expression of numerous genes by the plant and the bacteria.**

The formation of the root nodule and the small vacuoles in which the bacteroids live, requires plant cells to synthesize many new proteins and therefore requires the expression of many plant genes not expressed in normal root cells. The most important one of these proteins is the red protein leghemoglobin, a molecule that is closely related to muscle myoglobin and to red blood cell hemoglobin. Nodules contain so much leghemoglobin that they actually look pink. The function of leghemoglobin is to diminish the concentration of oxygen in the cells without making them totally anaerobic. The functioning of the bacterial enzymes that reduce atmospheric N_2 to ammonia requires that the level of oxygen be very low in the cells. Other proteins that are made by the plant cells include enzymes that use the ammonia made by the bacteroids for the synthesis of amino acids that can be transported in the vascular system to the rest of the plant, and special membrane transporters.

The bacteroids are contained in special vacuoles, and there is an enormous flux of metabolites across the membranes of these vacuoles: products of photosynthesis in one direction, products of nitrogen fixation in the other direction. Transporters that allow these metabolites to be carried across the vacuolar membranes that enclose the bacteroids are synthesized in response to the bacterial invasion. Many new genes are also activated in the bacteria as they form bacteroids capable of fixing N_2. Most important are the genes that encode the enzymes for N_2-fixation, as well as the proteins that regulate the activity of these genes.

One often publicized goal of plant genetic engineering is to transfer the ability to nodulate and to establish effective symbioses to nonlegume plants. This could save enormous amounts of nitrogen fertilizer, and therefore would also save energy because producing nitrogen fertilizers requires large amounts of energy. The problem of groundwater contamination by nitrogen fertilizers might also be avoided if more plants had symbiotic nitrogen fixers and if we relied less on inorganic nitrogen fertilizers. However, because so many proteins are needed to make functional nodules and because of the species-specific recognition between plants and *Rhizobia*, transferring the nitrogen-fixing capacity from legumes to nonlegumes will remain a distant goal for quite some time. In addition, it is also important to remember that nitrogen fixation by bacterial symbionts is not "free" but depends indirectly on photosynthesis. Photosynthate must be diverted to the roots to provide energy for nitrogen fixation, and is therefore not available for growth and seed production or other energy-requiring processes. Although reliable estimates are difficult to make, as much as 20% of the photosynthate in a nodulated legume plant may be used for nitrogen fixation.

6 | Free-living bacteria and fungi convert atmospheric N₂ into ammonia.

In recent years, there has been much interest in the many different free-living organisms in the soil that fix N_2 and that live in the rhizosphere in close association with the roots of plants or that thrive as a crust on the surface of the soil. The nitrogen-fixing microbes in the rhizosphere live off the exudates from the roots and contribute to the soil ecosystem by augmenting its store of fixed nitrogen. Some, such as the fungi that form root nodules on many trees, may contribute ammonia directly to the plant in the same way as the *Rhizobia* in the root nodules of legumes. Others, such as the bacteria that live in the sheaths that surround the roots of many tropical grasses, contribute fixed nitrogen indirectly. The bacteria grow and multiply, using the plant's secretions as food, and they fix nitrogen from the atmosphere. When they are consumed by grazers (amoebae, nematodes, or microarthropods) that eat bacteria, the excess nitrogen will be released into the soil as ammonia and is now available to the plant, as well as to microbes that do not fix nitrogen. The nitrogen fixers that live in a crust at the surface of the soil are cyanobacteria (blue-green algae) that are photosynthetic and fix nitrogen as well, or they form symbiotic associations with fungi that are called *lichens*. Measurements in the Sonoran Desert showed that such cyanobacteria-lichen crusts contribute 7–18 kg per hectare annually of fixed nitrogen to the ecosystem. Similar values are found for the nitrogen fixers that live in association with plant roots. The input of fixed nitrogen from a legume–*Rhizobium* symbiosis is much greater, ranging from 50–200 kg per hectare per year, depending on the legume species. Nitrogen-fixing bacteria growing in rice paddies have been shown to contribute anywhere from 5 to 80 kg per hectare per year. This enormous range shows that there is much scope for increasing nitrogen fixation by soil organisms. The challenge for agriculturalists is to encourage those microbes that fix nitrogen, without upsetting the delicate food webs in the soil.

Soil micro-organisms that fix N_2 depend on soluble root exudates as their main source of energy. However, they can also use the organic matter in the soil. It is well known that fungi, because they secrete cellulase and other cell wall degrading enzymes, can break down the cellulose of wheat straw. This, in turn, promotes nitrogen fixation when this straw is mixed with the soil probably because soluble carbohydrates are now available to the nitrogen fixers. It has, therefore, been proposed that wheat straw could be "upgraded" before adding it back to the soil. Upgrading would mean a partial decomposition by acids and inoculation with cellulose-degrading fungi and N_2-fixing bacteria, as well as with bacteria that will degrade the straw and secrete soluble carbohydrates. Experiments to look at the yield effects of such upgraded straw are now underway.

7 | Mineralization, the slow decay of soil organic matter, provides a steady stream of mineral nutrients for plants.

Organic matter in soils is derived from residues of the plants and animals that live in and on the soil. Leaf litter accumulates on the soil surface in many forests; as it decays, it is gradually mixed with the top layer of soil

Table 8.1	The organic residues in the roots of crops grown in central Ohio

Crop	Residue (kg/ha)
Soybeans	600
Wheat	830
Corn	1,270
Alfalfa	3,850
Kentucky bluegrass	5,000

Source: Data from U.S. Department of Agriculture.

(earthworms are especially important in this process) and carried downward by percolating water. In this way, it becomes distributed throughout the top-soil. Organic residues are also contributed by the roots of plants (see Table 8.1) and by the soil organisms. The organic matter in soil often gives it a characteristic brown color. Sandy soils and many tropical soils are usually light-colored, because they contain very little organic matter (1 to 2%), whereas heavy clays can vary from dark brown to black, because they contain much more organic matter (5 to 10%).

The proportion of organic matter in a soil depends on the rate at which it is being added to the soil and the rate at which it is being broken down by the soil organisms. The rate of decay is influenced by the prevailing temperature, by the availability of oxygen, and by the acidity of soil. The first phase of the decay process is carried out partly by earthworms and other soil animals, which ingest large amounts of leaf litter and dead roots. These materials are used as a source of food, and the animals excrete a black organic residue called **humus**. This process of transforming the organic residues into humus is aided by the soil micro-organisms that secrete cell-wall-degrading enzymes that in turn convert the insoluble complex carbohydrates into soluble products they can take up. Their growth also requires minerals, especially phosphate, nitrate, and sulfate. As they multiply, they are in turn eaten by grazers who serve as food for yet larger soil organisms. Thus, the continuous process of "eat and be eaten," in which the whole soil ecosystem is involved, gradually converts fresh organic matter into humus and also results in the slow decay of humus. Mineral nutrients that are taken up and therefore immobilized by one group of organisms, are released again when these organisms are eaten by the next trophic level. This complex process, which releases mineral nutrients, is called **mineralization**.

Decay is carried out by organisms that require oxygen for their respiration. As a result, it usually proceeds much more rapidly in well-aerated soils. The soil temperature and degree of soil acidity also affect the decay of organic matter, because they influence microbial activity. High temperatures, as in the tropics, speed up the growth of micro-organisms, which in turn speeds up decay (see Table 8.2). Acidic soils are usually rich in organic matter, because neither earthworms nor most bacteria can thrive in such soils; so decomposition of organic matter is slow. Although humus is more resistant to decay than fresh organic residues, it too is eventually broken down by micro-organisms. The plant nutrients contained in humus are then released into the soil and become available to the plants. The slow decay of humus provides the plants with a steady flow of nutrients and is of great importance to soil fertility.

Table 8.2	**Annual leaf production and turnover time for the organic matter in several ecosystems at different temperatures**		

System	Annual Leaf Production (kg/ha)	Residual Litter Accumulation (kg/ha)	Time for Decay of Organic Matter (years)
Rainforest (tropical)	14,000	9,000	1.7
Deciduous forest (temperate)	4,500	14,000	4.0
Conifer forest (northern)	2,700	40,000	14.0
Tundra (arctic)	900	45,000	50.0

Source: Data from various sources, quoted by C. Kucera (1973), *The Challenge of Ecology* (St. Louis: Mosby), p. 64.

To understand how the process of decomposition works, let us examine what happens when fresh organic matter, such as straw or a rich farm manure, is added to the soil. Immediately, the soil micro-organisms start working on this organic matter, and because the microbes are provided with a large source of food, they start to multiply rapidly.

In addition to an energy source (provided by the carbohydrates), these micro-organisms need nutrients to grow, especially nitrogen. If the organic matter has more nitrogen than the micro-organisms need for their own growth, nitrogen will be released into the soil by the decay processes. If the organic matter is poor in nitrogen, the micro-organisms will use up the nitrate already present in the soil, thereby robbing the plants of their source of nitrogen. Since farmyard manure contains nitrogen-rich organic matter (animal wastes) and many mineral nutrients, its decay provides the plants with a steady supply of nutrients. Straw and leaf litter are poor in nitrogen and minerals, and their decay may slow down plant growth, because plants and micro-organisms are competing for the same nutrients. The setback for the plants is only temporary; eventually all these plant nutrients will be released back into the soil when the decomposition is complete and the population of micro-organisms declines. Avoiding this temporary slowdown is the purpose of composting. **Compost** is organic matter that is partially decomposed by the soil fauna and micro-organisms. Basically, the first rounds of "eat and be eaten" have already taken place, thereby enriching the organic matter in important nutrients. The enrichment is relative, of course. There is no absolute increase in nutrients, only a decrease in carbon (from polysaccharides and lignin) that is released as carbon dioxide into the atmosphere as a result of the respiratory activities of the decomposers. Adding compost to the soil tends to boost plant growth more quickly than adding fresh organic matter, which must still be decomposed to release the minerals. Compost can also be used as a mulch to cover the soil underneath the plants (Figure 8.6). In that case, the nutrients are washed into the soil when the plants are watered.

Converting a natural system to an agricultural one usually decreases the amount of organic matter in the soil, for several reasons. Tilling the soil and mixing the crop residues into the topsoil increases microbial activity because it increases the aeration of the topsoil. Thus, cultivation accelerates the decay of organic matter by initially increasing microbial activity. Maintaining a healthy soil ecosystem therefore requires the continuous addition of fresh organic matter so that the level of humus in the soil can be maintained and so a vigorous population of microbial decomposers can release nutrients.

Figure 8.6

Home Gardener Applying Composted Mulch on a Flower Bed. Material, such as leaves, grass clippings, and stems and stalks from harvested vegetables were allowed to decompose in a corner of the yard. Compost conserves moisture, prevents erosion and adds nutrients and organic matter to the soil. Source: U.S. Department of Agriculture.

Organic matter improves the structure of the soil, its aeration, and the percolation of water, and these all contribute to healthy root systems that are not so readily invaded by the many pathogens present in the soil.

8 | Biological warfare goes on in disease-suppressive soils.

Take-all (isn't *that* an unusual name for a disease!) is a fungal disease of wheat roots caused by a pathogen with the unpronounceable Latin name *Gaeumannomyces graminis* var. *tritici*. It doesn't kill the plants, but the damaged root system makes the plants susceptible to drought stress. The plant matures early, and wheat yields can be severely reduced. In the northwestern United States, an important wheat-growing area, 5–10% of the wheat crop is lost to take-all disease. There are no commercially available chemical control agents, and there is no known resistance in related species that could be used for breeding resistant wheat strains. The fungal pathogen remains in the wheat stubble that stays on the field, and the disease gets steadily worse if wheat is grown year after year on the same land. The small time gap between the harvest of wheat in midsummer and the sowing of the next crop of (winter) wheat in the fall makes matters worse. The surviving fungus jumps from the decaying stubble to the new young plants ready for a new round of devastation. If a monoculture is continued for three to five years, the severity of the disease greatly increases, but beyond that time—after more than five years—the disease begins to decline again. Researchers found that factors in the soil were responsible for the decline in the severity of the disease, called "take-all decline," and that these factors could be transferred from one field to another

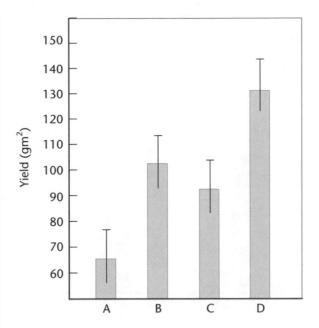

Figure 8.7

Yield of Spring Wheat in Soil Naturally Infected with Take-All and Inoculated with Bacillus pumilus.
(a) *Uninoculated control;*
(b) *inoculated at sowing;*
(c) *inoculated 6 weeks after sowing; and*
(d) *inoculated at sowing and 6 weeks after sowing.*
Source: *Modified from A. L. Capper and R. Campbell (1986), The effect of artificially inoculated antagonistic bacteria on the prevalence of take-all disease of wheat in field experiments,* Journal of Applied Bacteriology *60:155–160.*

if about 1% of the topsoil was transferred from a field in which take-all decline had occurred to a heavily diseased field.

Soils that decrease the severity of a disease are referred to as "disease-suppressive soils." Disease suppression is widely attributed to Pseudomonad bacteria that live and multiply in the lesions on the roots caused by the take-all fungus. The Pseudomonads probably produce antibiotics that inhibit the growth of the fungus. Once take-all decline has occurred, the bacteria survive both on and inside the roots and stubble of the wheat after the crop has been harvested. In the next growing season, the suppressive Pseudomonads are in the right place at the right time to colonize new wheat roots, as soon as the new seeds start to sprout, and suppress the growth of the take-all fungus. Further studies have led to the isolation of a strain of *Pseudomonas fluorescens* that produces an antibiotic and is able to control the take-all fungus.

Other bacteria have also been shown to give protection against take-all disease, and it is likely that different soil types (different ecosystems) have different micro-organisms that keep each other in check. An experiment showing the yield of wheat on soil naturally infected with take-all and the effect of inoculation with *Bacillus pumilus* is shown in Figure 8.7. The highest wheat yields were obtained when the seeds were inoculated at the time of planting and the soil was inoculated again six weeks later. Such experiments clearly show the value of biological control agents in disease control.

Thus, in a disease-suppressive soil, a new equilibrium is established between the pathogen and one or more organisms that inhibit its growth. The pathogen has not disappeared, and the roots and plants (and farmers) still suffer from its presence. However, severe crop losses are curtailed, and an ecosystem brought into disequilibrium by continuous monoculture has found a new equilibrium. The challenge for agronomists and biotechnologists will be to tip the balance just a bit more in favor of the plant without upsetting the soil ecosystem.

One way to find out which micro-organisms may be helpful to prevent diseases is to examine the rhizosphere of healthy plants in otherwise heavily infested fields. If beneficial strains can be identified in laboratory and field

Figure 8.8

Percentage of Healthy Cucumber Seedlings at Various Times After Planting in a Pythium ultimum *Infected Soil.*
The seeds were treated with the chemical fungicide thiram (T), with Trichoderma harzianum *(H) or were not treated at all (O). Seedling emergence was excellent with thiram and with* Trichoderma, *but thiram did not protect the seedlings during the next five days; nearly all the seedlings died. More than half the seedlings survived to day 8 when the seeds had been coated with* Trichoderma.
Source: *Data from Harman and Taylor (1988), Improved seedling performance by integration of biological control agents at favorable pH levels with solid matrix priming,* Phytopathology 78:520–525.

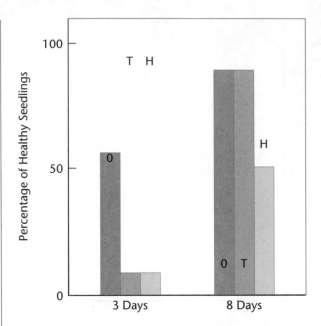

studies, the microbes can be grown in fermentors (an apparatus in which controlled fermentation takes place) and then applied to the infested fields as small pellets or as a coating on seeds. Seed treatments are already widely used by seed companies to enhance seed vigor or seedling performance by treating the seeds with certain chemicals.

Seed priming has been found to be an effective way of incorporating beneficial microbes into the planting material. Seed priming is a controlled or partial hydration of seeds that permits certain pregerminative metabolic activities to be initiated without leading to the actual germination of the seed. For example, it is possible to coat the dry seeds with the microbes and then add a powdered inert filler material with a high water-holding capacity, such as peat moss. The water in the peat moss lets seed priming occur and supports the growth of the beneficial microbes. When the seeds are planted, the microbes multiply vigorously and protect the seedlings from disease organisms already in the soil. This approach was found to be successful with cucumber seeds. Seeds coated with *Trichoderma* gave rise to seedlings that were better able to withstand infection by the fungus *Pythium ultimum* (Figure 8.8).

9　Plants and micro-organisms compete for scarce supplies of soluble iron.

The soil contains a huge amount of iron, but because a good soil is well aerated, the iron exists in the oxidized Fe^{3+} form, which often forms insoluble complexes. To ensure its own growth, each organism must convert a portion of the iron in its own immediate environment into a form that can be taken up. If that form of iron is inaccessible to other competing organisms, then the iron acquisition mechanism can function to deny this essential material to a competitor.

Plants use two strategies for iron acquisition. Most plants, with the exception of the grasses, have a special enzyme in the outer membrane of the root cells that converts Fe^{3+} to Fe^{2+} by a reductive reaction. The Fe^{2+} is soluble and can be taken up by the root hairs. The grasses use a mechanism that is also used by many bacteria and fungi: they synthesize and secrete organic molecules that complex the Fe^{3+} and then take up the iron complex. These organic molecules are called **siderophores** (literally "iron carriers"), and different species use different siderophores. Using different siderophores prevents organisms from parasitizing each other's iron uptake efforts. Clearly, the competition for iron is very keen down in the underworld.

Not all siderophores are equally effective, and not all micro-organisms make them. A research group at the University of California at Berkeley discovered that inoculating a soil with Pseudomonads that produce siderophores enhanced the yield of potatoes. The explanation for this phenomenon is that these siderophores scavenge iron, which is now not available to minor pathogens or other micro-organisms whose growth is not necessarily beneficial to plants. For example, many micro-organisms in the rhizosphere produce HCN (hydrogen cyanide gas) in the soil, and even extremely low concentrations of HCN slowly kill root cells. The siderophore-producing Pseudomonads successfully outcompete the HCN-producing bacteria when both are present in the soil.

10 | When they infect roots, nematodes cause serious crop losses.

Nematodes, also called *eelworms,* are tiny animals that live in the soil—as many as 500 in an average-sized scoop. Many are free-living grazers that eat bacteria, fungi, protozoa, and microalgae, but others cause severe damage to plants. Parasitic nematodes have a sharp mouthpiece with which they pierce the cell wall of the epidermis.

Having gained entry, nematodes have two different ways of parasitizing the plant. The **root knot nematodes** invade the roots and begin to multiply, causing huge, tumorous growth on the roots (Figure 8.9). The **cyst nematodes** remain on the outside of the root and are attached to a single plant cell, which becomes very large. The plant feeds the growing nematode by continuously transferring photosynthate to the giant cell and then to the nematode via the mouthpiece. The body of the nematode swells up until it consists of a giant cyst containing thousands of eggs. These eggs are released, and the cycle starts over again.

Most nematodes parasitize only one species of plant. As a result, their abundance in the soil, like that of pathogens, is greatly influenced by crop rotations. Continuous cultivation of the same crop leads to a buildup of nematodes in the soil that parasitize that particular crop.

On a worldwide scale, nematodes cause enormous crop losses, yet there are no chemicals that kill nematodes and are also safe to use. The nematocides produced by chemical companies are effective against nematodes because they are extremely toxic to many organisms. They are, in essence, biocides, and their use is being phased out in the United States and other developed countries.

Effective control of nematodes can be achieved in several ways. Cultivation practices and crop rotations can do much to keep nematode levels down by encouraging their natural enemies and pathogens. A healthy soil may harbor as many as 60 species of predatory mites, microarthropods that feed mainly on nematodes. The different species of mites are extremely well

Figure 8.9

When Root Knot Nematodes Infect the Root System of a Soybean Plant, They Cause the Formation of Galls. Such galls impair the function of the roots and depress yields. Source: U.S. Department of Agriculture.

adapted to different microhabitats in the soil. Conventional tillage (plowing the soil) upsets these microhabitats, disturbing the ecosystem, reducing the number of species as well as numbers of individuals. Much research is needed to understand the dynamics of these organisms in the soil. Perhaps if we learn to manipulate these mites in the laboratory and understand their feeding preferences, biotechnology can provide biological control methods. A return to farming methods that are ecologically sound could possibly achieve the same result.

11   The agronomic practices of sustainable agriculture promote a healthy soil ecosystem.

The soil is part of a complex ecosystem, and relations between the many different species may be harmed by modern agricultural practices, such as monoculture, chemicals used for weed control and pest control, frequent tilling that reduces the organic matter, irrigation that raises the salinity and the pH of the soil. The complexity of the root ecosystem makes it impossible to predict how and when the balance will be tipped in favor of pathogens. A pathogen does not become a pathogen until its numbers get out of hand. Before that, it lives saprophytically, helping to break down organic matter.

In the past 15 years, many studies have been done on the effect of agronomic practices on pathogens in the soil, and a few general principles are emerging. Clearly, the agronomic practices advocated by proponents of sustainable agriculture stabilize the soil ecosystem and ameliorate many of the problems we are now encountering.

Crop Rotation. The devastation caused by take-all disease can be attributed to continuous monoculture of wheat in the Pacific Northwest of the United States. The severity of many diseases increases when crops are grown continuously on the same land. Productivity is usually inversely related to the frequency with which that crop is grown on the same field within a three- to six-year rotation. Earlier (Figure 7.15) we showed the effect of rotating corn and soybeans instead of growing corn continuously. The yield differential is caused in part by the nitrogen fertilization of the soil by the legume crop (soybean), and in part by the general beneficial effect of crop rotation.

The most important benefit from crop rotation is a lower level of pathogens and pests in the soil. Crop rotation always reduces disease problems unless another crop in the rotation is also a host. With take-all disease, for example, barley can also serve as a host, but potatoes, lentils, and peas can be used as "break" crops.

Tillage. Tilling the soil can have positive and negative consequences for crop productivity. Tilling mixes the crop residues into the top layer of soil, causing organic matter to decay faster. Soils with lower levels of organic matter tend to have less stable ecosystems, because there is a lower overall level of microbial activity. Tilling destroys weeds, but it accelerates soil erosion by wind and water. Because of the potential detrimental effects, some farmers practice no-till agriculture. This involves drilling small holes in the soil in which the seeds are dropped, a method known as *direct drilling*. This method reduces erosion especially if the stubble is left on the field but increases the severity of weeds, requiring the application of more herbicides (Figure 8.10).

Direct drilling has also been shown to increase the severity of three root rot diseases of wheat in the Pacific Northwest of the United States. The likely reasons are (1) it allows more pathogens to survive in the crop residues (stubble) and (2) the seeds are then placed in the layer of soil that is richest in pathogens. The complexity of the situation is demonstrated by the fact that direct drilling of wheat in the same region greatly reduced damage from the cereal cyst nematode. A possible solution and one way to achieve sustainable agriculture is to abandon tilling in favor of no-till methods *and* to abandon continuous monoculture in favor of crop rotation.

Fertilizers. The type of fertilizer used (nitrogen as ammonium or as nitrate), the time of its application, and where it is placed in the soil all can affect pathogens. If seedlings get off to a slow start because the soil is nutritionally deficient, they are more prone to infection by pathogens. However, the growth of the pathogens as saprophytic organisms—at first they all grow saprophytically—is often limited by the low nitrogen content of the stubble. If no nitrogen is available, they may be outcompeted by other saprophytes that grow well on low-nitrogen stubble. The take-all fungus generally cannot survive on acid soils, so keeping the soil acidic will control take-all. However, legume–*Rhizobium* symbioses do not thrive on acid soils, and many crop rotations involve cereals and legumes. Liming acidic soils improves the growth of the legumes, but may increase problems with take-all and other pathogens for the cereal in the crop rotation. Pathogens that are most damaging in acid soils are usually favored by ammonium fertilizers, while pathogens that damage plants in neutral or basic soils are encouraged by nitrate applications. The solution may well be longer rotations—wheat every third year—and less inorganic fertilizer.

Figure 8.10

Winter Wheat Grown with Conservation Tillage to Protect the Soil from Erosion. Normal tillage reduces weed populations but when conservation tillage is practiced, there are more weeds (a). To eliminate the weeds, herbicides must be used (b). Thus, there is a tradeoff: conservation tillage reduces soil erosion but often requires more herbicide use. Other forms of cultivation, such as ridge tillage, kill weeds and protect against erosion at the same time. Source: *U.S. Department of Agriculture.*

(a)

(b)

Herbicides. In recent years, many instances have been recorded in which herbicides have increased the damage from root disease. In some cases, the damage can be avoided by a short period of fallow between the application of herbicide and the planting of new seed. With the direct drilling method, the weeds are first killed with a biodegradable herbicide such as glyphosate and the seeds are planted a few days later. Researchers in Australia found that this procedure increased pathogen damage to the crop, because the pathogens started multiplying in the root systems of the plants killed or weakened by the herbicides. By waiting two to four weeks instead of 2 to 3 days, the problem was avoided. However, the problem cannot always be avoided in such a simple way, and the farmer must weigh the benefits of the herbicide application—no competition from weeds—against these and other drawbacks of herbicides. Another experiment in Australia showed that the use of chlorsulfuron (Glean®) for weed control in wheat increased *Rhizoctonia* root rot in the barley crop that followed wheat, and reduced the yield of barley by up to 1 ton per hectare. The results show the complexity of interactions among herbicides, plants, and their pathogens. Controlling weeds by other methods, such as mulching or intercropping, would avoid the problem completely.

Summary

Life underground is tremendously complex. Twenty years ago, people knew very little about the rhizosphere; today we know only a little of what is needed for devising intervention strategies to tip the balance in favor of beneficial organisms.

At least three general classes of microbes benefit plants. Certain bacteria and fungi establish symbioses with plant roots. They either live completely within the root cells, or they grow in between the cells with their hyphae extending into the soil. Most important among these are the symbiotic *Rhizobia* and the mycorrhizal fungi. Other micro-organisms keep pathogens in check and therefore contribute indirectly to plant growth. A third group, the saprophytes, is involved in the mineralization or breakdown of organic matter, the fixation of lesser amounts of nitrogen compared with the symbiotic *Rhizobia,* and the solubilization of inorganic minerals such as phosphate and iron. Many pathogens grow initially as saprophytes and become pathogenic when conditions are right. The food for all these micro-organisms and for the entire rhizosphere ecosystem comes from the plant: the products of photosynthesis are transported to the roots, which secrete sugars, amino acid, and complex carbohydrates.

Agronomic practices greatly influence the rhizosphere ecosystem. Biotechnology can help by growing microbes in the laboratory that can be used to inoculate the soil. This can be done by coating the seeds just before they are planted, or later, when the plants are already established. It is likely that in the future micro-organisms will be genetically engineered to more effectively promote plant growth and crop yield. Adopting agronomic practices that contribute to the stability of the soil ecosystem (crop rotation, mulching, no tillage, and so on) would eliminate many of the problems caused by present practices.

Further Reading

Allen, M. F. 1991. *The Ecology of Mycorrhizae.* Cambridge, UK: Cambridge University Press.

Campbell, R. 1989. *Biological Control of Microbial Plant Pathogens.* Cambridge, UK: Cambridge University Press.

Elliott, L. F., and J. K. Frederickson. 1986. Plant–microbe interactions in the rhizosphere. In *Future Developments in Soil Science Research.* Madison, WI: Soil Science Society of America. Pp. 145–156.

Garbaye, J. 1991. Biological interactions in the mycorrhizosphere. *Experientia* 47:370–375.

Harman, G. E. 1992. Development and benefits of rhizosphere competent fungi for biological control of plant pathogens. *Journal of Plant Nutrition* 15:835–843.

Keister, D., and P. Cregan. 1991. *The Rhizosphere and Plant Growth.* Amsterdam: Kluwer.

Lynch, J. M. 1983. *Soil Biotechnology: Microbiological Factors in Crop Productivity.* Oxford, UK: Blackwell.

Lynch, J. M. 1990. *The Rhizosphere.* New York: Wiley.

Pankow, W., T. Boller, and A. Wiemken. 1991. Structure, function and ecology of mycorrhizal symbiosis. *Experientia* 47:311–322.

Chapter 9

The Molecular Basis of Plant Breeding and Genetic Engineering

There are few more striking examples of science in the service of humanity than the application of the principles of heredity to crop improvement and food production. To obtain high yields from their crops, farmers can use all the technological inputs available to them. However, sound agronomic practices, irrigation, fertilizers, and pest control will improve the yield of a given crop variety only to the point at which the plants are making maximum use of the inputs. When the yield plateau is reached for a given variety, further improvement in yield is not possible, because the variety has reached its inherent maximum yield. The major achievement of plant breeding has been to create new varieties that have pushed upward the limits of yield of the major crops, giving us our present high-yielding varieties.

In the future, increasing yield for traditional crops, improving nutritional or postharvest quality of these crops, adapting them to more stressful environments, domesticating new crops, and converting existing ones to plant factories that produce chemicals for industry will all flow from applying the principles of plant breeding, genetics, and molecular biology. Even "classical" plant breeders—those who perform mostly field work—are relying more and more on the tools of molecular biology to expedite the long process of producing a new variety.

1 | **Charles Darwin first identified the important role of natural variation in evolution and crop improvement.**

Charles Darwin, who has been called the father of modern biology, was a naturalist who made extensive observations on plants and animals. He observed that individual members of a given species are not exactly alike, but show a great deal of variation. He knew that "Like begets like," but noticed that subtle differences often appear between parents and their offspring. He proposed in 1858 that because of this variation, some individuals would be more suited than others to the environment in which they lived. Darwin concluded that species are not fixed—their characteristics change over time—

Figure 9.1

Frequency Distributions for Seed Weight of Beans of Plants Grown in a Single Field and Using Only a Single Variety of Bean Plants. Source: *D. L. Hartl (1991),* Basic Genetics *(Boston: Jones and Bartlett), p. 215.*

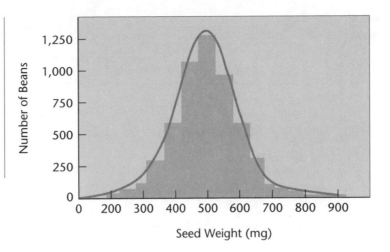

and suggested that the driving force behind this change is the number of offspring each individual leaves as a result of its fitness for the environment. Darwin called this process **natural selection**, and he proposed that it could cause a gradual shift in the characteristics of the population of individuals that constitute a species. Darwin's hypothesis was controversial because of the implication that the human race itself evolved from a nonhuman ancestor. Most scientists agree that natural selection is the driving force behind evolution. Indeed, evolution can readily be demonstrated in the laboratory and in the field with organisms that have a short life cycle.

The variation that occurs among individuals of a species can be all-or-none, or it can be continuous. Genetic diseases are clear cases of all-or-none variation: some people have hemophilia, others do not; some wheat varieties are susceptible to rust fungus, others are not. Most variation is continuous. For example, most plant populations, like human populations, show continuous variation in height. Many plants cluster around an "average" height, but other individuals are taller or shorter. The same is true for the seed yield obtained from a single plant or the weight of individual seeds, as illustrated in Figure 9.1. Data on variability of a trait, such as seed size, are often represented as a histogram. The researcher arbitrarily divides the total size range in a number of size classes, and then counts the number of individuals in each class. Plotting the results then produces a bell-shaped curve.

The causes of variability in the seed yield of individual plants or the seed weight shown in Figure 9.1, and of other traits that show such continuous variability, are complex. Both *genes* and the *environment* are involved, and discovering the contributions of heredity and the environment in determining a character is an important and difficult job. If the environment produced the observed variability because there are differences in soil fertility or soil moisture, then greater and more uniform yields may be obtained by optimizing the environmental factors (agricultural inputs) as much as possible. If genes are primarily responsible for the variability, then it should be possible to select for a variety that has a high seed yield per plant.

The answer usually lies somewhere between these two possibilities: both genes and the environment play a role in the observed variability. By growing different varieties of wheat in the same environment, and a single wheat variety under different environments, agronomists can assess the relative contributions of heredity and environment to the variability of different

	Table 9.1		Heredity and environment

Characteristic	Percentage Determined by	
	Heredity	Environment
Conception rate in cattle	5%	95%
Ear length in corn	17	83
Egg production in poultry	20	80
Yield in corn	25	75
Oil content in corn	65	35
Egg weight in poultry	60	40
Root length in radishes	65	35
Slaughter weight in cattle	85	15

Source: J. Brewbaker (1964) © *Agricultural Genetics.* Reprinted by permission of Prentice-Hall, Inc., Englewood Cliffs, NJ.

traits. Such assessments have been made for a number of important traits of plants and animals (see Table 9.1).

Technically speaking, the many characteristics by which people recognize an organism constitute its **phenotype,** or appearance. For example, we recognize a wheat plant by the shape and positioning of its leaves, by its size, by the presence of the ear, and by many other features. These characteristics constitute the phenotype of the wheat plant. All characteristics of the phenotype are inherited; that is, transmitted from the parent to its offspring. They are the result of the inheritance of specific genes, the interactions between those genes, and the degree to which the environment modifies the expression of these genes. In the example shown in Figure 9.1, the variability is caused entirely by environmental effects because the beans with which this experiment was done were highly inbred and genetically as identical as possible.

2 | **By studying the inheritance of all-or-none variation in peas, Mendel discovered how characteristics are transmitted from one generation to the next.**

More than a hundred years ago, Gregor Mendel, a Moldavian monk, discovered the basic rules that govern heredity as a result of a series of experiments in which he crossed two varieties of pea plants, one with round and one with wrinkled seeds. In addition, these varieties had a number of other characteristics that also show all-or-none variation. Remarkably, the characteristics he used can still be found today, precisely because they all have all-or-none variation. He published his observations in 1865 in a local scientific journal, but his findings went unnoticed until the same observations were made 35 years later by other scientists.

Mendel, who was very familiar with peas, knew that they always breed true, because peas are self-fertilizers and the pollen fertilizes the pistil (female reproductive organ) even before the flower opens. Mendel had many types of true-breeding peas growing in his garden, and he knew that certain lines of his peas had different characteristics and wanted to know what would happen if he cross-fertilized them. One line of pea always produced smooth, round peas, but another one produced wrinkled peas. When the latter peas dry out at the end of seed maturation, some of the inner tissues collapse, providing the seed with a wrinkled appearance (phenotype).

When Mendel crossed round and wrinkled peas, he observed that the first generation (the F_1 generation) consisted entirely of round peas. When these round peas were allowed to sprout, grow, and flower, and when the plants were allowed to set seed, most of the seeds (about 75%) were round and a minority (about 25%) were wrinkled (Figure 9.2). None were in between: just a little wrinkled or nearly round. Thus, a characteristic that disappeared in the first generation reappeared in the second.

Mendel repeated these experiments using seven other discontinuous characteristics (for example, green and yellow seeds) and found them to be generally true. Interestingly, he found that these seven different characteristics were not linked, but were transmitted to the next generation independently. Thus, when he crossed a round yellow pea with a wrinkled green pea, the first generation had only round yellow peas. However, in the second generation he found four types: the two original types and two additional combinations of the characteristics: round green, and wrinkled yellow.

From such experiments, Mendel drew two important conclusions. First, characteristics or traits are transmitted to the next generation as discrete units, now called **genes.** Second, an individual must contain two copies of each of these units, and each parent transmits only one copy to the next generation. That is the only way he could account for the disappearance and subsequent reappearance of a characteristic. The implication is that the unit (gene) is always present, but may not be *expressed,* as is the case with the wrinkled or green characteristics in the first generation of crosses of round and wrinkled or green and yellow peas.

Although Darwin and Mendel were contemporaries (Mendel published his work in 1865, fifteen years before Darwin's death), Darwin did not know of Mendel's work. Darwin knew that characters could disappear and reappear, but he did not understand their inheritance. In his book *On the Origin of Species by Means of Natural Selection,* Darwin discussed the notion that a specific trait can be inherited by one child but not by another and that traits of grandparents sometimes appear in the grandchildren although they were not present in the parents. He did not understand this skipping of a generation, which is so beautifully explained by Mendel's experiments.

The raw material that evolution works with is the variability in the population. As the environment slowly changes, individuals with certain characteristics are more fit and leave more offspring, and the changing environment is said to provide the **selection pressure.** Selection pressure can also be provided by humans, either knowingly or unknowingly. We already mentioned that early agriculturalists some 10,000 years ago may have unknowingly selected in favor of certain traits in their crop plants. Plant breeders generally select consciously in favor of certain characteristics, and crop improvement is therefore an accelerated form of evolution. This type of breeding depends on the same natural variability in the population that allows evolution to occur.

The emergence of pesticide-resistant plant pests—whether weeds, insects, bacteria, or fungi—is also the result of accelerated evolution. If the pesticide kills or greatly slows down the growth of the overwhelming majority of the pests when it is applied, then the selection pressure is enormous, and resistant offspring will quickly multiply and replace the original population that

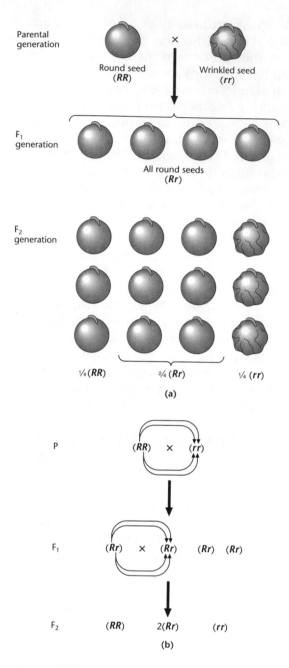

Figure 9.2

Inheritance of the Round and Wrinkled Characters in Peas. *(a) When plants that always produce round seeds are mated with plants that always produce wrinkled seeds, the plants in the first generation (F₁) all produce round peas. However, the F₁ plants still have the wrinkled gene. When the F₁ plants are crossed with one another, one-quarter of the seeds in the F₂ generation are wrinkled and three-quarters are round. (b) This pattern of inheritance can be explained by assuming that each plant carries two copies of the genes for each character, and that being round is dominant over being wrinkled. The original parents, which bred true, are either RR or rr, and when these plants are crossed, gametes are either R or r, and all offspring are Rr. When these F₁ plants are crossed, both can produce R and r gametes and there are more possibilities for the offspring: there will be two Rr plants for every one that is either rr or RR. However, because R is dominant, three-quarters will be round (RR and Rr) and only one-quarter will be wrinkled (rr). Source: Redrawn from J. L. Marx, ed. (1989),* A Revolution in Biotechnology *(Cambridge, UK: Cambridge University Press), p. 10.*

was sensitive to the pesticide. The presence of a resistant individual is not *caused* by the pesticide, but is the result of the variability that exists in any natural population. The pesticide simply supplies the selection pressure. This accelerated evolution of pests accounts for the emergence of so many pesticide-resistant insects and weeds in agricultural fields.

Another example of accelerated evolution is the rapid evolution of plant pathogens that attack genetically uniform crops grown as monocultures. After the plant breeders release a new variety of wheat that is resistant to the wheat rust fungus, the fungus has a temporary setback. It cannot infect the plants because it cannot penetrate the plant's defenses, so the fungus population is kept at very low levels. However, genetic variability in the fungus population ensures that individuals (mutants) will arise that can breach those defenses. Then the fungal population rapidly increases, because of the genetic uniformity of the wheat plants. Once the fungal mutant has arisen, all the wheat plants will be equally susceptible to the new fungal strain. In former times, when the genetic variability of crop plants was much greater and many plants were more or less susceptible, the selection pressure on the fungus was much less. The genetic uniformity has increased the selection pressure and the rate at which new pathogen varieties arise.

3 | **Mutations, heterozygosity, and independent assortment of genes are the sources of genetic variation.**

Why are there smooth and wrinkled peas? Biochemists have recently discovered that wrinkled peas lack one of the important enzymes for starch synthesis. During their development, peas import sucrose from the rest of the plant. This sucrose is quickly converted to starch, which makes up 60% of the weight of a mature pea seed. In wrinkled peas, one of the enzymes for starch synthesis is lacking, so when the seeds are mature they contain a lot of sucrose and water and much less starch than is normal. Then, when the seeds dry out they become wrinkled. The absence of the enzyme is caused by an alteration in the gene that encodes the information for the synthesis of this enzyme. All-or-none variation, as in the case of the smooth and wrinkled peas, is usually controlled by a single gene.

Many agronomically important characteristics show continuous variation, such as yield or protein content of seeds, and the inheritance of such traits is controlled by many genes. Such traits are referred to as **multigenic traits.** Thus, the ability to take up soil nutrients, to photosynthesize, to transport photosynthate to the seeds, to withstand drought all will affect yield. And each of these characteristics is controlled by many genes, making yield truly a multigene trait.

As already mentioned for the round and wrinkled peas, one source of genetic variation is that genes are not stable, but can mutate. Mutations occur rarely, but when they do they may be passed on to the next generation. Many mutations produce only minor changes in the gene, and this brings about a small change in the phenotype. However, when these small mutations accumulate over time, their combined effects can account for quite a bit of the observed variation.

Figure 9.3

Visualization of Chromosomes in Microsporocytes (the precursors of the gametes or sex cells) of Lily. The cells have been treated with a dye that binds to DNA and allows all DNA-containing structures to be seen clearly. In **(a)** the chromosomes have just started to contract and are visible as long thin threads. In **(b)** contraction is at its maximum, and each chromosome is clearly distinguishable. In **(c)** the chromosomes are being separated from each other during the second meiotic division, and in **(d)** the formation of the four gametes is nearly complete. Source: *D. Sadava (1993),* Cell Biology *(Boston: Jones and Bartlett), p. 487.*

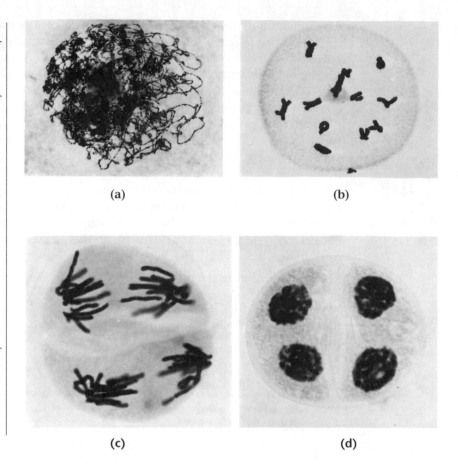

(a)   (b)

(c)   (d)

The second, and most important factor that creates genetic variation in a population is the existence of two copies of every gene, as postulated by Mendel. Genes are located on filamentous structures in the cell nucleus called **chromosomes.** The chromosomes are discrete structures that become visible in the light microscope when a cell undergoes the process of cell division (Figure 9.3). Most organisms have in their cells some 50,000–100,000 genes, but only a small number of chromosomes. Thus, thousands of genes are linearly arranged on each chromosome.

Higher organisms have two copies of every gene in every cell, except in the sex cells (sperm and egg cells), which only have one copy of each gene. In humans, each cell has 23 pairs of chromosomes, for a total of 46. Corn has 10 pairs of chromosomes, and *Haplopappus,* a plant that thrives in dry areas, has only 2 pairs. Because cells have two copies of every chromosome they also have two copies of every gene, one copy on each of the chromosomes that make up a pair of chromosomes. When these two copies are identical, the organism is said to be **homozygous** for that gene. If one of the two gene copies has mutated, they will be different, or **heterozygous.** These different forms of the gene are called **alleles.** Plants that normally self-fertilize, such as peas and beans, are homozygous for a very large number of genes, whereas plants that outcross, such as corn, may be heterozygous for most of their genes.

The chromosomes play a vital role in passing on genes, and therefore, phenotypic characteristics, from a mother cell to the daughter cells during cell division, and from the parents to the offspring during reproduction. Cell division, which in plants occurs in meristems, is preceded by **mitosis**, a process of chromosome duplication, and the subsequent separation of the two sets of chromosomes. Thus, for a short while, a cell contains four copies of every gene, but as the new chromosomes separate, each of the daughter cells contains again two copies of every gene (see Box 9.1). During the formation of egg cells and sperm cells, a different type of cell division occurs. A single cell undergoes one round of chromosome replication, followed by two rounds of chromosome separation and cell division. Four cells are formed in this process, called **meiosis**, and each cell ends up with a single copy of each gene. If the organism is heterozygous for that gene, then half the sex cells will have one allele of the gene and the other half will have the other allele.

When two sex cells fuse during fertilization to start a new organism, the cells of this new organism will again contain two copies of every gene. A plant homozygous for a given gene will produce sex cells, all of which have identical forms of that gene. That is, in terms of that gene, it produces only one type of reproductive cell. However, a plant that is heterozygous for a certain gene, with (for example) one normal and one mutant copy, will produce two types of reproductive cells. Half the sex cells will carry the normal gene, half will carry the mutant gene. The fact that many plants are normally heterozygous for many genes creates many possibilities for variation in the offspring, because random assortments of genes are brought together when the sperm cell fertilizes the egg cell.

A third factor that leads to variation in a population of organisms is that many genes affect a single phenotypic characteristic. For example, among the more important genetic characteristics that determine yield are disease resistance, photosynthetic efficiency, seed size, and ability to take up soil nutrients and respond to fertilizer. Each of these traits is determined by many genes. For example, responsiveness to nitrogen fertilizers involves genes that regulate the uptake of nitrate by the roots, the conversion of nitrate to ammonium in the plant body, the translocation of amino acids to the developing seeds, and the biosynthesis of seed proteins. To sort out the combinations of genes involved in agriculturally important characteristics is the challenge facing the plant breeder.

4 The genetic material is DNA.

Careful observation of the separation of the chromosomes during mitosis and the discovery that sex cells contain only half the number of chromosomes (one set instead of two), led scientists to postulate that Mendel's inherited units, later called *genes,* must be organized on the chromosomes, the filamentous structures found in nuclei. Biochemical analysis showed that nuclei contain primarily two substances, proteins and nucleic acids. For a long time, biochemists believed that genes were made up of protein because proteins were known to be large, complex molecules (long strings of 20 different amino acids) that showed a great deal of specificity in their activity (mostly enzymatic activity). Biochemists thought that only protein molecules could contain enough information to specify inherited traits.

Nucleic acids, in contrast, were thought to be small and relatively simple molecules. Decisive experiments carried out with bacteria about 50 years ago showed that genes are made of DNA (deoxyribonucleic acid) and that they

Box 9.1

*Mitosis and
Meiosis*

Cell division in plants serves two main purposes. First, adding new cells is the basis of growth, although in plants much growth in size occurs by cell expansion. Thus the growth of a new leaf from a microscopic leaf primordium and the addition of an annual ring in the stem of a tree both require much cell division. Each cell division is preceded by **mitosis** (see figure, **a**), a nuclear process in which each chromosome duplicates itself, the two halves separate and two new nuclei are formed. Second, cell division is also crucial to the formation of sex cells or gametes, and gamete formation is accompanied by **meiosis** (see figure, **b**), a process in which the chromosomes also duplicate and separate, but the cells end up with only half as many chromosomes as they originally had.

Let us consider a plant with a haploid chromosome number of 2. Its normal diploid cells will have two pairs of chromosomes in each cell. Such a cell is shown (figure, **a**) with two heterozygous genes: *Tt* located on the long chromosome and *Aa* located on the short chromosome. *A* and *a* denote two different alleles of the same gene, and so do *T* and *t*. Before mitosis begins, the cell contains in its nucleus four chromosomes: one pair of each type. The first step in mitosis is a duplication of each chromosome. This doubling actually occurs before the chromosomes are visible with a light microscope as discrete structures. Thus, when the chromosomes do become visible, they appear as double strands, already containing two copies of each of their genes. At that point, the cell contains four copies of each gene.

During the subsequent stages of mitosis, these doubled strands line up at the center of the cell and then separate longitudinally, a single strand of each pair going to one end of the cell, and its sister strand to the other end of the cell. Because the two strands are identical copies of the same, original genes, this means that the two new cells that form will contain the identical genetic complement of the cell from which they came. Each daughter cell will have two copies of every gene. If the two copies were different, then those differences would be passed on from the mother cell to the daughter cells.

The first phase of meiosis is identical to that of mitosis (see figure): the chromosomes duplicate themselves, and each one becomes a double strand. These double-stranded chromosomes now line up in the center of the cell in such a way that two homologous or look-alike chromosomes line up beside each other. In the next step, these two look-alike chromosomes separate and move to opposite poles of the cell, each chromosome in its double-stranded form. The resulting two daughter cells each have only half as many chromosomes as the mother cell.

This separation of chromosomes is the fundamental difference between mitosis and meiosis, for in mitosis the two strands of each chromosome separate and move to opposite poles of the cell. In meiosis, when these double-stranded chromosomes migrate to the ends of the cell, they do not split, as in mitosis. Instead, one doubled chromosome of each pair goes to each end of the cell. This process is followed by mitosis without an intervening duplication of the chromosomes. Thus, four sex cells are formed, each with a single copy of each chromosome and gene. In this way, a parent passes only one of its two copies of each gene to its offspring, and the offspring inherits one copy of each gene from its maternal parent and one from its paternal parent. This is obviously one way in which an organism can become heterozygous for a given gene.

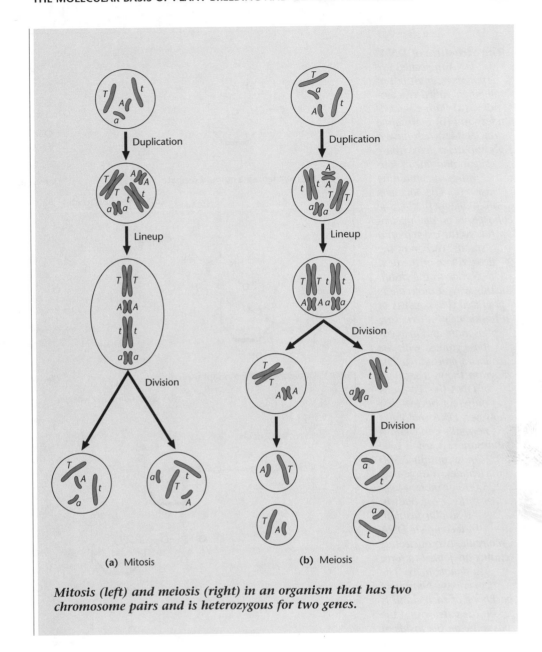

Mitosis (left) and meiosis (right) in an organism that has two chromosome pairs and is heterozygous for two genes.

have the necessary information to direct the synthesis of proteins. Mutation of bacterial genes was shown to result in the loss of specific enzymatic activities, indicating that genes are responsible for the presence of proteins in the cell. This discovery later led to the "one gene, one enzyme" hypothesis, which stated that a distinct gene controls the presence of each enzyme in the cell.

Chemical analysis showed that DNA is made up of four different nucleotides that are released when DNA is broken down with mild acid. More careful extraction of DNA from cells led to the realization that DNA is made of huge molecules—much bigger than proteins—but that DNA is easily degraded when extracted from tissues. The nucleotides—the building blocks of DNA—consist of a phosphate group, a sugar (deoxyribose), and a nitrogen-containing base (Figure 9.4). The four bases in DNA are adenine, guanine,

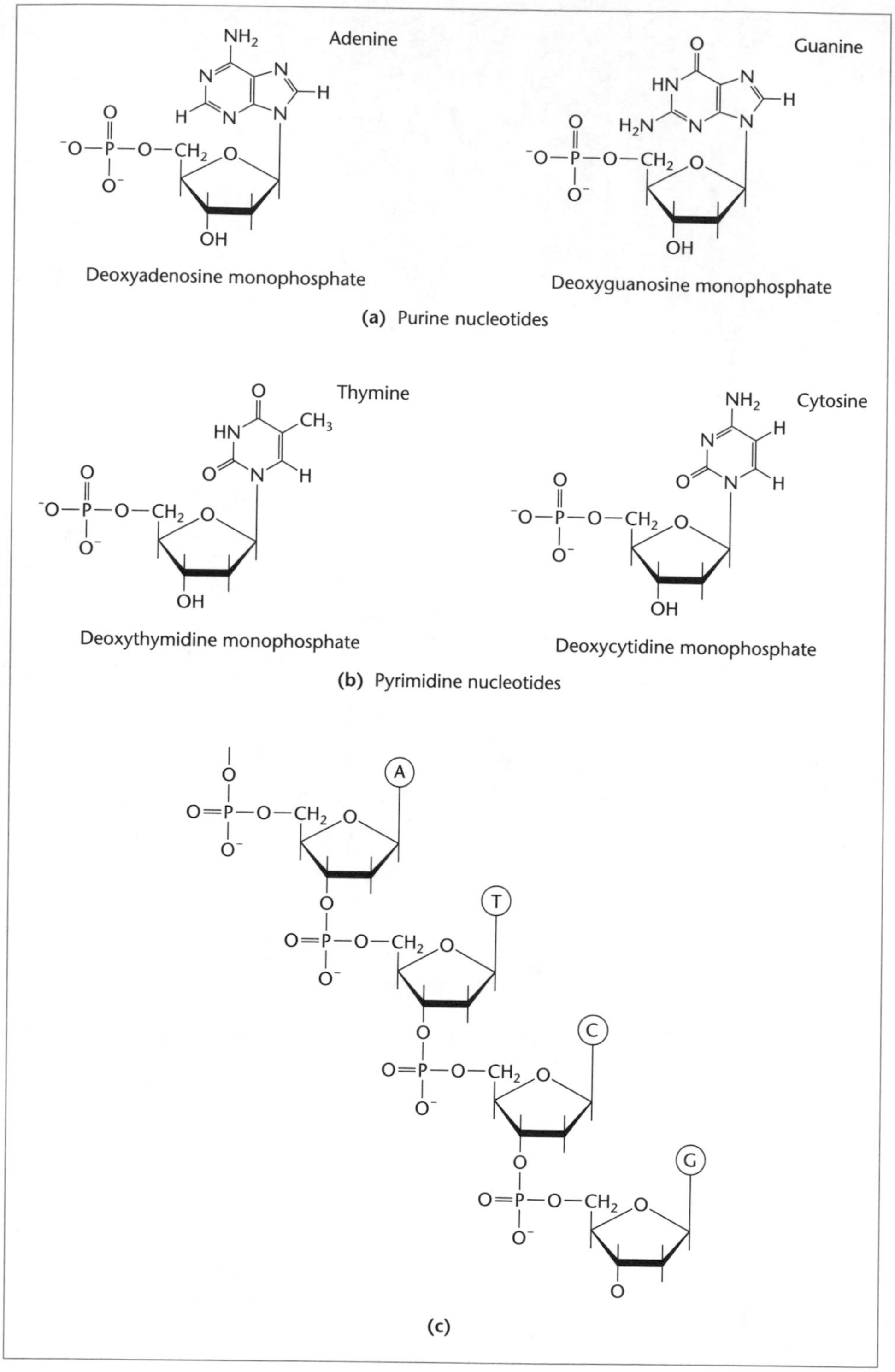

Figure 9.4

The Structure of DNA.
DNA molecules are
polymers, built out of
monomers called nucleo-
tides. **(a)** *Purine nucleo-*
tides, and **(b)** *pyrimidine*
nucleotides each have a
phosphate, a deoxyribose
sugar, and one of four
nitrogen-containing
bases. **(c)** *The polymers*
are made by linking the
phosphate on one nucleo-
tide to the oxygen on a
specific carbon of the
deoxyribose of the next
nucleotide in the chain.
In this way, a molecule is
produced that consists of
a backbone of alternating
phosphate and deoxyri-
bose groups, with the
bases protruding. The
diagram shows a segment
of DNA that contains
only four nucleotides;
however, DNA molecules
normally can contain
thousands. A small gene
may contain 2,500
nucleotides; a large one,
25,000 or more. Mole-
cules of RNA are built in
a similar fashion.
In a complete DNA
molecule, two nucleotide
chains are joined to form
a double helix. The
deoxyribose–phosphate
backbone of each chain is
on the outside and the
bases are on the inside of
the double helix. The
bases of one chain are
paired with those of the
other chain according to
the base-pairing rules:
thymines always pair
with adenines and
cytosines with guanines.

(a) Purine nucleotides

Deoxyadenosine monophosphate   Deoxyguanosine monophosphate

(b) Pyrimidine nucleotides

Deoxythymidine monophosphate   Deoxycytidine monophosphate

(c)

thymine, and cytosine, often written as their first-letter abbreviations A, G, T, and C. Thus, DNA has a backbone of alternating sugar and phosphate groups with bases attached to each sugar group. The information or specificity that DNA contains is to be found in the exact sequence of the bases.

The higher-order structure of DNA was elucidated by a research team whose best-known members were Francis Crick, Rosalind Franklin, and James

Figure 9.5

Replication of DNA. When DNA replicates, the two strands of the molecule separate and each serves as a template for the synthesis of a new complementary strand, here shown in gray. The obligatory pairing of adenines with thymines and guanines with cytosines ensures that the two daughter molecules will be duplicates of the original. Source: From D. L. Hartl, Basic Genetics (Boston: Jones and Bartlett), p. 97.

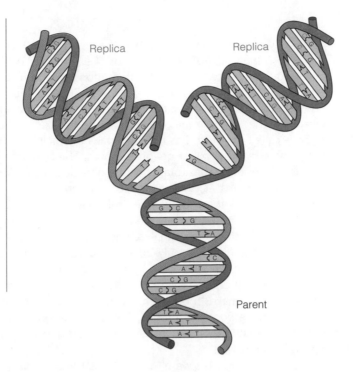

Watson. They concluded that DNA is a double-stranded helix with the adenine (A) in one strand always hydrogen-bonded to the thymine (T) in another strand; similarly cytosine (C) in one strand is always hydrogen-bonded to guanine (G) in the other strand. Together, the two strands form the double helix. The two strands of the DNA molecule are said to be **complementary** because if we know the base sequence of one strand, we can deduce the base sequence in the other complementary strand. If one strand contains the sequence ATTGCC, then the other strand must in the same region have the sequence TAACGG because of the base pairing rules (A opposite T, and G opposite C).

The beauty of the molecular design of DNA is that it provides a means of maintaining the information in DNA and for copying and then transmitting it to daughter cells. When DNA replicates the two strands, separate and complementary strands are synthesized using the existing strands as template. When the two new strands are synthesized, A is always opposite T and C opposite G, so that two identical molecules are formed (see Figure 9.5). Because the information is contained in the exact sequence of the bases, DNA replication allows for a new and exact copy of the information to be made.

5

> **Genetic information stored in DNA is used to assemble RNA, and this information is then used to synthesize protein.**

Proteins are linear molecules of 20 different amino acids—usually between 100 and 10,000 of them—folded in a three-dimensional configuration that is often globular in overall shape. The individual amino acids are linked together by linkages called *peptide bonds* (Figure 9.6a), and for this reason a protein molecule is often called a *polypeptide.* This string of amino acids is then wound

Box 9.2

*A Common Weed
with an
Uncommon
Genome*

Fifteen years ago, few researchers had ever heard of a small weed called *Arabidopsis thaliana,* but now hundreds of scientists all over the world are analyzing its genes. There is even an international program to determine the base sequence of each and every one of its 30,000 genes (about 100 million nucleotides total). Why such a sudden flowering of interest in a seemingly useless weed? In the 1980s, molecular biologists discovered that *Arabidopsis,* as the plant is called for short, has only five pairs of small chromosomes and that it is the plant with the smallest genome known. The term *genome* is used by geneticists to describe the sum total of all the genetic information or all the genes of a species. Other species, such as tobacco and wheat, probably don't have many more genes, but these genes are embedded in 15 and 60 times more DNA respectively. The presence of this large amount of "excess" DNA makes it much harder to find the genes, analyze them, and figure out how they work. The genome of *Arabidopsis* is only 20 times larger than that of the lowly intestinal bacterium *Escherichia coli,* yet the plant is incredibly more complex. With its small genome, it does all the same things that plants with much larger genomes do: *Arabidopsis* makes roots, stems, leaves, and flowers; photosynthesizes; takes up nutrients; responds to stresses; fights off pathogens; and synthesizes a large number of metabolites; and its development is governed by hormonal and environmental signals.

Molecular geneticists were the first to realize that many of the things we want to know about our important crops can be figured out much more easily by working with *Arabidopsis.* An international collaborative effort under the leadership of Chris Sommerville, now at Stanford University, is under way to map the *Arabidopsis* genome. Mapping involves physically mapping all the chromosomes, showing the location of each gene that is known from a phenotypic trait. Once a high-density map has been made, and the location of many genes is known, it becomes much easier to find and isolate genes. And once a gene from *Arabidopsis* has been isolated, it is usually quite easy to isolate the gene that encodes the same or a very similar protein from a crop plant. In other words, if scientists understand the molecular details of how *Arabidopsis* regulates the uptake of nitrate from the soil, then that information is immediately transferable to crops.

After the molecular geneticists showed the way, others interested in crop plants followed. They also realized that the genetic secrets of crop improvement may indeed be clarified by studying the genes of this diminutive common weed.

into a helix or can take the shape of a pleated sheet. When the entire chain is folded, some parts will be helical, other parts will be pleated, and yet other portions of the chain will form a random coil (Figure 9.6).

Proteins may interact with other molecules in the cell in very specific ways: as enzymes, as regulators of gene activity, or as transporters of ions. This specificity resides in their exact three-dimensional shape, which is in turn determined by the sequence of amino acids. Thus, each kind of protein, and cells probably have more than 25,000 different proteins, has a characteristic amino acid sequence. The information that specifies the order in which the amino acids must be assembled when a protein is synthesized is contained in the DNA, specifically in the sequence of bases that are attached to the sugar–phosphate backbone.

At first sight this might seem to be a problem, because DNA has only four different bases and proteins have 20 different amino acids. This problem is

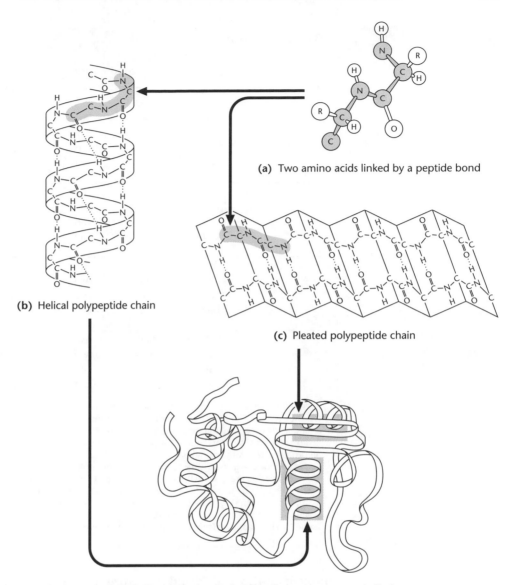

(a) Two amino acids linked by a peptide bond

(b) Helical polypeptide chain

(c) Pleated polypeptide chain

(d) Protein molecule showing that some parts are helical
and some are pleated; yet other parts form a random coil

Figure 9.6

Hierarchy of Protein Structure. (a) *Primary structure: peptide bond between N and C, linking two amino acids.* (b) *and* (c) *secondary structure: helical and pleated polypeptide chains,* (d) *tertiary structure: helices, sheets, and random coils in a globular protein. See also Figure 5.13b for a model of an entire protein.*

solved by having a sequence of three bases specify one amino acid. When four different nucleotides are arranged in groups of three, it is possible to make 64 different combinations (4 × 4 × 4), more than enough to specify 20 amino acids. In the 1960s, researchers figured out which three-base combination or **codon** specifies a particular amino acid. Most amino acids are represented by more than one codon. Of the 64 different possibilities, 61 codons specify amino acids, and 3 specify "stop"—signals marking the end of a protein-coding segment of the DNA (Table 9.2).

Cells have an elaborate machinery for translating the nucleotide sequences in the DNA into amino acid sequences in proteins. It would be

Table 9.2 | **The genetic code**

First Position	Second Position				Third Position
	U	C	A	G	
U	Phe	Ser	Tyr	Cys	U
	Phe	Ser	Tyr	Cys	C
	Leu	Ser	Stop	Stop	A
	Leu	Ser	Stop	Trp	G
C	Leu	Pro	His	Arg	U
	Leu	Pro	His	Arg	C
	Leu	Pro	Gln	Arg	A
	Leu	Pro	Gln	Arg	G
A	Ile	Thr	Asn	Ser	U
	Ile	Thr	Asn	Ser	C
	Ile	Thr	Lys	Arg	A
	Met	Thr	Lys	Arg	G
G	Val	Ala	Asp	Gly	U
	Val	Ala	Asp	Gly	C
	Val	Ala	Glu	Gly	A
	Val	Ala	Glu	Gly	G

Note: A sequence of three nucleotides forms the nucleic acid codon for a single amino acid. The four nucleotides C, A, U, G can produce 64 different three-nucleotide combinations. All the amino acids except methionine (Met) and tryptophan (Trp) have more than one codon. The "stop" codons UAA, UAG, and UGA do not code for amino acids but signal the end of a protein. All proteins start with methionine.

The codons are given as they appear in messenger RNA. The four bases in the nucleotides of ribonucleic acids are uracil (U), cytosine (C), adenosine (A), and guanine (G). The amino acids specified by the genetic code are alanine (Ala), arginine (Arg), asparagine (Asn), aspartic acid (Asp), cysteine (Cys), glycine (Gly), glutamine (Gln), glutamic acid (Glu), histidine (His), isoleucine (Ile), leucine (Leu), lysine (Lys), methionine (Met), phenylalanine (Phe), proline (Pro), serine (Ser), threonine (Thr), tryptophan (Trp), tyrosine (Tyr), and valine (Val).

simple if amino acids could recognize their own codon and just line up in the right order on the surface of the DNA to be linked together by an enzyme, but nothing so simple evolved. Rather, cells use a different class of nucleic acids, **RNAs (ribonucleic acids),** to help translate the information contained in the DNA. RNAs are also strings of nucleotides, and consist of a sugar–phosphate backbone with a base attached to each sugar group, except that the sugar is ribose. Three of the bases in RNA are the same as in DNA (C, A, and G) but instead of thymine (T), uracil (U) is substituted in RNA.

Transcription of a gene is the formation of a messenger RNA (mRNA) molecule. The mRNA carries the information that will specify the amino acid sequence, from the nucleus, where the message is contained in the nucleotide sequence of the DNA, to the cytoplasm of the cell, where this message will be translated and the protein synthesized. **Translation** of the message involves transfer RNA (tRNA) molecules as well as ribosomes, specialized structures on whose surface new proteins are assembled (Figure 9.7b). As the amino acid chain gets longer, it begins to fold up, first in a helical configuration, and then in a three-dimensional globular shape.

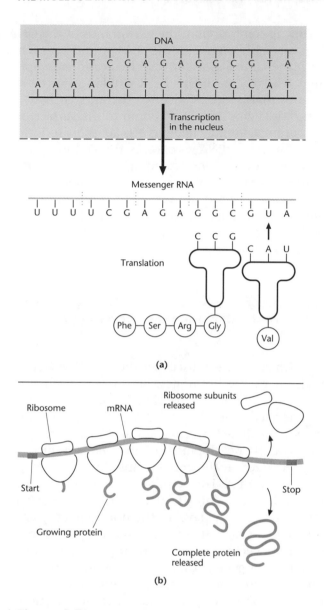

Figure 9.7

Protein Synthesis, **(a)** *and* **(b).** *Before protein synthesis can begin, the genetic information (sequence of bases) in the DNA must be copied into an RNA molecule. The two strands of the DNA molecule separate, allowing one strand to be copied into a messenger RNA (mRNA). This process, called* transcription, *takes place in the nucleus. The mRNA moves from the nucleus to the cytoplasm where protein synthesis occurs. This coincides with a maturation process in which parts of the mRNA that do not code for protein are removed. In the cytoplasm, specific enzymes join each of the amino acids to the appropriate transfer RNA. The messenger RNA and the transfer RNA for the first amino acid of the protein form a complex with the ribosome. Then the ribosome effectively moves along the messenger RNA molecule. As it moves, the "anticodon" on the transfer RNA for each successive amino acid recognizes the corresponding codon on the messenger RNA and brings the amino acid into position for joining to the growing protein chain. In this way, the information encoded in the linear sequence of nucleotides in the DNA of the genes is translated into the linear sequence of amino acids in a protein. Source: Redrawn from J. L. Marx, ed. (1989),* A Revolution in Biotechnology *(Cambridge, UK: Cambridge University Press), p. 10.*

During transcription, the two strands of the DNA separate and ribonucle-otides assemble on one of the DNA strands in such a way that the bases will pair in the same way as for DNA replication, except that A in the DNA will be paired with U (uracil) in the RNA being synthesized (Figure 9.7a). Thus, a sequence that is CGATC in DNA becomes GCUAG in RNA. When the mRNA reaches the cytoplasm, it will bind to the surface of a ribosome where tRNAs with amino acids attached serve as adaptors to line up the amino acids in the correct order. A tRNA recognizes the codon in the mRNA by means of an *anticodon,* which aligns according to the base-pairing rules. There is at least one species of tRNA for each amino acid, and there are several tRNAs for most amino acids specified in the genetic code. As the tRNAs with their respective amino acids line up on the mRNA one by one, the amino acids are linked together by an enzymatic reaction and the tRNAs are released to react with new amino acids (Figure 9.7a).

6 | **The regulation of gene activity is an important aspect of cell differentiation and plant development.**

In Chapter 5, we discussed the role that light and other environmental signals, as well as hormones, have in the normal development of the plant. Light causes leaves to expand and chloroplasts to develop, and the hormone ethylene accelerates flower fading and fruit ripening. The action of all these signals is mediated by enzymes and other proteins, all products of specific genes. Orderly development of an organism implies that the correct proteins are synthesized at the correct time during development and in response to the proper stimuli. For example, the enzymes that cause the softening of the cell walls in the ripening fruit, are only synthesized in the fruit, and as a result of an increase in the level of the hormone ethylene.

So far, we have described genes only in terms of the information they contain to specify a sequence of amino acids and the structure of a protein. This part of a gene is called the **protein-coding region.** However, a gene is more complex (much longer, really) than just a protein-coding string of nucleotides. First of all, the protein-coding portion may be interrupted from one to a dozen times by short stretches of DNA whose function is not yet clear. These short stretches are also transcribed when RNA is first made in the cell nucleus, but they are later removed when the initial product of transcrip-tion is processed to a mature mRNA that is transported to the cytoplasm. The interruptions in the coding sequence are characteristic and different for every gene. Second, there is a DNA segment on each side of the protein-coding region that specifies when the gene is to be activated during development, to which stimuli (hormonal or environmental) the gene should respond, in which cells it should be active, how much mRNA is to be made, and when the gene is to be turned "on" and "off." These **regulatory regions** are, of course, of great importance for the orderly development of a cell and an organism (Figure 9.8).

Thus, genes that encode enzymes for the synthesis of chlorophyll are "on" in the leaves, but "off" in the roots and the flowers. Furthermore, they are "off" in the dark and "on" in the light. When a root finds itself in a soil that is rich in nitrate, genes are turned "on" that encode proteins needed for the uptake of nitrate, its transformation into ammonia, and the use of this

A cell contains thousands of different proteins, many of which are the enzymes that carry out the biochemical functions that make up life's processes. The conversion of light to chemical energy, the assimilation of carbon dioxide, the oxidation of sugar, respiration, and the biosynthesis of fatty acids, proteins, and nucleic acids are all processes that each require numerous enzymes. Yet the cell is not an "enzyme soup," but a beautifully structured factory with subcellular structures called **organelles** such as the chloroplasts, mitochondria, nucleus, and vacuole, which all have unique sets of enzymes. The cell's biochemical reactions and enzymes are compartmentalized in these organelles. The organelles themselves also may have compartments, each with their own set of enzymes. Scientists presume that cells have evolved this way to improve the efficiency of life's biochemical processes.

Many organelles are surrounded by a lipid-rich membrane similar to the plasma membrane that surrounds the whole cell. This membrane separates the enzymes within from the enzymes in the cytosol, the soluble phase of the cytoplasm. This membrane also regulates what goes in and out of the organelle. Organelles have not only soluble proteins within them, but specific proteins are part of their limiting membrane and of their internal membranes. Some organelles, such as the microtubules and microfilaments that make up the cytoskeleton, do not have membranes. They are simple, proteinaceous rods to which other proteins can be attached. Given this subcellular complexity, how do proteins reach their correct destination after they are synthesized?

One solution to this problem would be for each organelle to synthesize its own specific proteins. However, that is not how things work. Only two organelles, the chloroplasts and the mitochondria, synthesize a few of their own proteins. Most of their proteins and all the proteins of the other organelles, as well as all the proteins that are secreted from the cell, are made on the ribosomes that are in the cytosol. After the proteins are made, they must reach their correct destination in a process that cell biologists call **protein targeting.** Protein targeting involves recognition between the newly synthesized protein and the organelle for which it is destined. This recognition is the function of a part of the polypeptide called the *targeting domain.* This targeting domain recognizes (binds to) a receptor protein in the outer membrane of the organelle, and this allows it to enter the organelle. Once the protein has entered the organelle, the targeting domain is usually removed from the protein, making the protein somewhat smaller.

This targeting of proteins to specific subcellular organelles has important implications for genetic engineering. If we want a protein to work properly in a genetically engineered plant, we must be sure that it reaches its correct destination within the cell. We now know that if we take from beans the gene for a chloroplast protein and we transfer that gene to wheat, the protein will be targeted to the wheat chloroplasts. This is also true for other organelle proteins, indicating that targeting domains are not species specific. But what happens when a bacterial protein is made in plant cells after a plant is transformed with the bacterial gene? Bacterial cells have no internal compartments, and their proteins do not have domains that will be recognized by the receptors on the organelles of plant cells. This means that we have to produce genes that encode chimeric proteins: the gene for the bacterial protein must be fused with a short piece of DNA that encodes a specific targeting domain known to work in plant cells. When such a gene is introduced into a plant, the protein made will be bigger than it normally is in the bacterium, and the targeting domain will let it enter the organelle.

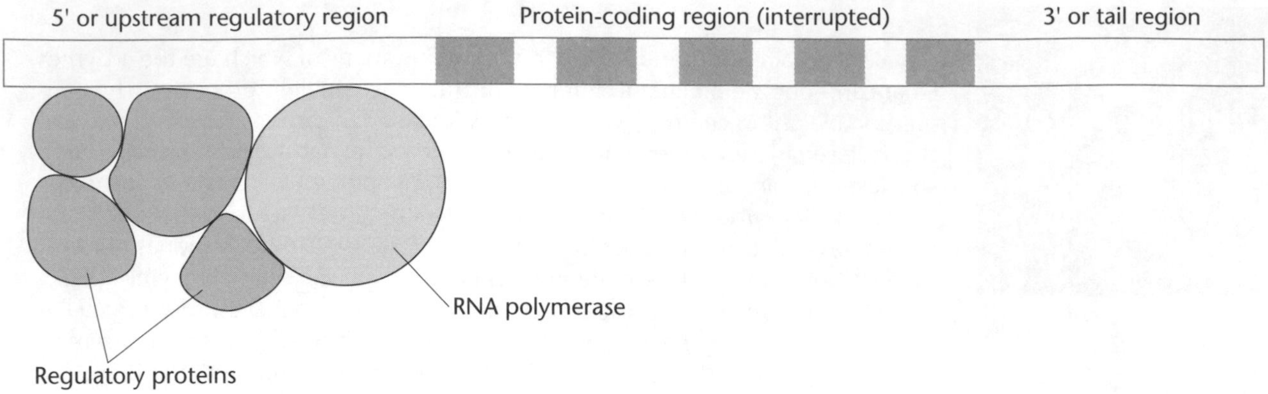

5' or upstream regulatory region   Protein-coding region (interrupted)   3' or tail region

RNA polymerase

Regulatory proteins

Figure 9.8

Schematic Representation of a Gene and Some Proteins That Can Bind to It. The protein-coding region is shown as being interrupted. RNA polymerase is the enzyme that is responsible for RNA synthesis and the RNA that is made first (initial transcript) will contain the entire protein-coding region with the interruptions. Regulatory proteins, called transcription factors, determine in which tissues or cells the gene is active, at what time of development, how much messenger RNA is made, or as a result of which signals the gene becomes active.

ammonia in amino acid biosynthesis. When the spore of a plant pathogen germinates on the surface of a leaf and tries to penetrate into the leaf cells, defense genes are rapidly turned on in the plant cells of resistant varieties. These defense genes encode enzymes that synthesize toxic compounds that will kill the invader. When light strikes a seedling that has been growing in the dark underneath the soil surface, hundred of genes that were completely off or barely on, are turned on and the cells start making hundreds of new proteins that allow the chloroplasts to develop, photosynthesis to start, and autotrophic growth to be initiated. These are but a few examples of the regulation of gene activity in response to specific stimuli.

The regulation of gene activity is the responsibility of proteins called **transcription factors** that bind to the regulatory regions of genes, and these transcription factors are themselves the products of genes. Thus, turning genes "on" and "off" is not such a simple matter and may involve a regulatory cascade in which the product of gene A activates gene B, whose product activates gene C, whose product activates the gene D by binding to its regulatory region.

Plant breeding and genetic engineering both involve the transfer of genes from one organism to another. In plant breeding, the breeder selects for an ultimate outcome or a phenotype, thus ensuring that the whole gene regulation pathway will operate correctly. Let us take the example of transferring resistance to a specific pathogen into wheat. A resistance gene may encode a protein that allows the plant to detect the invader quickly so that the plant can turn on its defenses. The gene that encodes this detector protein responds to a stimulus from the pathogen—perhaps a chemical or metabolite made by the pathogen—and this response involves several other genes. Thus, when the breeder transfers resistance to that specific pathogen from a wild variety of wheat to a domesticated wheat by selecting for pathogen resistance in the field, the entire regulatory cascade must be working correctly. If the new combination carries the resistance gene, but it is not correctly expressed, then the new variety would not be resistant to the pathogen.

Genetic engineers transfer one gene at a time and need to understand how the gene they transfer is regulated. First, they need to transfer not only the protein-coding part of a gene, but also its regulatory region. Alternatively, they can equip the gene with a new regulatory region that ensures correct expression. For example, regulatory regions of bacterial genes generally do not work in plants because the transcription factors of bacteria are different from those of plants. Similarly, a regulatory region of the gene from a monocotyledon (corn) may not work in a dicotyledon (bean). Thus, a full understanding of the regulation of gene expression is of great interest not only to plant biologists who want to know how plant development is regulated, but also to genetic engineers who transfer genes between unrelated organisms.

7 | **Enzymes that cut and rejoin DNA molecules allow scientists to manipulate genes in the laboratory.**

Progress in genetic engineering depends on a new technology that allows scientists to isolate, identify, and clone (produce multiple copies of) genes. This recombinant DNA technology was developed in the 1980s, but finds its origins in the discovery of plasmids and restriction enzymes in bacteria in the 1970s. **Recombinant DNA** is made by joining, or "recombining," DNA segments from different sources: plant and bacterial DNA, or plant and animal DNA. Since DNA strands from all organisms have the same structure, they can be cut into segments and the segments linked together again (ligated) in new and different ways.

Cutting the DNA is done with special enzymes called **restriction enzymes,** which occur in bacteria as part of a natural defense system against invading viruses. Restriction enzymes are highly specific to the nucleotide sequence, and each enzyme only cuts at one specific short base sequence in the DNA. For example, the enzyme *Eco*R1, found in the intestinal bacterium *Escherichia coli,* cuts only at GAATTC (the complementary strand, CTTAAG in this case, is assumed), and such a sequence occurs on the average only every 4,000 nucleotides of DNA. An interesting feature of most restriction enzymes is that they make a staggered cut across the two strands of DNA, producing what molecular biologists call "sticky ends." Another enzyme, **DNA ligase,** can ligate those sticky ends together again (Figure 9.9), and this can happen whether the DNA is from the same organism or from different organisms. Such manipulations of DNA in the laboratory are often referred to as gene splicing.

The job of cloning—that is, of producing multiple copies of a gene or DNA segment—is carried out with the help of a plasmid. **Plasmids** are (small) double-stranded circular DNA molecules that replicate in bacteria. An important property of plasmids is that they may be transferred from one organism to another. When a foreign gene is introduced into a plasmid, the plasmid can still replicate. By using a restriction enzyme, it is possible to cut the circular plasmid open. If DNA from another organism has been cut with the same restriction enzyme so that the same sticky ends were created, DNA ligase can be used to reform circular molecules that contain a segment of foreign DNA spliced into the site where the plasmid was cut. The new plasmid is transferred to bacteria that have been specially treated for this purpose; when the bacteria are allowed to multiply, the plasmids will be copied within the bacteria. Millions of copies of the plasmid, and of the inserted gene, can be reisolated from these bacterial cells.

Plant transformation with *Agrobacterium tumefaciens* (discussed in Chapter 3) also relies on plasmids. This plant pathogenic bacterium carries a large plasmid, and when the bacteria infect a wound site a portion of the plasmid DNA is excised and becomes integrated in a plant chromosome. This segment of plasmid DNA carries about 10 genes, some of which encode enzymes for

Figure 9.9

Restriction Enzymes Often Make "Sticky" Ends. A specific restriction enzyme—EcoR1, for example, as shown here—makes a staggered cut producing "sticky" ends. The single-stranded end of one fragment can therefore recognize and bind to the end of any other fragment produced by the same enzyme, even if the fragments originally came from the DNAs of different species. The two fragments can then be joined by the action of an enzyme called DNA ligase. Restricting and ligating DNA in this way forms the basis of recombinant DNA technology.

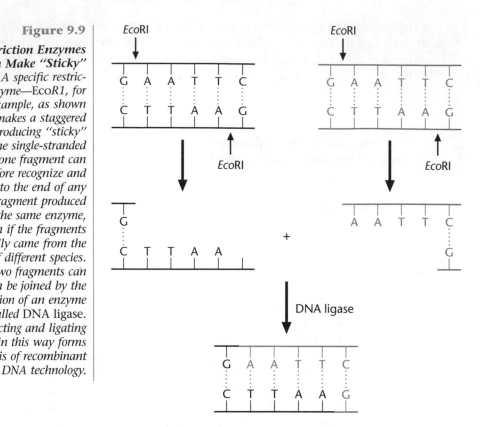

auxin and cytokinin biosynthesis. When these genes are active in the plant cell, the enzymes cause excessive production of these plant hormones and tumor growth ensues. By studying these plasmids and how the transfer of DNA occurs, molecular biologists learned that all genes in the DNA segment that is transferred to the plants can be deleted. If one then substitutes another segment of foreign DNA, this segment will be transferred to the plant (see Chapter 15).

8 | **Crop improvement through plant breeding and crop improvement through gene transfer have the same molecular basis.**

Plant breeders use a number of approaches to obtain plants with desired characteristics. Some of these are discussed later, but one—backcrossing—is discussed here to show what the method accomplishes in molecular terms. Breeders often wish to introduce a single important new trait, such as disease resistance into a high-yielding variety that is well adapted to a particular environment. The desirable characteristic may be encoded by a single gene and can be present in a closely related variety, a more distantly related variety (for example, from a different part of the world and therefore not well adapted to local conditions), a wild ancestor, or even a different species that is a close relative.

To start the process of crop improvement, the breeder makes a cross between the "native," well-adapted parent and the "foreign" parent. In the offspring of this cross, each parent will have contributed 50% of its genes.

However, agronomic characteristics will vary considerably. The breeder first selects among the offspring the individual plants that have the new desired characteristic (such as resistance to the disease) and also the best agronomic traits. Then the breeder crosses the selected plants back to the native, well-adapted high-yielding parent. In the progeny of this cross, 25% of the genes will be from the foreign parent and 75% from the native parent. This process of backcrossing is repeated over and over again, and after the second, third, fourth, and fifth backcrosses the foreign parent contributes on the average 12.5%, 6.25%, 3.12%, and 1.55% of the genes, respectively.

The purpose of this backcrossing program is to dilute out the gene pool of the wild relative and maximize the number of genes coming from the high-yielding variety while at the same time maintaining the gene for the desired characteristic. A breeder releases a new variety to the farmers when enough backcrosses have been made so that the new variety has both the new desirable trait and good agronomic characteristics. Ideally, the breeder would like to get the number of genes from the foreign parent down to the absolute minimum, perhaps one gene if the trait is an all-or-none characteristic governed by a single gene, or just a few genes. In practice, that is not possible because each backcross takes six to eight months, and a commercial breeder may be competing with other commercial breeders to release improved varieties.

At the molecular level, the breeder wishes to introduce as little foreign DNA as possible into the variety which he or she knows to be well adapted. However, with complex multigenic traits it is always necessary to introduce more than just a few genes. Because the species that can be crossed sexually are of necessity closely related, it is not necessary to transfer the regulatory genes as well as the gene that confers the specific trait. For example, the regulatory genes that are necessary for correct expression of a pathogen-resistance gene in goatgrass are closely related to the ones in wheat.

Plant genetic engineering achieves in a very quick and precise way what breeders try to accomplish by years of backcrossing: the introduction of only (1) that segment of the DNA that has the necessary gene(s) to confer the new trait and (2) the necessary regulatory region or genes to confer the correct expression of that gene. Although such introduced DNA is often called "foreign" DNA, one should realize that the crop varieties now in use have been improved with all kinds of "foreign" genes. In *Food Crops of the Future* (1988; Oxford, England: Blackwell Scientific, p. 88), Colin Tudge lists the following examples of the improvement of wheat by "foreign" genes.

- The wild wheatgrass, *Agropyron,* has contributed genes to Russian wheat varieties that improve resistance to frost, and increased yield of wheat has been obtained by crossing Russian wheat with wild species of rye, *Secale.*
- Wild and primitive species of wheat have contributed genes to cultivated wheats that make them resistant to the rust fungus.
- French plant breeders obtained genes for resistance against the fungal eyespot disease from the goatgrass *Aegilops ventricosa* in a backcross program with French domestic wheats. Special tissue culture techniques often need to be applied to rescue the embryos that come from the first cross because wheat is not closely related enough to its wild ancestors or to wheatgrass or goatgrass to be able to cross normally with them.

These examples should be sufficient to illustrate that present-day crops already have many "foreign" genes, largely as a result of the ingenuity and persistence of plant breeders.

Plant breeding can be considered "classical gene manipulation," whereas genetic engineering is based on "molecular gene manipulation" and tissue culture. It should be stated clearly, however, that genetic engineering will not simply replace plant breeding. After a single gene has been introduced into a crop by gene manipulation and tissue culture, several years of plant breeding are always needed to make sure that the new plant has the right agronomic characteristics. Genetic engineers can complement plant breeders, but cannot replace them.

Summary

Crop improvement through plant breeding can readily be explained at the molecular level. Plants have thousands of individual traits that are transmitted from one generation to the next, because the information for each trait is contained in one or more genes, short segments of DNA. Each cell contains two copies of each gene, and when a cell divides, its DNA is faithfully replicated and two copies of each gene go to the daughter cells. The accumulation of small mutations and the independent assortment of genes between plant generations result in continuous variation for traits that are encoded by many genes.

Complex traits such as yield are influenced not only by genes but also by the environment, and plant breeders must establish to what extent a specific characteristic has a genetic basis. Crop improvement invariably entails a change in the genetic makeup of the plant. This can be done by conventional breeding techniques or via molecular gene manipulation. The introduction of a new gene into a plant means that it has new information to synthesize a new protein that establishes the new trait. Understanding the molecular basis of heredity and progress in molecular biology have both made possible crop improvement through gene manipulation.

Further Reading

Beckmann, J. S., and M. Soller. 1988. Restriction length polymorphisms in plant genetic improvement. *Oxford Survey Plant Molecular Cell Biology* 3:196–250.

Bevan, H. 1991. Identification and expression of genes of potential interest in plant biotechnology. *Current Opinion in Biotechnology* 2:164–170.

Hartl, D. 1991. *Basic Genetics*. Boston: Jones and Bartlett.

Micklos, D., and G. Freyer. 1991. *DNA Science*. New York: Cold Spring Harbor Publications.

Peters, P. 1993. *Biotechnology: A Guide to Genetic Engineering*. Dubuque, IA: W. C. Brown.

Stalker, H. T., and J. P. Murphy. 1992. *Plant Breeding in the 1990's*. Wallingford, UK: CAB International.

Watson, J. D., M. Gilman, J. Wirkowski, and M. Zoller. 1992. *Recombinant DNA*. 2nd ed. New York: Freeman.

Ten Thousand Years of Crop Selection

In recent years, scientists have made a concerted effort to increase food production around the world by breeding genetically improved crop plants. This effort has led to a renewed interest in the origins of agriculture and crop domestication; that is, in how crop plants evolved from their wild ancestors. Plant geneticists are now searching the world to collect seeds from both the wild relatives and the many cultivated races of some of the important modern crop plants. They hope to incorporate some of the desirable genetic characteristics of these wild relatives (for example, resistance to certain diseases) into modern crop plants. A better understanding of plant domestication and of the relationship between the crop plants and their wild relatives may help plant breeders produce superior varieties of crop plants.

Another reason for the renewed interest in the origins of agriculture and in various "primitive" agricultural systems comes from the recognition that agriculture is part of the natural environment and that agricultural systems operate under the constraints of nature. The success of an agricultural system may depend on how well it can simulate the natural system that it replaces. Thus we need to know more about the natural systems in which the crop plants were domesticated, and about the way in which the agricultural systems gradually replaced the natural ones in many parts of the world.

1 | **Hunting and gathering have been the methods of food procurement for humans through most of their history.**

Human beings have been a distinct kind of animal for at least two million years and perhaps much longer. Archaeologists believe that this animal first evolved in Africa, as *Australopithecus*. Little is known about the diet of our earliest ancestors, but it is generally believed that our more recent ancestors were hunters and gatherers of food, and had a varied diet that included both plants and animals. Agriculture did not become the usual method of procuring food until about 10,000 years ago, or about 1% of the time humans have existed.

One way of inferring the prehistoric diet is to examine the buried garbage dumps of our early ancestors. These show a preponderance of animal bones—plant parts do not persist in these remains, having been broken down by micro-organisms. Some information on ancient people's food habits comes from examining fossilized feces. Botanists have been able to identify the remains of seeds from many plants in these fossils, thus confirming that early people ate both plants and animals.

Prehistoric people probably experimented with a variety of plant foods, learning by trial and error which were good to eat and which were not. Some of these experiments may have been fatal, because many plants contain toxic chemicals (see Chapter 12). Because these toxic chemicals can often be inactivated by heat, cooking was important in expanding the range of plants human beings used as food. Archaeological evidence suggests that the controlled use of fire for cooking had been developed by at least 100,000 years ago.

Recent studies suggest that early people did not have to hunt, fish, and search continually just to find enough to eat. They may have had considerable leisure time to make tools, decorate caves, or take part in various rituals and other social activities. Some of these societies of hunter-gatherers have survived into historic times (for example, the Inuit of northern North America, other Native Americans, Australian Aborigines, and South African Bushmen), and for these people food procurement usually takes only a few hours each day.

A common misconception is that all hunter-gatherers had primarily a nomadic way of life. Obviously, this may have been true of those who followed large game herds. Their material culture and social organization would have been adapted to the exploitation of their principal food source. The use of wild plants for food may have been ancillary in these highly specialized hunting economies. But most hunter-gatherers are thought to have lived a fairly settled life, living in small bands in territories they knew intimately, and moving about less than the specialized hunters. Occupying primarily riverbanks and lakeshores in wooded areas, they probably obtained their food from a wide variety of wild plants, small game, and fish (see Box 4.3, "Food of the Native Californians").

2 | Agriculture began at several places around the world.

About 10,000 years ago, a remarkable change occurred in the way people procured food. They domesticated plants and animals, and started practicing agriculture. The gatherer became a farmer. This transition has been called the agricultural revolution, for it heralded a fundamental change in human beings' material wealth, social organization, and cultural achievements. It also marked a change in how human beings related to their environment. The hunter-gatherer was very much a part of nature, competing with other organisms for the food supply. The farmers started to modify the ecosystem to suit human needs. They interfered with the normal flow of energy in the biosphere and diverted it to products they could eat. They decided which plants grew where, protected useful plants from diseases, and even altered the course of plant evolution, by helping to evolve new species that would not have survived without human care.

Finding out where, why, how, and when this transition occurred has taken the combined efforts of plant biologists, geographers, archaeologists, and many others. People have domesticated only about 100 or 200 of the thousands of plant species, and of these no more than 15 now supply most of the human diet. These 15 can be divided into four groups, as follows:

1 *Cereals:* rice, wheat, corn, sorghum, barley
2 *Roots and stems:* sugar beets, sugarcane, potatoes, yam, cassava
3 *Legumes:* beans, soybeans, peanuts
4 *Fruits:* coconuts, bananas

In addition, human beings have domesticated about 50 species of animals. Of these only the dog, pig, cattle, horse, water buffalo, goat, sheep, and chicken are of great economic importance. The list of plants and animals that can be domesticated is by no means exhausted. In recent times, some new crops, such as the rubber tree, have been domesticated. Also, studies have been made of plants that could be domesticated because they have some unusual and potentially useful characteristics.

To answer the question of **where** cropping began, the Russian geneticist and plant geographer N. I. Vavilov, who traveled and collected plants from all over the world, identified regions where crop species and their wild relatives live with great genetic diversity. This, he reasoned, would provide the first farmers with "raw materials" from which to select crops to grow. He found that there were eight regions where diversity existed and crops overlapped, and termed these "centers of origin": China, India, Central Asia, the Near East, the Mediterranean, East Africa, Mesoamerica, and South America. Subsequently, others, such as U.S. geographer Jack Harlan, challenged Vavilov's hypothesis because many cultivated plants did not fit Vavilov's pattern, and appeared to have been domesticated over a broad geographical range over a long period of time. Harlan proposed "noncenters" of agricultural origin.

Combining these two scientists' ideas, it is possible to delineate three major centers and three noncenters for agricultural origin and assign the major crop plants to each (Figure 10.1):

1 *Northern China center:* rice, soybean
2 *Southeast Asia noncenter:* banana, coconut, sugarcane
3 *Near East center:* barley, wheat
4 *African noncenter:* sorghum, yam
5 *Mesoamerica center:* bean, corn
6 *South America noncenter:* cassava, peanut, potato, sugar beet

The question of **why** people started digging the soil, sowing seeds, and cultivating fields is equally difficult. Some have suggested that before agriculture the food supply was unreliable. As soon as people discovered farming, they would immediately *perceive its advantages* for a more reliable and easy to obtain food supply.

Observations of current hunter-gatherers belie this idea. They are not primitive but have a sophisticated knowledge of their environment. Even in the harsh world of the Kalahari Desert, the !Kung Bushmen of Africa spend relatively little time gathering the plant foods that form much of their diet. A

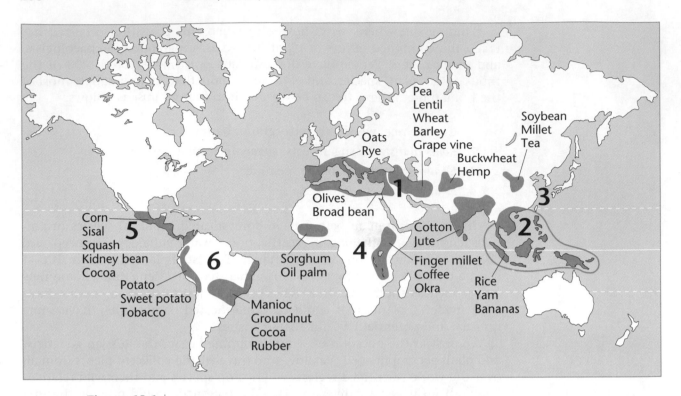

Figure 10.1

Regions of Plant Domestication. Areas 1, 3, and 5 are considered "centers," where crops originated in a narrow area at a specific time. Areas 2, 4, and 6 are "noncenters," where domestication occurred more broadly and took longer.

!Kung woman can easily gather enough of these foods in 6 hours to feed her family for three days. This lifestyle has been in existence for over 10,000 years and is clearly successful; there has been no compelling need to try agriculture.

Some scientists have argued that *changes in climate* forced people to become farmers. In the Near East, for example, when the last ice age ended there was a change from a cool, wet climate to a warmer and drier one. As this occurred, people were forced to gather near the remaining sources of water. Living near the plants that also grew there, these people would then "discover" agriculture.

After examining many archaeological sites in southwestern Asia, archaeologist R. J. Braidwood has put forward a plausible hypothesis about why people began farming in that region. He believes that the emergence of agriculture was part of humanity's overall *cultural evolution,* and that it occurred as a result of a slowly changing ecological relationship between human beings and the plants and animals they used for food. Agricultural food production developed "as the culmination of the ever-increasing cultural differentiation and specialization of human communities." Perhaps the emergence of agriculture was inevitable once people had become thoroughly familiar with the plants and animals that formed their major food source.

An answer to the question of **how** plant domestication occurred begins with the realization that many major crop plants are annuals that grow readily in disturbed soil. In this respect, the crops are similar to weeds. We usually define weeds as unwanted plants that are competing with the plants we do want. But weeds can also be defined as plants that are adapted to take advantage of disturbed or open habitats. Weeds quickly spring up where a disruption of the normal vegetation has left the ground bare.

The ancestors of modern crop plants may have been weeds that grew naturally in the disturbed and fertile soil surrounding the semipermanent settlements of humans. Such plants may have sought human beings out just as much as humans sought them out, because of their adaptations and their rapid growth in fertile soil. This hypothesis of the origin of agriculture has become known as the "rubbish heap" hypothesis, since it assumes that plants with weedy tendencies grew readily in rubbish heaps and other areas where the soil was disturbed and fertile.

People may have collected the seeds of the weeds in the same manner that they gathered the seeds of other plants. At first, they may have gathered them over great distances, but the abundance of these plants near human settlements may have increased as seeds were scattered on the ground from carelessness. Thus gathering may have given way to "harvesting." Finally, early agriculturalists may have sown seeds in fields or gardens in which the soil had been prepared.

Whatever the mechanism for starting farming, people did not do so because it was easier or more civilized than hunting and gathering. However, once farming and crop cultivation had started and populations had increased, humanity became dependent on farming. To avoid starvation, people had to cultivate the land and to store grain. In his 1988 book *Food Crops for the Future* (Oxford, UK: Blackwell), Colin Tudge suggests that early farmers may have resented this dependence on agriculture and the obligation to work so hard. And according to the account in the Bible, when Adam and Eve were banned from the Garden of Eden, God said to them, "Accursed shall be the ground on your account. With labor you shall win your food from it all the days of your life. It will grow thorns and thistles for you. . . . You shall gain your bread by the sweat of your brow" (Genesis 3:17–19).

3 | Wheat was domesticated in the Near East.

One way to gain a better understanding of the process of plant domestication is to study the available evidence for some of the major food plants. Wheat was originally domesticated in a hilly region of southwestern Asia called the Fertile Crescent (see Figure 10.2). The area is bordered on one side by the Tigris–Euphrates basin (ancient Mesopotamia) and on the other side by the mountains of what are now Iran, Turkey, Syria, and Jordan. Before 10,000 B.C., the hunter-gatherers who inhabited this region had followed herds of migrating animals, harvesting plants as they traveled.

But as the Ice Age ended, a cool climate was replaced by a warmer one, with enough rainfall for plant growth. Now the food supply was more diverse: plants such as emmer and einkorn (the ancestors of modern wheat), barley, peas, and lentils grew wild; and animals, such as wild sheep, pigs, goats, and deer, fed on the plants. As the hunter-gatherers developed a detailed knowledge of their varied foods, some of them moved into semi-permanent settlements, possibly because their growing population made it increasingly difficult to get enough food by hunting and gathering.

Some of these settlements have now been excavated by archaeologists, and several lines of evidence point to a beginning of agriculture there:

1 The earliest evidence is the presence of sickle blades in deposits dating back some 12,000 years. Pounding stones and mortars for grinding are even older, but they are not distinctly agricultural tools. The presence of these implements suggests that very

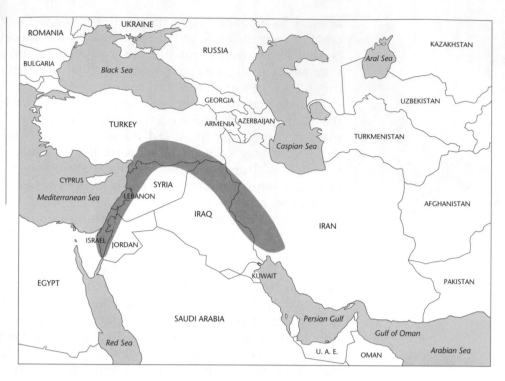

intensive gathering of grains was going on at the time. Human beings had undoubtedly entered the "harvesting" stage of agricultural development; whether they were also planting grains is difficult to ascertain.

2 There is a change in the animal bones found near the settlements. Bones from goats, deer, and gazelles decrease, whereas those from sheep, especially young sheep, increase. This suggests a shift from hunting wild animals to herding sheep. Herd animals are more easily killed than wild ones, and usually young rather than mature animals are slaughtered for food.

3 There are remains of three unmistakable progenitors of modern crops: two-row barley, emmer wheat, and einkorn wheat (Figure 10.3) in the villages, dating from 7500 B.C. These identifications were made from the imprints of grains in baked clay and charred grains found in fire sites. The plants from which these grains came differed from their wild relatives that grew in the same area and were also found among the plant remains in the excavations. This difference indicates that plant domestication had started.

Once people started to cultivate plants, harvest grains, and preserve a portion of the crop for planting at a later time, they also selected, probably unconsciously, in favor of certain plant characteristics; so the grains from cultivated plants began to differ from those of the wild ones.

Along with wheat, and later barley, people also domesticated peas, chickpeas, and lentils. These legumes also grew wild in that area, although evidence suggests that these plants were far less important than the cereal grains to the early farmers. The sites where the first signs of plant domestication

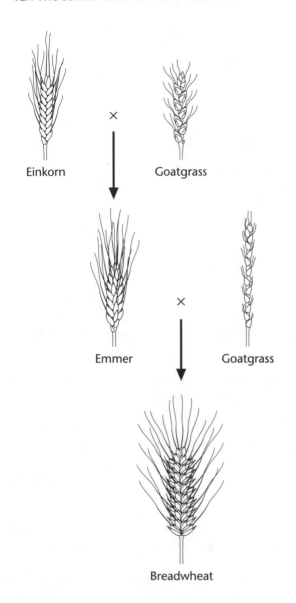

Figure 10.3

The Evolution of Wheat from Its Wild Relatives. Both einkorn and emmer wheats are found in the Fertile Crescent.

Einkorn   Goatgrass

Emmer   Goatgrass

Breadwheat

were found are believed to have been semipermanent settlements with populations of a few hundred people each. By 5000 B.C., such thriving agricultural communities were found in a wide area stretching from what is now Greece to Afghanistan. The agricultural way of life spread rapidly through Europe and western Asia, and by 1500 B.C., hunting and gathering as an exclusive way of procuring food was practiced only in the northern regions of this area. The early agriculturalists, like the subsistence farmers of today, probably supplemented their harvests with foods gathered from the wild.

4 | Corn was domesticated in Mesoamerica.

The developments taking place in southwestern Asia were not unique; similar changes were occurring in other parts of the world. In the southern part of what is now Mexico lie flat basins surrounded by steep mountain ranges. One of these, the Basin of Mexico, is the site of Mexico City. The next

Figure 10.4

Plants of Corn (left) and Its Ancestor, Teosinte (right). Both have an apical male part and seed-bearing, female parts along branches off the stem. The main difference is that corn produces a few large ears, while teosinte produces many small ones with seeds in a very hard casing.

basin to the south, at Tehuacan in the southern portion of the state of Puebla, contains many archaeological sites where a major transition to agriculture probably occurred around 7,500 to 5,000 B.C.

In this area, three crops were domesticated: corn, beans, and squash. Although this did not happen all at once, the result was fortunate from a nutritional point of view. Beans provide amino acids that are deficient in corn (see Chapter 4), and squash seeds provide additional protein as well as oil. Archaeological evidence indicates that squashes were domesticated first, around 7400 B.C., followed by beans, around 5000 B.C. In both cases, there are closely related wild relatives, so that it can be surmised that, as with wheat, the relatives were first gathered, and then selected over time causing the evolution of the crop.

The case of corn is more difficult to envision. The closest living relative of corn is teosinte, a weed that grows at many locations in Central and South America. Teosinte is similar in general appearance to corn, with multiple stalks that bear male flowers at the top and female flowers off the stem, and the two plants can interbreed in nature. In fact, botanists classify both in the same genus and species (*Zea mays*). But a look at the two (Figure 10.4) shows a major difference: while corn produces a few large ears with many seeds, teosinte expends its energy on producing vegetative shoots, and the ears are tiny in comparison. Obviously, this plant is not a good source of human food, as the quantity of seeds is simply too low. In nature, teosinte yields less than 100 kg per hectare, far below the 600 kg per hectare from the wild cereal grains in the Near East.

Genetically, teosinte and corn differ by only a few genes, and it is possible that a rapid evolution of the crop occurred. Corncob sizes found and dated at the Tehuacan sites bear this hypothesis out (Figure 10.5). From the tiny, teosinte-sized cobs at the earliest dates, yields must have improved to about 250 kg per hectare by about 1500 B.C. This would be enough to support human civilization, and from then on, the evolution of corn continued apace.

From Mexico, domesticated corn spread rapidly both northward and southward. It appeared in Peru as early as 2000 B.C. However, by that time

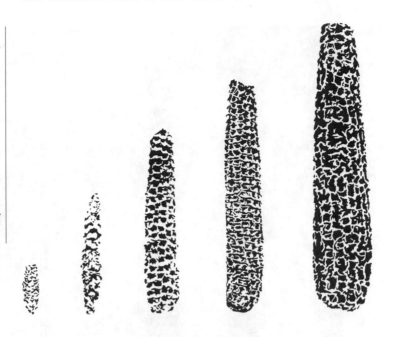

Figure 10.5

Increase in the Size of Corn Between 1500 B.C. (left) and A.D. 1500 (right) at Tehuacan, Mexico. The oldest cob is less than an inch in length, typical of those of corn's probable ancestor, teosinte. Source: D. B. Byers, ed. (1967), The Prehistory of the Tehuacan Valley, *vol. 1* (Austin: University of Texas Press).

agriculture was already well established in South America, having originated independently in the Peruvian highlands. Domesticated kidney beans and lima beans have been found there in deposits dated around 6000 B.C. By the time Columbus arrived in the Americas, corn was the most widely cultivated plant, growing from southern Canada to southern South America. There were literally hundreds of varieties, each adapted to their specific growth region and the customs of the inhabitants. The five major varieties of corn today—dent corn, flint corn, popcorn, flour corn, and sweet corn—were all cultivated by the Native Americans.

5 | Rice was domesticated in India and China.

Terraces that are unmistakably meant for rice cultivation have recently been discovered in northern India and dated at about 10,000 B.C. If this estimation is accurate, it means that rice may be the oldest cultivated grain. The exact origin of cultivated rice is less clear than that of wheat. On the basis of his search for wild relatives and the centers of diversity idea, Vavilov believed that rice originated in northern India. Yet there is strong evidence, from the existence of wild rices crossing with cultivated strains, that the crop originated in eastern China. Finally, biochemical studies have demonstrated considerable genetic diversity in wild rices growing in Southeast Asia.

The progenitor of *Oryza sativa*, the major cultivated rice in Asia, is clearly a wild species, *Oryza rufipogon*, that also grows in Asia. Several lines of evidence point to this conclusion:

1 The wild and cultivated plants can and do interbreed.
2 They occur in the same locations.
3 They have the same biochemical marker genes, although the frequencies of the genes in the species are different.
4 They have similar growth habits.

Figure 10.6

***Wild Beans Growing in Mexico Adjacent to the Cultivated Bean,** Phaseolus vulgaris.* *The wild bean plants are amid the natural vegetation below the road, and near the rows of planted beans.* Source: *Courtesy P. Gepts, University of California at Davis.*

There are two outstanding differences between the wild and cultivated rices. First, although the latter is an annual (flowers only once and then dies after seeds are produced), the progenitor can be a perennial (lasts many seasons and flowers repeatedly). The perennial strains have great genetic diversity, and these many characteristics doubtless formed the gene pool that farmers selected for planting. Second, this genetic diversity derives in part from the fact that the wild rices are cross-pollinated (one plant's pollen fertilizes another plant), while cultivated strains are self-pollinated. The mixing of genes in the wild allows for new combinations that will improve adaptability and survival.

Once it was domesticated, rice became very popular. Chinese manuscripts dating back 5,000 years describe rice as the most important food, and say the emperor alone was allowed to sow it. The ancient Indians called rice the "sustainer of the human race." A king of Nepal in the sixth century B.C. was named "Pure Rice." Rice spread throughout the world, as did the people who grew it. Indians brought it to north Africa and the Middle East; Arabs gave it to the Spanish; Greeks got it from the Persians.

The story of how rice was first cultivated in the United States illustrates how a crop can spread. In 1697, the governor of South Carolina had recently visited Madagascar, and observed (and eaten) the rice growing there. When a Dutch ship from Madagascar ran aground in a storm off the South Carolina coast, the governor asked the captain if he had any rice aboard. He did, and the governor planted some in the garden of his mansion. It was so successful that by 1850 rice cultivation was established over wide regions of the southern United States.

In this discussion of plant domestication, the seed crops, and primarily the cereal grains, have been emphasized. At least one cereal species and one **legume** species was domesticated by each of the independently emerging agricultural civilizations: wheat and lentils in southwestern Asia, maize and kidney beans in America, sorghum and cowpeas in Africa, and rice and soybeans in southeastern Asia. Once again, the presence of wild relatives of these plants indicates the site of domestication (Figure 10.6). Like cereals, legume seeds can be conveniently stored. Because of their low water content, they do not spoil easily.

Root crops (fleshy roots and tubers) are an important component of the diet in many countries. The plants that produce them are usually propagated vegetatively rather than by seeds. It has been suggested that human cultivation of plants may have started with such vegetatively propagated plants, because such cultivation is simpler than planting seeds. For example, the Australian Aborigines, who do not practice any kind of traditional agriculture, nevertheless always put some yams back into the ground whenever and wherever they dig them up in their food-gathering activities.

The archaeological record has not provided evidence that supports the suggestion that root crops were domesticated before seed crops, possibly because many root crops are grown primarily in the humid tropics, where prehistoric plant materials are much less likely to be preserved than in the much drier areas that have yielded important archaeological information about plant domestication. Root crops, being fleshy plant parts, are less likely to be preserved than seeds, which contain much less water.

6 | Domestication is an accelerated form of evolution.

In the relatively short span of 3,000 to 7,000 years, depending on the crop, human beings markedly altered the characteristics of their food plants. Many of these plants—corn, for instance—changed more in this period than they probably had in the million years before it. Primitive farmers, knowing nothing of genetics or plant breeding, seem to have accomplished much in a short time. They did so by unconsciously altering the natural process of evolution. Indeed, domestication is nothing more than directed evolution; as a result, the process of evolution is accelerated.

Evolution is based on two phenomena that can be observed in natural populations: variation and natural selection. Not all individuals of a species are alike; there is much variation. (The nature of this variability, and the way traits are inherited, were discussed in Chapter 9.) In a variable population, some individuals are better adapted to the environment, and so produce more offspring, than others. This illustrates the principle of "survival of the fittest," where nature allows the individuals best adapted to reproduce the species to survive.

In domestication and plant breeding, people select those individual plants that have the desired characteristics. Thus, natural selection is replaced by intentional selection by human beings. It is likely that the early farmers consciously selected in favor of certain characteristics, and by their interference with the natural course of events, they may also have unconsciously selected in favor of certain traits. Present-day subsistence farmers also often consciously select in favor of certain characteristics when setting aside a portion of one year's harvest for next year's planting.

This selection has resulted in profound differences between crop plants and their progenitors (Table 10.1). A few of these are discussed in detail below:

First, many wild plants have a **seed dispersal mechanism** that ensures that the seeds will be separated from the plant and distributed over as large an

Table 10.1	How plants change with domestication

Characteristic	Examples
Loss of dispersal mechanisms	Corn, wheat
Loss of dormancy	Wheat, oats, rice
Conversion from perennial to annual	Rice, rye, cassava
Loss of fruit production	Yam, sweet potato
Loss of seed production	Banana, citrus
Increase in size of	
Seed	Bean
Fruit	Squash
Storage organ	Cassava, carrot

Source: Adapted from J. Salick and L. Merrick.

Figure 10.7

Wild Varieties of Beans Scatter Their Seeds But Domesticated Varieties Do Not. The dry pods **(a)** of a wild bean (Phaseolus vulgaris) varieties open violently when they are touched, and curl up **(b)**, scattering the seeds on the ground. This characteristic makes harvesting difficult. The dry pods **(c)** of domesticated varieties must be opened by hand **(d)** to release the seeds that normally stay inside until harvest time. Source: *Photo courtesy of P. Gepts, University of California, Davis.*

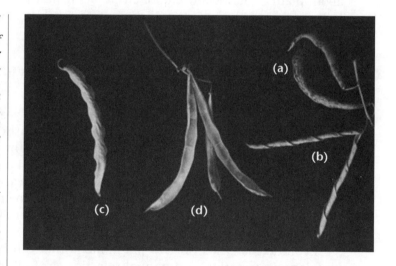

area as possible. In many grasses, the spikes that bear the grains become brittle when the grains are ripe, and disintegrate when the plant is hit violently (by wind or by humans or other animals) thereby scattering the seeds. This seed dispersal mechanism is found among the wild relatives of modern cultivated wheat and rice, and was undoubtedly characteristic of their ancestors. Many legumes have a different seed dispersal mechanism: when the seeds are mature and dry, the pods suddenly pop open, projecting the seeds in all directions. This happens spontaneously even if the pods are left undisturbed, but it happens more readily if anything touches the pods (Figure 10.7).

Seed dispersal mechanisms are obviously advantageous to the plant, for they scatter its offspring. But they are a disadvantage to the farmer who is trying to collect the seeds. The farmer would prefer a cereal grain with a tougher spike or a legume with pods that do not pop open. Once people started keeping a portion of the harvest for planting, they modified plant evolution by selecting against seed dispersal mechanisms. At each successive harvest, they collected a greater proportion of seeds from plants that had defective seed dispersal mechanisms. As a result, modern wheats and rices have tough spikes, corn does not lose its kernels at all from the cob, and

Table 10.2	Dormancy in wild rice collected in Taiwan as compared to cultivated rice	

	Percentage of Seeds Germinated	
Days After Seeding	Wild Rice	Cultivated Rice
5	14	7
10	19	79
15	23	98
20	26	98
25	32	98
30	36	98

Source: Data from T. Oka, National Taiwan University, and International Rice Research Institute.

modern legumes do not have pods that pop open. Seed dispersal in these crop plants depends entirely on human beings.

A second example of human influence on plant evolution is found in the disappearance of **seed dormancy mechanisms** from domesticated plants. Seeds from many wild plants do not germinate as soon as they are shed, but remain dormant for different lengths of time. This is beneficial for the plant, because it minimizes the risk that all the offspring will be wiped out by a spell of bad weather. But for the farmer this trait is a nuisance, for it means that the seeds will germinate throughout a long period of time, thus extending the harvest time. By harvesting a crop at a particular time (for example, autumn), early farmers selected only those plants whose seeds had germinated in the spring and grew to maturity by the fall. Plants that had germinated later in the season would not have produced seeds by harvest time and would not have been selected for harvesting and planting the next year. As a result, dormancy soon disappeared from crop plants. Now when a farmer sows seeds, they all germinate within a relatively short time.

The difference in seed dormancy between wild relatives and modern varieties is seen in an experiment conducted in Taiwan (Table 10.2). Wild rice seeds and cultivated rice were planted under identical conditions. After four weeks, only a third of the wild seeds had sprouted, and they did so at staggered intervals. In contrast, the cultivated variety all sprouted at once.

There were three important steps in the domestication process. The important point is not that people planted seeds, but that they

1 Moved seeds from their native habitat and planted them in areas to which they were perhaps not as well adapted

2 Removed some natural selection pressures by growing the plants in a cultivated field

3 Applied artificial selection pressures by choosing characteristics that would not necessarily have been beneficial for the plants under natural conditions

Moving plants from one area to another can have important evolutionary consequences. In a new environment, the plant may encounter new wild varieties of the same species or of other species with which it could interbreed. Such hybridizations are believed to have played an important role in the evolution of modern bread wheats (see Figure 10.3). However, when a plant was moved away from its immediate ancestor, intercrossing between these two was eliminated. This would have made the fixation of new characteristics, important to humans, much easier. Indeed, the fixation of a new genetic characteristic in a small population of cultivated plants is slow if there is continuous cross-fertilization between these plants and a much larger wild population of the same species.

Plants growing in a cultivated field need not compete with other species for sunlight, water, or nutrients. As a result, all kinds of individuals can survive that would not have survived under the conditions prevailing in the natural habitat. The removal of the natural selection pressures allowed more of this genetic variability in the plant population to be expressed in the field. People then imposed their own selection pressures on this variable population, causing certain characteristics to appear or disappear in a relatively short time.

It is clear that much more needs to be learned about the origins of agriculture before we will fully understand how and why people gradually changed their mode of subsistence from hunting and gathering to farming. But whatever these origins may have been, human beings and their crops are now united in a firm and mutually dependent partnership.

7 | The agricultural revolution irreversibly changed the course of human history.

As noted earlier, humanity's transition from fisher-hunter-gatherer to herder-farmer has been called the "agricultural revolution." Is *revolution* too strong a word to describe the domestication of a few plants and animals? After all, revolutions are supposed to cause profound changes in human existence. The agricultural transition can hardly be called a revolution if we consider that it took several thousand years to complete; that is, it was not a sudden upheaval in human existence. However, agriculture changed the course of history quite radically. The emergence of agriculture is the very foundation on which current civilizations rest. The transition to agriculture had several important consequences. First, it resulted in an *increase in the human population* (Figure 10.8), probably because it was easier to obtain more food, although the food supply was probably not more reliable nor the food necessarily more nutritious. Hunter-gatherers had relied on a wide variety of very small seeds (especially legumes) for their food supply. The cultivation of cereals—with much larger seeds—simply made it easier to obtain an adequate amount of food. Jack Harlan has shown that one person with a primitive flint sickle could easily harvest 1.8 kg of the primitive wheat grains in an hour. Thus, a primitive farmer could have harvested in a few weeks more than enough grain to feed a family for a year.

Agriculture changed the *people–land relationship.* Calculations by David Pimentel illustrate this. Assuming that a person needs 2,500 kcal per day, and 1 gram of seeds yields 3 kcal, then it would take about 300 kg of seeds per year to feed one person. Observations of the habitats used by current hunter-gatherers, as well as calculations of other natural ecosystems, indicate that this 300 kg would come from about 250 ha of land. This number—250 ha per

Figure 10.8

Increase in Population Density in Ancient Persia During the Transition to Agriculture. Note the gradual increase after farming was introduced, and the sharp increase after yields went up after the introduction of agricultural technology (irrigation). Source: K. V. Flannery (1969), Origins and ecological effects of early domestication in Iran and the Near East, in P. Ucko and G. Dimbleby, eds., Research Seminar in Archaeology and Related Subjects, *Duckworth.*

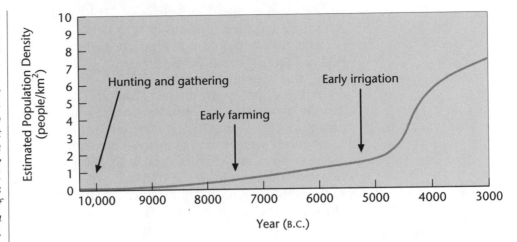

person—fell sharply after agriculture began, and fell further as yields rose. Currently, the number for the human population is about 1 hectare of cropland and pastureland per person.

Population growth led to an increase in the number of settlements, but also—and more importantly—to the formation of cities. Great civilizations arose in Asia, Africa, and the Americas, and in all of them the urban population was supported by a large agricultural population. Living together in small groups required people to experiment with new social and political structures. Urbanization also resulted in occupational differentiation. Some of the urban people became rulers, priests, soldiers, traders, builders, and artisans. Some farmers directed their efforts toward improving farming, making further increases in population possible. One such improvement, irrigation, seems to have had a dramatic effect on the population density in certain areas (Figure 10.8).

The emergence of *social stratification* was another consequence of the agricultural transition. Society now consisted of different classes: an upper class that owned the means of production (the land); a middle class of traders, artisans, and soldiers; and a laboring class of agricultural and other workers.

Archaeologist K. Flannery has suggested that this social stratification was not a result of the success of agriculture (fewer people being needed to produce the necessary food), but rather of the widening gap between the size of the population and the land area on which it was most dependent for its food (see earlier discussion). He studied the rise of agriculture in Iran, and found that 35% of the land was suitable for a hunting-gathering type of existence, the rest being uninhabitable deserts and lands of marginal productivity. The introduction of farming resulted in an increase in the population, even though probably only 10% of the area of Iran was suitable for farming. Irrigation boosted population densities even further, although it could only be applied to a very small proportion (less than 1%) of the land. Thus more and more people became dependent on less and less land for their food. Those who owned this land became the ruling class.

Finally, agriculture increased the *human impact on nature.* As natural ecosystems were replaced by agricultural ones, the landscape was irreversibly changed. Although every organism alters the environment in some way, ecosystems have evolved to assimilate these changes without great effect in the short term. The mechanisms that keep ecosystems in balance have been largely overwhelmed by agriculture. This made it impossible for people to return to their former means of subsistence even if they had wanted to.

The close relationship of people to farming has, ever since agriculture began, been a driving force of history. As the U.S. president Herbert Hoover once said, "The first word in war is spoken by guns—but the last word has often been spoken by bread." And the communist Chinese leader Mao Zedong once stated, "To grow one single additional grain of rice is to produce one more bullet and kill one more enemy."

A good example of the impact of agriculture is the U.S. Civil War of 1861–1865. In 1859, the southern states produced only one-sixth of the total U.S. wheat crop, and typically bought a lot of wheat from the northern states to supplement local production. When war broke out in 1861, President Lincoln blockaded the southern states. This created a dilemma for Britain, to which the southern states turned for help. Normally, Britain imported most of its cotton (the textile industry was an important one) from the southern states. The blockade virtually stopped this trade and left half a million British textile workers unemployed. Britain had every reason to help the southern states, except for the fact that wheat harvests in Britain and Central Europe had been poor and Britain had to import wheat from the northern states. Faced with the choice of cotton versus wheat, Britain chose the latter ("You cannot eat cotton!"). It is interesting to speculate what might have happened if the southern states had been able to grow enough of both cotton and wheat.

8 | High-input agriculture emerged as a result of voyages of discovery and modern science.

To measure the rate of agricultural progress in the thousands of years that followed the domestication of the major crop plants is an impossible task. Measured by present-day standards, progress was very slow. The introduction of irrigation, perhaps as early as 5000 B.C. in southwestern Asia, and the invention of such "simple" devices as the plow, the yoke, the wheel, and the waterwheel, were major steps forward toward making food production easier. The agricultural transition set in motion a worldwide population increase, accompanied by a steady expansion of the cultivated area. It is thought that this expansion came about as the early farmers migrated with their crop plants and domesticated animals into areas previously occupied by hunter-gatherers.

By A.D. 1500, the agricultural way of life had spread throughout the world, and hunting-gathering as a mode of subsistence was practiced by only a minority of the population. However, food procurement remained a major preoccupation. Although some people had been freed from the land, it is generally estimated that two-thirds of all people were engaged in food production. The efficiency of agriculture was low, and one family could not grow much more food than it needed for itself. Contrast this with the situation today, when, according to the FAO, only about one person in five is a farmer, and 80% of these are in Asia, engaged in growing rice. Freeing most people from farming has been a major contribution of modern agriculture to society.

Much of the high-input agriculture now practiced has its roots in the European Renaissance (1400–1600). Two things that happened during this period of cultural revival contributed to this development. The first was the voyages of discovery by the European explorers. Columbus and others set in motion an intercontinental exchange of crops, domesticated animals, and farming practices. This exchange greatly altered the global distribution of certain species of plants and animals, and increased the world capacity for food production. Crops that had been domesticated in one region often found an equally or even more suitable growing area on an entirely different continent.

By 1492, the Incas living in the highlands of South American Andes were planting more than 3,000 varieties of potatoes. They had a high degree of sophistication, constructing canals for irrigation and planting on terraces. They even invented a method of drying potatoes for storage. When they arrived in 1531, the Spanish conquistadors had little interest in the crop, but later that century potatoes reached Europe. For 200 years, potatoes remained somewhat of a curiosity until peasants discovered that potatoes grew easily in many types of soils, and in prodigious quantities. In fact, a farmer could grow far more potatoes than grain on a plot of land, and could have plenty left over to sell. This fact was important in the industrial revolution of the eighteenth and nineteenth centuries: surplus potatoes found a ready market in the cities, where sons and daughters of farmers moved to work in factories. The reliance on the potato as a staple was so great that two successive crop failures in Ireland in the 1840s led to widespread famine. Although the Irish at that time also grew enough wheat, they were too poor to buy it themselves and were forced to export it to England.

On his first voyage to the Americas, Columbus noticed, and was duly impressed by, the corn cultivated by the Native Americans on what is now Cuba. Indeed, at that time corn was grown from Canada to Chile, in both cold and warm weather, and in various conditions of moisture. As with the Incas for potatoes, a complex agricultural system had been developed, with hundreds of varieties. Columbus brought back some corn on his return voyage to Spain, and it first became a staple in Portugal, then spread as far as Turkey, China, and Africa. People soon realized the value of corn as an animal feed, and this ready supply of feed grain played an important role in supplying the increasing population with meat and animal products. When the Pilgrims landed in North America in 1620, they had little food and local natives taught them to grow corn. These Europeans soon found many uses for the grain, and it spread as they did through the plains of North America.

By and large, the Native Americans used simple tools for planting and harvesting of their crops. When Columbus arrived, there were no draft animals, such as horses and oxen. On his second voyage, Columbus brought the first horses to the Americas, astonishing the local inhabitants. Horses quickly spread throughout the Americas, and were important in opening up new lands for cultivation, for travel, and, of course, for war.

Contact among the different civilizations also allowed the exchange of agricultural practices. When Columbus arrived in America, European agriculture was in some ways more advanced than that of the Native Americans. But the latter used some important techniques not widely practiced by the Europeans. The Native Americans practiced intercropping (growing two or more crops in the same field) and planted seeds individually in rows or in small clumps. The Europeans practiced crop rotation (growing one crop one year and another one the next on the same field) and spread the seeds by broadcasting them over the land. The Native Americans intercropped cereals (corn) and legumes (common bean), whereas the Europeans often rotated cereals (wheat, barley, or other small grains) and legumes (clover, for fodder for their animals). Modern farming uses both crop rotation and an elaboration of the Native Americans' planting method. Crop rotation results in better pest control by not allowing pest populations to build up, and planting seeds in rows allows easier weed control.

The second major aspect of Europe's cultural revival in the fifteenth and sixteenth centuries that affected agriculture was the rise of modern scientific investigation, some of which was soon turned into agricultural technology. A renewed interest in medicine first caused an increase in the publication of "herbals," treatises that described and catalogued plants and listed their medicinal values. The study of plants became a science, and the invention of the microscope in the seventeenth century allowed botanists to examine and describe plant structure in detail. Later botanists, as we saw in earlier chapters, gained extensive knowledge of photosynthesis and plant nutrition, and these studies could now be put to use.

The research of Justus von Liebig, who formulated "the law of the minimum," which states (as noted earlier) that the growth of a plant is limited by the one nutrient that is in shortest supply, led directly to the manufacture and use of inorganic fertilizers. Knowledge about soil characteristics and about the physiological requirements of plants allowed farmers to manipulate the environment more intelligently to coax the most food from their land. Knowledge of the laws of heredity led to a more rational approach to plant breeding. At the same time, crop production was improved by the use of better farm implements and machines. For example, in 1831 Cyrus McCormick invented the mechanical reaper, which harvests grain as it moves across the field. As a result, one person could harvest much more grain than ever before.

These developments in the last 400 years have enabled people to produce food much more efficiently. In the process, agriculture has become completely dependent on science and technology. The output of agriculture has greatly increased, but so have the "inputs." Until about 200 years ago, the inputs were quite simple: they consisted of a portion of last year's crop, some manure, perhaps an ox and a plow, and a hoe, as well as a lot of human labor. Today, the inputs are provided by a host of agricultural industries virtually unknown 150 years ago. Chemical plants annually produce millions of tons of fertilizers, pesticides, and herbicides. Factories provide a whole inventory of specialized equipment for tilling the soil, applying fertilizers, planting, hoeing, harvesting, and storing the crops. Plant breeders constantly try to obtain better varieties that produce yet more food. Much of this food is not sold fresh to the consumers, but must first be processed in factories. Thus food production in industrialized countries has become a highly complex process.

In Chapter 2, we noted that there are many different farming systems in the world. The high-input agriculture practiced in developed countries is but one system. Nearly all systems have undergone dramatic changes as a result of the scientific discoveries and advances in technology. In some systems, these changes came early, in others they came much later, as a result of the Green Revolution (discussed in the next chapter).

9 | Selection led to major improvements in certain crops.

By the end of the eighteenth century, the informal processes of selection by farmers everywhere led to the creation of thousands of different varieties or **landraces** for each major crop species all over the world. Thus, when a Native American farmer in the Gila River area of Arizona collected bean seeds

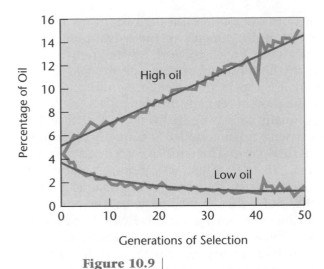

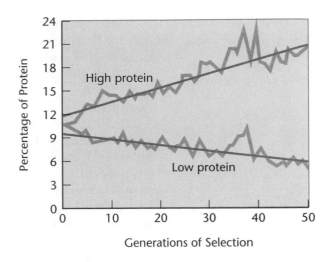

Figure 10.9

Selection for Oil Content and Protein Content in Corn. The solid thick lines are actual data; the thinner lines, trends. In a comparatively short time, people can use genetic variability to select strains of a desired phenotype. Source: *Data from U.S. Department of Agriculture.*

every year from her field for replanting the next year, she ended up with a landrace that was slightly different from that of her sister-in-law who lived 10 km away. Each landrace was exquisitely adapted to its own local environment.

The landraces of crops that self-fertilize (such as the small cereal grains) differ in terms of the characteristics of their gene pools when compared with crops that cross-fertilize (such as corn). Since the former are the result of repeated self-matings, they are relatively homozygous and uniform in their agronomic characteristics. Plant breeders often call them "pure lines." In contrast, the cross-pollinated plants are comparatively heterozygous, arising as they do from the mating of two different plants. Instead of pure lines, the landraces of cross-pollinators are heterogeneous genetically and form a relative continuum of agronomic types in the field.

Selection breeding involves the breeders choosing those plants for reproduction that have the most desirable genes from an agronomic viewpoint. Creating a better crop does not depend on the introduction of new genes. With self-pollinating crops, the process is quite simple: a landrace is chosen and propagated for many generations. As these plants are inbred, they become more homozygous than their progenitors. But as the desired genes become homozygous, so may undesirable genes, and characteristics that were masked in the past by being heterozygous may now be expressed. This inbreeding may lead to a reduction in yield termed inbreeding depression. Such inbreeding depression among humans is also well known, and is a rationale behind laws prohibiting incest in many cultures.

Selection in cross-pollinating crops is more difficult, because genetic homogeneity is not reached. Nevertheless, such selection can be done, and results in a trend toward a chosen phenotype, or "appearance." Modern science has made it possible to select not only for visible characteristics (appearance), but also for properties that can be measured biochemically (Figure 10.9).

Despite these drawbacks, selection was the main method of crop plant breeding until early in the present century. For many crops and by many peoples, special lines—often with specific desired characteristics—were selected from naturally occurring variants. Literally thousands of wheat and corn varieties were known in North America by the early 1900s. And plant collectors continue to find new variants of our major crops.

One variant of considerable nutritional interest is high-lysine corn. Selection for high protein contents (Figure 10.9) increased the overall nutritional value of this staple as a food and animal feed. However, there are two problems with this approach. First, the high-protein strains exhibit yield depression. Second, because corn protein is low in the essential amino acid lysine (see Chapter 4), the quality of the protein remains problematic.

In 1963, during screening of many corn varieties for protein quality, U.S. scientists found that grains carrying a gene called "opaque-2" had more lysine and tryptophan than normal corn. Their relative contents of other amino acids were about normal, so the variety was termed "high-lysine corn." Later studies by molecular biologists showed that the opaque-2 mutation dramatically reduces synthesis of a class of seed proteins, the *zeins,* that are low in lysine. This results in the grain having a relatively high content of the other classes of proteins, which have an adequate lysine content.

Realizing the importance of their discovery, the scientists fed the corn as a meal (supplemented with vitamins and minerals) to baby rats. The rats grew four times faster than those fed with normal corn. Similar results were obtained with poultry and pigs. Soon the opaque-2 cornmeal was being used in South America to cure children who had protein–calorie malnutrition.

Breeding programs were begun to develop new lines with opaque-2 and a similar allele, floury-2. However, the strains had a low-yield, soft grain with a chalky appearance that was unacceptable to consumers and was more susceptible to ear rot, a fungal disease. So far plant breeders have not succeeded, in spite of considerable effort, in transferring this interesting and valuable trait to a high-yielding commercial corn variety. It may be that the gene that specifies this trait is closely linked in the chromosome to genes that have detrimental effects.

10 Hybridization was the first systematic application of genetic principles to crop breeding.

In 1908, the U.S. plant breeder G. H. Shull reported the results of crossing corn plants from two different inbred lines. The cross between these two highly homozygous lines produced heterozygous offspring, because the lines were homozygous for different sets of genes. The result was astonishing. The two inbred lines had each produced about 20 bushels of corn per acre in the last crop. Their outbred offspring quadrupled this yield, to 80 bushels per acre! As noted in an earlier chapter, this unanticipated strength in the heterozygous outcross was called **hybrid vigor**, and such hybrids have played an important role in increasing yield. By 1920, the first commercial hybrid corn was available in the United States, and two decades later nearly all the corn grown commercially was hybrid, as it is to this day.

As with most scientists, Shull's achievement followed several important advances by others. Although early botanists had observed increased growth when unrelated plants of the same species were crossed, it was Charles Darwin who did the first experiments in this regard. He showed that crosses of related strains did not exhibit the vigor of hybrids. In 1879, William Beal demonstrated hybrid vigor in corn by using two unrelated varieties. Corn normally does not self-fertilize (the male part of one plant produces pollen that fertilizes the female part of another plant), so Beal detasseled (removed the male parts)

Figure 10.10

G. H. Shull's Hybrid Corn Growing in 1906 at the Cold Spring Harbor Laboratory, New York. Note the bags covering the tassels (male organs) of some plants. Source: *Cold Spring Harbor Laboratory Archives.*

of one strain to ensure cross-fertilization. Shull then took these experiments a step further by deliberately creating inbred lines by repeated self-fertilization. He ensured this by physically collecting pollen from a plant and putting it on the female part of the same plant (Figure 10.10).

One reason for this unanticipated yield increase was that the inbred lines Shull started with suffered from severe inbreeding depression. In other words, they yielded less than the old open-pollinated varieties. As noted earlier, crop plants consisted of hundreds of landraces, each one adapted to its own local area. Plants that normally self-fertilize (inbreeders) produce fairly uniform crops under such conditions, but plants that normally cross-pollinate each other (outbreeders, such as corn) produce quite heterogeneous stands. When plants cross-pollinate, the genes of individual plants can be quite different, resulting in variability among the plants and in variable yield. The main benefit of hybridization is to eliminate this variability and to produce seeds that produce plants that all give high yields.

A major problem with the application of Shull's experiment on a massive scale was that the cross had to be repeated every year. In contrast to pure line selection, the farmer could not use the seeds for next year's crosses. The reason for this follows the principles of genetics. Suppose the gene combination *AaBb* is a desirable combination for a hybrid. This can be achieved by crossing two inbred lines: *AAbb* and *aaBB*. In such a cross, all the progeny will be *AaBb*. However, crossing *AaBb* plants among themselves gives only one-quarter of the plants as *AaBb*. Most of the plants have some homozygosity, with its resulting yield depression. This heterogeneity is bad news for the farmer, but good news for the company that supplies seeds to the farmer each year.

After its introduction in the 1920s, hybrid corn spread rapidly. By the 1940s, virtually all corn grown in the U.S. Corn Belt was hybrid. Yields increased fourfold between 1920 and 1990, with most of that increase

Table 10.3	How a hybrid seed line is developed from 100 crosses

Year	Events	
1	100 crosses made	
2	200 rows of progeny planted and selfed	
3	2,000 rows of progeny planted and selfed	
4	8,000 rows of progeny planted and selfed	
5	1,200 hybrids tested	
6	80 hybrids tested	
7	12 hybrids tested	Testing must be done in multiple locations.
8	4 hybrids tested	
9	1–2 hybrids tested	
10	1 hybrid tested	
11	0–1 hybrid marketed	

Source: Modified from *Seeds Greenbook* (1990), Basel: CIBA-GEIGY Seeds.

occurring after 1940 when the spread of the hybrids had already been completed. Some have argued that 60–75% of the increase in corn productivity can be ascribed to introducing hybrid corn, and that the remainder can be accounted for by other technologies (fertilizers, pesticides, and mechanization). It is more accurate to say that after hybrids were introduced all breeding efforts were focused on producing hybrids that were compatible with the new technologies that became available after 1940. For example, after agronomists realized that corn responds to nitrogen fertilizers with increased growth and crop production, hybrids were bred specifically for this purpose. This resulted in hybrid lines that respond to high levels of nitrogen fertilizers. Very likely plant breeders could also have selected open-pollinated strains of corn that respond to fertilizers. However, there was no financial incentive to do so.

The introduction of hybrid corn in the United States was paralleled by the development of a corn-breeding industry, and its prosperity depends on the continued use of hybrids by the farmers. Most corn breeding is done by plant breeders associated with this industry, and they have no reason to improve varieties of corn that are not hybrids, as is normally done with wheat. Wheat farmers do not have to go back to the seed company every year to buy new seeds for planting, but corn farmers have no choice: they are dependent on the commercial breeders.

The development of new hybrid varieties is an ongoing process. But it is also long and tedious (Table 10.3), with hundreds of crosses needed to yield one new strain. One problem with this technique has been that it was labor intensive. To force corn plants to outbreed, Shull and later breeders had to manually detassel the female parents to prevent self-fertilization. The inbred lines were planted in such a way that one row of pollen-producing plants alternated with four rows of seed-producing plants. The four rows of plants that were to produce the seeds had to be detasseled to keep them from producing any pollen. The other line produced the pollen that fertilized the silks of the detasseled plants, and in this way pollination and reproduction could be controlled.

Figure 10.11

Production of Hybrid Rice Is an Elaborate Procedure. To produce hybrid rice, the seed producer first crosses a female sterile parent with a fertile male maintainer line. The seeds from this cross are planted in small plots (a). If allowed to cross with each other, they would produce sterile plants that have no grain. However, when surrounded by rows (b) or a restorer line, pollen from the restorer will fertilize the sterile plants in (a), so that these plants will set seed. This seed, the hybrid seed, is sold to the farmers and yields 15–20% more than conventionally bred rice varieties. Source: IRRI Reporter (Los Baños, Philippines), February 1992, p. 2.

Later, mutants of corn were discovered that were male-sterile: they did not produce fertile pollen. This discovery proved to be a boon for the hybrid corn seed industry because it eliminated the need to detassel the corn plants by hand. The practical application of this discovery depended on a second finding: that corn also has genes that restore male fertility by counteracting the genes that cause male-sterility. When making hybrid corn, it is possible to have a gene for male-sterility in line *A,* and a fertility-restoring gene in line *B.* When line *A* is planted, it produces plants that have no viable pollen. Line *B* produces the pollen, and the hybrid *AB* kernels will mature on the line *A* plants. When these *AB* kernels, the hybrids, are planted in the field, they will produce pollen because of the fertility-restoring gene in line *B.*

An unexpected consequence of using this system to produce all hybrid corn arose in the United States in 1971. By this time, most of the corn was grown from the so-called Texas cytoplasm male-sterile line. It turns out that this allele confers on the corn plant not only an inability to form pollen, but an enhanced sensitivity to the toxin produced by the fungus responsible for southern leaf blight. As a result, the U.S. corn crop was severely reduced by a blight infestation. Different genes that confer male-sterility are now used.

The production of hybrids in other species depends heavily on such male-sterility and fertility restorer genes, either naturally occurring or introduced by genetic engineering (see Box 10.1). Cross-pollination of plants that normally self-pollinate is only possible when pollen production can be efficiently eliminated. This can be done by hand, removing the pollen-producing anthers with tweezers, and it is a difficult and laborious task. In spite of the labor costs, China has moved ahead with the production of hybrid rice for large-scale rice production. However, large-scale production of hybrid seeds in developed countries requires more efficient mechanisms to suppress pollen production.

At the International Rice Research Institute (IRRI) in the Philippines, an elaborate technique has been developed for producing hybrid lines of rice that show a 20% increase in yield over current varieties planted locally (Figure 10.11). Since the discovery of male-sterile and restorer genes in rice,

Box 10.1

*Male-Sterile
Plants Through
Genetic
Engineering*

The efficient and cost-effective production of hybrid crop seeds requires strains of the crop that are male sterile (do not produce viable pollen) and strains with genes that restore male fertility, when crossed with the male-sterile strains. Although we do not yet understand how and why the natural genes that cause male sterility and restore it again work, gene transfer technology has allowed scientists to recreate such a system. This was accomplished by hooking the gene for an enzyme that causes cell death to a regulatory region that causes the expression of the enzyme, only in those cells that produce the pollen grains.

All plant cells express many of the same genes, but there are genes that are expressed only in specific organs or even cell types. The specificity of this expression resides in the control region of the gene. Robert Goldberg, from the University of California at Los Angeles, has isolated a number of control regions that direct the expression of genes in specific parts of the flower: the sepals, petals, stamens, and carpels. He found a control region that caused expression of a gene in the anther, that part of the stamen that produces the pollen. In a collaboration with scientists from Plant Genetic Systems in Belgium, he fused this control region to a gene that encodes the enzyme ribonuclease. Ribonuclease, when present in the cells, causes the uncontrolled breakdown of the ribonucleic acids, the molecules that carry the genetic information from the nuclear DNA to the protein-synthesizing machinery in the cytoplasm. When this chimeric gene was introduced into wheat, the transformed plants appeared healthy except that they produced no pollen.

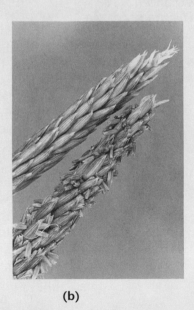

(a)   (b)

(a) *Male Sterile (right) and Male Fertile Normal (left) Corn Plants.* **(b)** *Close-up Photos of the Tassels of a Male Sterile (upper) and a Normal (lower) Corn Plant.* Note the absence of anthers from the male sterile tassels. Source: *Photos courtesy of J. Leemans, Plant Genetic Systems, Ghent, Belgium.*

Next they hooked the same control region to a gene that encodes an inhibitor of ribonuclease. When both the enzyme and the inhibitor are made together, in the same cell, the inhibitor binds tightly to the enzyme, preventing it from doing its nasty work. Plants transformed with the inhibitor looked perfectly normal. When the pollen of these plants was used to fertilize the plants containing ribonuclease gene, fertile seeds were produced. When these seeds were planted, they yielded normal plants that produced pollen. Thus, the ribonuclease gene functioned as a male-sterile gene, and the inhibitor gene as a restorer gene.

agronomists have designed a method to produce viable hybrids. The original female line has the male-sterile allele, and it is crossed with a fertile "maintainer" line. The seeds from this cross are sown, and the resulting plants are sterile. However, they do produce seeds when crossed with a third rice line that contains the restorer allele.

11 | Crossing allows specific genes to be introduced.

Besides selecting plants with useful characteristics, breeders also arrange "marriages" between plants with different traits and hope for fertile offspring carrying both traits. By understanding the reproductive capacities of plants, plant breeders can do more than hope: they can manipulate these crosses in such a way that fertile offspring *will* result, carrying traits from both parents.

In thinking about mating partners and barriers to reproduction, it is useful to adopt the terminology of gene pools (GP). We noted earlier that crop domestication resulted in the establishment of many different landraces. Such landraces can be crossed readily with each other because the plants all belong to the same species. Normally, the landraces do not interbreed because they are isolated and grow in different places. In addition, many plants do not interbreed readily because they routinely self-fertilize.

These landraces, along with the various wild ancestors of domesticated plants, belong to the **primary gene pool (GP1)**. When individuals from GP1 are crossed, they usually produce fertile offspring. Plants of related species, with which crossing is possible and where crossing results in the production of fertile progeny, are said to be in the **secondary gene pool (GP2)**.

Beans provide an example of these gene pools. Native Americans cultivated hundreds of landraces of beans (*Phaseolus vulgaris*). Although most of these landraces have now, unfortunately, disappeared, we still have many different varieties, such as green beans, wax beans, pinto beans, kidney beans, dry white beans, and so on. All these beans, in spite of their different appearances, belong to the same species and the same GP1 gene pool. In addition to this species, some other types of beans belong to other species in the same genus: lima beans (*Phaseolus lunatus*), tepary beans (*Phaseolus acutifolius*), and scarlet runner beans (*Phaseolus coccineus*). These species belong to GP2. Crosses can be made, but this takes some effort and fertile offspring are rare. Embryo rescue via tissue culture is sometimes needed.

To make crosses with other beans, which are more distantly related (mung beans or blackeyed peas) may be impossible, although techniques of tissue culture and embryo rescue make it possible in some cases. These distantly related plants are in the **tertiary gene pool (GP3)**. Thus, introducing genes from GP1 into a crop is relatively easy (although time consuming and requiring extensive backcrossing) while introducing genes from GP2 and GP3 is progressively more difficult. Genes from GP4, the gene pool of all the organisms that cannot possibly be crossed with the plants in GP1, can only be introduced via techniques of genetic engineering.

Transfer of a characteristic (one or more genes) cannot be accomplished by one simple cross, but requires extensive **backcrossing**. In this way, desirable characteristics of one strain can be combined with those of another strain, and the breeder circumvents the problem of trying to select simultaneously for many traits in the same strain. The introduction of genes for short

Figure 10.12

Hypothetical Scheme for Backcrossing. The objective is to introduce the gene for shortness from strain B into the high-yielding wheat strain A. From the first cross of A × B, plants (A') are selected that resemble A but are shorter. This selection is done after the progeny of the first A × B cross have been allowed to self-fertilize to make sure that the gene for shortness is homozygous. These are then crossed to the initial strain, A. Once again, plants (A") are selected that resemble A but are shorter. These in turn are crossed back to strain A. After repeated crossing and selections, the gene for shortness of strain B becomes part of strain A.

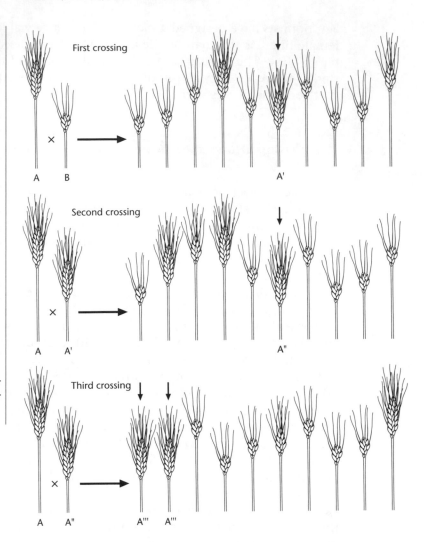

stems into traditional long-stemmed wheat provides a good example of backcrossing. When tall wheat plants are heavily fertilized with nitrogen, they fall over at maturity because the slender stalk is unable to bear the heavy load of grain. As a result, much of the harvest can be lost, especially if the wheat is harvested by combine. For this reason, plant breeders have tried to breed shorter strains of cereals.

Suppose one wants to introduce, by backcrossing, a gene for "shortness" into a normal, high-yielding, tall wheat strain (see Figure 10.12). The breeder does this in several steps:

1 A suitable wheat variety that is short is found. It may not have any other desirable characteristics, such as high yield, but that does not really matter.

2 The short variety is crossed with the normal tall wheat and the seeds that are produced are planted out to produce new plants. These plants are allowed to self-fertilize so that the gene for shortness will be homozygous. The seeds produced by these plants are planted out again, and now the breeder will select those plants that resemble the high-yielding parent and that are also short.

Table 10.4	The evolution of wheat-breeding methods in the United States (percentage of wheat planted derived from different genetic techniques)			
Year	Collected Strains	Selections	Crosses	Other
1924	57	26	7	10
1934	50	32	14	4
1944	25	27	45	2
1954	6	11	82	1
1964	6	8	84	2

Source: U.S. Department of Agriculture.

3 These plants are then crossed again with the high-yielding variety, the seeds from this cross planted, and again one selects plants that resemble the original high-yielding variety but are also short.

4 This backcrossing selection procedure is repeated 5 to 8 times, until a new variety of wheat emerges, one that has all the desirable characteristics of the original strain and also has the gene for shortness.

The purpose of backcrossing, in genetic terms, is to maintain as much as possible the genes of the high-yielding plant, and to introduce the fewest possible number of genes from the plant that has the gene for shortness. With each backcross, half the remaining genes from this plant are eliminated (see Chapter 9).

Crossing has been very valuable to plant breeders, because it allows some measure of control over the phenotype of a plant. Not surprisingly, crossing is responsible for most new varieties developed in recent years (Table 10.4). The story of modern North American wheat provides an example. In 1842, a Scotsman, David Fife, emigrated to Canada and brought some wheat seeds with him. He planted them on his homestead in what is now the province of Ontario, and the high yield and the good milling and baking qualities of the crop were remarkable. Soon Red Fife wheat had spread throughout the region. But it was susceptible to the early frosts that plague eastern Canada. In 1892, this wheat was backcrossed with varieties from the Himalayas, to make it more frost resistant. A result of these crosses in 1907 was Marquis wheat, which had all the good traits of Red Fife, with the addition of frost resistance. Marquis was sent to the Canadian and U.S. prairies, and was an instant success. By 1917, a decade after the first seeds were sown, its North American crop was 7 million tons.

Many useful genes that can be added to commercial cultivars by backcrossing are found in the GP1 gene pool among the wild relatives and the landraces, and that is the main reason why these plants should not be allowed to become extinct. Unfortunately, the adoption of commercially produced crop varieties has led to the disappearance of nearly all landraces in developed countries and in Third World countries where the Green Revolution has occurred. Similarly, the spread of agriculture onto lands that were previously not cultivated is causing more and more wild relatives to disappear. The saying "Extinction is forever" applies to common-looking relatives of our crops as well as to interesting mammals, beautiful birds, and pretty flowers. When the plants disappear, the genes disappear with them.

12 | X rays and chemical mutagens can create useful mutations.

Mutations create genetic variability, but the natural rate of permanent genetic change is very slow. Indeed, it has taken millions of years to accumulate the variability observed in natural populations of crop plants. Within the last half-century, geneticists have found ways to speed up this mutation process greatly. Both radiation (such as X rays or ultraviolet light) and certain chemicals can change DNA and therefore induce mutations. When large numbers of seeds are treated with these agents, mutations are induced at random. Most of these gene changes are harmful, but a very small number may be beneficial. Thousands of seeds must be treated, planted, and screened to detect one useful mutation, which can then be bred into the crop plant by backcrossing.

By the 1960s, some scientists were suggesting that mutation breeding would solve the food problems of the future. There have been a few notable successes, and the creation of mutations has on the whole had a minor impact in crop improvement. Mutant genes that create semidwarf varieties of rice have been made in Japan and for durum wheat in Italy. Indian scientists were able to create a variety of castor bean that matures in 120 days, rather than the usual 270, allowing two crops to be grown per year on the same land area.

An interesting X-ray-induced mutation is the "topless" field bean created in Sweden. Field beans and broad beans (*Vicia faba*), are strong-tasting beans not much in demand in the United States, but widely grown in Europe, partly for human consumption and partly for animal feed. They are legumes with an indeterminate growth habit, producing a stalk with a number of flowers and then continuing to produce leaves as they grow 2 feet tall. The leaves at the top apparently do not contribute to the yield and farmers can mechanically top the plants without affecting yields. Topping individual plants is a difficult and time-consuming task, but is worthwhile because it makes harvesting the beans much easier. The problem has been solved by an X-ray-induced mutation that produces a topless plant. It is a minor victory in the struggle for food production; minor, because although field beans have great potential as a cool climate legume, at present they are a relatively unimportant crop.

13 | Natural and artificial polyploids create new possibilities for plant breeding.

Most higher organisms are diploid, with two sets of chromosomes (and therefore, genes). If there are more than two sets, the organism is termed a **polyploid.** This condition can arise spontaneously in nature or can be induced by specific chemicals. In nature, polyploids usually arise from sex cells that have extra sets of chromosomes. For example, chromosomes may fail to migrate to the ends of the cell during meiosis; if one end of the cell receives all the chromosomes, and the other end none, this will result in the formation of a sex cell that has two sets of chromosomes. When this sex cell combines with a normal sex cell that has a single set of chromosomes during fertilization, the resulting organism will have three sets of chromosomes (triploid).

Box 10.2

Hot Chile Hybrids

Landraces that were the sole source of seed of all farmers have mostly disappeared in countries that practice high input agriculture, but among traditional farmers they may still be found. Gary Nabhan, an ethnobotanist from Phoenix, Arizona, has written engagingly about the use of landraces by Native American farmers and about the need to conserve this precious resource. He argues that although seeds can be stored in seedbanks as a conservation measure, their gene pools are not in dynamic equilibrium with the wild relatives that grow in their surroundings. Indeed, although crop domestication requires some degree of isolation of the crop that is being domesticated from its wild relatives, it does not exclude continuous hybridization with wild relatives. In his book *Enduring Seeds* he captures the essence of this gene flow beautifully in a conversation with Juan Ignacio Humar, a village farmer who grows chiles (hot peppers) on the subtropical edge of the Sonoran desert, in an ancient village of the Pima Indians.

"The wild chiles of the countryside inoculate those in my garden with their piquancy," Juan said, somewhat understating the case, considering the pain those natural hybrids have brought to his friends.

"Well," I asked, "does that injection of heat make them bad?"

"Oh, no. I eat the chiles, hot or not. There's great variability in the spiciness of cultivated chiles around here, especially if the wild ones have inoculated them. I never know what one will be like until I bite into it."

"When you get a lot of the hot ones, can you sell them?" I asked, wondering if Juan may have a viable hybrid variety on his hands. "Do you save the seed?"

"Well, I save the seed because I like eating them enough to plant them again. It's my own seed, not a selected commercial variety. But I believe they're too hot to sell to the buyers who come around for the market in the city . . . I simply keep them for my own pleasure."

Source: Nabhan, G. P. (1989), *Enduring Seeds.* North Point Press, San Francisco, p. 37.

Polyploids have played a major role in the evolution of modern crop plants. Plant strains, which early farmers selected for their vigorous growth or high yield, often turned out to be polyploids. In some cases, an increase in the number of chromosome sets is correlated with increases in the sizes of leaves, stems, fruits, and flowers. This is especially true for ornamentals, such as petunias, where series of both wild and cultivated varieties show both additional sets of chromosomes and larger flowers.

In addition to the extra chromosome sets in a single species, polyploids can arise from the combination of chromosomes from two different species. For example, the modern plum tree is diploid, with two sets of 24 chromosomes; 8 of these chromosomes represent the haploid complement of the cherry plum, and the other 16 represent the haploid complement of the blackthorn. The additional chromosomes in the modern plum lead to larger fruit and a higher yield. In a similar manner, modern tobacco, a diploid with two sets of 24 chromosomes, is believed to have arisen from two lower-yielding species with 12 chromosomes each.

Most modern wheat strains are hexaploid, with six sets of chromosomes per cell, and the evolution of modern wheats from their wild diploid relatives parallels, in terms of chromosomes, their agricultural selection (see Figure 10.3). Einkorn wheat has 14 chromosomes (two sets of 7), which are designated *AA*. This species is believed to have hybridized with a wild grass species

Figure 10.13

*Triticale (right), a
Polyploid Hybrid
Between Wheat (left)
and Rye (middle).*
Source: *U.S. Department
of Agriculture.*

to produce emmer wheat. The latter, now common as durum or pasta wheat, has the *AA* chromosome sets and the *BB* sets of the wild grass. This tetraploid, *AABB,* now mated with yet another wild grass species, which had the chromosome complement designated *DD,* to yield *AABBDD,* our modern hexaploid wheat, which has 42 chromosomes. The two wild grasses that contributed the *B* and *D* chromosomes to this evolutionary scheme have been identified as goatgrasses endemic to Turkey and the region of the Fertile Crescent, where wheat was domesticated.

Polyploids with an odd number of chromosome sets cannot reproduce sexually and do not produce seeds. This results from the biology of sex cell formation, which involves a division of a diploid cell such that each sex cell receives one chromosome set of the two in the diploid (see Chapter 9). In a triploid, this machinery breaks down, because three sets of chromosomes cannot be divided evenly into two cells. This sterility can be an advantage, as in the production of seedless fruits such as watermelons. But it is a disadvantage in that the plants must be propagated asexually and bred by grafting. Many apple, pear, and citrus varieties are triploid, and thus must be propagated asexually.

An important application of polyploid breeding has been the development of the wheat–rye hybrid called *triticale* (Figure 10.13). This resulted from an attempt to combine the high yield and high seed-protein content of wheat (*Triticum*) with the adaptability to adverse environmental conditions (such as cold and drought) and high-lysine content of rye (*Secale*). This is an example of GP1–GP3 breeding, since the plants are only distantly related. The initial cross was made in 1875, but the plants were sterile because the hybrid had one copy of tetraploid wheat chromosomes (*AB*) and one copy of diploid rye (*R*). Thus, when the resulting plant (*ABR*) attempted to form sex cells, there was no way for the gametes to receive one copy of every gene and half the copies of the parent plant.

Triticale remained a botanical curiosity until 1937, with the discovery of colchicine, a molecule from the seeds of the autumn crocus that essentially blocks the movement of chromosomes to opposite poles of the cell during cell division. When the triploid wheat–rye offspring embryos were treated with this drug as cells divided, the replicated chromosomes did not migrate to the poles. As a result, the new cells had double the number of chromosomes as the parents, and were hexaploid (*AABBRR*). With an even number of chromosome sets, these plants were fully fertile.

An intensive breeding effort in Canada and Mexico has been focused on the hexaploid triticale. Strains have been developed that have high yield, high protein content, resistance to moisture stress, early maturity, and disease resistance. Worldwide, it is now grown on well over 1 million hectares, mostly in eastern Europe, Canada and Mexico. It is currently used predominantly as an animal feed, but shows promise as a source of human food.

14 | Mapping the genomes of crop plants offers new possibilities for crop breeding.

We noted earlier (Chapter 9) that genes are located on chromosomes. Actually, each gene has a very precise location on a particular chromosome. Characteristics that are encoded by a single gene, such as flower color, can readily be identified in the progeny of a cross, and when two such characters segregate independently we know they are on different chromosomes. For example, the peas of Gregor Mendel were always round or wrinkled, and green or yellow. In the progeny of his crosses, these two characters behaved independently.

When two traits do not behave ("assort") independently in the progeny of a cross, they are said to be linked on the same chromosome. By keeping track of the linkage of many different single gene traits, geneticists can construct a linkage map of each chromosome. When the chromosome is drawn as a line, the location of each gene can be marked with respect to all other known genes. Such linkage maps have been constructed for many different crops, such as tomatoes, lettuce, corn, and beans.

The tools of molecular biology allow the construction of a totally new kind of linkage map in which the markers are not traits that can be recognized on the plant, but specific sequences of nucleotides that can be found when the DNA is analyzed in the laboratory. Scientists can also merge the two kinds of maps, so that they know which nucleotide sequences are close to which genetic traits.

Think of a chromosome as the Interstate highway (Motorway or Autostrade) between San Francisco and Chicago. If you have a map, and all the exits are marked (by number or location) you always know where you are. The more locations are marked, the more precisely you will know where you are at any one time. If you know the town you want to visit is between exits 41 and 42, then you look for one of these exits, and you know you are close. On chromosomes, the "exits" are represented by specific nucleotide sequences called **molecular markers**.

Not unexpectedly, given the knowledge that DNA is a long strand of nucleotides (see Chapter 9), the markers are small sequence differences. These can be recognition sites for restriction enzymes. For example, in one plant strain if a restriction enzyme cuts DNA at the sequence

(*rest of DNA*) . . . A -A-C-G-T-T . . . (*rest of DNA*)
(*rest of DNA*) . . . T -T-G-C-A-A . . . (*rest of DNA*)

But in another strain, the same location on the chromosome has the sequence

(*rest of DNA*) . . . A -A-C-A-T-T . . . (*rest of DNA*)
(*rest of DNA*) . . . T -T-G-T-A-A . . . (*rest of DNA*)

Thus we know that the enzyme will not cut the second strain's DNA at this location. Because this place on the genome is a site that a restriction enzyme recognizes (normally), and because it has become different (polymorphic), this is called a **restriction-fragment-length polymorphism (RFLP).** "Riflips," as these are colloquially named, can be detected after separating DNA fragments according to size. They are inherited as simple Mendelian alleles.

Making an RFLP map is a time-consuming and labor-intensive procedure (molecular biologists are very impatient people), and recently a new method called *RAPD* ("rapid") was devised. *RAPD* stands for "randomly amplified polymorphic DNA," and in this procedure, short, random segments (RAPDs) of DNA are mapped out, so that they can be used to locate nearby genes. By using RAPDs, the breeder need not rely on the natural differences in the DNA. RAPD sequences are also inherited as simple Mendelian alleles.

These small, heritable sequences of DNA (RFLPs or RAPDs) are usually not themselves in important genes. But if they can be mapped at intervals all over the genome (see Figure 10.14 for an example of such a map), they will undoubtedly be near genes of agricultural importance. Often such important alleles are inherited together with a RFLP, and this indicates that they also lie closely together on the chromosome. So localizing the RFLP means that the important gene is nearby.

The first major effort at making a molecular map was done with the tomato because so many tomato mutants had already been placed on a genetic map. This molecular map, with 800 molecular markers, serves as the model for generating maps of other important crop plants. Over 300 RFLPs have been mapped for barley, the third leading seed grain in the United States (Figure 10.14), and related to genes for disease resistance, nutritional quality, and malting quality (barley is used to make beer). Similar mapping efforts are underway for all other major crops.

The importance of such maps is that they will let molecular biologists find and isolate specific genes—a gene for disease resistance, for example—and they will let plant breeders identify among the progeny of a cross, those individuals that have desired characteristics that are otherwise not so easily identified. This is especially important for so-called quantitative traits, such as yield or drought resistance, that are encoded by many genes. Transferring such multigene traits from one line to another poses daunting challenges for crop breeders who want to be able to identify the genes associated with them. Without the molecular markers to guide them, breeders have found the task of identifying these genes next to impossible.

15 | Plant breeders have a long wish list for crop improvement.

The aim of crop improvement is to create new varieties that yield more food of better nutritional value than their predecessors. In previous chapters, we described the physiological bases for food production by plants, as well as

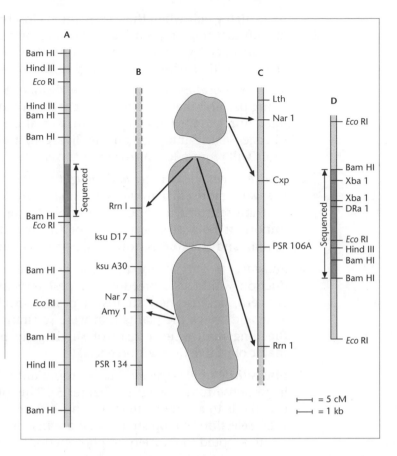

Figure 10.14

Mapping the Chromosomes of Barley. Lines B and C show several mapped genes on a barley chromosome. Line A is a detail of the region on B called Nar 7. The recognition sites for a number of restriction enzymes are shown. The same is true for line D, which is a map of the Nar 1 region on line C. The nucleotide sequence of some of the barley chromosomal DNA has been determined in regions shown on lines A and D. Source: The World and I, *vol. 6 (1992), p. 238; photo courtesy A. Kleinhofs, Washington State University.*

the environmental requirements for the optimal functioning of the plants. Now, with the tools of the plant breeder at hand, agronomists can decide what desirable attributes crop plants should have.

1 *High primary productivity.* Primary productivity is almost invariably related to crop yield. High productivity can be achieved by ensuring that all the light that falls on the field is intercepted by the leaves and that photosynthesis itself is as efficient as possible. Greater efficiency in photosynthesis could perhaps be achieved by selecting against photorespiration (Chapter 6).

2 *High crop yield.* In a world with a rapidly growing population, people need crop plants that produce more food per cultivated area. Plants must be selected that turn a high proportion of their total primary productivity into the plant parts that people eat (whether seeds, roots, or stems). The plants must also be strong enough to bear the crop.

3 *High nutritional quality.* Higher crop yield is more beneficial to people if the nutritional qualities of the crops can be maintained or improved. The proportions of essential amino acids and the total protein in cereal grains should be increased to improve their nutritional quality. The same should be done with root crops, such as potatoes, sweet potatoes, and cassava.

4 *Adaptation to intercropping.* So far, nearly all plant breeding efforts have been directed at crops that are grown in monocultures, and

have high yields when they are planted closely together. The realization that intercropping has many ecological benefits will redirect breeders' efforts toward strains that perform well when grown together with another crop in the same field.

5 *More extensive and efficient nitrogen fixation.* Strains of nitrogen-fixing bacteria that are more efficient at providing the plant with fixed nitrogen are being developed. Cereal grains that have more nitrogen-fixing micro-organisms living in the soil around their roots will require less nitrogen fertilizer.

6 *More efficient use of water and drought resistance.* Crop plants that use water more efficiently (have a smaller transpiration ratio) and that are resistant to drought are now being developed. This is important because many areas of the world have insufficient rainfall and the strains now in use produce only small crops.

7 *Resistance to pests.* If crop plants were resistant to pests, food production would rise. Breeding for pest resistance is an ongoing concern, because new strains of pests, especially fungi, always arise and attack the plants that were resistant to the old strains. Breeding multigene instead of single-gene pest resistance is a major objective of plant breeders.

8 *Insensitivity to photoperiod.* Many crop plants flower and set seed in response to day length (Chapter 5). This phenomenon limits the plants to a certain climatic zone and limits harvest to once a year. Selection of crop strains that are insensitive to photoperiod or are adapted to a variety of photoperiods would allow multiple cropping during a growing season.

9 *Plant architecture and adaptability to mechanized farming.* The number and positioning of the leaves, the branching pattern of the stem, the height of the plant, and the positioning of the organs to be harvested are all important to crop production. They often determine how well plants will intercept light, how closely they can be planted, and how easily the crop can be harvested mechanically.

10 *Elimination of toxic compounds.* Toxic compounds could be eliminated by plant breeding. For example, cassava produces cyanogenic glucosides, and potatoes produce toxic solanins (when exposed to light). The grass pea, *Lathyrus sativus,* a small legume that is widely cultivated and eaten in northern India, contains an unusual amino acid that causes lathyrism, a crippling condition of the legs.

Clearly, crop improvement means an attack on all these fronts. We have described some of the progress in and prospects for improving productivity, nitrogen fixation, and disease resistance. Recently, dramatic advances have been made, and are continuing, in the improvement of quality and yields.

Summary

The transition of hunting and gathering to planting and herding occurred between 12,000 and 8,000 years ago, independently in several parts of the world. At that time, most of our major crops were domesticated.

Agriculture was not "invented," but resulted from a gradual change in the relationship between humans and the ecosystem, and presupposed an intimate knowledge of the food plants.

Domestication, which is best described as "directed and accelerated evolution," quickly changed the plants because of the severe selection pressures that were applied by the first farmers. The emergence of agriculture dramatically changed human life, both quantitatively and qualitatively.

The European voyages of discovery and the application of science and technology to agriculture in the eighteenth and nineteenth centuries profoundly affected food production all over the world. Applying genetic principles to crop improvement created new strains of crops that could take advantage of the new inputs that technology offered. In the process, food production and industry became completely interdependent, at least in the developed countries.

Plant breeders have an extensive repertoire for improving crops. These include creating hybrids, backcrossing to introduce genes from other landraces or wild ancestors, creating variability with chemical mutagens or X rays, or using polyploids. Transferring multigenic traits by backcrossing is extremely difficult, but is aided by the new science of RFLP and RAPD map making, which necessitates mapping the entire genome of all the important crops.

Further Reading

Fehr, W. R., and H. H. Hadley. 1980. *Hybridization of Crop Plants.* Madison, WI: American Society of Agronomy.

Flannery, K. V. 1973. The origins of agriculture. *Annual Review of Anthropology* 2:271–310.

Galinat, W. 1992. Evolution of corn. *Advances in Agronomy* 47:203 –229.

Harlan, J. R. 1971. Agricultural origins: Centers and noncenters. *Science* 174:468–474.

Heiser, C. B. 1981. *Seed to Civilization: The Story of Food.* 2nd ed. San Francisco: Freeman.

Huke, R., and Huke, E. 1990. *Rice, Then and Now.* Los Baños, Philippines: International Rice Research Institute.

International Rice Research Institute. 1992. *Rice Genetics: I and II.* Los Baños, Philippines: IRRI.

Kloppenberg, J. 1988. *First the Seed.* Cambridge, UK: Cambridge University Press.

Kuckuck, H., G. Kobabe, and G. Wenzel, 1991. *Fundamentals of Plant Breeding.* Berlin: Springer Verlag.

Reed, C. A., ed. 1977. *Origins of Agriculture.* The Hague: Mouton.

Russell, W. 1991. Genetic Improvement of Maize Yields. *Advances in Agronomy* 46:245–280.

Schwanitz, F. 1966. *The Origin of Cultivated Plants.* Cambridge, MA: Harvard University Press.

Sleper, D. A., ed. 1991. *Plant Breeding and Sustainable Agriculture.* Madison, WI: Crop Science Society of America.

Ucko, P. J., and G. W. Dimbleby, eds. 1969. *The Domestication and Exploitation of Plants and Animals.* New York: Aldine.

U.S. National Research Council. 1989. *Triticale.* Washington, DC: NRC.

Viola, H., and C. Margolis. 1991. *Seeds of Change: A Quincentennial Commemoration.* Washington, DC: Smithsonian Institution Press.

Chapter 11

The Green Revolution and Beyond

The application of science and technology to crop production in the first half of this century resulted in yield takeoffs for rice, wheat, and corn in the developed countries. These efforts were diffuse, and many researchers, universities, and industrial companies contributed to them. The final result was that a new type of agriculture—high-input or chemical-genetic agriculture—replaced a more traditional system.

In 1970, Norman Borlaug, a U.S. plant breeder working in Mexico at an international plant-breeding institute, was awarded the Nobel Prize for peace. He was in charge of an organized international effort to breed wheat varieties that could use technological inputs and were well adapted to the tropical and semitropical wheat-growing areas of the world. A parallel effort in the Philippines resulted in the breeding of new high-yielding varieties of rice. The increased cereal production that these varieties made possible in the 1960–1980 period accounts for much of the steady rise in food production that was discussed in Chapter 1. In several Third World countries also, high-input agriculture has now replaced the traditional system. The results of this so-called Green Revolution are mixed: increased food production on the one hand, but a decrease in the sustainability of agriculture on the other; increased self-sufficiency and economic benefits for some, but increased landlessness and poverty for others.

1 High-yielding semidwarf varieties of wheat were developed in Mexico.

On June 30, 1794, the Hartford, Connecticut, newspaper ran an article describing a variety of wheat that had been selected several years before, in Virginia. Compared to conventional varieties, this Forward Wheat variety was remarkable in several respects:

- It matured two to three weeks faster.
- It was resistant to the fungus that causes wheat stem rust.

Figure 11.1

Origin of the New Wheats. Each cross represents many actual crosses to fix the new characteristics into the strains. The first two sets of crosses, resulting in Norin 10, were performed in Japan. The next set, producing Gaines, was done in the United States. The last set, producing the new wheats, was done by N. Borlaug and his colleagues in Mexico.

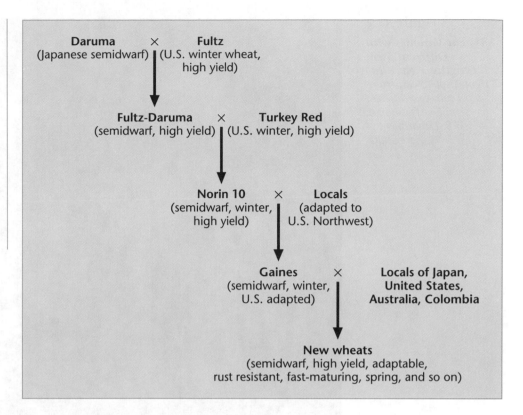

• It yielded more grain.
• It had a shorter stem, and so produced less straw.

By the turn of that century, this variety was widespread in the wheat-growing areas of the new country, but for some reason it did not spread to the Midwest and West as they were colonized.

Meanwhile, the Japanese had also been experimenting with short (dwarf), high-yielding varieties of cereal grains. They had noted that the tall landraces selected elsewhere and used over the centuries often fell over when they had high yields, simply because the spindly stems were not strong enough to carry the extra weight of grain. Although this helped the plant reproduce (the seeds fell to the ground), it was disastrous for the farmer. When the head of grain was on the ground, it was extremely difficult to harvest; moreover, moisture on the ground encouraged the growth of fungal spores on the grain.

By the middle of the nineteenth century, **semidwarf** wheat varieties—usually less than 90 cm tall, compared with traditional varieties that were well over 115 cm tall—were widely used in Japan. Even in the most fertile soils and even when manure was maximally used for fertilizer, the high-yielding varieties would not fall over. One of them, Daruma, became the ultimate source of dwarfing genes for the new wheats, which had worldwide impact (Figures 11.1 and 11.2).

To improve the yield of their semidwarf varieties, the Japanese decided to cross them with higher-yielding U.S. varieties. In 1917, a series of crosses was made between the Japanese Daruma and U.S. Fultz varieties. The offspring of these crosses, selected for both Daruma dwarfness and high yield, were again crossed with the U.S. strains in a type of backcross. By 1935, one product of

these crosses, Norin 10, was made available to farmers, and greatly increased yields were reported.

After 1945, the agricultural adviser to the U.S. occupation army in Japan saw the success of Norin 10 and brought some seeds back to the United States. Soon, workers in the U.S. Pacific Northwest were breeding the Japanese semidwarf with their local wheats, in order to confer on it genes that could adapt it to their local conditions. A result was Gaines wheat: short, strong, high-yielding, and well adapted to the heavy rainfall of that region. Again record wheat yields were reported.

Two problems with Gaines wheat made it unsuitable for tropical and semitropical climates:

1 Like the original Japanese semidwarfs, Gaines was a winter wheat: planted in the fall and harvested in the spring. Winter wheats need a period of cold during early growth, and if they do not get it they will not produce grain. This, of course, precluded using Gaines wheat in tropical Asian countries.

2 Gaines was susceptible to a rust fungus. Although the original U.S. semidwarf, Forward Wheat, resisted rust, the Japanese had not selected for this trait. What is more, the fungus had evolved to get around the older defense mechanisms of the wheat plant. Expensive fungicide dusting programs were needed to prevent infection, and even so, during rust epidemics yields plummeted.

In 1953, some Gaines seeds and others from similar Norin 10 crosses were sent to Borlaug in Mexico. Working under the auspices of the Rockefeller

Table 11.1	Some wheat varieties introduced in Mexico, 1950–1964

Year	Variety Name	Plant Height (cm)	Maximum Yield (kg/ha)
1950	Yaqui 50	115	3,500
1960	Nainari 60	110	4,000
1964	Sonora 64	85	5,580

Source: Data from D. Dalrymple (1985), *Development and Spread of High-Yielding Wheat Varieties in Developing Countries* (Washington, DC: U.S. Agency for International Development), p. 18.

Foundation, he and his colleagues set up breeding criteria for the wheat strains, traits that would make them adaptable to growing conditions in the poor countries. (These criteria included many of the aims of crop improvement listed at the end of Chapter 10.) By collecting many varieties of wheat seeds from all around the world, they amassed a "gene bank" from which they could breed into the semidwarfs desirable characteristics, such as adaptability to different climates. These crosses were made and grown at two locations in Mexico. One, Toluca, is on a high plateau near Mexico City (2,500 m above sea level); the other, Ciudad Obregon, is in the (hotter) state of Sonora. Choosing these two sites did two things for the breeding effort: (1) it allowed two growing seasons a year, thus speeding up experimentation; and (2) it allowed selection of strains adaptable to both cool and hot climates.

By 1962, the Mexico breeding project had produced new, high-yielding, semidwarf varieties with important new characteristics. They were spring wheats, meaning that they were planted in spring and harvested in summer. Where winters are mild, however, such wheats can also be planted in fall and harvested in spring, thus potentially allowing two crops a year. The new varieties matured quickly and were insensitive to photoperiod, two traits also conducive to multiple cropping. They were rust resistant, and they were well adapted to a variety of warm climates.

These first varieties were an instant success in Mexico (the Mexican government was a cosponsor of the breeding work). When used with adequate fertilizers, yields were twice as great as for traditional varieties (see Table 11.1). In 1963, after visiting India, Borlaug returned to Mexico and sent 100 kg of each of four varieties to M. S. Swaminathan at the Indian Agricultural Research Institute. These plants were grown at seven locations around the country in 1963–1964, and their success led to further requests to Borlaug—for 250 tons in 1965, and 18,000 tons in 1966! On his way back from India in 1963, Borlaug had visited Pakistan, and then sent the Pakistanis 205 kg of seed. Again, the plants delivered increased yields, and the government requested increased deliveries of seed stocks in planting: in 1965, they bought 350 tons, and by 1967 the request was for 42,000 tons! By 1980, 80% of the wheat crop in South and Southeast Asia was planted with high-yielding varieties.

2 | High-yielding rice varieties were developed in the Philippines.

At about the same time that the Mexican wheats were being developed, plant breeders were hard at work improving the yields and properties of the other main staple of the world, rice. Much of this research was done in India and Japan, but it culminated with the development of new, high-yielding rice varieties at the International Rice Research Institute (IRRI) in the Philippines.

The story of rice breeding has many parallels with that of wheat. Rice has two subspecies within the single species *Oryza sativa*, and each is adapted for specific climates and cultural preferences:

1   **Japonica** rice is grown in temperate regions such as Japan, Taiwan, the lower Yangtze valley of China, and Korea. It requires rather precise water, weed, and insect control and so is not well suited to the humid tropics. It has a somewhat short stalk, early maturation, and responds well to added fertilizer by giving a good yield. The cooked rice tends to stick together, and the people have adapted many dishes to this characteristic.

2   **Indica** rice is grown in the humid regions of South and Southeast Asia, as well as much of China. It is naturally well adapted to the monsoon climate, requiring little weed or water control. The plants are typically tall and leafy, and respond to fertilizer by producing more vegetative parts rather than grain. When they do produce a high yield, they tend to fall over. When cooked, the grains remain separate.

The Chinese had selected landraces for early maturation as early as A.D. 1000. The so-called Champa variety not only matured faster than traditional varieties, but also was resistant to drought and produced relatively high yields. By the twentieth century, the majority of the japonica rice grown in China was derived from the Champa lines. In Japan, success in breeding a short-stemmed and rapidly maturing rice led to the yield takeoff during the first half of this century (see Chapter 2). This prompted agricultural scientists to try to do the same for indica rice.

But, as noted, the indica landraces had adaptations to their environment (strong vegetative growth, tall and tending to fall over, sensitive to day length to flower, and so on) that were the opposite of what was wanted in high-yielding varieties. High-yielding varieties need to be semidwarf and come to maturity early in the season, and flowering must be independent of day length.

As early as the 1950s, significant steps were taken to improve the indica strains. In Indonesia, outcrosses to local varieties produced Peta, a higher-yielding, relatively drought-resistant variety. Meanwhile, the Taiwanese had long experience with semidwarf varieties, having developed the first such one, a japonica, Dee-geo-woo-gen, in the late nineteenth century. In 1956, this strain was crossed with a local disease-resistant indica and produced Taichung Native 1 (TN-1), a semidwarf (85 cm tall) that responded well to fertilizer by increasing yields, and since it had a short stalk, did not fall over. But TN-1 was somewhat late maturing and susceptible to the rice blast fungus.

In the late 1950s, the wheat-breeding program in Mexico was well underway, and its sponsors, the Ford and Rockefeller foundations, turned their attention to rice. A major breeding station, the International Rice Research Institute, was opened in 1962 near Manila, the Philippines, with the specific aim of doing for rice what Borlaug and colleagues had just done for wheat. A U.S. plant breeder, Robert Chandler, was put in charge of the program.

Of the 38 crosses made at IRRI in its first year, 11 involved semidwarf varieties. One of the crosses was between Peta and Dee-geo-woo-gen, and the

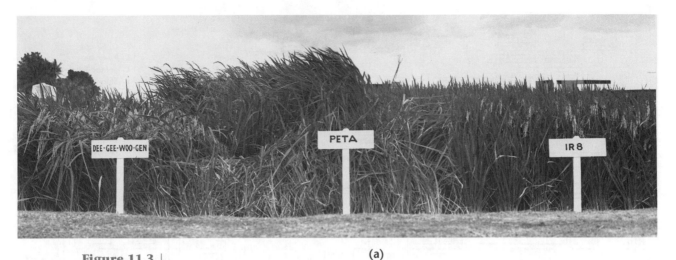

(a)

Figure 11.3

Green Revolution Rice Varieties. (a) The first Rice of the Green Revolution (IR-8) and Its Parents, Peta and Dee-geo-woo-gen. (b) The Effect of Nitrogen Fertilizer on Yield by the Tall-Stemmed Peta and the Semidwarf IR-8. The semidwarf does not fall over when the heads are heavy (high yield), and can be harvested. The harvest from Peta is lost when the plants fall over. Source: International Rice Research Institute.

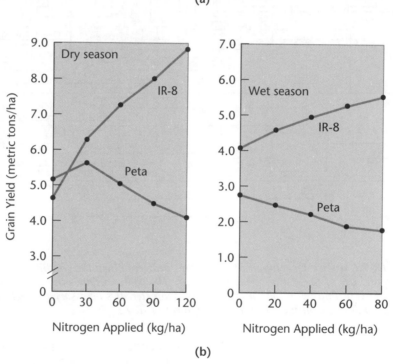

(b)

resulting variety, named IR-8 (Figure 11.3), was sent out a year later to breeders in Bangladesh, India, Thailand, and Colombia. IR-8 had many of the characteristics breeders and farmers were seeking:

- High yield (twice that of other strains)
- Early maturing (125 days, compared to 210 days for other strains)
- Flowering independent of day length
- Semidwarf growth habit
- Resistance to blast fungus

No wonder the variety was called "miracle rice"! Further crosses between TN-1 and Peta yielded another variety, IR-20, which was released for use in 1969.

But IR-8 was not perfect. Its grains tended to break when milled, and were chalky when cooked, making it less than acceptable for consumers. To

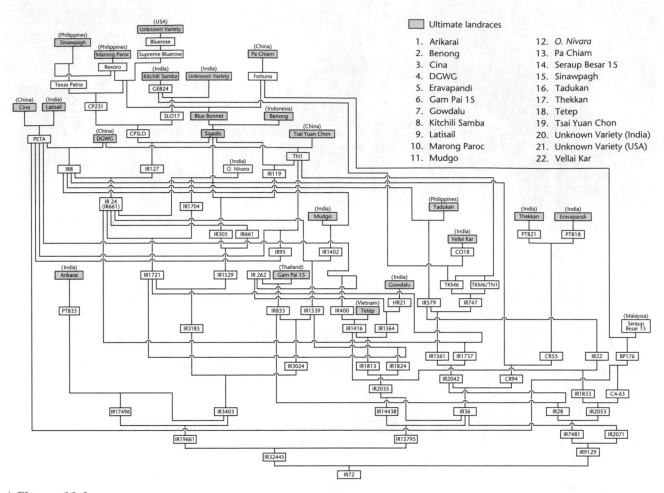

Figure 11.4

The Pedigree for IR-72, a Rice Developed at IRRI. *This pedigree illustrates the interdependent nature and importance of the conservation and use of the world's genetic resources. The pedigree shows the crucial role that landraces from all over the world play in contributing to the range of genetic characteristics necessary to meet future food needs. As indicated by the asterisks, 22 landraces contributed to the development of IR-72.*

overcome this problem, breeders crossed IR-8 with Sigadis, a variety whose grains are soft after cooking, and in 1971 the result, the more acceptable IR-24, was released. Because until then the high-yielding varieties were resistant to blast but susceptible to many insect pathogens, a cross was made between IR-24 and a disease-resistant TN-1 relative, and in 1974 IR-26 was released for planting. This strain combined the best characteristics of high yield, consumer acceptability, and disease resistance. But in 1976 a new genetic type of the brown plant hopper, an insect pathogen, attacked this variety, so further crosses led to IR-36, released later that year. IR-36 added not only plant hopper resistance but also drought resistance and tolerance to adverse soil minerals (such as salts). By the late 1970s, half of all Asian rice-growing land was planted with semidwarf varieties.

The rice-breeding effort at IRRI and other locations continues. The objective is not just to improve yields and consumer acceptability, but also to make the plants adaptable to their local environment, with disease resistance. The pedigree of IR-72, recently developed, illustrates the genetic complexity involved in producing a new variety (Figure 11.4). As farmers throughout Asia

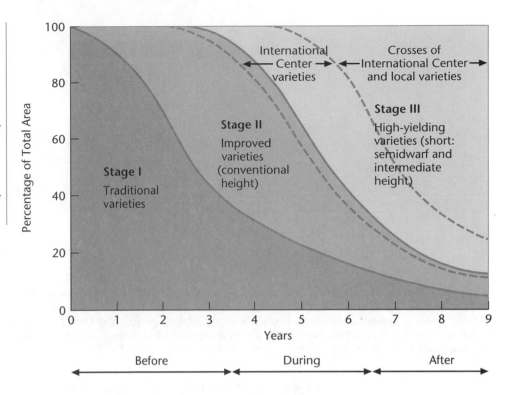

Figure 11.5

The Sequence of Adoption and Use of Crop Varieties Before, During, and After the Green Revolution. The graphs show the decline in the use of each set of varieties as they are replaced by new varieties. Source: *Data collected by D. Dalrymple, U.S. Department of Agriculture.*

have adopted the new varieties, they have abandoned the landraces that sustained them for centuries. But the landraces are still important, because they provide the source of genes for disease resistance and other agronomic characteristics, which are and will be needed again and again for crossing into the new strains. A look at Figure 11.4 shows that some two dozen landraces contributed to the development of this strain.

3 Adoption of high-yielding wheat and rice, along with the technologies needed to grow them, led to the Green Revolution.

The adoption of high-yielding wheat and rice varieties followed similar patterns in many countries of the Third World (Figure 11.5). Initially, the farmers primarily grew traditional varieties, most selected from landraces (stage I). These plants were often well adapted to their specific environment, but had problems such as lack of resistance to new diseases or low yield. To improve these varieties, local plant-breeding stations crossed in genes from other strains and landraces (stage II). But these improved varieties were still not yielding enough to feed the expanding population or to meet market demands.

At this stage, international research centers entered the sequence by breeding the semidwarf varieties (stage III). Their high yields and fast maturity made them attractive to many countries, but they were not as exquisitely adapted to local conditions as had been the original, traditional varieties. Local breeders have been extensively crossing these plants with local varieties, to the point that it is hard to distinguish local varieties from the strains that arrive from the international stations.

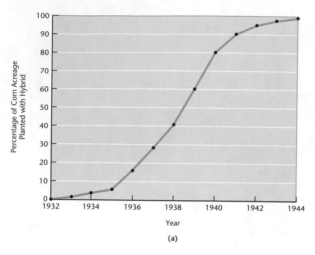

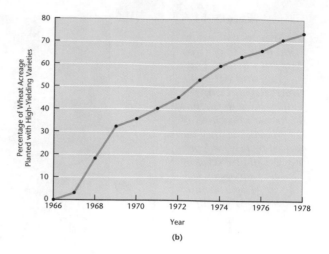

(a)

(b)

Figure 11.6

(a) *Cycle of Adoption for Hybrid Corn in the United States.* (b) *Cycle of Adoption for High-Yielding Varieties of Wheat in Nine Southern Asian Countries.* Sources: **(a)** *Data from Z. Grilches (1957), Hybrid corn: An exploration in the economics of agricultural change,* Econometrica *(October), p. 502.* **(b)** *Data from D. Dalrymple (1985),* Development and Spread of High-yielding Rice Varieties in Developing Countries *(Washington, DC: U.S. Agency for International Development), p. 110.*

Experience shows that the adoption process for a new agricultural technology, be it seeds, fertilizers, or tractors, often follows an S-shaped curve (Figure 11.6). After a new advance (such as a variety of seed) is introduced, the first farmers to adopt it take the greatest economic risk. If the new method fails on their farms, they could lose their entire crops, or at least produce less. However, if the technology is successful, they also reap the greatest benefits.

The first U.S. farmers to adopt hybrid corn thus received the highest increase in income, which more than offset the additional price they had to pay for the seed corn. New technologies are adopted first on the biggest and most prosperous farms; these are managed by farmers who keep abreast of the new developments in agriculture, and can obtain or have the capital needed to profit from such developments. The later a farmer adopts a new technology, the less will be the economic benefit.

In the case of the high-yielding varieties of wheat and rice, the farmers had to adopt not only the new seeds, but an entire technology package, that included fertilizers, insecticides, weed killers, equipment for irrigation, and tractors to till the land. Indeed, the new varieties made it possible to grow two or even three crops per year, thereby potentially increasing the demand for labor. Thus, labor-saving technologies (tractors and herbicides) had to go hand in hand with the new strains. Furthermore, nitrogen fertilizer was needed so that the new varieties would yield up to their potential. Without fertilizer, they produced the same yields (or less) than the landraces they replaced.

Cycles of adoption quickly succeeded each other in different countries, starting with the adoption of improved wheat in Mexico. In March 1968, the director of the U.S. Agency for International Development (U.S. AID), William Gaud, used the term **Green Revolution** to describe the great gains in production that the semidwarf varieties were achieving. In Asia, adoption of new rice varieties started in the late 1960s and was completed 12–15 years later. The gains in overall yield were quite impressive (Table 11.2), and the net

Table 11.2	The result of the Green Revolution: yields of wheat and rice, in metric tons per hectare		

Country	Crop	1963	1983
India	wheat	0.9	1.7
India	rice	0.9	2.2
China	wheat	1.0	2.5
China	rice	2.0	4.7

Source: Data from Food and Agriculture Organization.

improvement in food production certainly justified the awarding of the Nobel peace prize to Borlaug.

But the yield increases that were realized on farms were never as great as those observed in research stations. Scientists at the International Rice Research Institute (IRRI) showed that the new varieties could outproduce the old ones by a factor of 3 to 4. Yet most farmers in Asia realized increases of only 1.5. This discrepancy in production is not surprising and also occurs in developed countries, where yields obtained in experiment stations are seldom matched on the farm. Surveys of farmers who were using the new technology showed that the constraints on higher productivity were poor water control, inadequate nitrogen fertilizers, and diseases. To put it another way, the major constraint was inadequate capital to purchase the input package needed to obtain the high productivity of which the seeds are capable. The fundamental fact of the Green Revolution is that the farmers had to manipulate the environment to get the most out of the plants, and too often they have not been able to afford it.

Forty years after the first adoption cycles of the Green Revolution began in Mexico, and 15 years after they came to completion in Asia, we see that the world is not much better off. A similar percentage (10–15%) of the world population that was undernourished in the 1950s and 1960s is undernourished in the 1990s. The increases in agricultural production, while impressive, have kept just ahead of population growth, and not led to a more even distribution of food to all people. In addition, many problems plague sustainability of the high-input system, in the Third World just as well as in the developed countries. These problems are discussed in Chapter 16.

The solution to the problem of hunger that coexists with a high level of food production is not agronomic or scientific, but political. The importance of this political side of the problem is very much in evidence when we consider the relationship between land reform and the introduction of the high-yielding rice varieties in Asia. Before introducing the new varieties, Japan, China, and Korea implemented radical land reforms and redistributed to peasants land previously owned by large landlords.

But in other Asian countries the influence of the landlords was undiminished. When the high-yielding rice varieties were introduced, these rural landlords were in a strong position to reap the benefits. The prevalence of share rents for tenant farmers (such as 50% of the crop), as opposed to fixed money rents, greatly benefited the landlords. The need for capital to buy the input package also benefited the landlords, who had greater access to banks and became, in effect, the middlemen in a distribution chain. They lent money to the tenant farmers and sold them the technological inputs that were produced outside their rural districts.

Some countries implemented moderate land reform. For example, the Philippines converted shared tenancies to fixed-rent tenancies in the 1970s, with an option to buy the land over a 15-year period. Still, the small farmers were barely able to improve their lives, because other sources of income were scarce. In the United States in the 1930s–1950s, and in Japan, Taiwan, and Korea in the 1950s, marginal farmers who were squeezed out as new technologies were adopted found jobs in the cities because the industrial base was rapidly expanding.

This has not been the case in most of Asia, although many governments initially had high hopes that this scenario, which had played itself out in a number of industrializing countries, could be repeated. The labor-saving technologies introduced with the new strains in Japan, Korea, and Taiwan freed people to take jobs in industry. The result was a rise in the standard of living for everyone. But in other countries, such as Bangladesh, where poor women, without access to land, provided the labor force for planting rice in rice paddies and weeding the crops, the living standard of the poor remained low. Many subsistence farmers lost their land and became laborers on farms owned by others. Their needs were also not met by the Green Revolution.

4 | The Green Revolution was accompanied by the establishment of a network of international agricultural research institutes.

Until the 1950s, most agricultural research was conducted in temperate zone countries on temperate zone crops. Although the major schools of agriculture in Europe had sections focused on tropical agriculture, their major objective was to train managers for plantations in the tropics. The agriculture of the inhabitants of the tropics was generally viewed as primitive, and not really worth studying. Few efforts were made to improve the food crops of the tropics and subtropics, and what little research was done concerned cash crops such as cotton and sugarcane, rather than food crops.

The efforts by plant breeders in the United States and in Japan to produce high-yielding semidwarf varieties of wheat and rice resulted in the establishment of two major research institutes: IRRI in Los Baños, the Philippines, in 1960, and CIMMYT (International Center for the Improvement of Corn and Wheat) near Mexico City in 1966. Their founding and financing was the result of consultation and cooperation of major international donors, such as the Rockefeller Foundation and the Ford Foundation. The success of the Green Revolution, and the perceived need for more research on tropical and subtropical food crops, led to the establishment of other research institutes in different countries of the Third World.

The first institutes were sponsored by the Ford and the Rockefeller foundations. These international donors subsequently coordinated their activities with the World Bank and governments of developed countries, and subsequent research institutes were sponsored by the Consultative Group on International Agricultural Research (CGIAR). This group, based in Rome, now coordinates the research for all these international research institutes. Each institute has responsibilities for particular crops

Table 11.3	Research institutes sponsored by the Consultative Group on International Agricultural Research (CGIAR)

Institute	Location	Major Research Focus
IRRI, International Rice Research Institute	Philippines	Rice
CIMMYT, Centro Internacional de Mejoramiento de Maíz y Trigo	Mexico	Wheat, corn, triticale
CIAT, Centro Internacional de Agricultura Tropical	Colombia	Beans, cassava
IITA, International Institute of Tropical Agriculture	Nigeria	Legumes, roots, tubers
WARDA, West Africa Rice Development Association	Côte D'Ivoire	Rice
CIP, Centro Internacional de la Papa	Peru	Potato
ICRISAT, International Crops Research Institute for Semi-Arid Tropics	India	Sorghum, millet, legumes
ICARDA, International Center for Agricultural Research in Dry Areas	Syria	Barley
IIMI, International Irrigation Management Institute	Sri Lanka	Irrigation
ICRAF, International Council for Research on Agroforestry	Kenya	Agroforestry
ILCA, International Livestock Center for Africa	Ethiopia	Cattle, sheep
IBPGR, International Board for Plant Genetic Resources	Rome	Gene conservation

(see Table 11.3). These institutes have a triple mandate: (1) to improve the crops assigned to them, (2) to study farming systems for these crops, and (3) to create gene banks in which the landraces of the crops under study can be preserved (in the form of seeds stored under conditions that ensure their longevity). Initially the institutes focused on the approach that was successful with the high-yielding wheat and rice varieties in the 1960s and 1970s. That is, they concentrated on monoculture and on improving crop production by a package of improved varieties and associated technologies to modify the environment. More recently, this focus has begun to change, showing a greater concern for integrating agriculture more with the natural environment.

For example, scientists at ICRISAT in India have identified strains of groundnut (a drought-resistant legume) that produce more than 1,000 kg grain per hectare, even when stressed by drought, compared to the 500–800 kg per hectare produced by the normal varieties. Similarly, they have found strains of sorghum that produce two to three times more grain than the present commercial types. One of their goals is to select drought-resistant crop plants that will be used not only in Asia, but also in the Sahel region of Africa.

Scientists at CIAT in Colombia have identified genes in wild varieties of beans that enable them to be resistant to attack by bruchid beetles. These beetles do enormous damage to the seeds after harvest because they multiply in the dry, stored beans. By crossing a wild variety with a cultivated variety, CIAT scientists were able to introduce the bruchid-resistance genes into the cultivated variety.

The philosophy that underlies the CGIAR system of research institutes has been that biological and technological research problems are separable from political and social issues. The research institutes generate the necessary innovations and offer these as packages to various national and regional extension services or development authorities. Fine-tuning these packages to local conditions, and solving social and political problems resulting from implementation of new technologies, are the responsibilities of the implementing agencies or the governments of the countries in question. In this system, agricultural development is seen as an activity that comes from the top (the CGIAR institutes) and trickles down to the bottom (the often landless subsistence farmers).

This approach has worked reasonably well in some countries or areas, and poorly in others. Agricultural development has not often been a high priority for many Third World governments, except for developing cash crops that earn foreign exchange. In many African countries, national agricultural research efforts have been weak, in part, because governments mistakenly assumed that the CGIAR institutes would solve all their agricultural problems, and also because these high-profile institutes attracted all the international funds and the best scientists.

The Green Revolution has done more than raise overall food production. It has also caused the rapid disappearance of the landraces (see later discussion) and of indigenous farming systems that may have had much to offer to agricultural researchers. The top–down approach of development implies that agricultural researchers can learn little of value from subsistence farmers. For example, although the institutes have often stressed monoculture, many farmers have been growing more than one crop on the land (either crop rotation or intercropping) for centuries. These practices have led to a more sustainable agricultural system than monoculture.

Scientists at the CGIAR institutes now realize that they must do more than create new strains to hand over to the extension agents. They need to get involved in understanding the tropical agroecosystems at the farm level. For example, IRRI is located at a place where rainfall is reliable and adequate. But much rice in Asia is grown on land where rainfall is intermittent and unreliable. So how can a technology developed at IRRI be useful to a farmer who has very different needs? This type of problem has recently led to the formulation of the second mandate for these institutes: to study the farming systems and make sure that proposed development uses the strengths of those systems and does not replace or destroy them.

With a centralized system of research, reaching the farmers is difficult and expensive. The programs generally do not operate in villages lacking access roads suitable for motorized vehicles. The alternative, it has been argued, is to institute a completely decentralized system of agricultural (food crop) research. The new emphasis in several CGIAR institutes concerning on-farm research and agricultural microecosystem research should do much to counter the criticism that these institutes do not really solve problems, but just produce technology.

BOX 11.1

A Tale of Two Institutes: IRRI in the Philippines

As director of agricultural research for the Rockefeller Foundation, George Harrar, a plant biologist, launched the wheat-breeding program in Mexico led by Norman Borlaug. Then, in 1958, Harrar turned his attention to rice, the staple food for over 2 billion people in Asia. Yields on typical Asian farms in the 1950s had hardly improved over the past century, and the populations were growing rapidly. Harrar reasoned that the scientific approach that was succeeding on Mexican wheat farms could be used for rice.

With additional funds from the Ford Foundation and the government of the Philippines, the International Rice Research Institute (IRRI) was organized. A site was chosen adjacent to the University of the Philippines College of Agriculture in Los Baños, near Manila. By early 1962, the institute, with research laboratories and over 200 irrigated acres, was operational. It began as a showplace, attracting tourists as well as kings and premiers from all over Asia. It still is a showplace, but a scientific one, as the most impressive institution in the world primarily devoted to the physiology, genetics, and agronomy of a single crop.

The Green Revolution in rice that occurred during the 1960s and 1970s was largely led by the development of high-yielding varieties at IRRI. In addition to breeding for increased yield, IRRI scientists have extensively researched rice diseases and their control; water and nutrient requirements; responses of the plant to stress; and farming methods. Most importantly, they have worked with social scientists and extension officers to bring these developments to the individual farmer and shared their findings with the international community through publications and training programs. It is not surprising that when scientific exchanges with the People's Republic of China resumed in the 1970s, one of the earliest participants was IRRI.

Although the "IR" varieties developed during the 1960s and 1970s thrive on irrigated land, many rice farmers live on land where such water control is either impossible due to periodic flooding, or not affordable. IRRI is now making an effort to develop rice varieties that can grow in nonirrigated environments. For instance, some rice varieties in Thailand normally grow in 4 m of water. Their stems elongate rapidly as the water in which they are growing rises. The inheritance of this trait is being studied so that it can be transferred to high-yielding varieties. This would let farmers switch from costly transplanting methods to less costly broadcast sowing of the improved seeds.

IRRI has become very active in technological approaches to improving the rice plant. These range from molecular biology to engineer nitrogen-fixing bacteria that live in association with rice roots, to remote sensing by satellites to obtain data on crop growth and physiology.

5

Landraces have disappeared and the genetic base has narrowed as a result of modern plant breeding and the Green Revolution.

As crop plant breeding produced "miracle" strains over the past half century, farmers were encouraged to switch from the genetically diverse strains they were using, many of them landraces, to the new, high-yielding varieties. Moreover, as agriculture expanded into previously uncultivated areas, farmers

In Ibadan, Nigeria's second largest city, 150 scientists have been working since 1967 at the International Institute of Tropical Agriculture to solve the seemingly intractable problems of food production in Africa. In contrast to the spectacular successes in rice production catalyzed by IRRI, progress in Africa has been slow, and can hardly be called a revolution. Although the average poor person in Asia and Latin America has more food available in the 1990s than in the 1970s, in many African regions there has been an absolute decline in food availability. One reason for this decline has certainly been rapid population growth in Africa, which is pushing the people onto more marginal lands unsuited for agriculture. But there are other reasons as well:

1 Ecological conditions for agriculture are diverse, and in general not suited to intensive cultivation. The topsoil is precariously thin.

2 Long droughts, first in the Sahel and East Africa in the 1980s, then in southern Africa in the 1990s, have degraded the soil and made water for agriculture scarce.

3 Government policies often have encouraged big projects, such as irrigation dams, instead of farm-level reforms such as fertilizer subsidies. Civil unrest has wreaked havoc on agriculture.

4 Insufficient research has been done on crops that Africans use under local conditions.

IITA was set up to remedy this last problem, and has made some progress. High-yielding cassava and cowpeas, and innovative farming systems such as alley cropping (planting trees between crop rows to enrich the soil and reduce weeds), have been developed and put into practice. But the most important results have come from research on plant diseases. In the early 1970s, a mealy bug that attacks cassava arrived in Africa accidentally from Latin America, and soon threatened this vital staple. IITA scientists went back to Latin America and found a parasitic wasp that attacks the mealy bug. As a result, cassava production, and people dependent on it, did not suffer. Because of maize streak virus, corn has not been a successful crop in much of Africa. IITA scientists have bred a corn variety that is naturally resistant to this virus.

Norman Borlaug, who developed high-yielding wheat in Mexico, has been using a similar approach in Africa since 1986, with private funding from a Japanese philanthropist. Borlaug's team has essentially been going from farm to farm applying available information on African crops—including fertilizer, pest control, and water needs. Yields of wheat, sorghum, and corn have gone up as much as 250 percent in the Sudan, Ghana, and Tanzania. Making these inputs available to the farmers requires government intervention, and countries in the midst of privatizing have been reluctant to provide subsidies. Former U.S. president Jimmy Carter has been lobbying these governments to change their policies.

cleared away the wild relatives that grew in the marginal lands around farms and often interbred with the crops. The loss of these plants is potentially disastrous, because landraces and wild relatives have valuable genes that will be needed for plant breeding in the future.

As noted earlier, landraces are varieties developed by local farmers by using informal selection processes at a time when each farmer kept part of the harvested seeds for planting. Each landrace was adapted to the soil type and to the microclimate, and was more or less resistant to the pests and disease organisms that thrived in that particular locale. With respect to diseases, the plant population as a whole may not have been resistant to a particular disease organism, but it probably included individuals that carried resistance genes.

Landraces have a very broad genetic base—in genetic terms, they are highly *heterozygous*—in part because they may still occasionally be crossbreeding with the wild relatives that grow in their vicinity. Crop plants produced by breeders are much more *homozygous,* because variability is an unwanted trait. Farmers want every single plant to produce an equally abundant crop of seeds. This is especially true of self-pollinating (inbred) crops such as wheat and rice, and of hybrids such as corn.

The genetic potential of the landraces and wild relatives is tremendous. Here are some examples:

- The durum wheat Kubanka, which set the standard for wheat yields in the northern U.S. plains for over 50 years, was originally collected as a landrace in southern Russia in 1900.

- One wild corn species collected in Mexico during the 1970s offers a number of possibilities for plant breeders. It is a perennial (as opposed to modern corn, which is an annual), opening up the possibility of a farmer not having to plant seeds every year. Moreover, the wild plant has resistance to a spectrum of viral diseases that together account for a 1% per annum drop in corn yields worldwide (this is worth $500 million per annum). Finally, this strain of corn grows at higher elevations than modern corn, raising the possibility of planting the crop on land that is now considered unsuitable.

- In 1948 Jack Harlan, then at the U.S. Department of Agriculture (USDA), collected a small sample of a wild wheat variety in Turkey. This wheat variety was not high-yielding or resistant to leaf rust, but was conserved nevertheless, under the assumption that it might contain genes of interest to later breeders. Fifteen years later, the fungal disease stripe rust became epidemic in the U.S. Northwest. On testing, Harlan's variety was found to have a gene for resistance to this disease, as well as to several others. This wild variety's genes were promptly crossed into the cultivated wheats, with great success in terms of disease resistance. In fact, most wheat in the U.S. Northwest now has the Turkish variety in its pedigree.

The dangers of a narrow genetic base in modern crops are particularly well illustrated by the epidemic of southern corn leaf blight that lowered U.S. corn production by 15% and by more than 50% in some southern states, where the weather conditions favored the rapid spread of the blight fungus in 1971. We noted in the previous chapter that the discovery of a gene for male sterility and restoration of fertility led to a convenient way to produce hybrid corn. This method was so convenient that within a few years all corn planted in the United States had this particular gene. But the same gene that conferred male sterility also conferred susceptibility to corn leaf blight. In a year when conditions were right for the rapid spread of the disease (a warm, wet summer), the corn crop was devastated. This was a result of its narrow genetic base. Similar disease epidemics (for example, tungro virus and plant hoppers) attacked successive rice varieties bred by IRRI.

The problem is that when breeders select specific strains, they know what they are selecting for (for example, high yield or early maturity), but these are complex characteristics, usually determined by many genes. When a series of crosses is made, it is hard to focus only on the genes of interest; breeders often do not know which other, possibly harmful, genes are coming along for the ride. That may only become evident when a certain set of conditions prevails, as with the corn leaf blight.

6 | **Conservation areas and gene banks have been set up to preserve precious genetic resources.**

To breed resistance to corn leaf blight and all other plant diseases, and to breed plants that are more adaptable to particular environmental conditions that are present on the farm, breeders need access to the entire germ plasm of the species. This includes the cultivated varieties as well as all the primitive cultivars and landraces, and the wild and weedy relatives. So plant genes must be preserved. There are two main approaches to the conservation of plant genetic resources:

1 *In situ* conservation involves locating plants of interest, and then setting aside the land where they grow permanently so that these plants can thrive. This approach is well known for large animals, such as the preserves on the plains of Kenya, where an entire eco-system is being protected from human intervention. Less well known are two sites in Ecuador, one in the north and one in the south, that have been specifically set up to preserve the wild relatives of crop plants. An inventory by plant biologists found the wild relatives of potatoes, tomatoes, walnuts, beans, and papayas growing there. The land has been purchased by a private organization, the Nature Conservancy, to be maintained as a preserve.

    There are many advantages of this approach. Because the plant continues to be in its natural setting, it is not a "fossil" in some distant collection, but a continually evolving organism. New adaptations will arise; for example, resistance to pests that are present and are themselves evolving. Because predators of pests are also there, information on biological control is available. The interactions of the target plant with other plants, both competing and beneficial, are also preserved. Finally, if the target plants continue to be grown by indigenous farmers, their farming systems are kept intact for ongoing study.

2 *Ex situ* conservation involves collecting plants in the field, transporting them to a central location where they are stored, and continual testing of the stored material to make sure that it is viable. This is the "gene banking" approach.

Collecting plant genetic resources at one place and using them somewhere else has a long history. As noted in Chapter 10, crop plants originated in both centers and noncenters of diversity and then spread over a wider area as people traveled. The European explorers exchanged crops among continents (for example, they brought corn and potatoes from the Americas to Europe). In 1827, U.S. president John Quincy Adams sent a message to all U.S.

Table 11.4	Number of samples in gene banks for major crops, and estimate of the total diversity collected			
Crop	Samples (× 1,000)	Strains (× 1,000)	Percentage of Diversity Collected Landraces	Wild Species
Wheat	410	125	95	60
Rice	220	90	80	13
Corn	100	50	95	15
Soybean	100	18	60	30
Barley	280	55	85	20
Sorghum	95	30	80	10

Note: Not all samples collected by the gene banks are from separate strains; there is some duplication.

Source: Data from International Board for Plant Genetic Resources.

embassies in foreign countries instructing the consular officers to send cultivated plant seeds and cuttings back home.

By the end of the nineteenth century, the U.S. Department of Agriculture had set up a plant exploration section. When plant explorers brought a new plant to the United States, it was first test-grown on a special 200-ha farm near Washington, DC (on what is now a parking lot at the Pentagon!). Out of this effort came many new and valuable crops. For example, the importation of soybeans from China led to what is now a multibillion-dollar soybean oil and protein industry.

As concern about the conservation of genetic resources mounted, it was a natural progression from exploration and collecting for immediate use, to collecting and then storing living plant parts for future use. In addition, the aim was not only to collect popular cultivated varieties, but also the landraces and wild relatives of crops. Initially, this idea was practiced by the great plant geographer, N. I. Vavilov, who set up a large gene bank in Leningrad (now St. Petersburg) to further his studies of the origins of crop plants. The Soviet dictator Stalin did not favor Vavilov's approach, and greatly hampered his work just as the collection was building up. Meanwhile, in the United States, plant introduction stations were set up at four locations to receive, store, evaluate, and make crop strains available to farmers.

By the 1980s, there were dozens of gene banks in 70 countries around the world, some with general collections of many species, and others with specialized collections of one or two crops. Some of these collections are housed at the international research institutes (Table 11.3). All are under the general coordination of the International Board for Plant Genetic Resources (IBPGR), headquartered at the FAO in Rome.

The extent of these collections is considerable, but there is a long way to go before the bulk of the genetic heritage of people's crops is represented (Table 11.4). The cultivated crop plant strains are generally abundant in the collections, because farmers are encouraged by agricultural extension officers to send new seeds to the bank. In fact, a fair number of "books" in the gene "library" may be duplicates—two farmers may be growing a wheat variety that is new to them, and when they are sent to the gene bank each is given a separate identification number (called an *accession number*). The only way for the bank to know that these are the same is by growing both of them, and/or comparing their characteristics as noted by the farmer. This is a challenge

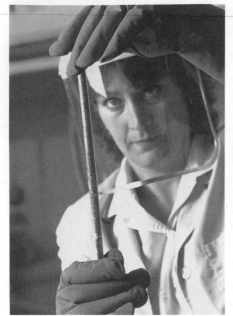

(a)

(b)

Figure 11.7

Storing Seeds in a Gene Bank. **(a)** *At the National Seed Storage Laboratory (United States), most seeds are stored frozen in liquid nitrogen tanks.* **(b)** *Fleshy and oily seeds are stored in bags in a refrigerated room.* Source: *U.S. Department of Agriculture.*

for managing data, and information scientists and computers have been used for the task rather successfully.

IBPGR has put a high priority on collecting and storing landraces and wild relatives of crops. For example, in 1991–1992, a special effort was made to collect wild species in northern Africa and West Asia. In Morocco, collectors found a perennial chickpea; in Algeria, a perennial oat; and in Namibia, wild sorghums. In all, over 2,000 new accessions were collected and described in this one effort. Meanwhile, during the same year, new wild relatives of potatoes were collected in Ecuador, maize and sesame in India, and okra and eggplant in Nepal.

For most important crops, the plant part that is stored in a gene bank is the seed. Of course, when a crop does not produce seeds or reproduces vegetatively (such as with many fruit trees), cuttings or, more recently, tissue culture cells are stored. Clonal propagation of stored tissue culture cells is the way forest genetic resources are stored.

A good example of a gene bank is the National Seed Storage Laboratory at Fort Collins, Colorado, in the United States. Initially established in 1959, this government organization moved into a new, ultramodern facility in 1992. At present, there are about 250,000 accessions of over 400 species, with an ultimate capacity for 1 million accessions. As befits a bank, there is tight security and the building is designed to be fireproof and weather resistant (in case of floods, tornadoes, and earthquakes). Inside is not money, which is replaceable, but an agricultural genetic heritage that is irreplaceable.

When a package of seeds arrives at the laboratory, the seeds are first given a computerized identity card (with a bar code like that used in supermarkets) and evaluated for germination and moisture content. They are then stored. Because seeds do not remain viable indefinitely at room temperature (the record appears to be a lotus seed from Manchuria, which germinated 700 years after it was formed), they are stored in the cold. First the seeds are dried down to about 6% moisture content in a cold room at low relative humidity.

Seeds are stored in one of two ways (Figure 11.7):

1 Seeds with a high oil content or large size are usually stored in bags in vaults at −18°C. These seeds must be tested for germination at intervals (once every 10 years), because they tend to deteriorate over time. When the germination percentage becomes low, seeds are grown in the field to regenerate the accession for storage.

2 Small seeds with less oil (this includes the cereal grains and most other crops) are stored in liquid nitrogen tanks at −196°C. This method is superior for long-term storage, and current research indicates that the time period for germination testing could be several decades, thus reducing this laborious task considerably.

Currently, research is underway to determine whether other plant material besides seeds can be stored using the second method (ultralow temperature). Success has been obtained with pollen, tissue culture cells, and buds of vegetatively propagated species.

A computerized data base, the Genetic Resources Information Network, connects the computers of the many gene banks around the world. When a breeder wants a gene for resistance to a particular variety of rice blast, for example, locating that strain in a gene bank can be done by computer and modem. The bank is then contacted, and a sample of seed is thawed, tested, and sent to the breeder.

7 | The Third World has contributed most of the genetic bases of global crop production.

We noted in Chapter 10 that crop plants originated, and agriculture began, in centers as well as broader areas ("noncenters") of genetic diversity. From this information and from crop production data, we can calculate to what extent each area is independent of, or dependent on, the genetic resources of other areas. For example, soybeans are a major food and feed source in the United States, but were domesticated in China and only brought to the United States in the late nineteenth and early twentieth centuries. Therefore, the United States depends on the genetic resources of China for its soybean lines.

Calculations for the food crops of the world, made by J. Kloppenburg, Jr., and D. L. Kleinman and shown in Table 11.5, document the enormous interdependence of the world's regions with respect to genetic resources. These authors divided the world in 10 regions of production and calculated the percentage of the 20 most important food crops grown in each region that were also domesticated there. Thus, 70% of the food crops grown in West Central Asia were also domesticated there (principally wheat and grain legumes), whereas only 12% of the crops grown in Africa were domesticated in Africa (principally sorghum and cowpeas). Crops that originated in one region are now grown all over the world and the most important contribution of one region to another is shown in the last column of Table 11.5. Thus, 82% of the crops grown in Australia originated in West Central Asia (principally wheat and other small grains), whereas 52% of the crops grown in Africa came from Latin America (principally corn, beans, and cassava). None of the world's 20 most important food crops are indigenous to North America or Australia.

| Table 11.5 | Food crops and their regions of origin |

Region of Production	Percentage of Crops Grown in This Region of Production That Also Originated Here	Origin and Percentage of Crop Area Devoted to Crops from That Area in This Region of Production	
China-Japan	37%	Latin America	41%
Indochina	67	Latin America	32
Australia	0	West Central Asia	82
Hindustan	51	West Central Asia	19
West Central Asia	70	Latin America	17
Mediterranean Basin	2	West Central Asia	46
Africa	12	Latin America	56
Euro-Siberia	9	West Central Asia	52
Latin America	44	China-Japan	19
United States and Canada	0	West Central Asia	36

The genetic resources of West Central Asia and Latin America have made the largest genetic contribution to feeding the world. Crops originating in these two regions account for two-thirds of all the crops that are produced. These two regions have given the world six of the seven major crops (corn, potato, cassava, sweet potato, wheat, and barley).

If these data are considered from the standpoint of the Third World–First World relations, it is clear who is indebted to whom: the Third World regions have contributed 95% of the genetic resources for global food production, and the advanced industrial nations (minus Japan) have contributed only 5% of the genetic resources. Thus, financially poor but genetically rich countries have given their genetic resources to financially rich but genetically poor countries.

In recent years, a controversy has developed over the ownership of genetic resources and over the need for proper compensation for the use of genetic resources from one country by another. The Food and Agriculture Organization of the United Nations has declared that genetic resources are the shared and common heritage of all humanity. Included in these resources are not only the landraces but also the elite genetic stocks developed by plant breeders and biotechnology. According to this principle, there is no real difference between landraces developed by peasants and hybrids developed by breeders. In the developed world, companies—some of them quite powerful politically—have been set up to breed, generate, and sell seeds to farmers. These companies believe that new varieties of crops should be patented, much like industrial processes, and should not be made freely available to anyone.

Third World countries generally applaud the idea that all breeding lines should be freely available. They say, "We will share our landraces, if you will share your elite breeding lines." Developed countries object, for they wish to patent and sell genetic stocks. They want free access to the landraces of the Third World, to produce new varieties that will then not be freely available to the Third World. Yet Third World countries feel that they have already made an enormous and uncompensated contribution to the economies of the developed countries. The developed countries pay for extracting natural resources such as coal and oil, but do not wish to pay for the landraces they so desperately need to maintain their food production base.

Yet crop plant genetic resources are not necessarily gifts of nature, but in some cases have been developed with thousands of years of human efforts. They also have potential for the development of new products by biotechnology. Why should these developed products be the common heritage of humanity, if oil and coal are not? It is ironic that in this germplasm controversy, each side in the debate wants to define as "common property" what the other side owns. At the U.N. Environment Conference in Rio de Janeiro in 1992, an attempt was made to solve some of these problems by a proposed treaty on biodiversity, which asserted that the country of origin would be entitled to compensation not only for genetic resources themselves but also for products made from them. Under prodding from some of the biotechnology and pharmaceutical industries who stood to lose the most to the poor countries, the United States initially refused to sign the treaty. It would appear that the deadlock will continue, and the intransigence on both sides makes it more likely that the poor countries will try to benefit financially from their ownership of unique genetic resources.

8 | The Green Revolution in Third World countries has had profound social impacts.

The adoption of Green Revolution strains and technology certainly has had positive effects on food production (see Table 11.2). But in some cases, these benefits have been achieved at the expense of a social structure that had evolved over a long period of time. First with the adoption of semidwarf wheat in Mexico, and then of wheat and rice in Asia, the requirements for the revolution—not only strains but also expensive technologies—made it likely that the gap between rich and poor would widen.

A major impact has been a sharp increase in **landlessness**. Landowners who had tenant farmers living on their land—some tenant families, for generations—often evicted these people and began to operate the farms themselves because money was now available to use the strains and technology to reap profits. In Pakistan, the number of tenant farmers, and the amount of land they farmed, was halved during the 1960–1980 period. Farmers were given subsidies to buy tractors, for example; this reduced the need for tenant farmers and drivers for the bullocks that had been used to till the fields. In fact, each tractor was estimated to replace four tenants. Landlessness doubled during this period. In Sudan during the 1980s, the government allocated large landholdings to owners willing to invest in mechanization. The result was greatly reduced farm employment. The poor got even poorer, because no alternative jobs were available. Childhood malnutrition increased, because the former tenant farmers had lost their access to the land where they used to grow their own food crops. This greatly increased the pressure on the common areas, which are open to everyone. In many cases, the government decreased the lands held in common by leasing them out to farmers willing to farm them with tractors, new strains, and fertilizers.

A second impact of high-input agriculture has been **disruption of social systems**. In Mexico, the conversion of lands from mixed culture to monoculture of semidwarf wheat upset a system that the local inhabitants had developed since before the Europeans arrived. As cultural practices and jobs

Figure 11.8

Women Provide Most of the Labor for Crop Production in Developing Countries. Source: *U.N. Educational, Scientific, and Cultural Organization.*

changed, family life and roles were altered. Whole communities were destroyed, and the people moved to the cities. It is not surprising that Mexico City is the largest urban area in the world. In communities that relied on barter and reciprocity, the change to cash-based agriculture virtually eliminated the old style of economics, as well as the cultural practices on which they were based.

Perhaps the most striking effect of the Green Revolution was the increasing **marginalization of women**. In many agricultural societies, women have traditionally been the primary food producers (Figure 11.8). Time studies show that women work more hours and expend more energy in food production than men. A study in Java showed that women worked an average of 11.1 hours per day and men 8.7 hours per day. (Housework was not included in this survey.) A study in India calculated the relative contributions to agriculture of men, women, and children in rural villages as a percentage of their respective energy inputs, and obtained the following figures: men, 31%; women, 53%; and children, 16%. In some African cultures, plowing is the only activity in which men do more than 50% of the work. Other activities such as planting, weeding, transporting, storing, processing, marketing, and husbandry are disproportionately the work of women (Table 11.6).

The important point is not that women work more than men in production, but that they have the major responsibility for maintaining the agroecosystem. They maintain soil fertility with green and animal manures, select seeds for planting, remove weeds, and know and understand the subtleties of their own farming system. Indeed, their lives and those of their dependents require it. Although they generally do not share ownership of the land, they share with men access to the land and to the benefits that the land produces, whether measured as food for consumption or food for marketing.

| Table 11.6 | Male and female share of agricultural work in Africa |

Task	Male	Female
Plowing	70%	30%
Planting	50	50
Hoeing and weeding	30	70
Transporting	20	80
Storing	20	80
Processing	10	90
Marketing	40	60

The Green Revolution has split food production into two sectors:

1 A highly visible, centrally planned, and state-subsidized sector, in which crops are produced for markets and profits
2 A less visible decentralized sector, usually carried out on more marginal land near the home, which produces food for local consumption

The first sector, dependent on bank loans, trade agreements, and purchases of chemicals and technology, has become the province of men, while the second sector is now almost exclusively the province of women. Thus, as high-input agriculture has spread, women's control over food systems has greatly diminished (Table 11.7, p. 323), while their responsibilities as the main providers for their dependents have increased. In the cash economy that has evolved, the compensation for women's work has declined, because women are not involved in the money-oriented production of crops, except as hired laborers. They have lost their access to the land and, as a result, to the food production process itself.

The decline in the status of women is well documented in a U.N. "Report of the Committee on the Status of Women in India." For example, between 1951 and 1981 the proportion of all women who were cultivators (food producers with access to the land) declined from 45 to 33%, while the proportion of all women who were hired agricultural laborers rose from 31 to 46%. As is usually the case the world over, employed women are paid less than their male counterparts.

When technology displaced these laborers, women were often the first to lose their jobs. For example, in India most weeding has traditionally been done by hired women. As herbicide use has increased, so has these women's unemployment. In Bangladesh, rice polishing has traditionally been done with a foot-operated pestle and mortar, almost always operated by women. As much more productive and efficient mechanized rice mills have replaced this method, these women have become unemployed. What is more, the few people who get jobs in the mills have been men.

Govind Kelkar, a sociologist who has written extensively about the role of women in the Third World, studied the effects of the Green Revolution in three villages in India and found that women lost decision-making power when there was a shift away from food crops for home use and toward cash crops for the market. Men decide what to do with cash crops, and in their decisions

Box 11.3

*A Farm Woman's
Work Is Never
Done*

About half the world's population lives in the subsistence economies of the Third World. In such economies, people do not have enough income to meet their basic needs and therefore rely on their own labor to get food, fuel, and water from their surroundings. Subsistence farmers grow their own food, provide their own clean water, and gather their own fuel. And in such farming systems, the women work longer hours than men in support of their families.

It is normal for women in the rural villages of India to work three shifts. A typical day for such women starts at 4 A.M. They light fires, milk the buffaloes, draw and carry water from the well to their homes, sweep floors, and feed their families. From 8 A.M. till 5 P.M., they work their own fields or those of a landlord for a meager wage. In the evening, they gather firewood, gather food from wild plants to feed their families, and collect grass to feed the buffalo. Still later they return home to prepare food for their families and do other domestic chores. These women work twice as many hours as their husbands to support their families. Their work is truly "never done," and they therefore may regard a child not as another mouth to feed, but as another pair of hands to lighten the burden of work.

The roles men and women have in society are in part self-assumed and in part imposed by society. Generally speaking, women have fewer choices than men, especially in rural societies with subsistence economies. Women assume and are assigned the role of primary caretakers of the family. As a result, women and men have different agricultural strategies and objectives. Men are wage earners if jobs are available, or raise cash crops on a landlord's farm or on their own land. They see money as the key to a better life. Women may also be wage earners, although their work is always paid less well, but their concern is primarily for their family. They raise food crops on their own land or on the communal land of the village, and they gather food from edible wild plants as well as fuel and animal feed. Most of these contributions to the family economy also come from communal lands. These common lands that are open to all villagers have always constituted an important resource in subsistence economies, a resource that is primarily cared for and used by women.

The contributions that women make to the household are often undervalued because they cannot easily be expressed in monetary terms. For example, a study of villages in Nepal showed, comparing only wages, that women contributed only 20% to the household, and men contributed 80%. Yet when nonmonetary contributions were counted, it became clear that men and women contributed money, food, and services of equal value. But even when money alone is considered, all is not what it seems. A study in Mexico found that although the women earned less than the men, women contributed 100% of their earnings to the family, while men contributed at most 75%, and often less.

The aim of most development projects is to increase the well-being of the families living in the area, but the experts generally ignore the problems just discussed because the contributions that women make do not show up in government statistics. Therefore, no importance is attached to the resources needed to generate those contributions. Specifically, development projects have ignored the value of common lands to the well-being of village families, precisely because those common lands are used by women in nonmonetary transactions.

Developments such as the Green Revolution have generally increased the land available for cash crops at the expense of common areas that are used for subsistence crops. Increased populations are putting greater pressures on these communal lands at the same time that these lands are shrinking in area. The women who use these lands now must overexploit them and/or go farther afield to meet their families' needs. The net result is environmental degradation of the common lands.

Development also often entails the replacement of human labor by tractors, and this increases the competition among the laborers and lowers wages. As the women subsistence farmers sink deeper into poverty, they come to value children more as a ready source of labor and a hedge against poverty in old age. Their daughters will face the same predicament. It is perhaps not surprising that a study of work patterns in India found that 75% of females over age 5 are working, as opposed to 63% of the males.

financial gain is paramount; setting aside grain to feed the family is less important. In an economy dominated by men and money, women have lost control over, but not the responsibility for, the survival of their families. Vandana Shiva, a physicist, philosopher, and feminist, writes in her 1988 book *Staying Alive: Women, Ecology and Development in India* (London: Zed Books): "This devaluation, combined with increased work burdens, reduces women's entitlement to food, nutrition and even life itself. As women carry more burdens for society, they are increasingly seen as becoming a burden on society, and can be dispensed with, through discrimination, dowry deaths and femicide."

Both international and national organizations have belatedly recognized these social impacts and are trying to repair the inadvertent social damage caused by the Green Revolution. Broader availability of inputs to smaller and smaller farmers, and subsidies to tenants, have begun to stem the tide of landlessness. Some CGIAR institutes have begun to hire social scientists to evaluate the effects of the new seeds on the lives of poor people. The United Nations has an active Commission on the Status of Women, whose focus is on greater involvement of women in economic development through empowerment and greater self-reliance. Hopefully, these efforts will not only continue but will also prompt larger efforts to assess the impacts of new agricultural technologies on the people who are encouraged to use them.

9 Increased food production has led to concern about food stocks.

In addition to biological and social impacts, the Green Revolution has had effects on patterns of local and international markets for food and feed. The internationalization of most industries includes agriculture. The

| Table 11.7 | Rice-growing technology and women farmers in Gambia (1985) |

Variable	Traditional Methods	High-Input Farming
Cost of inputs ($/ha)	$20	$294
Yield (t/ha)	1.3	5.9
Percentage of fields farmed by women	91%	10%

Source: Data from H. Binswanger and J. von Braun (1991), Technological change and commercialization in agriculture: The effect on the poor. *World Bank Research Observer* 6:57–80.

interdependence of different countries was felt strongly in the early 1970s, when the conjunction of several events around the world had profound implications:

1 Because of an El Niño current (see Chapter 2), the anchovy fish catch off Peru plummeted. Since anchovies were an animal feed, demand for alternatives (such as soybean meal) increased.

2 The Soviet Union had inadequate snowfall to cover its winter wheat. This led to much reduced yields and a world wheat shortage. The United States responded by selling wheat to the Soviet Union for the first time.

3 The monsoon in Asia did not bring expected levels of rainfall. This placed additional stress on the global grain markets.

The result of all these events was steep rises of prices for food and feed on the world markets. Indeed, economists feel that this was a major cause of the inflation that plagued the world during this period. Many poor countries could not afford to buy grain on the open markets and went into debt to finance purchases. In some cases, these debts remain a burden, inhibiting development. To satisfy demand, producers (for example, U.S. farmers) put crop acreage back into production that had been idled due to soft demand and low prices. This placed more stress on the land. In addition, high prices encouraged producing farmers to sell as much of their grain as possible, rather than keep some as carryover stocks for the next year.

The concept of **carryover stocks** is an ancient one (for example, in the Bible, Joseph advised the Egyptian pharaoh to store grain for a coming famine). In addition to protecting a country, and the world, against food shortages in those inevitable years when production does not achieve its goals due to weather and other factors, stocks protect farmers and purchasers against fluctuating prices. When the supply is low, stocks can be put on the market to lower the tendency of prices to rise. When the supply is high, grain is removed to stocks and thus taken off the market, so prices do not fall too far.

World grain stocks have fluctuated over the years as producers have tried to adapt to the market and keep it stable (Figure 11.9). No international agency regulates stocks, so policies are the sum of those of the individual stockholders. Over 80% are held by six regions: the United States, Canada, China, countries of the former USSR, India, and the European Community. In China and India, stocks are only for local use and are not available for the world market. In these two countries, stocks have been very valuable in keeping the countries almost self-sufficient for grain (although China continues to import). In the former USSR, stocks have traditionally been low, and have not prevented substantial imports.

In the United States, which has the largest share of the world's grain stocks, the peak levels of the 1980s have been reduced. Reasons for this include the droughts of the late 1980s and increases in demand for U.S. exports. An important cause is government policy: the large stocks were considered a potential drag on prices, depressing them. For this reason, cropland was idled through subsidies to farmers not to grow crops. Although it appears paradoxical that cropland is unused in one place and excess food stored, while in another place people are hungry, this only emphasizes again that the food problem is an economic and social one, and not, in the near future, a biological one.

Figure 11.9

World Cereal Grain Stocks After Each Year, Plotted as Days of Consumption. For 1993, the stocks were 330 million metric tons, which was about 19% of total annual use. When the stocks fall below the line of instability, prices may rise considerably. Source: Data from Food and Agriculture Organization.

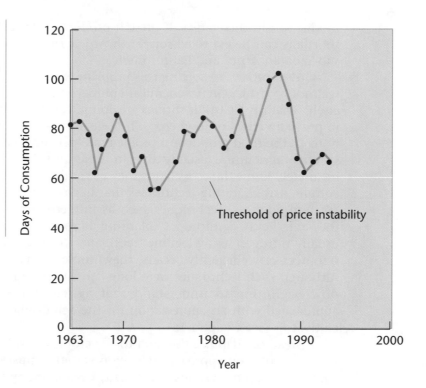

10 | What lies beyond the Green Revolution?

By 1985, the Green Revolution—the introduction of high-yielding varieties of wheat and rice—had about run its course, not only in Third World countries, but also in the developed world, where semidwarfs greatly boosted wheat production. Countries such as India, which once faced chronic food shortages, today have substantial grain reserves. Indonesia, once the leading rice importer, became a rice exporter.

Yet the benefits of the Green Revolution were not felt everywhere. The high-yielding varieties were adopted by farmers who had access to markets, and who could afford fertilizers, pesticides, irrigation machinery, and small tractors. These inputs are required to reap the most benefits from the Green Revolution strains. According to a report by the Worldwatch Institute of Washington, DC, high-yielding varieties were being grown on 36% of the grain-producing lands in Asia and the Middle East, 22% of Latin America's grain lands, and only 1% of Africa's grain lands. Whereas per capita food production has increased in Asia and Latin America, it has declined in Africa.

Africa gained the least from the Green Revolution, in part because it has large areas with sufficient rainfall but poor soils, and other areas with reasonable soil but insufficient rainfall. In addition, its population cannot afford the necessary inputs. The Green Revolution emphasized wheat and rice, but the staples of Africa are maize, sorghum, millet, cassava, sweet potatoes, and yams—all crops that do well on marginal lands. On a worldwide scale, as many as 1.5 billion people live on such marginal lands, where they practice a rainfed subsistence agriculture, in which food crops are raised for local consumption and not for export. Many of these areas have high population growth rates (see Chapter 1), and feeding the 7.8 billion humans expected to inhabit the earth by the year 2020 will require substantial increases in food production in these areas. If present production trends continue, substantial shortfalls can be expected for North Africa and the Middle East, along with sub-Saharan Africa and Latin America.

One approach to solving this problem is to extend the Green Revolution technology (the genetic–chemical approach) to other crops, and that is precisely what the CGIAR institutes are doing. Together, they are now working to improve a variety of food crops. It is clear, however, that more creative solutions to the food production problem are needed. Farmers who practice subsistence agriculture usually live in areas unsuited to monoculture. They intimately know the ecosystems they use and have evolved a sustainable agriculture, using farming techniques that have often been underappreciated by Western advisers. Techniques such as intercropping (planting two crops in the same field), agroforestry (planting a crop underneath a canopy of trees, usually nitrogen-fixing legume trees), and shifting agriculture all use natural resources very efficiently because they mimic the natural ecological processes. Although such techniques were long considered "primitive," people are just now beginning to understand that agricultural development must be approached with an appreciation for the achievements of these indigenous cultivators (see Chapter 16).

Instead of adapting the land to the crop, which has been the top–down, Green Revolution approach, new crop varieties must be adapted to the very complex cropping systems that already exist in many areas. Clearing the land in Asia or in Africa to use cropping systems that work well in Kansas or in France has invariably led to disaster. This does not mean that traditional breeding and biotechnology have no role to play. Quite to the contrary. Many experts believe that biotechnology will help create crop strains that will fit into these subsistence agriculture systems. Agricultural experts are meeting to discuss crop by crop what the major constraints to production are and where biotechnology can help. For some crops, the constraint may be the slow clonal propagation of elite specimens (tissue culture can help), for others it may be specific fungal or virus diseases (transgenic virus-resistant plants can help), and yet others may have toxic substances (breeding can help). Virus diseases are particularly devastating to tropical crops, and biotechnology is particularly well suited for making virus-resistant plants.

The biotechnology companies of the developed world are unlikely to make a real contribution in this area. Their goal is not to develop crops for Third World countries. It is therefore imperative that international organizations and the governments of wealthy regions help such countries develop their research infrastructure so that the developing countries can themselves generate the new crop strains adapted to their own environments.

Summary

The Green Revolution is the breeding of high-yielding varieties of wheat and rice and their use in Asia and Latin America to raise food production. These new varieties have performed well with the agricultural inputs (fertilizers, pesticides, and mechanical energy) that typify agriculture in developed countries and quickly replaced traditional crop varieties. The breeding of these high-yielding varieties was a planned international effort supported by U.S. foundations and governments of developing countries. Quick adoption of these varieties led to dramatic increases in grain production. Grain importers became grain exporters. The export of grain was possible, even necessary, because the local people did not always have the purchasing power to buy the available food. Thus, in spite of the Green Revolution, hunger persists.

Other social dislocations have been associated with the Green Revolution. Two of the most notable are the increasing marginalization of women and the creation of more landless peasants. Green-Revolution-type agriculture requires access to capital, and only men have access to capital. Thus the women in much of the Third World have lost control over the food production system although they remain responsible for the nutrition of the family. Another drawback of the Green Revolution is the disappearance of landraces. Developed countries are trying to address this problem with *in situ* conservation and seed banks.

Further Reading

Binswanger, H. P., and J. Von Braun. 1991. Technological change and commercialization in agriculture: The effect on the poor. *World Bank Research Observer* 6:57–80.

Brown, A. H. D., ed. 1989. *Use of Plant Genetic Resources*. Cambridge, UK: Cambridge University Press.

Brown, L. R. 1985. *The Changing World Food Prospect: The Nineties and Beyond*. Washington, DC: Worldwatch Institute.

Carr, M., ed. 1991. *Women and Food Security*. London: Intermediate Technology Publications.

Chang, T. T. 1984. Conservation of rice genetic diversity. *Science* 224:251–256.

Conway, G., and Barbier, E. 1990. *After the Green Revolution*. London: Earthscan Press.

Dalrymple, D. 1986a. *Development and Spread of High-Yielding Rice Varieties in Developing Countries*. Washington, DC: U.S. Agency for International Development.

Dalrymple, D. 1986b. *Development and Spread of High-Yielding Wheat Varieties in Developing Countries*. Washington, DC: U.S. Agency for International Development.

Doyle, J. 1985. *Altered Harvest: Agriculture, Genetics and the Fate of the World's Food Supply*. New York: Penguin.

Greeley, M. 1991. *Postharvest Technologies: Implications for Food Policy Analysis*. Washington, DC: EDI Case Studies, no. 7. World Bank.

Harlan, J. R. 1976. Genetic resources in wild relatives of crops. *Crop Science* 16:329–333.

Juma, C. 1989. *The Gene Hunters*. Princeton, NJ: Princeton University Press.

Kelkar, G. 1991. *Gender and Tribe: Women, Land and Forests in Jharkand*. (New Delhi, India: Kali for Women).

Lipton, M. 1989. *New Seeds for Poor People*. Baltimore, MD: Johns Hopkins University Press.

Mier, M., and V. Shiva. 1993. *Ecofeminism*. London: Zed Books.

Shiva, V. 1988. *Staying Alive: Women, Ecology and Development*. London: Zed Books.

Tinker, I. 1990. *Persistent Inequalities*. Oxford, UK: Oxford University Press.

U.S. Department of Agriculture. 1990. *World Grain Stocks: Where They Are and How They Are Used*. Economic Research Service, Bulletin 594. Washington, DC: U.S. USDA.

Wolf, E. C. 1986. *Beyond the Green Revolution*. Worldwatch *paper 73*. Washington, DC: Worldwatch Institute.

Wright, A. 1990. *The Death of Ramon Gonzalez*. Austin: University of Texas Press.

Chapter 12

Pests and Pathogens

The yields produced by crop plants can be substantially reduced by weeds, by organisms such as insect larvae or nematodes that feed on plants, and by disease organisms—fungi, bacteria, and viruses—that attack the plants. Most of the organisms that can reduce crop yields are present at all times, but crop losses only occur when the numbers of these pests and predators exceed certain thresholds, resulting from rapid population increases.

Such population increases occur more easily in agricultural systems characterized by continuous monoculture of genetically uniform plants. Plant diseases spread rapidly in monocultures when weather conditions favor the growth of pathogens, or stress the crop plants. Many predators or disease organisms are host specific, meaning that they attack only one or a few species of plants. Monocultures provide such pests with abundant food sources, or ecological niches in which to multiply. When crops are rotated from year to year, and/or when crops lack genetic uniformity, predators and pests are likely to be less prevalent. The absence of natural enemies often caused by applying pesticides may further aggravate the situation, making sudden pest population increases possible and crop losses likely. For a single crop, such as rice, there may be a large number of organisms that can attack it (Table 12.1).

Table 12.1	Rice diseases	
Agent	**Plant Organ Attacked**	**Number of Diseases**
Virus	Leaf	12
Bacteria	Leaf	4
Bacteria	Grain	3
Fungus	Leaf	11
Fungus	Stem, root	10
Fungus	Seedling	5
Fungus	Grain	10
Nematodes	Root	11

Source: From S. Ou (1972), *Rice Diseases*, Commonwealth Agricultural Bureau.

In this chapter, we first briefly discuss some of the biological characteristics of the plant pests that reduce yield, and then survey the chemical defenses that plants have evolved to protect themselves against insects and pathogens. A better understanding of these defense mechanisms allows biotechnologists to transfer genes from one plant to another or from bacteria or fungi to crop plants, so that the crop plants will be better protected against the pests.

Agricultural Pests and Plant Pathogens

1 | **Weeds compete with crop plants for nutrients, water, and light.**

Farmers define weeds as plants that compete with crop plants and diminish crop yield. Weeds reduce crop productivity by an estimated 12% in spite of all efforts made to control them. Converting a natural system to an agricultural one usually involves clearing the land, plowing it, and fertilizing it. These conditions are suitable for the growth not only of the crop plants, but also of many other plants that grow readily in disturbed fertile soil. These plants become weeds. Some of these weeds may have been present in the natural system, but their numbers were low because conditions were not particularly favorable for their growth. The same conditions that favor crop growth also favor weed growth. Many weeds are not indigenous to the area where they are found. Indeed, the spreading of crop plants from their centers of origin was accompanied by the spread of weeds. Thus, many weeds commonly found in North America originated in Europe, and were not found in North America in pre-Columbian times (see Box 12.1).

Many weeds have efficient mechanisms for reproduction and seed dispersal: some produce only a few hundred seeds; others may produce several million. The dandelion and the tumbleweed (see Box 12.1) are good examples of weeds with an effective seed dispersal mechanism. Seed dormancy mechanisms often ensure that not all weed seeds will germinate immediately; some seeds may remain dormant for several years before germinating. For example, the cocklebur, a broadleaf weed, has prickly burrs, each containing two seeds. One seed germinates in the spring following the maturity of the plant, whereas the other seed usually germinates a year later. Many seeds require light for germination, and every time the field is tilled a new batch of weed seeds is brought to the surface and germinates if the conditions are suitable. Some weeds do not reproduce by seeds, but instead have horizontal underground stems that send up aerial shoots at intervals. Such weeds often form mats that are difficult to break up, and when broken up each segment of the plant can give rise to a new plant through vegetative propagation. Serious weeds share certain botanical characteristics:

- Their seeds germinate over a long period of time and are long-lived.
- They grow rapidly and produce seeds throughout the growing season, often reseeding themselves several times.
- They have a high seed output.
- If not self-fertilizing, then they are wind fertilized or fertilized by unspecialized insects.

Box 12.1

*The Weed That
Won the West*

Sometime in the late 1870s, local farmers in South Dakota noticed the appearance of a new weed bouncing across their fields. This tumbleweed, as it was called, spread its seeds as it tumbled along the rangelands, and in 30 years it invaded the entire U.S. West. Plant taxonomists soon identified it as *Salsola kali* or Russian thistle, and determined that it had been imported from Europe in 1877 as a contaminant of flax seed. The weed's spread put fear in the hearts of farmers and politicians alike. It was not just a nuisance, but it spread range fires as the flaming weeds blew across the firebreaks, and most importantly, it severely depressed wheat yields. A legislator from North Dakota proposed to fence the state to keep the weed out, and the state later passed a tax to compensate the farmers for their losses. However, the losses incurred by the farmers were much greater than all the taxes collected by the state.

Well known and oft repeated are the success stories of the rapid spread of certain crops and how these migrations changed food production: potatoes from South America to Europe, corn from Mexico to Africa, and soybeans from China to North America. Less is said about the noxious weeds that accompanied this crop migration. The instant success of the Russian thistle was caused by its many adaptations to the climate, terrain, and conditions of the U.S. West.

The Russian thistle has a highly branched main stem; the entire above-ground portion of the plant is a ball-shaped body, and a special layer of cells in the main stem allows it to break smoothly when the seeds mature and the plant dries out. Gusts of wind therefore set the weeds tumbling. The seeds—as many as 250,000 in a good-sized plant—are embedded in bracts, and they do not scatter immediately when the weeds start to tumble. Rather, they are released gradually as the plants roll along the fields. Like other successful weeds, they are opportunistic and can take advantage of the disturbed soils that characterize agriculture. Russian thistle competes poorly against established range plants, but grows quickly in plowed soil and other areas where there are no competitors. Finally, the plant is well adapted to the dry western ranges and does well in salty soils. All these adaptations account for its rapid spread from the initial site of introduction over the entire West.

The farmers basically had to live with this successful pest until the introduction of chemical pesticides around 1950. The herbicide 2,4-dichlorophenoxyacetic acid (2,4-D) brought relief to the wheat farmers because this chemical kills dicot plants without killing the monocot grains. Tumbleweeds are still troublesome for highway departments and irrigation districts, which every year spend millions of dollars cleaning up the Russian thistles that accumulate along roadside fences and in irrigation canals.

It is likely that this drought-resistant weed could be developed into a new fodder crop. Its fresh growth is high in good-quality protein and could be harvested at intervals and fed to cattle. A water-starved West may in the future be forced to abandon its irrigated alfalfa fields and replace them with Russian thistle.

- If they are perennials, they can regenerate entire plants from small fragments (vegetative propagation), and are hard to uproot.
- They may form mats or otherwise choke out plants.

Weeds diminish crop yield by competing with the crop plants for water, light, CO_2, and nutrients. The actual effect of weeds on a crop may not be to reduce the crop's net productivity, but instead to reduce the proportion of that productivity that goes into the seeds. In an experiment at the University

| Table 12.2 | Production of dry matter from corn and foxtail grown in competition |

Foxtail Spacing	Foxtail	Corn Grain	Cobs	Stalks	Total
1 inch	1,020	3,710	850	2,750	8,330
2 inches	820	3,880	860	2,690	8,250
4 inches	600	4,020	880	2,600	8,100
12 inches	240	4,090	900	2,810	8,040
24 inches	90	4,280	920	2,860	8,150
No foxtail	0	4,430	940	2,940	8,310

Note: All figures are pounds (dry weight) produced per acre.

Source: E. Knake and F. Slife (1962), Competition of *Setaria faberii* with corn and soybeans, *Weeds* 10:28.

of Illinois, competition between a weed (foxtail) and corn was tested throughout a range of weed densities. The results (in Table 12.2) showed that the increase in the density of the weeds reduced the total net productivity of the corn plants and specifically lowered grain production. Total biomass production remained the same.

Weeds can also diminish crop yield by interfering with the harvest. The morning glory, which grows as a vine, can be a troublesome weed if it is allowed to remain in the field until the crop reaches maturity. Its twining stems make mechanical harvesting difficult, and this contributes to harvest losses.

The purpose of many agricultural practices is to try to eliminate weeds so that they do not cause crop losses. If weeds are not controlled, yields may be reduced by 50%; however, most farmers make considerable efforts to reduce weeds. Strategies to keep weeds in check include mechanical cultivation, applying mulches that cover the bare soil between the plants, intercropping with a second crop that grows quickly but remains low to the ground, hand hoeing, and spraying chemical herbicides.

Like other plant pests, weeds can evolve rapidly to adapt themselves to specific agronomic practices. If weed seeds can survive the winnowing process in which farmers separate the crop seeds from the chaff and the weed seeds, they will be stored with the grain and planted again the next season. Scientists have recorded several cases of seed mimicry in which the seeds of a weed rapidly evolved to have the same shape and size as the seed of the crop. Not uncommonly, both weed and crop belong to the same family or genus. For example, wild rices are major weeds in the rice fields of India and Africa and sometimes reduce yields of the crop rice by 50%. Their resemblance to cultivated rice makes hand weeding very difficult. These rice varieties interbreed with the cultivated rice, so that new genes introduced in the crop plants quickly spread to the weeds.

2 Insects diminish crop yields by damaging roots, leaves, or seeds.

From the locusts mentioned in the Bible to the boll weevils that attack cotton of this century, insects have plagued agricultural production. In nature, the population of a particular insect species is kept in check by the

Figure 12.1

Scanning Electron Micrograph of the Head of an Aphid, Showing the Large Stylus Pointing Downward. Aphids suck out plant juices and damage plants. An aphid inserts its stylus into the phloem, which transports the sucrose and amino acids it uses as food. In addition, aphids are a major vector for transmission of viruses. Source: *U.S. Department of Agriculture.*

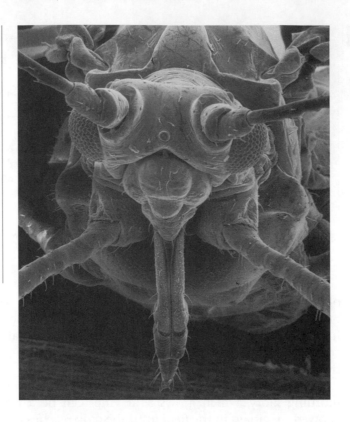

limitation of its food sources and by its natural enemies. Some insects, such as the locust, feed on a broad range of plants, but many insects have very specific food preferences and only feed on one or a few species of plants. These food preferences are the result of specific adaptations by the insects to the chemical defenses used by plants to defend themselves against insects. In the course of evolution, plants started producing toxic chemicals that prevented all insects from eating them. However, when a particular insect species evolved the necessary detoxification mechanism, it could use this plant as food whereas the other insects could not. In an agricultural system, there are usually fewer plant species, and a few species (the crops) predominate over all the others (the weeds or the plants along the periphery of the field). Insects that feed on crop plants find an abundant food source, carefully cultivated by the farmer, and as a result they can multiply rapidly. Their numbers are kept in check only by their natural enemies or by the farmer's pest control programs.

Insects can damage crop plants in several ways. Sometimes the damage is caused by the adult insects, such as locusts and grasshoppers, which eat the leaves of the plant, thereby decreasing its ability to photosynthesize. Other insects, such as aphids (Figure 12.1), suck out sap from the phloem, the conductive tissue that transports the products of photosynthesis, thus starving the plant organs of nutrients. Much damage is caused by insect larvae,[1] an immature form of insects. Some insect larvae live in the soil and feed on young roots, thereby diminishing the plant's ability to take up nutrients; others remain in the foliage or actually tunnel within the plant tissues. Some

---

[1]Insect larvae are commonly called "worms" or "grubs," hence the common names *cutworm, wireworm, corn earworm,* and so on.

feed mainly on the stem and leaves, and others, such as the corn earworm,
eat the seeds (Figure 12.2). Some larvae, such as the cutworm, live in the soil
but come to the surface at night. Cutworms can chew completely through the
stem of a young seedling, cutting it off at ground level. Bruchid beetles lay
their eggs on dry seeds, and their larvae develop within the dry seeds. They
infest the crop before harvest but continue their damage while the seeds are
being stored. A new generation of beetles appears every 30–35 days, and each
female lays 50–100 eggs. Thus, the bruchid population may mushroom during six months of crop storage, causing extensive damage.

Finally, over 200 microbial and viral diseases are transmitted from plant
to plant by insects, such as leafhoppers or aphids. When an aphid inserts its
stylus into a plant cell, it sucks out the juice together with the virus particles,
and when it flies to the next plant to feed again, it will inoculate this plant
with virus particles. The transmission of these plant diseases is similar to the
transmission of certain human diseases, such as malaria, which is spread by
mosquitoes. The best way to combat such microbial and viral diseases is to
cultivate resistant crop varieties, and if these do not exist, to control the
insects that serve as vectors for the pathogens.

Many insects overwinter either in the soil or in the crop residues that
remain on the land. The damage caused by such insects is often magnified if
the same crop is grown year after year on the same field, because doing so
allows the populations of such insects to build up, a problem that can be
avoided by crop rotation. However, many insects are fairly mobile, and can
easily move from one field to another.

Despite the increasing use of pesticides, insects do considerable damage
to crops and depress food production. The FAO estimates that about a sixth of
all plant food grown in the world is consumed by insect pests. Even in the
developed countries, where insecticides are used heavily, damage by insects is
considerable. The USDA estimates that 3% of the annual U.S. wheat crop is
lost to insects. This represents enough grain for 2 billion loaves of bread, or
food for 2 million people. Note that the heavy use of insecticides has not

decreased the proportion of the food crop that is lost to insects. The percentage of crops that is lost to insects and other pests has remained fairly constant over the past 50–100 years.

Like most pathogens and pests, insects can evolve rapidly as a result of particular agronomic practices or the use of chemical insecticides or biological control agents. Selection pressure, whatever its source, causes rapid changes in the insect population, encouraging the survival and reproduction of those individuals that can cope with the new stress. An interesting example is provided by the northern corn rootworm, an important pest of corn in the U.S. Midwest. In areas where corn is grown every year, the adults produce eggs that remain in the soil for one winter and then hatch and feed on the corn roots the next spring. The farmers can combat this problem by introducing a corn–soybean rotation. When the rootworm larvae emerge in the spring in a soybean field, they do not find the necessary food and very few will grow into adults to lay eggs for the next year. However, after a few cycles of this rotation, the rootworms can become a problem again because the agronomic practice selects in favor of a new variant (a mutation) that allows its eggs to remain dormant in the field for two winters, not just one. This population takes over and replaces the population that overwintered only a single season.

3 | Soil nematodes cause extensive damage to root systems.

There are several thousand species of nematodes, tiny animals with transparent worm-shaped bodies that measure 0.25 to 2.5 mm in length. Most live freely in fresh or salt water and in the soil, where they feed on microscopic plants and animals. Numerous species can attack and parasitize humans, and several hundred species feed on living plants—usually the roots—causing serious plant diseases and substantial crop losses. The adults lay eggs that hatch in the soil, and after several larval stages—each one terminating in a molt—the males and females differentiate, mate, and reproduce. If no males are around, the female produces her own sperm or can reproduce parthenogenetically (without sperm).

Plant-parasitic nematodes may be ectoparasites that feed on the surface of the root, or endoparasites living within the tissues of the plant. All nematodes have a sharp stylus or mouthpiece that is used to pierce through the outer cell wall of the plant epidermis. Some nematodes secrete enzymes that digest the contents of the plant cells as they invade the roots, and the larvae grow in a cavity within the root tissues. Most nematodes show host specificity; a particular species of nematode attacks only one (or a few) species of plants. Once the infective larval stage develops, a nematode must find its host plant or starve to death. It cannot simply go on eating organic matter in the soil. In some nematode species, the larvae can dry up and go into a resting stage, to wait until a suitable host plant appears.

Nematodes cannot move very far under their own power, and in a single season they may travel only 1 meter. However, nematodes are spread by anything that moves and carries soil particles, including people and animals, farm equipment, irrigation water, runoff from rain, dust storms, trucks, and even airplanes that transport crops and nursery stock.

Two major groups of plant-parasitic nematodes are the root knot nematodes (*Meloidogyne*) and the cyst nematodes (*Heterodera*). **Root knot nema-**

Figure 12.3

A Root Knot Nematode Enters a Tomato Root. Once inside, the larva establishes a feeding site and causes formation of galls that impair root function (also see Figure 8.9). Source: U.S. Department of Agriculture.

todes occur throughout the world, but are most prevalent in areas with warm or hot climates. They are universally present in greenhouses unless the soil is sterilized with nematocides. They can attack more than 2,000 different species of plants, including most crops (Figure 12.3). Root knot nematodes feed by piercing the epidermal cell wall with their stylus and then injecting saliva into the plant cell. The surrounding plant cells disintegrate, creating a giant cell, and the vascular system of the plant continues to unload the products of photosynthesis into this giant cell. These products are then sucked up by the growing nematode that lives inside this giant cell, gradually causing the root to develop a gall where the nematode is developing.

The presence of root knot nematodes causes the roots to become grossly deformed and root growth to diminish. As a result, plant growth is also reduced, although the plant never dies from this invasion. If the crop is a root crop (sugar beets, carrots, potatoes), market value will be adversely affected. The mechanical damage caused by the nematodes provides points of entry for other pathogens (bacteria and fungi) that live in the soil, and these often cause root rot. In addition, nematodes may serve as vectors for viruses that cause diseases in the above-ground parts of the plant.

Cyst nematodes cause serious losses of temperate region plants, such as soybean, potato, and sugar beet. These nematodes also invade the roots, causing the formation of large cells that continue to serve as a source of food because the plant continues to transport more photosynthate to these cells. When the nematodes reach adulthood, the males leave the root and fertilize the females, who produce several hundred eggs. These eggs cause the body of the female to swell up, and this gelatinous egg mass later acquires a hard outer shell and forms a cyst. Cysts can be seen attached to the outside of infected root systems. Even plants resistant to nematodes may still be attacked, and the first stage of nematode development is completely the same as in susceptible varieties. However, the young larvae do not mature into adults, no eggs are formed, and the damage to the plant is minimal. Heavy infections of susceptible plants result in smaller root systems that are less able

Figure 12.4

Tobacco Mosaic Virus Forms Rod-Shaped Particles. The single RNA molecule within the rod is completely coated with protein molecules. Source: *U.S. Department of Agriculture.*

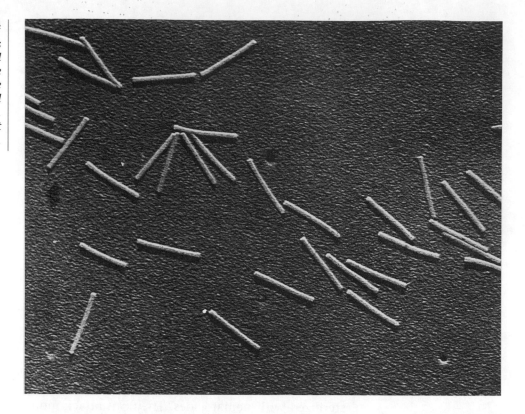

to support plant growth. The soybean cyst nematode interferes with the normal development of the nitrogen-fixing root nodules.

4 | Viruses and virusoids may be present in all plants at all times.

Viruses and virusoids are the smallest infectious agents that cause diseases. They are so small that they can only be seen with an electron microscope. Viruses and virusoids are all parasitic: they can only live within other cells, whether plant, animal, or microbial. Unlike bacteria, viruses are not cells, but consist only of a molecule of nucleic acid (RNA or DNA) wrapped up in a coat of protein (Figure 12.4). The nucleic acid has only enough genetic information for a few proteins. Virusoids are even smaller and consist of only a relatively small RNA molecule. Both viruses and virusoids use the cellular machinery of their host to reproduce themselves. Normally, they never quite kill their hosts, but because they use the cellular functions of RNA and protein synthesis for their own ends (replication), they weaken the plant and diminish crop yields.

We noted in Chapter 3 that meristem culture is used to produce plants that are free of viruses because viruses cannot live in meristematic cells. All other tissues of a plant seem to be always infected by one or more viruses. That they diminish the vigor of the plant is shown by the fact that the virus-free plants produced via meristem culture usually grow more vigorously than normal (virus-infected) plants. Since viruses are everywhere, most people eat plant viruses every day.

The nucleic acid molecule of a virus encodes only a few genes. Its DNA or RNA contains the necessary information to make an enzyme that allows the nucleic acid molecule to replicate itself, to make the viral coat protein, and to

BOX 12.2

*The War Against
the Hessian Fly*

The Hessian fly is the most destructive insect pest of wheat in the world, and the war against it has been waged for 30 years by wheat breeders. Its Latin name *Myetiola destructor* is well chosen, for it has destroyed millions of bushels of wheat all over the world. The black two-winged fly, about the size of a mosquito, is thought to have arrived in North America during the American Revolution. Its larvae or eggs may have been present in wheat straw brought by Hessian mercenaries (from Germany). The adult female lays 200 to 300 eggs on the leaves of a young wheat plant, and the larvae eat their way into the leaves and stem, completely stunting the growth of the plant.

Winter wheat, which is planted late in the summer and grows about 20 cm tall before winter arrives, is often killed outright, because the weakened plants do not survive the winter. If 20% of the plants are affected, the farmer will suffer economic loss, and before the 1950s up to 70% of the wheat was infected in bad years. Cultural practices were the farmer's only defense against this pest.

In the late 1940s and early 1950s, plant breeders from Purdue University and Kansas State University began releasing the first resistant wheat varieties. These were obtained by crossing domestic wheats with wild relatives of wheat that resist the Hessian fly. The percentage of wheat plants affected by Hessian flies soon dropped below 20%. However, after a few years the new varieties proved less and less resistant, so new resistance genes had to be found. Researchers screened the large wild wheat collection of the University of California in Riverside, which contains more than 2,000 varieties of *Triticum* and *Aegilops* species that were originally collected in the Middle East. Many of the resistant varieties apparently came from a region around the Caspian Sea in Northern Iran and Azerbaijan, the area believed to be the original home of the Hessian fly. New resistance genes are continuously being transferred from the wild varieties to common wheat, to keep up with the continuing evolution of the Hessian fly.

Hessian Fly

make a "movement" protein that lets the virus move from cell to cell. Viruses usually enter the plant through a small wound, are often transmitted by insects such as aphids or leafhoppers, and multiply within this wounded cell. The thin channels (plasmodesmata) that connect plant cells exclude viruses, but the movement protein modifies these channels so that the viral nucleic acids can move from one cell to another.

By usurping the cell's metabolism and directing it toward their own growth and multiplication, viruses diminish the plant's capacity for photosynthesis. Wherever they occur in large numbers, they cause yellow spots on leaves, and these characteristic coloring patterns often give their name to the viral disease. Tobacco mosaic virus, cucumber mosaic virus, barley yellow dwarf virus, and papaya ringspot virus all are important plant diseases that affect not only the plants after which the virus is named, but other plants as well.

Tristeza citrus virus illustrates the devastation that can be caused by a single virus disease. Most plant viruses require insect vectors to spread, and the tristeza virus is spread by aphids, especially the tropical citrus aphid. The virus also spreads when citrus trees are grafted. Citrus trees are normally propagated in nurseries by grafting twigs from elite trees that produce good fruit onto a rootstock (a young seedling cut off near the ground) of an undomesticated citrus variety that will produce a strong root system. Citrus trees need to be replaced at regular intervals, so millions of new grafted trees are needed every year. It is important, therefore, to be able to certify that the trees from which the twigs are taken to make the grafts, are free of virus, and that the rootstocks that are used are virus resistant. Tristeza citrus virus lives in the conductive system of the stem, and its spread in the tree causes a rapid decline in growth and productivity. In the state of São Paulo in Brazil, 75% of the orange trees were killed within 12 years of the appearance of the tristeza virus in the state. The same virus also decimated the California citrus industry in the late 1950s and early 1960s.

Like the viruses that affect humans, plant viruses exist in many strains and evolve rapidly. In former times, they caused less devastation because the crops were not genetically uniform. This point is dramatically illustrated by the loss of much of the Philippine rice crop in 1971. In the 1960s, the Philippines had reduced the genetic diversity of its rice crops by the new Green Revolution high-yielding rice strain IR-8. This strain was susceptible to the leafhopper-spread virus disease called *tungro*. In 1971, a tungro epidemic severely decreased the rice crop, turning the Philippines from a rice exporter to a rice importer. Since that time, a new rice variety has been developed that is resistant to the tungro virus, but new, virulent virus strains will, of course, arise again.

5   Plant-pathogenic bacteria cause many economically important diseases.

Some bacteria, such as those that fix nitrogen, transform nitrogen and sulfur compounds in the soil, or contribute to the decay of organic matter and the process of mineralization, play vital roles in crop production. However, many species of bacteria cause important diseases and contribute substantially to the 10–15% of crop losses worldwide (Figure 12.5). Because they are microscopic in size, bacteria can enter the plant through natural openings

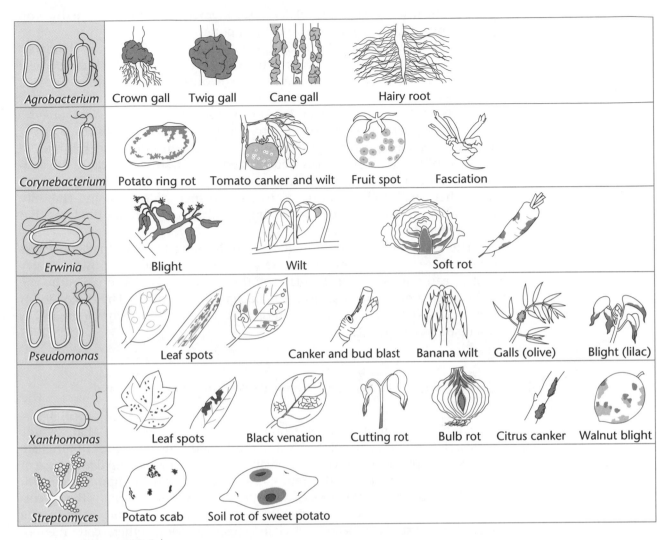

Figure 12.5

Symptoms and Diseases Caused by Six Different Types of Bacterial Pathogens.
Source: *G. N. Agrios, Plant Pathology (San Diego: Academic Press), 2nd edition p. 438.*

or wounds. In some cases, they secrete enzymes that help dissolve the protective barrier of the epidermal cell wall and penetrate the plant in this way.

Many bacterial diseases show up as spots on fruits, leaves, and stems. These spots, local infestations of the bacteria, diminish the photosynthetic capacity of leaves and disfigure fruits, thereby decreasing their commercial value. Other bacteria cause softrots by secreting enzymes that digest the cell walls of the plants and cause the plant tissues to become very soft or even liquid. Anyone who has left vegetables in the bottom drawer of the refrigerator for a month has witnessed the destructive action of *Erwinia* bacteria. Several other members of this genus cause postharvest losses of fruits and vegetables.

Bacteria that invade the conductive tissues can cause these tissues to become clogged so that water and minerals cannot get from the roots to the leaves. Such bacteria cause wilting of young plants, and the diseases are referred to as *wilts*. These bacteria also secrete enzymes that dissolve the cell walls of the conductive tissues, and the disease then spreads to other cells, quickly killing the plants. The most economically important wilt disease of plants is caused by a *Pseudomonas* that can infect plants from 44 different genera. The disease occurs especially in warmer regions and affects many crop plants, including bananas, tomatoes, and potatoes.

Bacteria also cause blights and cankers of fruit trees. Blights are characterized by rapidly spreading infections that cause the death of the infected cells and tissues. Fire blight of apples and pears is caused by a species of *Erwinia*. The disease can kill young trees in one season, and devastated U.S. pear orchards in the 1930s. Citrus canker, a disease caused by *Xanthomonas campestris* cv. *citrii* devastated the citrus industry in the 1920s. In 1984, it allegedly reappeared in some of Florida's largest nurseries, and millions of citrus seedlings and young trees were immediately destroyed because the agricultural authorities were afraid there might be a repeat of the earlier devastation. There are no chemical sprays to contain such outbreaks, and the only way to keep the disease from spreading is to destroy (burn) all the trees that are suspected of being infected.

The genus *Agrobacterium* includes soil-dwelling bacteria that can cause serious diseases that manifest themselves as galls or other outgrowths (hairy root disease). Infection of the plant by these bacteria is accompanied by the transfer of genetic material from the bacteria to the plant. The discovery of this process was of paramount importance for plant biotechnology and is discussed in Chapter 15.

6 | Pathogenic fungi cause serious losses all over the world.

Fungi are a group of microscopic heterotrophic organisms that include molds, mushrooms, and yeasts. Most of the 100,000 known species of fungi are strictly saprophytic, living on dead organic matter that they help decompose. A small minority—about 100 different species—cause diseases of humans and animals, but more than 8,000 species cause plant diseases. Some fungi—the mycorrhizal ones—live in close association with plant roots and help in the acquisition of scarce minerals, especially phosphate, while others, certain endophytes, produce substances that protect the plant against grazers or bacteria.

The vegetative body of a fungus, called a *mycelium*, consists of a mass of long filaments or hyphae. Fungi propagate by means of spores that may be formed asexually (vegetatively) or as a result of a sexual process that involves the fusion of two nuclei from two different mating types that correspond to the male and the female found in higher organisms. Spores are microscopic and easily carried away by water and wind, people, and animals.

Almost all plant-pathogenic fungi spend part of their lives on their host plants and part in the soil or on plant debris on the soil. Some live continuously on their hosts, and only their spores may land on the soil, where they remain inactive until a new host appears. Some need to spend part of their lives on dead host tissues in order to complete their life cycles. Their life cycle includes a parasitic (pathogenic) and a saprophytic stage, but both stages require the same host. Once these host tissues decay, they cannot survive. Yet other fungi have a saprophytic stage and can live on any organic matter in the soil. These pathogens, which often have a wide host range, can survive in the soil for years, even in the absence of their hosts.

Fungi, like bacteria, can enter plants through natural openings or wounds or by "eating" their way into the plant. All fungi secrete enzymes that break down the large macromolecules of the cell wall, and if the plant's defense is

not quick enough, their hyphae will quickly grow and spread from cell to cell. Once inside, they can affect the plants in many ways. Some produce toxins that alter the permeability of the cell membrane, thereby destroying its ability to regulate what goes in and out of the cell. Many fungi secrete slime that accumulates in the vascular tissues, thereby preventing the transport of water and nutrients from the roots to the shoot, causing the plant to wilt and die. Diseases that are caused by such fungi are called *wilts* (like similar diseases caused by bacteria). Some fungi produce plant hormones, and as a result the plant loses control over its own developmental processes. Some soil-borne fungi attack seedlings as soon as they germinate and kill the root system. Seedlings that look healthy one day are gone the next, and many seedlings are killed even before they emerge from the soil. This disease is known as "damping off." Many fungi simply cause necrotic (dead) spots on leaves, stems, fruits, and seeds of the plants they attack. They decrease the vigor of the plant, render the seeds and fruits less fit for human consumption, or decrease the market value.

The plant **rusts**, caused by different *Basidiomycetes*, are among the most destructive diseases known to humanity, having caused famines and economic depressions. These fungi attack primarily cereals such as wheat, oats, corn, and barley, but they can also do severe injury to beans, soybeans, cotton, and many other plants. They grow primarily at the surface of leaves and stems, causing the appearance of strictly localized rust-colored spots. Various species of *Puccinia graminis* infect cereal grains and produce long, narrow

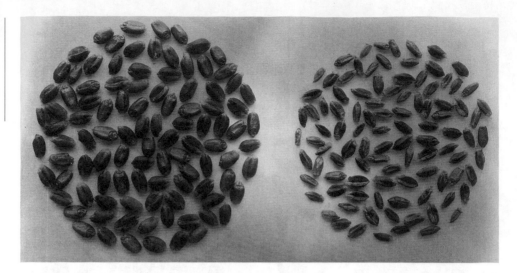

blisters on the leaves and stems (Figure 12.6)—hence the name *stem rust*. Heavy infections greatly diminish plant growth and seed size (Figure 12.7). Losses from this disease vary from year to year, but in the more severe epidemics the wheat crop may be reduced by a quarter. The only practical control of wheat stem rust (*Puccinia graminis tritici*) is through the breeding of varieties resistant to infection by the pathogen. Resistance genes have been identified in the wild relatives of wheat and then introduced in domesticated wheat by crossing and backcrossing. Since the genes of this rust fungus, and therefore the pathogenicity, keep changing all the time, the work of these plant breeders is never finished.

Rice blast, a serious disease that manifests itself as lesions on the leaves and rot of the stem, is caused by the fungus *Pyricularia oryzae*. This fungus produces a toxin that kills the plant cells wherever it grows. Rice blast was recorded as a disease as far back as the Chinese Ming Dynasty in 1637 and is widespread in Asia. The blast fungus thrives on plants that have been fertilized with nitrogen, and the high-yielding varieties introduced as part of the Green Revolution are therefore especially vulnerable. These rice varieties require the nitrogen fertilization to reach their high yields. The blast generally reduces by 20% the yield increase that nitrogen fertilizers induce in rice.

The Chemical Defenses of Plants

7 | **Secondary metabolites that deter herbivores, kill pathogens, and attract pollinators improve the evolutionary fitness of plants.**

If plants are such good food sources, and if insects can be so voracious—witness the devastation caused by locusts—why do plants still dominate the landscape? P. Feeny, an expert on the coevolution of plants and their insect herbivores, says that the most conspicuous nonevent in the history of the flowering plants is the failure of insects and other herbivores to eat the plants out of existence. At least part of the answer to this apparent contradiction is

Table 12.3	Major classes of secondary plant compounds involved in plant–animal interactions		
Class	**Number of Structures**	**Distribution**	**Physiological Activity**
Nitrogen Compounds			
Alkaloids	5,500	Especially in root, leaf, and fruit	Many toxic and bitter tasting
Amines	100	Often in flowers	Many repellant smelling; some hallucinogenic
Amino acids (nonprotein)	400	Especially in seeds of legumes but relatively widespread	Many toxic
Cyanogenic glycosides	30	Sporadic, especially in fruit and leaf	Poisonous (as HCN)
Glucosinolates	75	Eleven plant families	Acrid and bitter (as isothiocyanates)
Terpenoids			
Monoterpenes	1,000	Widely, in essential oils	Pleasant smells
Sesquiterpene lactones	600	Mainly in Compositae	Some bitter and toxic, also allergenic
Diterpenoids	1,000	Especially in latex and plant resins	Some toxic
Saponins	500	In over 70 plant families	Hemolyze red blood cells
Limonoids	100		Bitter tasting
Cucurbitacins	50		Bitter tasting and toxic
Cardenolides	150		Toxic and bitter
Carotenoids	350	Universal in leaf, often in flower and fruit	Colored
Phenolics			
Simple phenols	200	Universal in leaf, often in other tissues as well	Antimicrobial
Flavonoids	1,000	Universal in angiosperms, gymnosperms, and ferns	Often colored
Quinones	500		Colored
Other			
Polyacetylenes	650		Some toxic

Source: Adapted from J. B. Harborne (1982), *Introduction to ecological biochemistry*, 2nd ed. (San Diego: Academic Press), p. 3.

to be found in the synthesis by the plants of chemical compounds that make the plants both repellent and toxic to most insects and other grazing animals. About 10 percent of the dry mass of some plants is made up of chemicals designed for defense against predators. Herbivores have developed mechanisms to overcome some of these defenses, allowing them the limited feeding that we see today. Plants and insects have evolved together, the plants always "inventing" new chemicals to ward off the grazers, the insects evolving new enzymes to detoxify the chemicals. As a result of this coevolution, plants synthesize some 10,000 different **secondary metabolites** (Table 12.3). Ecological biochemists refer to these compounds as *allelochemicals*, molecules that regulate the interactions between different species. Most of these molecules defend plants against herbivores, but some have other functions, such as attracting insect pollinators or larger animals that eat fruits and in this way disperse the seeds.

When we examine the biochemistry of these defense compounds, and the manner in which they are synthesized, we see that many are derived from primary metabolites such as sugars or amino acids. The syntheses of these

Figure 12.8

Scanning Electron Micrograph of an Alfalfa Weevil Larva Trapped by the Hairs on the Surface of a Leaf. The chemicals produced by leaf hairs help trap the larva. Source: *U.S. Department of Agriculture.*

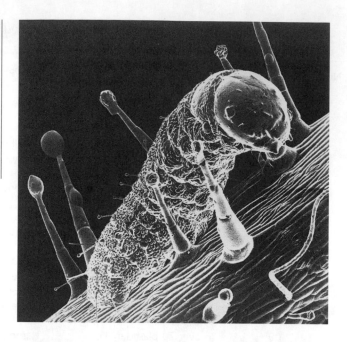

complex compounds probably evolved gradually, one enzymatic step at a time. Synthesizing a particular chemical so that it accumulates in the plant to a significant level has an associated "cost": photosynthate must be diverted for this purpose. However, if the cost of producing a defense compound is minimal and allows the plant that produces it to leave more offspring, then that plant has a greater evolutionary fitness than its nondefended colleagues. This can readily be demonstrated by partially defoliating plants or giving them a mild bacterial, fungal, or viral infection. Such plants grow much less vigorously and produce fewer seeds. Thus the appearance of defense chemicals or chemicals that attract insect pollinators provided a strong selection pressure and as long as mutations in the genetic material caused new enzymes to appear, the evolution of defenses against herbivores and pathogens was inevitable.

Plants have developed many different ways to defend themselves—so many that only a few can be discussed in this chapter. The waxy cuticle that covers leaves, flowers, stems, and fruits is the first line of defense of the plant. To be able to infect a plant cell, a pathogen must penetrate the cuticle and the underlying cell wall. Alternatively, a pathogen may enter through a natural opening (stomate) or through a wound.

The plant also has **constitutive** and **inducible** chemical defenses. The term *constitutive* means that the defense is present whether the predator attacks or not. In Chapter 4, we discussed the presence of cyanogenic glucosides in plants; these compounds are a good example of a constitutive defense system. Such systems are present in certain species at all times. In the following few sections, we discuss other constitutive systems, such as terpenoids, alkaloids, and phenolics. Many constitutive defense chemicals are produced by epidermal hairs that can trap and kill insect larvae (Figure 12.8). *Inducible* systems are those that are absent before a predator or fungal attack, but are induced when the attack occurs. This is how many defense mechanisms against fungi, bacteria, and viruses work. The plant will mount its defense as soon as it detects the attacker, and the attacker will try to escape detection so that it can invade the plant.

8 Secondary metabolites affect human metabolism in mostly unknown ways.

It should come as no surprise that many of the defense compounds that are effective against herbivores and pathogens are also effective against humans who are part of this evolutionary game. Agriculturally important food plants contain lower levels of these compounds. Their partial elimination was part of the crop domestication process. For example, wild lima beans have high levels of cyanogenic glucosides, but cultivated lima beans have very low levels. Cultivated plants are therefore more susceptible to attack by insects and pathogens, and this, in turn, necessitates the use of insect and pathogen control measures. Such control measures can take the form of chemical control (pesticides), biological control, or the genetic transfer of defense mechanisms from one species to another.

Our foodstuffs have the secondary metabolites that give plants greater evolutionary fitness. In some cases, these are removed after harvesting by elaborate procedures, such as the leaching of cyanogenic glucosides from cassava pulp discussed in Chapter 4. But in most cases, we eat these defense compounds. If they are proteins, they are inactivated by cooking or roasting, but most secondary metabolites survive those processes. The plants we use as stimulants (coffee, cacao, tea, and so on) or as spices are especially rich sources of these chemicals, and they affect us in ways that are largely unknown. For example, many secondary compounds (about half of those that have been tested) cause cancer in laboratory rodents when tested in the same way that we test industrial chemicals. Indeed, more than 99% of all the cancer-causing chemicals that we ingest are made by nature or arise during cooking and are not of industrial origin. The effects of these chemicals on the human body are only now being investigated. Not a month goes by without a report in the popular press that a certain vegetable is good for us because a specific secondary metabolite affects one of our metabolic processes. Recent reports show that celery extract lowers cholesterol in mice; that indole, carbinol, a chemical found in broccoli, inhibits the development of breast cancer; and that one clove of garlic each day lowers blood fat! It is important to remember that our normal diet includes hundreds of secondary plant metabolites that affect the biochemical reactions in the human body in many different ways with positive and negative outcomes simultaneously. We are largely ignorant of these outcomes and of the interplay of these hundreds of chemicals interacting with the thousands of biochemical reactions in our bodies.

9 Terpenes deter herbivores in many different ways.

Plants synthesize three major classes of secondary metabolites: terpenes, phenolics, and nitrogen-containing compounds such as alkaloids. Within each of these classes, there are many different chemicals and they all deter herbivores in different ways. Terpenes are a class of lipids that have as their building block a 5-carbon molecule similar to a small hydrocarbon called *isoprene*. The terpene molecules have 2, 4, 6, 8, or more of the isoprene units. Small terpenes are often synthesized and secreted by hairs on the epidermis of leaves and stems, and in this way they already advertise the toxicity of the plant to the insects that come close. **Essential oils**, the aromatic substances that give peppermint, basil, and sage their characteristic smells, are typical terpenes. These compounds are important commercially as food flavorings and for making perfume. The aromatic substances present in the resin that is exuded by wounded conifer trees are also terpenes. The synthesis of these terpenes is triggered when conifers are attacked by insects such as bark beetles that burrow between the bark and the wood.

Pyrethrins are another class of terpenes. Produced in the leaves and flowers of *Chrysanthemum* species, they show striking toxicity to insects and have negligible effects on mammals. Natural and synthetic pyrethrins are popular commercial insecticides because they have no effect on warm-blooded animals. Sterols and compounds with sterol-like structures are also terpenes. For example, some plants contain **phytoecdysones**, steroids that have the same effect on insect larvae as the insects' own molting hormone, α-ecdysone. Insect larvae that eat plants containing phytoecdysone do not molt properly, so they die. **Cardenolides** are extremely toxic molecules that consist of a terpene and a sugar linked together. These molecules, also called *cardiac glycosides*, are prescribed to millions of patients with heart disease. In small doses, they are useful drugs; in large amounts, they are strongly toxic. The extreme toxicity of oleander leaves and stems is caused by the cardiac glycoside oleandrin (Figure 12.9). In spite of their toxicity, oleanders are grown as ornamentals in many areas with a Mediterranean climate.

Yet another class of molecules that have a sugar half and a terpene half are called **saponins**. As their name implies, they are extremely good detergents, and they are toxic to mammals because they get in the bloodstream and cause rupture of the red blood cells. Structurally, saponins are so much like human steroids that saponins are used as the starting material for synthesizing steroids used for making birth control pills. **Gossypol**, another terpene, protects cotton plants against insects (Figure 12.9). It has been found to have excellent contraceptive properties for humans when taken at the right dosage. This is yet another example of the unexpected ways that these plant defense chemicals affect human metabolism.

10 | Tannins make plants unpalatable and indigestible.

The second major class of plant defense chemicals are the **phenolics**. Their building block is a ring of six carbon atoms with a hydroxyl (OH) group attached. They are derived from certain amino acids, through the loss of the amino (NH_2) group. We discussed phenolics earlier (Chapter 8), because they leak out of the roots and signal the *Rhizobium* bacteria that they have found the correct partner for a symbiotic association. **Tannins** are polymeric phenolics that have been used for centuries to preserve animal hides. When raw hides are soaked in solutions of tannins, usually made by extracting bark, the tannins react with the skin proteins (collagen). As a result, the skin proteins are impervious to attack by the microbes that otherwise would cause the hide to rot. Tannins can react with all kinds of proteins, and this reaction makes the proteins much less digestible because they are not as easily degraded by digestive enzymes. High levels of tannins in forage crops deter feeding by cattle. When cattle are browsing, they avoid plants with a high tannin content. Similar observations have been made with monkeys in their natural

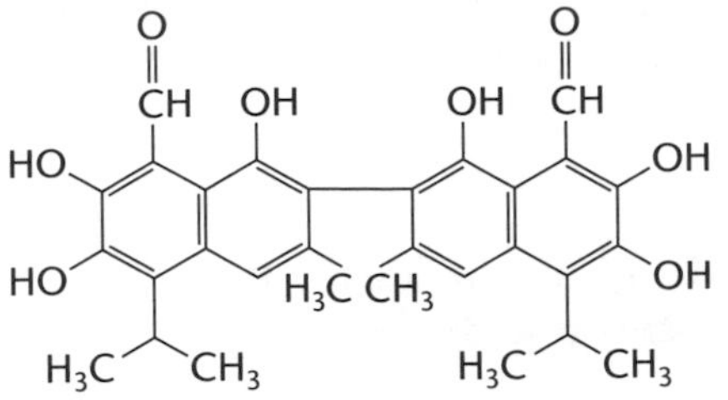

Gossypol

Gossypol is a terpene that is produced by hair cells on the epidermis of cotton plants. It protects cotton against insects and pathogens.

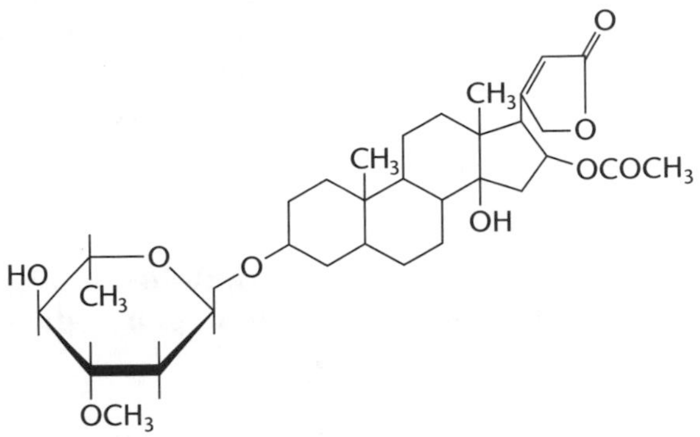

Oleandrin

Oleandrin, a cardiac glycoside present in the leaves and stems of oleander, a widely grown ornamental shrub.

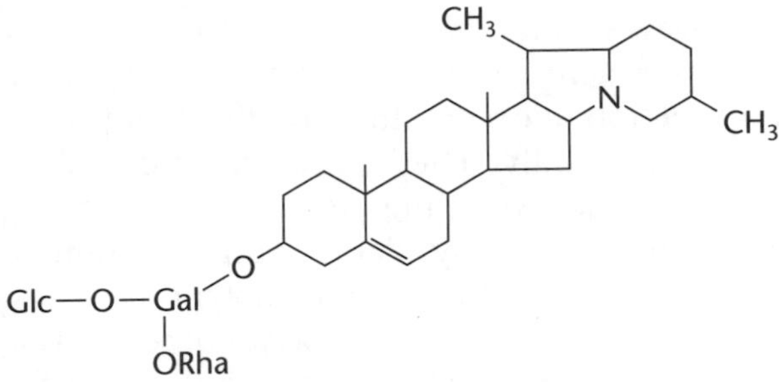

Solanine

Solanine, a toxic alkaloid produced by potato tubers. Potato tubers exposed to light produce substantial amounts of solanine.

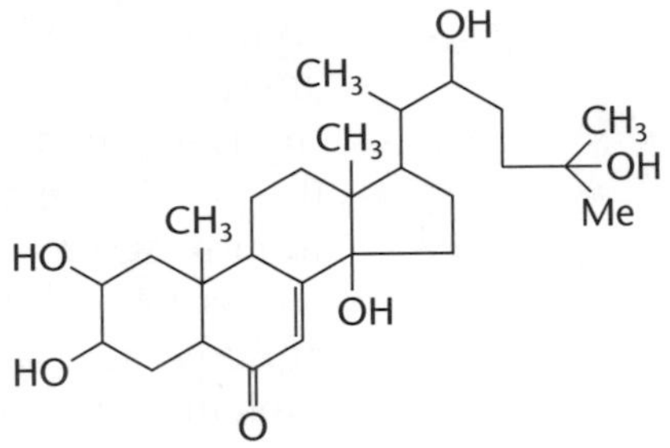

α-Ecdysone

Ecdysone, a molting hormone required by insects for normal larval development and synthesized in large quantities by certain plants.

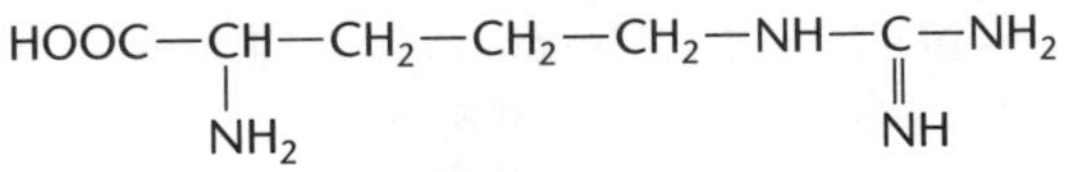

Arginine

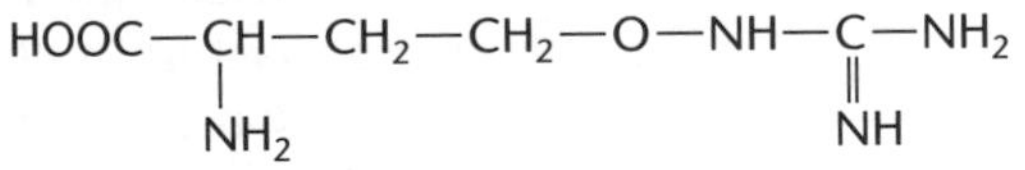

Canavanine

Arginine, a protein amino acid; and canavanine, an amino acid analog.
When this analog replaces arginine in proteins, the proteins are nonfunctional.

Figure 12.9

Some Plant Defense Molecules.

habitats. Although tannins are not necessarily toxic, they give the leaves an unpleasant, highly astringent taste. Some hunter-gatherers (see box on food sources of the Native Californians in Chapter 4) adopted practices to remove tannins from valuable foods (such as acorns) to make them edible.

Tannins are chemically related to **lignin**, another polymerized form of phenolics. Lignin is the material found in the cell walls of the conductive tissues of plants, that gives cell walls more strength; it is also responsible for holding water within the conductive tissue cells. Cellulose allows water to pass freely through the wall, but once the walls are impregnated with lignin, water can only pass through the holes that connect different cells, not through the walls themselves. When plants are invaded by pathogens (fungi and bacteria), they respond by making more phenolics, which are secreted into the wall and polymerize to form lignin. This accumulation of phenolics limits the infection because phenolics are toxic and the walls are reinforced by lignin.

11 | Alkaloids, unusual amino acids, glucosinolates, and cyanogenic glucosides help defend plants.

The third major group of plant defense chemicals all contain the element nitrogen (N); they are synthesized starting from the common amino acids used to make proteins. In this class of organic chemicals belong the **cyanogenic glucosides** that liberate toxic hydrocyanide gas when the cell structure is destroyed by chewing insects or invading fungi (see discussion in Chapter 4). **Mustard oil glycosides** (also called *glucosinolates*) are closely related to the cyanogenic glucosides. They are activated in the same way. The chemical compound and the enzyme that cleaves it and produces the active ingredient are stored in different cellular compartments. The disruption of cellular integrity is required for the activation reaction to occur. Mustard oils occur primarily in wild members of the mustard family. They cause irritation of the mouth, severe gastroenteritis, and diarrhea when ingested.

As part of a novel defense mechanism, many plants synthesize unusual amino acids that are not normally present in proteins. About 300 of these nonprotein amino acids are known to be present in different species of seeds. They usually closely resemble one of the 20 amino acids in proteins (Figure 12.9). When the seeds are eaten by insects, the protein-synthesizing machinery in the cells of the insects cannot tell the difference between the two types of amino acids, and these cells synthesize proteins in which the nonprotein amino acid is substituted for the "real thing." This substitution results in proteins that do not function properly, and as a result the insects die. The plants that synthesize such unusual amino acids are more discriminating; their own protein-synthesizing machinery can tell the difference between the two types of amino acids and only uses the appropriate amino acid for protein synthesis.

The most familiar class of plant toxins are the **alkaloids**, of which more than 5,000 different ones have been described. Because they affect the central nervous system, they are widely used in medicine (for example, morphine, codeine, atropine, and ephedrine), but also as stimulants or sedatives (caffeine, cocaine, and nicotine). Alkaloids have been used since ancient times for poisoning people—an extract of hemlock is said to have killed Socrates—and their toxic effects are well known.

Less well known are the teratogenic effects of alkaloids that are less obviously toxic. About 20% of all plants synthesize alkaloids and a significant number of deaths among range cattle are caused by browsing on plants that contain alkaloids. A number of alkaloids have recently been implicated in birth defects in cattle. These alkaloids are not toxic enough to kill, but if ingested over a long period of time, may cause congenital defects in the offspring. In humans, ingestion of solanine (Figure 12.9), the main alkaloid present in potato tubers, has been linked to a birth defect known as spina bifida; whether overeating of potatoes actually causes this condition has not been established.

12 | Inhibitors of digestive enzymes of insects constitute a major class of plant defense proteins.

Although the majority of defense chemicals are small molecules (secondary metabolites), plants also use proteins as part of their defense strategies. Some of these proteins are incredibly toxic, and a single molecule is enough to kill an entire cell. (There are about 10^{19}—10 with eighteen zeros after it—molecules in a gram of protein.) Other proteins, called *lectins*, bind to proteins in the gut and interfere with the absorption of nutrients in the small intestine (see Chapter 4).

An interesting class of defense proteins are the inhibitors of digestive enzymes that occur in many plants. They are present in many seeds, and are induced in leaves when insects start to chew on the leaves. The two main classes of inhibitors discovered so far are the amylase inhibitors and the protease inhibitors. Amylases are enzymes that digest starch in animals as well as in plants, and proteases digest proteins. When the action of these enzymes is inhibited, nutrition is impaired. Without digesting the proteins in their food, animals cannot obtain the amino acids necessary to synthesize their own proteins; without digesting starch, they will be short on energy, especially if starch is their major energy source.

Although the various amylases and proteases present in different organisms generally have the same function—splitting bonds between sugar residue in starch, and between amino acids in proteins, respectively—the individual amylases and proteases of different organisms are not all exactly alike. The protease and amylase inhibitors work by making a one-to-one complex with their respective enzyme. This tight binding of the inhibitor to the enzyme prevents the enzyme from doing its job in digestion. However, because the various amylases and the various proteases differ from one another, a given inhibitor may not inhibit all the amylases or all the proteases. Generally, the enzymes found in plants are not inhibited by the inhibitors found in plants. These inhibitors function (have evolved) to inhibit the digestive enzymes of insects and other animals.

The **amylase inhibitors** present in beans show an exquisite specificity for the amylases of certain bruchid beetles. The domesticated common bean (*Phaseolus vulgaris*) contains an amylase inhibitor that inhibits the digestive amylases of humans and other vertebrates as well as certain insects. This inhibitor provides an excellent example of the coevolution of animal digestive enzymes and plant defense proteins. Researchers believe that the primary function of the amylase inhibitor is to defend bean seeds against bruchid beetles (seed weevils); these insects lay their eggs on the dry seeds, and the larvae develop in the dry seeds. Different species of bruchids are found all over the world. For example, the mung bean weevil is an Eastern Hemisphere bruchid that thrives on cowpeas, mung beans, and other Eastern Hemisphere legumes. The Mexican bean weevil is the Western Hemisphere bruchid that thrives on the common bean, a Western Hemisphere plant. The amylase inhibitor found in the common bean does not inhibit the amylase of the Mexican bean weevil, but completely inhibits the amylase of the mung bean weevil. When purified amylase inhibitor obtained from beans is mixed in with otherwise acceptable cowpea flour, the mung bean weevil larvae fail to develop. The amylase inhibitor probably evolved from another protein to help protect the bean against the Mexican bean weevil and other Western

Hemisphere weevils. Subsequently, the amylase of the bean weevil evolved so that it was not inhibited any longer, resulting in a feeding preference of the Mexican bean weevil for beans (hence its name).

Protease inhibitors, the second group of proteins that inhibit digestive enzymes, are found not only in many seeds but also in fruits, tubers, and leaves. They may be continuously present at high levels, or they may be induced by wounding. Clarence Ryan of Washington State University discovered that when the leaves of tomato plants and potato plants are wounded by an insect that is feeding on the leaves, the plant cells start to synthesize copious amounts of several inhibitors that block the plant digestive proteases of the insect larvae. Interestingly, it is not just the wounded leaf that responds, but all the leaves, suggesting that a signal is being transmitted from the wounded leaves throughout the plant.

13 | Insects exploit the defense chemicals of plants to their advantage.

If plants are so good at defending themselves with chemicals, how can insects eat any plants? It is quite paradoxical that some insects actually thrive on high levels of defense compounds; that is why plant scientists have been slow in accepting the idea that secondary metabolites are defense compounds. The answer to this riddle is to be found in evolutionary adaptations and counteradaptations. Insects can evolve mechanisms to detoxify, sequester, or otherwise neutralize harmful chemicals. Once such a mechanism has evolved, the insect will be able to feed preferentially on the plant species that makes these chemicals, because other insects remain excluded.

The insect even sometimes uses the defense chemical as a food source. An excellent example of such an evolution is the feeding preference of a South American bruchid beetle for the seeds of a tropical legume vine called *Dioclea megacarpa*. This bruchid, *Caryedes brasiliensis*, is the sole predator on the seeds, which are loaded with the unusual amino acid L-canavanine. As explained earlier, such amino acids are analogs of the normal protein amino acids; after they have been taken up by the insect larva, the protein-synthesizing machinery of the larva will confuse the surrogate amino acid for the real one, with disastrous results. The *C. brasiliensis* bruchid has evolved several new enzymes that allow it to quickly break down canavanine, first into urea and then into ammonia, and to use the ammonia for synthesizing its amino acids. Thus a chemical that is poisonous to all other insects that might feed on this plant now becomes a source of dietary nitrogen for just one insect, which is now specifically adapted to this plant.

In some cases, the defense chemicals are even known to have become feeding attractants. Their odors tell insects where their food sources are. The glucosinolates already mentioned that are present in the mustard family, and give mustard its pungent flavor, are extremely toxic to some insects (and can cause discomfort in people), but insects that normally feed on the plants of this family require them, to "spice up" their food. Larvae of the cabbage butterfly that are caught in the wild will not eat from a bland artificial diet unless it is spiced up with glucosinolates.

Some insects store the bitter-tasting and toxic chemicals they eat in special organs, to defend themselves against their predators. The best-researched

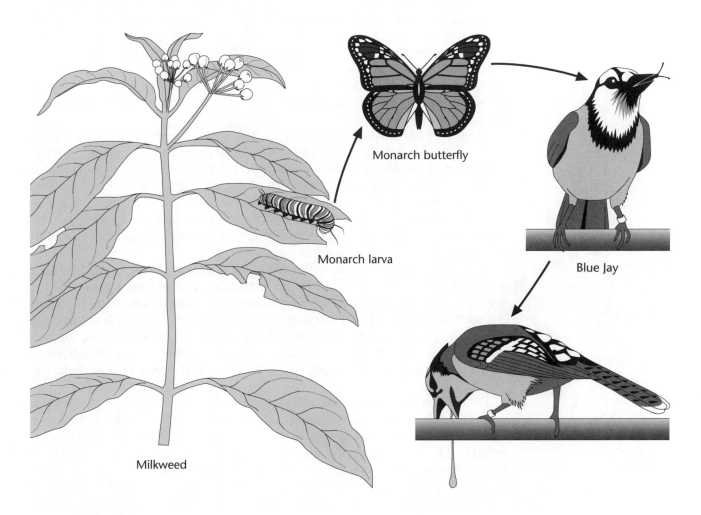

Monarch butterfly

Monarch larva

Blue Jay

Milkweed

Figure 12.10

Milkweed. Milkweed synthesizes cardiac glycosides that can be stored by monarch butterflies that feed on this plant.

example concerns that of the caterpillars of the monarch butterflies, which feed on milkweeds that produce toxic cardiac glycosides (cardenolides). In the course of evolution, the monarchs have become adapted to milkweed and use it as their sole food source. The caterpillars store the cardenolides, and when the adults develop and fly away, they still contain the bitter-tasting cardenolides. When blue jays eat the adult butterflies, they get a mouthful of the cardiac glycosides and vomit. The blue jays quickly learn not to eat these butterflies (Figure 12.10).

14 | **The hypersensitive response is a clever defense against pathogens.**

Pathogens such as viruses, bacteria, and fungi have various ways to breach the outer defenses—the cell wall—of the plant. They may infect a wound site, grow through a stomate, or simply try to dissolve the cell wall by producing enzymes that degrade the wall. Having gained entry, they will elicit a response from the plant, and the swiftness of that response determines if the invasion will be successful. When the response is swift, the cells that have been infected and the cells immediately adjacent to the infection site synthesize **phytoalexins**, chemicals that will kill the invader, as well as the plant cells in this localized area. This killing of the cells results in a small brown spot, called a *necrotic lesion*, on the leaves. Thus, the plant sacrifices a small group of cells to save itself. The generic term *phytoalexin* describes different chemical compounds. Some plant species use phenolics, other species use terpenes as phytoalexins.

The hypersensitive response includes a "strengthening" of the cell walls by the deposition of lignin and new structural proteins. In addition, a series of proteins are made that are secreted into the intercellular spaces and accumulate in the vacuoles. This set of proteins, called the "pathogenesis-related proteins," is produced regardless whether the pathogen is a fungus, a bacterium, or a virus. The production of these proteins appears to be a general response of the plant to infection. These proteins all inhibit the growth of pathogens in one way or another.

How does the plant know that it has been invaded? Fragments from the cell walls of the invader or of the plant can trigger the hypersensitive response. For example, if the invader secretes enzymes that degrade the plant cell wall, then the cell wall fragments immediately trigger the plant's defense response, unless the invader has found a way to neutralize the detection system. In that case, the plant mounts the same response, but much more slowly, and the invader (fungus, bacterium, or virus) spreads throughout the plant tissues.

The cell wall fragments are called **elicitors** because they elicit the defense response. What this means, in molecular terms, is that they activate the genes that encode the enzymes necessary for synthesizing phytoalexins and phenolics. The elicitor probably first interacts with a single protein in the plasma membrane, and this interaction sets off a cascade of events leading to the defense response. Being able to detect the invader rapidly enough hinges on a single gene in the plant and a single gene in the invader. The properties of the proteins specified by these genes determine whether a potential pathogen is indeed a pathogen and whether a plant is susceptible or resistant to its invasion.

The involvement of single genes explains why it is so easy for a crop that was made resistant to a specific pathogen through plant breeding, to become susceptible again. Breeding for disease resistance introduces a resistance gene, one whose protein allows the plant to detect the pathogen more rapidly so the plant can mount an effective response. However, a mutation in a single gene in the pathogen, whose pathogenicity was foiled by the breeder, will restore its pathogenicity. Now it will again escape detection and once more ravage the crop. If genes are the basis of resistance to pathogens, and if we understand how plants defend themselves, it may be possible to genetically engineer such defenses. Can we transfer the chemical defense system from one plant to another? Can we engineer a plant in such a way that the defenses are triggered by a chemical spray rather than relying on the early warning system of the plant? Can we isolate the resistance genes from wild relatives and quickly introduce these into crop plants by genetic engineering rather than slowly by plant breeding? Asking these questions already indicates which direction plant biotechnologists are moving.

15 | Many different plant proteins inhibit the growth of pathogens.

Plants respond to pathogen invasion by producing pathogenesis-related proteins. However, plants already contain many proteins toxic to pathogens. One group of proteins that inhibits the growth of many fungi are the chitin-binding lectins. Chitin is a complex carbohydrate polymer that makes up the exoskeleton of insects and the cell walls of many fungi. Many different plants, such as potatoes, tomatoes, stinging nettles, and rubber trees for example, contain proteins that bind to chitin. Because chitin is a carbohydrate, such proteins are called **lectins**, a general name for carbohydrate-binding proteins. When insect larvae are fed an artificial diet that contains such lectins, their development is inhibited. When these same lectins are added to the culture medium of fungi, the growth of the fungal hyphae is inhibited. The lectin appears to take effect at the tip of the hypha, where growth normally takes place and where chitin is being laid down to form the new cell wall. Chitinases—enzymes that break down the chitin cell walls of the fungi—also inhibit fungal growth. Several chitinases are among the pathogenesis-related proteins that are induced when a pathogen invades a plant (see earlier discussion).

The thionins are another class of proteins that inhibit the growth of fungi. These proteins were first found to be abundantly present in grains of wheat, and their function remained a mystery for quite some time. Now that scientists have begun looking for proteins that inhibit the growth of pathogens, it is likely that many more will soon be found.

Summary

A number of organisms thrive in agricultural ecosystems and diminish the yield of the crops people want to harvest. Weeds compete for nutrients, water, and light; insects and nematodes feed on plants; and fungi, bacteria, and viruses multiply within tissues and cells, often greatly diminishing the capacity for growth, and sometimes spoiling the crop itself.

The fact that most predators and pathogens only attack, or even prefer, certain species of plants implies that all other plant species have defenses. These defenses are mostly of a chemical nature. Chemicals called *secondary metabolites* are present in many plants and defend against predation and disease. Some 10,000 different secondary metabolites have been identified, and many more remain to be discovered.

Plants also use another approach to defend themselves. When attacked by a predator or invaded by a pathogen, a whole battery of genes is activated and the proteins encoded by these genes help defend the plant. In some cases, defense means death of a small group of cells, as in the formation of small lesions that limit the spread of fungi, bacteria, and viruses. A micro-organism becomes a pathogen when it can breach the defense perimeter of the plant *and* escape detection after it has invaded. Detection is followed by an immediate response that limits the spread, while lack of detection permits spread. It is nature's unending game of "spy versus spy," and we can exploit it to make plants more resistant to their predators and pathogens, using plant breeding, gene transfer technology, or a combination of the two.

Further Reading

Ashton, F., and T. J. Monaco. 1991. *Weed Science: Principles and Practices*. New York: Wiley.

Fox, R. T. 1992. *Principles in Diagnostic Techniques in Plant Pathology*. London: CAB International.

Gould, F. 1991. The evolutionary potential of crop pests. *American Scientist* 79: 496–507.

Johnston, A., and C. Booth. 1983. *Plant Pathologist's Pocketbook*. London: CAB International.

Parry, D. W. 1990. *Plant Pathology in Agriculture*. Cambridge, UK: Cambridge University Press.

Warwick, S. I. 1991. Herbicide resistance in weedy plants. *Annual Review of Ecology and Systematics* 22:95–114.

Zillinsky, F. J. 1983. *Diseases of Small Grain Cereals*. Londres, Mexico: International Center for the Improvement of Corn and Wheat (CIMMYT).

Chapter 13

Strategies for Pest Control

Farmers want to control the pathogens that diminish plant growth and therefore crop yield, as well as the predators that feed on the plants. There are several major strategies to achieve this goal. The most direct strategy is to kill the pests and the pathogens with chemicals. Although this approach was hailed by industrialists and agriculturalists alike, its many disadvantages are now becoming apparent. A more subtle approach is biological control: bring the population of pests back into equilibrium within the agricultural eco-system. A third approach is to make plants resistant to pests and pathogens. This approach necessitates gene transfer from other varieties, from more distant relatives or from unrelated organisms. Plant breeding and genetic engineering are strategies to achieve this goal. No single approach is likely to work best in the ever-changing interactions that occur in ecosystems, and integrated pest management (multiple methods) is rapidly gaining acceptance to keep the levels of pests and pathogens below the economic injury level.

Chemical Control

1 **Worldwide use of pesticides has increased dramatically since 1950.**

A wide variety of pesticides are used in agriculture in developed countries. For example, in the United States over 45,000 formulations with 600 different active ingredients are registered as pesticides. However, only about 50 active ingredients account for 80% of all pesticide use.

Contrary to popular opinion, chemicals have been used for hundreds of years to control pests. After learning of its use by farmers in the Far East, Marco Polo introduced pyrethrum, a "natural" insecticide produced by Asian chrysanthemums, into Europe. In the mid-1800s, it was discovered that Paris Green, an arsenic compound, effectively controlled the Colorado potato beetle on potatoes, and by the early 1900s, rotenone, pyrethrum, nicotine, kerosene, and compounds containing lead, arsenic, mercury, and sulfur were in common

Figure 13.1

The Amounts of Synthetic Pesticides (insecticides, herbicides, and fungicides) Produced in the United States from 1945 to 1985. Approximately 90% of what is produced is sold in the United States. The decline in total amount produced is in large part due to the 10- to 100-fold increase in toxicity and effectiveness of some new pesticides. Source: D. Pimentel and others (1991), Environmental and economic effects of reducing pesticide use, Bioscience 41:402–409. © American Institute of Biological Sciences.

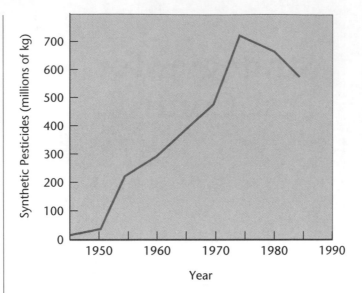

use. However, large-scale spraying of crops was not carried out until after 1950. Until that time, natural compounds extracted from plants (the secondary metabolites discussed in the previous chapter) and simple inorganic chemicals were the only chemical pesticides. For reasons that are not entirely rational, such chemicals are still approved for use in "organic" farming.

A significant breakthrough in chemical pest control occurred during World War II when the insecticidal properties of DDT were discovered and the herbicide 2,4-D (2,4 dichlorophenoxyacetic acid) was first manufactured. The spectacular successes of these two chemicals—in eradicating malaria (DDT kills the mosquitoes that carry the malarial parasite) and killing broadleaf (dicot) weeds in cereal crops—led to a search for more and better pesticides.

In their enthusiasm, agriculturalists gave little thought to what might eventually happen to the agricultural ecosystems in particular and the global environment in general because of the increasing number and volume of chemicals released in the environment. The book *Silent Spring,* written by U.S. biologist and science writer Rachel Carson and published in 1962, first awakened public opinion to the unintended effects pesticides were having on the environment. Most scientists now recognize that their colleagues were overenthusiastic in supporting the use of chemicals to eliminate pests and that a more sophisticated approach to pest control is needed.

In the past 40 years, total pesticide use increased enormously, especially in the United States (Figure 13.1) and other developed countries. The increase lasted until 1975, and then pesticide use declined somewhat. The decline was not caused by a reduction in the area of crop and rangelands on which pesticides are used, but by the use of chemicals that have greater specificity and are more toxic for their target pests. Some new pesticides can be used at one-hundredth the dosage per hectare that was used for some older pesticides. However, using small dosages of potent pesticides does not necessarily reduce the environmental and public health hazards compared to larger dosages of less toxic chemicals.

In spite of their many drawbacks, chemicals will continue to be used, if only because they appear effective to the individual farmer, who may be reluctant to try another approach. Worldwide preharvest crop losses have been estimated to be 13.8% from insects and other arthropods, 11.6% from

Figure 13.2

Annual Harvest Lost to Pests in the United States. Source: *Data from U.S. Department of Agriculture.*

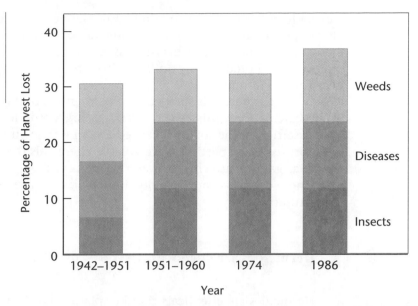

disease, and 9.5% from weeds, or a total of more than one-third of the crop. Pesticide applications do not reduce these values to zero, but nevertheless there are substantial benefits. The economic estimates of pesticide use in the United States suggest a return of $3 to $5 for every dollar invested in pesticides. Champions of chemical agriculture sometimes falsely pose the problem as a choice between chemicals and famine, and have data to prove how effective chemicals are. The data invariably compare yields with and without pesticides, but do not compare different ways of achieving similar levels of pest control. According to the estimates of David Pimentel (Cornell University), crop losses caused by various groups of pests (weeds, diseases, and insects) have remained more or less constant in the last 50 years, in spite of enormously increased pesticide usage (Figure 13.2). The reason is that we are substituting pesticides for good agronomic practices, such as crop rotation, that decrease the need for pesticides. In a study published in 1991, Pimentel calculated that reducing U.S. pesticide use by 50% would only increase food prices by a very small fraction (1%).

The dollar return value of pesticides cited earlier is the return to the farmer. However, the farmer does not pay the societal costs. The societal costs of pesticide usage—such as loss of fish, damage to water supplies, health care costs for the 20,000 people who suffer from pesticide poisoning, and especially the cost of regulating and monitoring pesticides—are borne by society at large and are estimated to be between $4 billion to $10 billion per year in the United States.

The destructive effects pesticides can have on the natural environment are well documented, and a number of pesticides are now banned in the United States, Canada, Western Europe, and other developed countries. These include DDT, dieldrin, endrin, lindane, and aldrin. Yet these same chemicals continue to be used in Third World countries, where the contradictions between the benefits and drawbacks of pesticide use are less acutely felt. When food is in short supply, or hard currencies must be earned from exports, people care less about contaminating their environment. In addition, they may have little information about the effects of pesticides on themselves or on the agricultural systems that produce their food. Although pesticide use in the Third World is still relatively low compared to that in the developed countries, it is increasing rapidly, and environmental disasters have already occurred. Nevertheless, this is a clear case where the need to increase food production conflicts with ethical concerns and the need to protect the environment.

When a pesticide is sprayed on a field, much of it (up to 99% according to some calculations) misses the intended target (the leaves of the plants) and either settles on the soil or is carried away by the wind to nearby fields, lakes, and forests. Many pesticides are volatile chemicals that move through the air, either as gases or attached to water droplets or dust particles. The net result is that these chemicals are now distributed all over the earth and are present in all natural waters.

2 | Herbicides are chemicals that kill plants.

Herbicides are chemicals that kill plants, and in the United States herbicides are used on about half the cropland (90 million ha), with corn and soybeans accounting for three-quarters of herbicide use. Herbicides kill plants either selectively or nonselectively. **Nonselective herbicides** kill all plants and some, such as chlorate, are used along railroad rights-of-way to keep tracks free of plants. Others, also referred to as pre-emergence herbicides, may be used to kill weeds on agricultural fields just before a new crop is planted. The former class is most useful if they persist and are not degraded, while the latter class must rapidly degrade if it is not to harm the crop that is planted a few days later. Degradation is accomplished, in part, by soil microbes, and the rate of degradation may depend on how often the herbicide is used. Constant use of a herbicide encourages the growth of those bacteria that degrade it.

Selective herbicides can be sprayed after the crop has emerged. To be useful, they must kill the weeds without affecting the crop. This selectivity is based on differential uptake of the herbicide by the leaves, differential inactivation of the chemical after it has entered the cells of the plant, differential binding at its site of action such as an enzyme, or by controlling the timing of application. The most useful herbicides kill specific weeds or classes of weeds (for example, all grasses) or protect specific crops (for example, all weeds are killed in a field of corn). Such exquisite specificity does not exist, however, and herbicides also affect crop plants and reduce yields.

The herbicidal qualities of specific chemicals usually emerge from screening procedures conducted by chemical companies. Chemicals that affect biochemical processes of plants that do not exist in animals often turn out to be herbicides. The biochemical processes affected by herbicides are usually essential processes in most plants. For example, the atrazine-type herbicides inhibit the light-induced flow of electrons in chloroplasts. Several herbicides inhibit the biosynthesis of amino acids in the chloroplast and the assimilation of ammonia. Some herbicides, such as 2,4-D, mimic the action of the plant hormone auxin and thus disturb normal growth control processes. The selectivity of herbicides is based on the finding that these processes are less severely affected in some plants than in others. The development of genetically engineered crop plants that are selectively tolerant of herbicides provides a new way of achieving a goal that industrial organic chemists could only dream about in the past: to kill all plants but one.

During the more than 40 years that herbicides have been used on a wide scale, a number of weed species have evolved resistance. For example, more than 20 weeds are now resistant to the herbicide triazine. In most cases, resistance is due to a single dominant gene, and results from a mutation in an enzyme that is the target of the herbicide. Herbicides often kill plants through inhibition of key enzymatic reaction(s); the herbicide must bind to the enzyme. If a mutation occurs so that enzyme binding does not occur, the enzymatic reaction will not be inhibited. If the mutated enzyme can still perform its normal metabolic function, the plant will resist the herbicide. The second way plants can become resistant is by developing a mechanism to detoxify the herbicide: the herbicide may become linked to a common metabolite (such as a sugar), or it may be modified slightly. The enzymes that carry out these detoxification reactions often arise as mutations from enzymes that catalyze already existing reactions.

It is particularly significant that herbicides affect bacteria in the same way as they affect plants, and that bacteria quickly develop the same strategies for coping with the herbicides: the target enzymes mutate, or detoxification mechanisms become operational. In bacteria, detoxification can mean breakdown of the herbicide, or partial modification, with the modified herbicide remaining in the environment. This is what happens to most herbicides and other chemicals that enter the ecosystem: they are partially or completely degraded by bacteria.

In addition to selectivity, certain herbicides have the added advantage of low persistence. In annual cropping systems, low persistence is important. If the field has a cereal–legume rotation, residual herbicide in the soil from the cereal crop could harm the legume if enough is taken up by the plants. Several studies have shown that 2,4-D disappeared within days after its application to a crop, and that 2,4,5-T lingered several weeks. Breakdown is caused both by the plants and by the soil micro-organisms. The breakdown products (various chlorinated hydrocarbons) may remain in the soil for a much longer time however, and their effects on plants, animals, and micro-organisms are unknown. Some herbicides, such as picloram and simazine, can remain in the soil for more than a year. Most desirable are those herbicides, such as glyphosate, that can be easily broken down, by the plants or in the soil, into products that can enter the normal metabolism of the cell. Society's recent concern with the protection of the environment is causing the chemical companies to search for such herbicides.

3 | Chemicals that kill insects, nematodes, fungi, and bacteria are often toxic to humans.

Chemicals that kill insects, nematodes, fungi, and bacteria are widely used in the United States and other industrialized countries, and their use has increased dramatically in the past 40 years. With the banning of certain long-lived pesticides and the appearance of new ones, there is a general shift toward pesticides with lower persistence in the environment and greater target specificity (more toxic to the organism a pesticide is intended to kill).

Fungicides are usually applied as "dusts" to crops growing in the field, or as a coating on the seeds before they are planted. To be effective, fungicides generally must be present on the plant before fungal spores land on that

Box 13.1

Reporting the Benefits of Pesticide Control Programs

How should the benefits of pest control programs be reported? To show how useful pesticides are in preventing crop losses, the makers of pesticides often compare yields with and without pesticides. Such figures always show that pesticides are extremely effective. For example, in a rotation of corn, soybeans, and wheat, the yields of these three crops were measured with or without herbicide application. The following yields (in kg per hectare) were obtained:

	With Herbicide	Without Herbicide
Corn	7645	6270
Soybeans	3025	2420
Wheat	2800	2750

So even when crop rotation was used to minimize the effect of weeds and other pests, herbicides still increased corn production by 22% and soybean yields by 24%. The effect on wheat yields was minimal. Although this is a sound experiment, no attempts were made to find out if other methods of weed control (such as intercropping) might have been equally effective.

In a study conducted at the University of California on weed control in cauliflower, cucumbers, lettuce, and peppers, one single hoeing in cauliflower, lettuce, and cucumber, combined with half the usual dose of herbicide, gave the same weed control (and yields) as a full dose of herbicide without hoeing. In this case, an alternative was tried, but no cost estimates were included in the report to advise readers of the economic effect.

In an integrated pest management program, reducing costs is often the goal. The "experiments" are often not quite so scientific, and the data often consist in a comparison of an IPM farm with other farms in the same area that use conventional pesticide treatments. As an example, the production of processing tomatoes in California on the Kitamura farm is cited in a study by the U.S. National Academy of Sciences. The Kitamuras use pest-resistant tomatoes and plant them only in fields where tomatoes have not been grown for six years. They withhold water during the last 40 days before picking to minimize the spread of pathogens (fungi). They monitor pest populations themselves by collecting 100 leaves at weekly intervals and counting the eggs of tomato worms (*Heliothis zea*). If there are more than three to four eggs per 100 leaves, they make a single pesticide application. The "bottom line" is that their 70 hectares achieves the same yield as on other farms in the county, the percentage of tomatoes that show damage is as low or lower (1–2%) as on other farms, and they save about $7,000 each year on pesticides (they spend $1,500, whereas others spend $8,500 for a farm of the same size). Because of their high management skills, they can grow processing tomatoes with a minimum of pesticides, and do it more cheaply than others can with pesticides.

plant, for they act by preventing spore germination and the initial growth of the fungus. Therefore, as new plant organs form, the crop must be redusted to coat these new organs with fungicide. This need for reapplication can make the use of fungicides quite expensive. The most widely used fungicides are relatively simple compounds containing copper, sulfur, or mercury. Applied to cereal crops, they could prevent such diseases as wheat rust and rice blast.

However, because these dusts are toxic to people, the crops must be thoroughly washed before marketing to remove the fungicide, and this is impractical for cereal crops. Most fungicides are therefore used on fruits and vegetables. For example, 95% of all grapes and potatoes are treated with fungicides, but fungicides are not used on corn or wheat.

Nematocides kill nematodes, but these chemicals are often so toxic that they kill most living organisms. Several nematocides, such as methyl bromide, are gases. They can be used quite effectively in greenhouses, where nematode infestations are a real problem. When they are used in open fields, the ground must be covered with thin plastic to keep the gas from escaping, and the crops (usually vegetables) are planted in holes in the plastic sheets. These plastic sheets also prevent water evaporation from the soil and keep weeds to a minimum. Because of their toxicity and other environmental effects—methyl bromide contributes to the destruction of the earth's ozone layer—nematocides currently in use will be phased out in the United States in the near future. So new ways will have to be found to control nematodes.

Since the insecticidal properties of DDT were discovered during World War II, synthetic **insecticides** have played a major role in saving millions of human lives. By killing insects that carry human disease agents, such as those that cause malaria and schistosomiasis, these pesticides have considerably lowered the number of deaths caused by these diseases. The realization that pesticides have saved so many lives was largely responsible for early enthusiasm for chemical pest control.

Most synthetic insecticides fall into two categories, according to their chemical structure: (1) *chlorinated hydrocarbons* (such as DDT, aldrin, and chlordane) or (2) *organophosphates* (such as malathion and parathion), which are chemically similar to nerve gases. Both are usually sprayed on the crop plants and are absorbed into the plant. When the insect pest starts to feed on the plant, it ingests the pesticide and is killed. It may also absorb the poison from the plant surface. Because chlorinated hydrocarbons are not easily degraded, multiple applications are not necessary, and one spraying often protects a crop for the entire growing season. Both types of insecticides interfere with the normal functioning of the nervous system of the insects, causing convulsions, paralysis, and death, but do not harm the plants.

Chlorinated hydrocarbons have two properties that cause them to accumulate in the organisms at the top of the food chain. They are much more soluble in fatty tissues than in water, and they are not readily broken down. As a result, they are neither excreted nor metabolized once they have been taken up into the body, but continue to accumulate in the body if the food source is contaminated. Studies conducted before 1972, when DDT was banned in the United States, showed that the concentration of DDT in the tissues of different organisms increased as it moved up the trophic levels of the food chain. For example, the DDT content of plankton is 250 times greater than that of the lake water in which the plankton lives. The small fish that feed on the plankton have 10 to 50 times greater concentrations of DDT in their tissues, and the larger fish and the aquatic birds that feed on the small fish have 20 times as much again. Thus, the concentration in their tissues is 100,000 times as great as that in the lake water.

These high concentrations of DDT have detrimental effects: the pesticide interferes with the calcium metabolism of some birds, resulting in fragile eggshells and reproductive failure, and in salmon it affects the nervous

Figure 13.3

Spraying Pesticides Without Protective Clothing. Although pesticides protect plants from insects and pathogens and may increase yield, there are also definite hazards associated with their use: contamination of the soil or the water in the rice paddies—as shown in this photograph of a Liberian field worker spraying rice plants—and toxic side-effects resulting from handling and spraying of these chemicals without protective clothing. Source: *Courtesy United Nations, FAO/Tortoli, Jr.*

system. Although long-term effects on humans have not been studied, the known environmental side effects of DDT led to a severe restriction of its use in the United States in 1972 and to its replacement by the nonpersistent organophosphates. However, although the organophosphates are not persistent, they are extremely toxic to humans and this is their principal drawback.

Considerable human **health risks** are associated with the use of these potent synthetic insecticides. Although many consumers are concerned about pesticide residues on the produce they buy in the store—at least in developed countries—that is probably not the major health problem associated with pesticide use. Every year nearly 1,000 cases of pesticide poisoning are reported in the United States alone, and two-thirds of these are caused by organophosphates. The actual incidence is undoubtedly much higher. One study of field workers in Tulare County, California, found that only 6% of the workers showing symptoms of pesticide poisoning had actually notified health officials. These data suggest that many more people than reported suffer some noticeable pesticide effects each year. The people most likely to be affected are the workers who mix insecticide solutions, load them into airplanes, and serve as spotters on the ground to guide the airplanes. Field hands who work in the sprayed fields and greenhouse workers are also likely to contract pesticide-related illness. In developing countries, the number of people who suffer pesticide-related illnesses is equally high and probably rising rapidly as the use of pesticides grows (Figure 13.3). Although pesticide containers carry clear warning labels, these are not always printed in a language the users can read (if they can read). Safety standards are often less enforced, and safety precautions less adhered to, by pesticide users in developing countries.

4 | **Resurgence, secondary pests, and resistance are problems associated with chemical pest control.**

The **resurgence** (bounding back) of the pest population, after the crop has been sprayed, occurs naturally because the food supply—the crop—is still present. Resurgence is often fast because the pesticide killed the predators of the target pest along with the target pest itself. For example, when potatoes are sprayed with guthion or methoxychlor to kill the green peach aphid, the pesticides also kill lady bird beetles and lacewing larvae, two organisms that prey on green peach aphids.

In some cases, the causes of resurgence are even more complex. When the population of a pest species falls to a very low level—as a result of spraying, for example—the population density may be too low to maintain the normal populations of its pathogens (viruses, bacteria, fungi). As the remaining individuals start to multiply, they are not held in check by their normal diseases, and the population increases faster than it otherwise might have.

Secondary pests are created when the populations of insects that did no damage start to increase because their natural enemies have been destroyed. For example, spraying of pesticides on cotton fields to kill the destructive boll weevil also killed the natural enemies of the cotton budworm and the cotton bollworm, which now became major pests of cotton. Nearly all the species of insects now considered serious pests in California were not pests before the days of chemical pest control. They became pests as a result of chemical pest control measures. Sometimes insects are kept in check by fungal diseases, and fungicide spraying can cause insect pest numbers to rise, as happened with velvet bean caterpillars and cabbage loopers in soybean fields.

The emergence of **resistant pest strains** is another major consequence of chemical pest control, although resistant pest strains also arise regularly when genetic pest control (new resistant crop strains) is used. Any regimen that causes extensive mortality and acts on a large population of organisms is a strong selective agent. Already more than 600 species of insects, weeds, plant pathogens, and nematodes have evolved resistance to chemical pesticides (Figure 13.4). In many cases, the pests have evolved resistance to more than one pesticide. Development of resistance is unavoidable and as natural as evolution (see Chapter 10). The emergence of pest resistance adds considerably to the financial cost of developing chemical pesticides. In many cases, effective pesticides are not immediately available, and crops are devastated by the resistant pest.

The history of cotton production in the Cañete Valley in Peru gives a good insight into the dangers of the chemical pesticide treadmill. In the 1940s, farmers started controlling insects with arsenic- and nicotine-based insecticides. By 1949, the emergence of resistant insect strains caused crop failures. The introduction of DDT allowed effective control at first, but resistant insects and record crop losses occurred in 1956. With government help and advice, the farmers drastically reduced pesticide applications and introduced many new cultural practices that control insect pests.

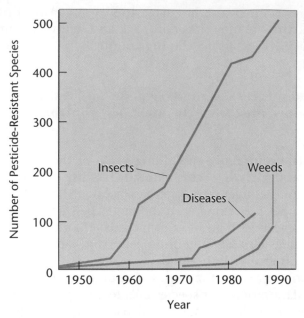

Figure 13.4

Pest Species Resistant to Pesticides. *Hundreds of crop pest species have become resistant to pesticides since synthetic chemicals were first used on a large scale for controlling agricultural pests in the 1940s. Insecticide resistance became a problem first, and today more than 500 insect species are known to be resistant to one or more insecticides. Resistance to pesticides has also developed in plant pathogens, primarily in fungi exposed to systemic fungicides. Genetic resistance to herbicides began to appear among weeds in the 1960s and has been reported in at least 84 species to date.* Source: *After G. P. Georghiou (1986), The magnitude of the resistance problem, in E. H. Glass, ed.,* Pesticide Resistance, Strategies and Tactics for Management *(Washington, DC: National Academy Press), pp. 14–43.*

Biological Control

5 | **Biological control relies on the use of predators, parasites, and pathogens of pests.**

   A crop field is an ecosystem in which various organisms interact, and the population density of a pest is determined not only by its own biology (rate of reproduction) and its food source (the crop), but also by the predators, parasites, and disease organisms that attack it. The aim of biological control is to lower permanently the population density of the pest by introducing natural enemies (Figure 13.5). If the strategy is successful, the population of the pest will fluctuate around a mean that is below the economic injury level of the pest population. Such a lowering can often be achieved with perennial crop plants that are part of a stable agricultural ecosystem (see Chapter 16), or in other established managed systems (forests, rangeland), but is more difficult with annual row crops. Other tactics, such as the periodic liberation of natural enemies reared in the laboratory, or the conservation and encouragement of natural enemies already in the field, may be more appropriate.

   Many pest species have been introduced from other geographical regions and have become pests because they had no natural enemies in their new habitat. Classical biological control is therefore based on exploration of foreign lands to find natural enemies. Once the enemies are identified, they are

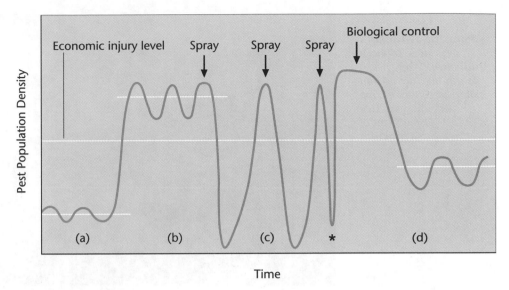

Figure 13.5

Pest Population Density of an Agricultural Ecosystem. (a) *Before agriculture, the population fluctuates around an equilibrium level.* (b) *After agriculture begins, the additional food for the pest that is provided by the crop allows the pest to establish a higher equilibrium level. Occasionally the pest population rises above the economic injury level, causing damage to the crop. When the crop is harvested, the pest populations fall again.* (c) *When pesticides are introduced, they kill the pests but also many of their natural enemies; the population of the pest thus rises again until the pesticide is reapplied. Occasionally, pesticide-resistant pests appear (*), causing the pest population to rise more rapidly than before.* (d) *When integrated biological and chemical pest control is introduced (arrow), a new equilibrium well below economic injury is established.* Source: *Adapted from R. Smith and R. van den Bosch (1967), in W. Kilgore and R. Doutt, eds.,* Pest Control *(New York: Academic Press).*

cultured in the laboratory and released into the environment with the expectation that they will suppress the pest population and that a new equilibrium will be reached (Figure 13.5). Such control measures can be compared to inoculations against diseases. They are meant to control the disease and prevent new outbreaks. The natural enemies include predators (insects that eat other insects), parasites (nematodes that parasitize insects), and pathogens (fungi, bacteria, and viruses).

Biological control has actually been practiced for centuries, although the term itself was first introduced in 1914, by a German plant pathologist. Its growing popularity derives from its record of safety. Although biological control relies heavily on introducing "foreign" (nonnative) organisms into the environment, no such introduced organism has ever become a pest itself or caused a measurable negative effect on the ecosystem into which it was introduced.

The classical example of a biological control strategy is the eradication of the prickly pear cactus in Australia. After this cactus was introduced in Australia in 1890, it grew unchecked because it had no natural enemies. By 1920, the cactus had become a serious pest, covering more than 60 million acres of arable land and rendering it useless. A natural enemy, an insect from Argentina, was introduced in 1925 and cactus populations were decimated within ten years. A one-time introduction was sufficient to establish a new equilibrium.

Some 10–15 species of other weeds are now controlled partly or completely by insects or by plant pathogens that are released because of their ability to decimate weed populations. This approach appears to be most

Figure 13.6

A Parasitic Wasp Lays Eggs in a Tobacco Budworm. Wasps, such as this one, are used in biological control of crop pests. Source: *U.S. Department of Agriculture.*

successful on rangeland as well as in other natural ecosystems that have been invaded by weeds. The target weeds for biological control are generally introduced plant species that are not consumed or parasitized by indigenous insects or micro-organisms. Such introduced plants often come to dominate the natural ecosystems they invade, and exclude native plants and animals from their habitat. By introducing insects or pathogens from other countries or regions, it is often possible to control the weed. When the new biological agent is introduced, its population will skyrocket, but as the weed is exterminated the population of the introduced pest will decline again and the two will remain in balance.

Natural enemies of insects are often other insects (Figure 13.6). In Barbados, the larvae of the sugarcane borer caused considerable losses of the sugarcane crop, the backbone of the island's economy. Introducing an insect parasite from India reduced the annual losses, and the problem became manageable (see Table 13.1).

A dramatic new example of biological control comes from sub-Saharan Africa where a "new" pest, the cassava mealybug, attacked cassava, the primary staple for 200 million Africans. In the early 1980s, the mealybug caused enormous crop losses. Researchers at the International Institute for Tropical Agriculture in Nigeria, searching for enemies of the mealybug in its South American

Table 13.1 — **Decrease in damage by sugarcane borer in Barbados after a parasite was introduced in July, 1966**

Percentage of Cane Infected	Year
15.0%	1966
12.7	1967
9.7	1968
8.0	1969
5.9	1970
4.8	1971

Source: F. J. Simmonds (1972), Biological control of the sugarcane borer in Barbados, *Entomophaga* 17:251–264.

homeland, found a small parasitic wasp that kills mealybugs by laying its eggs in the mealybug's body. Periodic distribution of the wasp in the 28 African countries of the "cassava belt" is now the responsibility of the Institute, and the dollar return value of the program has been calculated at $150 to $1.

Entomopathogenic fungi (fungi that cause diseases of insects) provide an excellent example of the potential for biological control. At least 750 different species of such fungi have been identified and attempts to use them as biological control agents of insects began in the late nineteenth century. Some research has been done in the past 20 years on these fungi, but few are used extensively for biological control. Variable test results have plagued the development and use of such **mycoinsecticides**, as the formulations of entomopathogenic fungi are called. Their effectiveness appears to be very dependent on the physical conditions of the field (such as humidity and intensity of the sunlight) and the density of the insect pest. The U.S. Department of Agriculture is trying to develop the fungus *Entomophaga grylli* as an effective control agent for grasshoppers. Grasshoppers and locusts have plagued humanity since the beginning of recorded history. They are the primary group of herbivorous insects that lower the productivity of rangelands worldwide.

A well-established biological control method involves the use of the bacterium *Bacillus thuringiensis,* or *Bt*, which kills the larvae of lepidopteran insects (butterflies and moths), such as the cabbage looper. Different formulations of Bt are widely used in the vegetable fields of Florida and are available to home gardeners. Overuse of Bt has already led to the emergence of pests that resist Bt, showing that the emergence of resistant pest strains is a problem associated with all types of control strategies.

Nematodes and viruses that attack insects also have considerable potential as biological control agents, but much more research is needed to produce cost-effective, usable formulations that can be marketed to farmers. Because biological control is species-specific rather than general, the commercial market for any one method is much smaller. A chemical that kills all or most weeds before corn is planted can be sold all over the world, but a biological agent that kills one particular species of weed may only be needed in that region where the weed thrives.

6   Biological control may involve the use of the pest against itself.

Knowledge of insect biochemistry has permitted the development of highly specific insecticides and attractants. Like other organisms, insects produce **hormones** that regulate their growth and development, and each hormone usually acts at one stage in the insect's life cycle. If a solution containing a few parts per million of the hormone is sprayed on the insect at another developmental stage, it will die. For example, a juvenile hormone is present in the insect while it is immature. This hormone can be synthesized in the laboratory, and when it is applied to an insect in the middle of its normal development, the insect remains "juvenile" and dies. The use of such hormones is specific to insects, and therefore they are very potent, biodegradable, and extremely specific pesticides. It was thought at first that insects could not and would not evolve resistance against their own hormones. However, this has proven false: insects can evolve so that they do not recognize their own hormone when it is used as an insecticide.

A **pheromone** is a volatile substance secreted in very small amounts by one animal, to regulate the behavior of other individuals of the same species. For example, in many species of insects a virgin female signals her readiness to mate by emitting a small amount of a species-specific chemical sex attractant. The males in the area detect this substance and follow it to its source.

Over two dozen insect sex attractants have been isolated and identified in the last decade, and research is now in progress to find out if these natural chemicals can be used to control these insects. One method is to bait traps with small amounts of the pheromones to try to catch the males before they copulate. The problem with this approach is that one must catch almost all the males if the program is to be effective. Another approach is to spray a large area with pheromone so that the smell of virgin females fills the air, leaving the males unable to track down a female. These insect control methods are species-specific, and only minute amounts of natural, readily degradable chemicals are used; as a result, they cause a minimum of environmental perturbation. One of the main uses of these chemicals is in pheromone-baited traps, to determine the population density of a pest. The number of individuals caught in the trap over a 24-hour period is an indication of the population density. This information is used in integrated pest management programs (discussed in the next section).

The use of **sterile mating partners** is a method of biological control in which an insect is used against itself. This technique has been used successfully to eradicate the screwworm (the parasitic larva of the screwworm fly) from the U.S. Southwest. These larvae are a serious cattle pest, and were causing as much as $40 million damage per year. The females of this fly mate only once during their lifetime, and if they mate with a sterile male, they do not produce fertile eggs. The USDA initiated a massive program to rear the flies in the laboratory, irradiating them to render the males sterile, and releasing the sterile males into the natural breeding grounds of the fly. Within a few years the screwworm problem was brought under control in the target areas. Approximately 150 million sterile males were released each week along an 1,800-mile front from the Gulf of Mexico to the Pacific Ocean, at an annual cost of $5 million, which was far less than the economic damage caused by the screwworm.

A program is underway in Africa to use the sterile-male technique to control the tsetse fly, which is a serious pest to cattle and humans. Chemical control of the tsetse fly was reasonably successful in the past, but the fly populations have become resistant to insecticides.

A major limitation to expanding these sterile insect release programs is that there is no way to rear only males in the laboratory. Males and females are reared together and released together after the males have been sterilized by X-irradiation of all the insects together. This doubles the cost of rearing the insects, and diminishes the effectiveness of the sterile males when they are released together with females. Using gene transfer technology, it should not be too difficult to generate a mutant insect line in which a gene can be activated that disrupts the development of females. This would allow only males to be released into the environment.

Another way of using a pest against itself is to "**immunize**" plants against a virus disease. The mechanism of protection through immunization is very different in plants and in animals, although the general principle is the same. When plants are exposed to a nonvirulent strain of a virus, or to a virus that is pathogenic on a different plant, they become resistant to subsequent

Table 13.2	Strategies of an IPM program

Cultural and Physical Control Tactics
Cultivation or tillage
Sanitation (removal of crop residues)
Planting and harvesting dates
Resistant varieties
Crop rotation

Chemical Control Methods
Pesticides
Behavior-disrupting pheromones

Biological Control Tactics
Mass release of natural enemies
Conservation or augmentation of natural enemies
Release of sterile mating partners
Release of pathogens or nematodes

Monitoring
Monitoring crop damage and pest populations is an important component of an IPM strategy

infection with a virulent virus. This phenomenon, called *cross-protection,* was discovered in 1929, and is used commercially. In Brazil, as noted earlier, tristeza virus is controlled in citrus trees by inoculating the trees with a mild virus strain. This approach protects the trees against more severe strains. This phenomenon has direct applications in plant genetic engineering.

Integrated Pest Management, Resistant Plants, and Cultural Practices

7 | **The goal of integrated pest management is to minimize the use of chemical pesticides.**

Born out of conflicting views of biological control advocates and chemical control specialists, **integrated pest management (IPM)** is an ecology-based strategy that relies primarily on natural factors such as pathogens, predators, parasites, weather, cultural practices, and resistant varieties. Chemicals are used, but only as a last resort (Table 13.2). It can be viewed as a pesticide minimization strategy that aims to reduce the use of pesticides, or as a pesticide preservation strategy that aims to prolong the useful life of pesticides. The goal is to optimize pest control economically, as well as ecologically and environmentally. So far, IPM has been applied primarily to insect pests.

Although many IPM programs have as their primary objective to maximize economic profit on the farm, proponents of IPM see many other goals. These include, among others, protecting the environment by minimizing pesticide use, increasing regional cooperation among farmers, substituting imported inputs (pesticides) with local skill and inputs (especially in Third World countries), developing new products, and prolonging pesticide life span.

The key to IPM is an understanding of the population dynamics of the pest in a particular region, given a particular set of conditions. There is no general recipe for pest control, but the pest level is determined at specific intervals. Data gathered in the field (age of plants, temperature, humidity, pest level, levels of certain pathogens that attack the pest, and so on) and other data concerning cost of the pest control intervention, price of the crop, and expected crop gain are all fed into a computer program that then

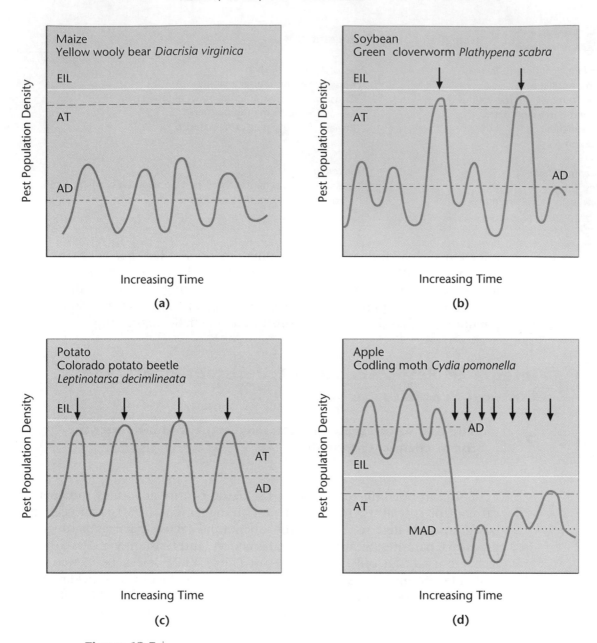

Figure 13.7

Economic Injury Level (EIL) and Action Threshold (AT) for Representative Pest Densities. AD = average density, MAD = modified average density, and arrowheads = pest control intervention. Source: After Stern and others (1959), The integrated control concept, Hilgardia 29:81–101.

prescribes a treatment based on past experiences. The real question is whether the pest, at this level of population density and in this environment, is likely to cause economic injury. The level of pest density at which action is taken is called the **action threshold** and is usually just below the economic injury level. Examples of fluctuations in pest density around an average density (AD) and its position relative to the action threshold (AT) and the economic injury level (EIL) are shown in Figure 13.7.

That the understanding of population dynamics is crucial to pest control is shown by the work of Larry Pedigo and his collaborators from Iowa State University, who studied outbreaks of the green cloverworm in soybean fields in the state of Iowa. This pest, like many others, does not overwinter in Iowa but arrives each year as an immigrant from the southern United States. If green cloverworm arrives too late in the spring, its population never builds up and the soybean crop suffers no damage. If it arrives too early in the spring, a

Table 13.3	Relative yield of three vegetables with different types of weed control		
		Cucumber	Lettuce	Cauliflower
Full dose of herbicide		82	74	79
Half dose of herbicide		73	77	90
Half dose of herbicide plus one hand hoeing		99	110	—
Hand hoeing all season		100	100	100
No treatment at all		50	—	89

local fungus will attack it and cause the cloverworm population to collapse before it can cause harm. However, in the years when it arrives at just the right time, green cloverworm can cause damage. This example makes clear that pest control programs should be fine-tuned to take into account each particular situation, and that there is no general prescription even for controlling a single insect species.

Ideally, the farmer should determine the population density level of an insect and this can often be done, especially with flying insects, with hormone-baited traps. However, with other insects this may be next to impossible or prohibitively expensive. For example, corn rootworm does considerable damage to corn grown in continuous culture in the upper U.S. Midwest. However, counting rootworms in the same year as the corn is growing is too expensive, and using the previous year's damage data is unreliable. Thus farmers must base their strategies on the assumption that damage will occur.

Integrated pest management is not only applicable to insect pests, as discussed, but also to weeds. Most vegetable growers control weeds by mechanical cultivation, the spraying of herbicides, or hand hoeing, or a combination of these methods. Each of these treatments controls weeds for a while; however, when should a farmer repeat the treatment, and which method is the best? Integrated pest management asks a different question: "What is the economic injury level, and how many weeks must a field be weed free after the plants emerge from the soil?" In a recent study carried out at the University of California, yields of cucumber, lettuce, and cauliflower were measured in plots that were treated with herbicide, weeded with a hand hoe at biweekly intervals, or treated with a reduced level of herbicide and hoed only once (Table 13.3). The results showed that weed control in cauliflower was almost unnecessary. Cauliflower grows so fast that it covers the ground quickly and the weeds don't stand a chance. Lettuce and cucumber need two weed-free weeks. After that time, weed control is not necessary anymore; once the plants touch each other and cover the ground, weeds can not grow. Some vegetables that do not grow as fast, such as bell peppers, need eight weed-free weeks or production is reduced.

8 Classical breeding has kept the crop one step ahead of the disease.

Some 85 years ago, the first rust-fungus-resistant commercial wheat was produced by crossing a resistant wheat with a susceptible wheat. This success began the practice of introducing pest and pathogen resistance genes into

Figure 13.8

Alternate Rows of Beans That Are Resistant and Susceptible to the Rust Fungus.
Source: *U.S. Department of Agriculture.*

crop plants, first exclusively by breeding and most recently through genetic engineering in combination with plant breeding.

The most important sources for disease-resistant crop varieties are the relatives of modern crop plants that grow wild in nature, because some of these plants must be disease resistant for the species to survive. The resistance characteristic can be bred into crop varieties by backcrossing. Unfortunately, new pest strains can arise that can infect the resistant crop plants, so the search for resistant plants is never finished. Pests evolve in parallel with their hosts, and a given pest-resistant variety of plant seldom remains resistant for more than three or four years. The shortness of the useful life of a pest-resistant variety is, in part, the result of using continuous monoculture (the same crop, year after year). This approach to agriculture greatly increases selection pressure on the pest, and resistant pest strains evolve sooner than in an agricultural system that uses crop rotations. Because of their economic importance, a major effort in plant breeding has been to develop pest-resistant varieties of the three important cereal grains—corn, wheat, and rice—and of food legumes (Figure 13.8).

The economic importance of pest resistance is illustrated by wheat rust. Plant breeders have managed to keep wheat resistant to rust for over 50 years by continuously developing new rust-resistant varieties. As the rust fungi evolved and managed to overcome the defenses of their hosts, the wheat plants, it was necessary to introduce new resistance genes. In 1991, the wheat harvest in the United States was decreased by 18% as a result of the rust fungus. At the moment, the pest is ahead once again. Researchers are now trying to breed a type of resistant wheat in which resistance is conferred not by a single gene, but by multiple genes. Resistance will then last much longer. It appears that the 20-year effort at CIMMYT in Mexico to achieve this goal may finally have paid off.

This susceptibility to diseases is caused, in part, by the genetic uniformity of crops. When crops are genetically heterogeneous, it is unlikely that

Figure 13.9

The Dangers of Genetic Uniformity, Illustrated by Rapid Spread of a Southern Corn Leaf Blight Epidemic in 1973. The picture shows ears and husks blackened by the fungus. Source: *U.S. Department of Agriculture.*

everyone will suffer crop losses when the pest evolves the ability to overcome plant resistance. The dangers of genetic uniformity are best illustrated by the corn southern leaf blight epidemic that swept the United States in 1971. This incident, discussed in Chapter 12, clearly illustrates the dangers inherent in genetic uniformity. If all the corn plants have similar genetic characteristics, they will all be susceptible to the same diseases, and epidemics may devastate the crops in very large areas (Figure 13.9).

9 Good agronomic practices keep pest populations down.

Good agricultural practices and timely cultivation are among the cheapest and most effective methods of pest control.

Crop Rotation. As we noted earlier, many kinds of pests thrive only on specific plants or groups of plants. Populations of such pests will tend to build up if the same crop is planted year after year on the same field. Thus, *crop rotation* is one of the best ways to ensure that pest populations will not build up to harmful levels, and it is probably the most widely used method of pest control. Crop rotation also controls weeds, because different crops require different agricultural practices, thereby decreasing the chances that weed populations will build up.

Recent experiments have shown that intercropping also reduces pest problems. When corn and peanuts were grown together in the same field, corn borer damage was reduced 80%. The biological reasons for this reduction are not yet understood, but it is possible that the peanuts harbor organisms that prey on the corn borer. Researchers have observed that insects often have a shorter residence time in fields that have mixed plant populations (intercropping), as compared to monoculture. If encounters with nonhost plants (those the insect cannot eat) are frequent enough, the insects go elsewhere!

Hoeing. *Hoeing* is undoubtedly the oldest method of weed control, and it is still widely used not only in the developing countries but also in the technologically advanced countries. Farmers normally cultivate their fields, before planting crop seeds, to kill the weeds that have germinated. This procedure brings to the surface another crop of weed seeds that will germinate at about the same time as the crop seeds. Thus the field must be hoed again to kill these weeds. It is important to hoe when the weeds are still small, since mechanical hoeing does not kill the plants directly, but simply breaks up the soil, partially dislodging the root system. The small weed seedlings lose part or all of their capacity to take up water, and quickly dry out on bright and sunny days. Perennial weeds, such as various grasses with underground runners, cannot be dealt with so easily, and must be treated before the crop is planted. Although weeds do not start competing with crop plants for water, nutrients, and sunlight until they are several weeks old, it is more difficult to kill them once they have taken hold.

Increasing Crop Plant Density. Another important way to suppress weed growth is by *increasing crop plant density.* In an experiment, decreasing the distance between rows of soybeans from 1 m to 0.5 m allowed the soybean canopy to cover the interrow area in only 47 days instead of 67. Once the canopy is closed and the interrow area is shaded, weeds cannot establish themselves. Thin stands of crop plants generally favor a heavier population of weeds; intercropping may reduce weeds. Recent tests have shown that corn yields increased by 20% when mung beans were grown with the corn, because the beans suppressed weed growth and provided less competition for the corn than did the weeds.

An effective method to control a number of pests and pathogens is to increase *sanitation* by open-field burning of crop residues after the harvest. Sanitation works the same way for crop diseases as for human diseases: it helps tremendously if the sources of infection are eliminated. Many disease organisms and insects overwinter in crop residues, and they are ready to infect the next crop as soon as it is planted. Burning the crop residues at the end of the growing season prevents this reinfection. However, burning also has negative effects. It pollutes the air and contributes to the depletion of soil organic matter that normally accompanies continuous agriculture. Although burning returns most nutrients to the soil, nitrogen is always lost. In some parts of the United States, open-field burning is now strictly controlled, or banned.

Manipulating Planting and Harvesting Dates. It is also possible to minimize damage from insects or pathogens by *manipulating planting and harvesting dates.* Planting dates can be chosen to coincide with periods of low pest density. A somewhat earlier harvesting date or the use of an earlier maturing variety can be used so that the pest has less time to complete its life cycle. Planting can even be legally restricted to create a time period when there are no host plants and the pests cannot multiply.

Genetic Engineering Strategies for Pest Control

Plant breeding—the introduction of new genes—is a powerful way to make plants resistant to pests and predators. This can be done by traditional breeding techniques or through genetic engineering. The methods used to introduce genes into plants via genetic engineering technology, and the possible problems that may be associated with genetically engineered crops (for example, is such food safe to eat?) are all explored in Chapter 15. Here, we discuss some of the different approaches that scientists are taking to create pest-resistant crops.

10 | **Genetic engineering can produce herbicide-tolerant plants.**

In discussing herbicides, we noted that certain herbicides that kill many or most plants are still quite useful because they can be used as *pre-emergence* herbicides to kill weeds before the crops are planted. They would be even more useful if crop plants could be made resistant to them because they could then be used as *post-emergence* herbicides. The manner in which plants and bacteria cope with herbicides and become resistant to them points the way to a strategy for developing herbicide-tolerant plants. Some plants or bacteria are resistant because they have an enzyme—and therefore a gene that encodes the enzyme—that detoxifies the herbicide. Transfer of this gene to a crop plant should allow the crop plant to do the same. Other plants or bacteria become resistant to a herbicide because of a mutation in the target enzyme that renders it no longer sensitive to the herbicide. The enzyme can carry out its function even when the herbicide is present. Transfer of this gene to a crop plant should also let the crop grow in the presence of the herbicide. Two successful approaches to creating herbicide-tolerant plants are discussed below.

Glyphosate is a herbicide that acts by inhibiting one of the enzymes that is necessary for the synthesis of amino acids in the chloroplast. Glyphosate, initially produced and marketed by Monsanto under the trade name Roundup®, is a widely used nonselective herbicide; it effectively kills 76 of the world's 78 worst weed species. Researchers at Monsanto isolated a gene for an enzyme involved in amino acid biosynthesis (EPSP-synthase) glyphosate from resistant *E. coli* bacteria, modified the gene in such a way that it could be expressed in plants, and then used it to generate transgenic tobacco, tomato, and soybean plants. Expression in the plant required a control region that would direct the expression of the gene in the plant (bacterial control regions do not work in plants). In addition, the gene had to be modified in such a way that the enzyme would be transported into the chloroplasts after it had been synthesized in the cytoplasm. When using bacterial genes for genetically engineering plants, the scientist must make sure the proteins encoded by these genes go to the right cellular compartment of the plant cells. The gene was introduced into soybean (Figure 13.10), and the transgenic soybean plants proved fully resistant to normal glyphosate applications. Crop yields were within a few percentage points of yields of untreated (non-transgenic) control plants.

Phosphinotricin is a herbicide that acts by inhibiting another enzyme (glutamine synthetase) necessary for amino acid biosynthesis and nitrogen assimilation by plants. Inhibiting the activity of this enzyme leads to rapid

Figure 13.10

Figure 13.10

Genetically Engineered Herbicide-Tolerant Soybeans. This field of soybeans containing plants that are transformed with the EPSP synthase gene has been sprayed with glyphosate (Roundup®) at twice the normal application rate. Notice the complete absence of weeds. A row of control non-transgenic soybeans has also been killed. The transgenic soybean plants produce the same yield of beans as unsprayed control plants. Source: Courtesy of Sherry Brown, Monsanto Company.

accumulation of ammonia within the plant cells. Because ammonia is toxic, except at very low levels, the plants are killed. Phosphinotricin, produced and marketed by Hoechst AG under the trade name Basta®, is also a very effective nonselective herbicide. It is related to an antibiotic that is also a herbicide, produced by the fungus *Streptomyces hygroscopicus*. Researchers at Plant Genetic Systems in Belgium obtained a gene from this fungus that encodes an enzyme that converts phosphinotricin to a nonherbicidal derivative by combining it with a common cellular metabolite. When this gene, called the *bar-gene*, was introduced into tobacco and potato plants, the crops were fully resistant to applications of Basta® at the normal levels used for the control of weeds. The transgenic plants were first tested in greenhouse experiments and later in field trials. The yields of tobacco and potatoes—several commercial cultivars were used—were similar for the transgenic plants and the control plants.

11 | Genetically engineered potatoes are resistant to potato virus X.

On a worldwide basis, potato is one of the most important agronomic crops. Because it is a tetraploid (potato cells have four copies of each chromosome instead of two), breeding is much more difficult than with other crop plants. Important commercial cultivars—such as the cultivar Bintje, grown in western Europe—are susceptible to major pathogens and pests, including cyst nematodes, an important fungal disease (potato blight), and several virus diseases. Potatoes are vegetatively propagated, so material for planting must be certified virus free, to help control the spread of viruses. Potatoes suffer from three important virus diseases called *potato virus X, potato virus Y,* and *potato leafroll virus.*

As noted earlier, an intriguing and not yet understood aspect of a plant's defenses against viral infection is that plants can be "immunized" against viruses in almost the same way that people are immunized against bacterial

Figure 13.11

The Bintje Variety of Potato, Popular in Europe, was Transformed with the Coat Protein of Potato Virus X to Produce Virus-Resistant Plants. Individual transformants were propagated vegetatively, then planted out in the field and harvested. There were marked differences in the yield and the shape of the potatoes. **Top left:** control Bintje potatoes. **Top right:** transgenic potatoes, same yield, same shape. **Bottom left:** transgenic potatoes, same yield, odd shape (elongated). **Bottom right:** transgenic potatoes, low yield, odd shapes. Source: Courtesy of Mogen International, Leiden, The Netherlands.

diseases. This phenomenon is called *cross-protection* in plants. When a plant is inoculated with a mild strain of a virus that causes almost no disease symptoms, it will be partially or fully protected against a subsequent inoculation with a virulent form of the virus. This cross-protection is in some way related to the synthesis of the coat protein by the plant cell. When viruses infect plant cells, a considerable portion of the plant cell's protein synthetic machinery is diverted to making coat protein instead of plant proteins.

In 1986, Roger Beachy and his collaborators at Washington University introduced the gene that encodes the coat protein of tobacco mosaic virus into tobacco plants, so that every cell in the transgenic plant made viral coat protein. These plants showed considerable resistance to infection by tobacco mosaic virus. The virus apparently was unable to multiply as rapidly in the cells that already contained some coat protein, and the number of virus particles per cell remained much lower in the transgenic plants compared to the control plants.

Scientists at Mogen International in The Netherlands, used the same approach to make potatoes resistant to potato virus X. The gene encoding the coat protein of potato virus X was introduced into two commercial potato cultivars that are highly susceptible to this disease. Control and transgenic regenerated plants were infected with a standard amount of virus, and the appearance of virus particles in the leaves was measured. In the control plants, virus particles started accumulating rapidly in the cells, six–eight days after the inoculation. In the transgenic plants, accumulation of virus particles was delayed and two weeks after the inoculation the transgenic plants had 100 times less virus than the control plants. Most of the transgenic plants had normal agronomic characteristics and would be suitable for commercial use. The experience of the Mogen scientists shows the importance of carefully studying the properties of the transgenic plants. They found that some transgenic lines of potatoes carrying the potato virus X coat protein gene had a reduced yield, whereas other lines had the same yield, but produced many elongated potatoes (Figure 13.11). Such elongated potatoes may not appeal to the consumers.

12 | Genetic engineering produces insect-resistant plants with Bt-toxin genes.

Several species of bacteria produce copious quantities of proteins that kill insect larvae that ingest these bacteria with their food. The most widely studied of these bacteria is *Bacillus thuringiensis,* or **Bt** for short, a species that lives all over the world and is itself divided into a large number of subspecies and strains. When these bacteria form spores, they also form a large crystal-like structure, in the bacterial cytoplasm, that is made out of protein. One of the proteins in this crystal-like structure is called the *Bt protoxin.* When insect larvae eat the bacterial cells, the spores, and the crystal-like structures containing the protoxin, are released in the larval gut, where the digestive enzymes cleave the protoxin, producing an active toxin. The toxin binds to the membrane of the cells that line the gut, inserts itself into that membrane, and creates a channel through which other molecules can freely pass. Punctured by many holes, the gut cells cannot survive long, so the insect larvae starve for lack of nutrition.

As a result of the coevolution between insects and their pathogens, there is an unusual specificity in this interaction between the Bt toxin and the membranes of the gut cells. For example, the Bt toxin of a particular *B. thuringiensis* strain will bind to the gut of lepidoptera larvae, or only some species of lepidoptera, but not to others. When the toxin does not bind, there is no effect on the cells that line the gut, and the larvae do not die. Thus, some Bt toxins will kill lepidoptera (butterflies and moths), others coleoptera (beetles and weevils), others diptera (mosquitoes), and yet others have been found that kill nematodes.

How could these toxins be used for insect control? A relatively simple way is to grow the *B. thuringiensis* bacteria, dry them out, and prepare the heat-killed and dried bacteria in such a way that they can be sprayed or dusted on crops. These preparations are initially highly effective, but the Bt protoxin is not stable after the product has been sprayed on the plants. The Bt protoxin crystals are released from the bacteria, and the protoxin quickly disappears from the plants.

Scientists at Mycogen, a biotechnology company in San Diego, California, introduced a Bt protoxin gene in a different bacterium (*Pseudomonas fluorescens*). These bacteria can readily be grown in large fermentors, killed, and then formulated as a spray. With this bacterium, the protoxin crystals remain in the bacterial cells, and as a result they are stable even after they have been sprayed on the plants. We do not know whether such genetically engineered bacteria would be hazardous to the environment if they were released alive, but by using dead bacteria Mycogen circumvented this difficult regulatory issue. Since *P. fluorescens* bacteria are normally found on the leaves of most plants, the use of a live spray could be dangerous, since the Bt toxin-producing bacteria might quickly spread the gene to other strains of *Pseudomonas.* An example of potatoes protected from devastation by Colorado potato beetles by applying such a spray is shown in Figure 13.12.

The approach just outlined will work quite well with insect larvae that live on the surfaces of leaves, but would be less effective with insect larvae that live in the soil and eat roots, or with larvae that live inside the plant and eat mostly internal tissues. To control those insects, scientists have expressed Bt toxin in cotton, potatoes, and other crop plants. They found that the Bt

Figure 13.12

Potato Field Infested by the Colorado Potato Beetle. The rows in the back were treated with M-Trak®, a formulation of Bt toxin encapsulated in Pseudomonas fluorescence *bacteria that have been heat-killed.*
Source: *Courtesy of Mycogen Corporation, San Diego, California.*

protoxin gene had to be extensively modified to bring about the production of Bt toxin in plants.

Scientists from Monsanto Company in St. Louis, Missouri, introduced the gene from the Kurtsaki strain of *B. thuringiensis* into tomato plants and checked the resistance of these plants to three lepidopteran insects: tobacco hornworm, tomato fruitworm, and tomato pinworm. Only the last two are actual pests in tomato fields. The results of field tests showed that feeding damage done by tobacco hornworm to the transgenic plants was minimal and damage done by the two tomato pests was reduced, but not dramatically so. This experiment emphasizes the importance of finding the right Bt toxin gene for the right insect pest. Tobacco hornworm is particularly sensitive to the Bt Kurtsaki toxin, but tomato fruitworm and pinworm are less sensitive. This sensitivity probably results from the affinity between the toxin and its binding site in the gut of the larva.

If coevolution has indeed produced a wide variety of Bt strains, each very effective against some insects, less effective against other insects and ineffective against most insects, then it will indeed be a challenge to match the strains with the pests so that the Bt genes can be used for genetic engineering experiments. However, if the overwhelming majority of one particular insect species is killed in this way, there will be enormous selection pressures, and the few resistant individuals will quickly increase in numbers. As with chemical pesticides, resistant strains will evolve rapidly. What is needed is a strategy in which several resistance genes are introduced into the plant at once. This approach is likely to result in longer-lasting resistance.

Scientists from CIBA Research in Triangle Park, NC, have succeeded in transforming corn plants with a Bt toxin gene that makes the corn resistant to the European corn borer. This insect pest normally arrives in the spring and lays its eggs on the leaves. The larvae hatch and eat the corn leaves, and after about four weeks new adults emerge. These adults lay eggs again, and it is this second generation of larvae that does most of the damage. The larvae bore

Table 13.4	Companies that have conducted field tests or have had field tests approved by USDA for transgenic plants containing insecticidal genes derived from *Bacillus thuringiensis* (through 9 February 1993)

Company	Crop
Monsanto	Potato, Cotton, Tomato, Corn
Calgene	Cotton, Tobacco, Potato
CIBA-GEIGY	Tobacco, Corn
Agrigentics	Canola (Rapeseed)
Campbell Institute R&T	Tomato
Rohm & Haas	Tobacco
Rogers NK Seed	Tomato
Frito-Lay	Potato
Dekalb	Corn
Northrup King	Corn
Dow Gardens	Allegheny Serviceberry

Source: S. Krimsky and R. Wrubel (1993), *Agricultural Biotechnology: An Environmental Outlook,* Tufts University, Department of Urban and Environmental Policy, p. 29.

into the stalk, eating out its tissues, thereby retarding the growth of the plant and reducing its yield. The transgenic plants made with the aid of a DNA gun (see Chapter 15) have proven resistant to the European corn borer. The corn plants have been crossed with five different elite lines of corn, and yield trials are now under way to select lines of corn that produce yields that are just as high as the parent lines.

Scientists from CIBA and from other companies (Table 13.4), such as Monsanto and Pioneer, are also attempting to produce corn that is resistant to the corn rootworm, another major insect pest. The objective of these research programs is to produce pest-resistant lines of corn. If these programs are successful, we are likely to see less use of pesticides, but more continuous cultivation of corn with less crop rotation. Many of these pathogens can also be controlled by sound agronomic practices that include crop rotation. Thus, these developments do not necessarily push agriculture in a sustainable direction (see Chapter 16). Rather, they will make it dependent on a different set of inputs: on transgenic plants instead of on pesticides. There will be environmental advantages to this substitution, but they will not be as great as could be achieved by researching and adopting those technologies that contribute to sustainability.

13 | Resistance to insects can be engineered with lectins and inhibitors of digestive enzymes.

Yet another approach that can be taken to protect plants from insects is to introduce genes that encode lectins or the inhibitors of digestive enzymes (amylase and protease). These are part of the natural defense mechanisms of plants (see Chapter 12). Because they are all proteins, their genes can be readily transferred from one plant to another. This approach is not as far

Figure 13.13

Comparison of the Damage Done by Insect Larvae to Tobacco Leaves of Control Plants (left) *and of Transgenic Plants Containing a Protease Inhibitor Gene* (right). *The leaves were inoculated with three small larvae in each case. Note the larger size of the larvae on the left. Source: Courtesy of C. Ryan, Washington State University.*

along as genetic engineering with Bt genes, in part because of the difficulty in matching insect pests with inhibitory proteins.

Wheat germ agglutinin is a chitin-binding lectin that inhibits the growth of insect larvae that feed on artificial diets, to which the protein has been added. Efforts are now underway to introduce the gene that encodes wheat germ agglutinin into corn to see if the protein is expressed at levels high enough to retard the development of various pests. Because wheat germ agglutinin and other lectins also inhibit essential processes in human cells, it must be established that the levels present in the transgenic plants do not affect humans. The problem can be avoided by making sure that the lectin gene is not expressed in the plant part consumed. To inhibit the corn rootworm, it should be possible to use a root-specific gene control region, so that wheat germ agglutinin would only be made in roots.

There have been several reports of the introduction of different protease inhibitor genes into tobacco and potato plants. In both cases, the transgenic plants were much more resistant to tobacco hornworms than the control plants. The resistance took the form of a failure of insect larvae to grow and develop on the transgenic plants (Figure 13.13). Tobacco and potato were used in these experiments because they are easily transformed plants.

14 Genetic engineering produces plants resistant to fungi and bacteria.

The first attempts to make plants resistant to fungal and bacterial pathogens involve introducing into the plant genes that encode enzymes that degrade cell walls. The targets are the cell walls of fungi and bacteria. These cell walls are completely different from the plant cell walls, and enzymes that degrade fungal and bacterial cell walls do not harm the plant.

Fungal cell walls consist of two polymers, glucan and chitin, that can be broken down by glucanases and chitinases (two classes of enzymes), respectively. However, the chitin of one species differs in subtle ways from the same polymer in another species. A particular chitinase may be effective against one species of fungus, but not against another. Thus the effective use of chitinase genes for genetic engineering again depends on the exact matching of the gene with the target pathogen, just as with the Bt toxins or the digestive enzymes.

Bacterial cell walls are made up of unusual polymers called *peptidoglycans*, which have both protein and carbohydrate components, and which can be digested by the enzyme, lysozyme. Lysozyme is found abundantly in chicken eggs. It is also found in cow stomachs, where its role may be to digest bacteria, which themselves are multiplying because they are helping degrade the cellulose and other components of the plant cell walls. Cow lysozyme can also degrade the cell walls of plant-pathogenic bacteria and can therefore be used for biotechnological approaches to the control of bacterial diseases. Scientists at SIBIA, a biotechnology company in San Diego, California, have introduced the gene that encodes cow lysozyme into a yeast, *Pichia pastoris*. When these transgenic yeast cells are grown in large fermentors, they produce substantial quantities of lysozyme, which is secreted into the culture medium. By treating various seeds with very dilute solutions (25–100 parts per million) of lysozyme, the scientists have been able to completely kill the bacteria on the seeds.

Many bacterial diseases are propagated by seeds: the bacteria stick to the seeds and when the seeds are planted, they infect the soil and subsequently the crop. The only effective remedy is to treat the seeds with compounds containing mercury or copper. Once an infection is established, copper is again the most commonly used bacteriocide. However, many bacteria have become resistant to copper and there are at present few alternatives. Although the lysozyme treatment is currently much more expensive than the old copper or mercury treatments, it is also much more effective because it produces complete sterility of the seeds. There may be a problem, however. Lysozyme is a stable enzyme, and it could do considerable harm to the soil microbes that live in the rhizosphere. Research will be needed to show how long the enzyme survives in the soil and what its effects are.

Summary

In this chapter we described several entirely different ways to control pests: with chemicals, through biological mechanisms, by using resistant plants, and through cultural practices. The chemical route has been around for some 50 years and has now been shown to have several serious drawbacks. The chemicals can create serious health hazards for many agricultural workers and pollute the environment, and the pests may develop resistance to them. Chemicals affect crop plants adversely in ways that remain largely unexplored. They also may have a negative impact on the rhizosphere. Breeding resistant plants is a tried and true method that has yielded positive results for 85 years. Because organisms evolve, the plant breeders will always be busy keeping the crops one step ahead of the pests and pathogens. Genetic engineering is greatly expanding the sources from which resistance genes can be obtained. Once the biology of plant–microbe and plant–insect interactions is understood, scientists should be able to identify different genes that could be used to make plants resistant to pests and pathogens. Although biological control is an equally old concept, and has been applied successfully many times, it remains grossly underused worldwide. The reason is a lack of interest by commercial companies, and until recently also by university scientists.

Sound agronomic practice requires the farmer not to choose among these strategies, but to use all available methods in a program of integrated pest management. Applying the principles of integrated pest management has decreased the use of pesticides on some crops, cotton being a prime example. Crucial to such a program is knowledge of the pest and its level of economic injury. Pest levels must be measured accurately, which often requires much sophistication on the farmer's part. In the future, we will see the integration of the biological and the genetic approaches (traditional breeding and genetic engineering) to keep pests under control, as society reconsiders the benefits of the chemical approach.

Further Reading

Brunke, K. J., and R. Meeusen. 1991. Insect control with genetically engineered crops. *Trends in Biotechnology* 9:197–200.

Croft, B. A. 1990. *Arthropod Biological Control Agents and Pesticides*. New York: Wiley.

Duncan, L. W. 1991. Current options for nematode management. *Annual Review of Phytopathology* 29:469–490.

Gatehouse, J. A., V. A. Hilder, and A. M. R. Gatehouse. 1991. Genetic engineering of plants for insect resistance. *Plant Biotechnology* 1:105–135.

Gill, S., E. A. Cowles, and P. V. Pietrantino. 1992. The mode of action of *Bacillus thuringiensis* endotoxins. *Annual Review of Entomology* 37:615–636.

Harms, C. T. 1991. Engineering genetic disease resistance into crops: Biotechnological approaches to crop protection. *Crop Protection* 11:291–306.

Hedin, P., J. Menn, and R. Hollingworth, eds. 1988. *Biotechnology for Crop Protection*. New York: American Chemical Society.

Julien, M. H., ed. 1987. *Biological Control of Weeds*. Tucson: University of Arizona Press.

Keen, N. T. 1992. The molecular biology of disease resistance. *Plant Molecular Biology* 19:109–122.

Lambert, B., and M. Peferoen. 1992. Insecticidal promise of *Bacillus thuringiensis*. *BioScience* 42:112–122.

Lavabre, E. M. 1991. *Weed Control*. New York: Macmillan.

Owen, M., and R. G. Hartzler. 1990. *Agricultural Weed Control*. Ames: Iowa State University Press.

Oxtoby, E., and M. A. Hughes. 1990. Engineering herbicide tolerance into crops. *Trends in Biotechnology* 8:61–65.

Rappaport, R. 1992. *Controlling Crop Pests and Diseases*. New York: Macmillan.

Tjamos, E. C., G. C. Papavizas, and R. J. Cook, eds. 1992. *The Biological Control of Plant Diseases: Progress and Challenges for the Future*. New York: Plenum.

Zalom, F., and W. Fry. 1992. *Food Crop Pests and the Environment: The Needs and Potential for Biologically Intensive Integrated Pest Management*. St. Paul, MN: American Phytopathology Society.

Chapter 14

Valuable Chemicals from Plant Cell and Tissue Culture

One branch of plant biotechnology, already mentioned in Chapter 3, proposes to use plant cell cultures to manufacture, on an industrial scale, some of the unusual and complex chemicals that plants produce and that have found uses in our lives as pharmaceuticals, fragrances, flavor compounds, dyes, and insecticides. The production of chemicals by microorganisms via fermentation is a well-established industry. Plant cells can be grown in similar cultures in which the cells are kept in suspension by aeration, and scientists in many countries are trying to find the right culture conditions that will allow them to establish plant cell cultures that produce valuable chemicals, thereby creating an entire new industry.

1 Plants produce valuable secondary metabolites.

Plants produce a complex array of organic chemicals that do not seem to play any direct role in their growth and development. These substances, the *secondary metabolites,* are quite unlike the *primary metabolites* such as sugars, amino acids, nucleotides, or chlorophyll. They do not seem to be required for the normal processes of plant growth: photosynthesis, nutrient uptake and assimilation, respiration, and the biosynthesis of protein, RNA, or DNA. Furthermore, there are probably more than 50,000 of these chemicals, which are all different and have a restricted distribution in the plant kingdom. Some are made only in one plant family or perhaps just one species. In contrast, virtually the same primary metabolites are found in all plants.

The secondary metabolites protect plants against predation by insects and mammals and against pathogens (see Chapter 12). In addition, they may be attractants for pollinators and fruit-dispersing animals. Biological chemists often refer to the chemicals as "natural products," to distinguish them from synthetic products that are produced in the laboratory. Many secondary metabolites can also be synthesized in the laboratory, but the yield is often quite low because a number of synthetic steps are involved. Plant cells use

Although plants contain thousands of chemicals, the effects of which on the body remain largely unknown, the medicinal properties of many plants and plant extracts have long been known. To this day about 75% of the world's population relies entirely on herbal medicines, administered by physician-herbalists. In the United States, where herbalism is thought of by some as quackery, most people associate it with primitive cultures that do not have access to "real" medicine. Yet 3 million Americans with heart conditions daily take a molecule extracted from *Digitalis purpurea* (foxglove) to help them stay alive. This molecule, the cardiac glycoside digitoxin, increases the force of heart muscle contraction without a concomitant increase in oxygen consumption. This makes the heart a more efficient pump, better able to meet the demands of the circulatory system.

But this is not the only drug of plant origin that meets the standards of "real" medicine. Worldwide, 120 other clinically useful prescription drugs are derived from plants. About three-quarters of them came to the attention of drug manufacturers because of their use in traditional medicine.

Modern physicians do not study botany, and most of them are only interested in drugs whose efficacy is proven in clinical studies. Therefore, they cannot appreciate the value of traditional remedies. Herbalists, in contrast, study with other herbalists and pass down their knowledge from generation to generation, adding their own experience in the process. An apprentice of a Navajo medicine man will learn to use medicinally nearly 200 different plants to treat both physical and mental afflictions. Although such medicines may be sold in the United States, the manufacturer is not allowed to make any claims as to their efficacy, because they have not been tested in clinical trials. In the United States, very few traditional herbal remedies ever become "real" medicines because the cost of testing in developing a new drug is around $100 million. Drug manufacturers are seldom willing to invest such sums of money to develop a natural product for clinical use. Yet the efficacy of the herbal medicine may be known to thousands of patients and hundreds of doctors. In western Europe, in contrast, expensive clinical trials are not always necessary to get a drug approved. Reports from general practitioners that a drug is effective are just as important to gain approval. Furthermore, natural products that have been used for a long time are generally presumed to be safe.

There is perhaps no better example than the worldwide use of extracts from willow bark (*Salix alba, Salix nigra,* and *Salix lucida*) to relieve headaches and fevers. For thousands of years, the Greeks as well as the Native Americans used such extracts to relieve aches and pains. The active ingredient in willow bark is now known to be the glycoside salicin, which decomposes in the stomach to give salicylic acid. Methyl salicylate, another derivative of salicylic acid, was found in plants that were being used in antirheumatic drugs.

Salicylic acid is a relatively simple organic compound; German pharmaceutical chemists readily synthesized it and made a variety of derivatives in the middle of the nineteenth century. The derivative acetylsalicylate was found to be more effective and had none of the side effects of salicylic acid itself. Fifty million tablets of this derivative are now sold daily in North America usually under the name *aspirin.* The development of this drug therefore went through three stages: (1) widespread use in herbal medicine, (2) identification of the active ingredient with continued use of the best plant extracts, and (3) organic synthesis of the same chemical and more effective derivatives.

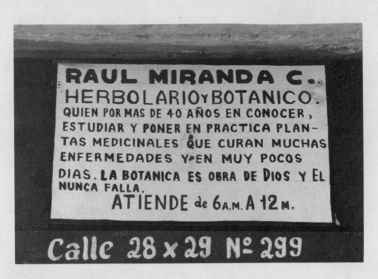

Sign on the Door of a Herbalist Physician in Mexico

Although aspirin may seem a relatively commonplace drug there are other, similar success stories. For example, plant extracts that were known to relieve hypertension or reduce the pain caused by glaucoma led to the discovery of important, clinically proven drugs. This does not mean that all herbal remedies will be found to be effective. For example, all cultures have attributed aphrodisiac and male-potency-enhancing properties to scores of plants, but so far none has proven effective. Unfortunately, of the estimated total of 250,000 plant species, only 5,000 have been examined for their possible medical application. As the earth loses its plant biodiversity and as the herbalists who understand the medicinal uses of plants disappear, humanity is losing an enormous cultural treasure that took a million years to accumulate.

batteries of enzymes to synthesize these important compounds, which sometimes accumulate to high levels in specific tissues or cells. Thus, what the organic chemists do with difficulty, the plant cells do with ease.

Many of these chemicals have found their way into our lives as flavorings, dyes, insecticides, and pharmaceuticals, and some have been used since ancient times. The curative properties of many plants are described in early medical texts dating back to the ancient Chinese, Egyptians, Greeks, and Romans, and were known to peoples without written histories, such as Native Americans. Today, if substantial amounts of the chemicals are used, the plants are often grown on plantations, usually in the tropics or subtropics, and the chemicals are extracted from the plants. For example, a full 25% of all our medicines come from plant extracts. Many other pharmaceuticals originally also came from plants, but extraction from plants has been replaced by organic synthesis.

Many of these extracted chemicals are so difficult to synthesize that it is unlikely that they will ever be produced via organic synthesis. Furthermore,

Table 14.1	Estimated world market for selected plant products

Compound	Estimated Retail Market Value (millions)
Medicinal	
Ajmalicine	U.S. $ 5
Codeine	90–100
Corticosteroids	300
Ephedrine, pseudoephedrine	100
Quinine	50
Vinblastine, vincristine	50–75
Flavor and fragrance	
Cardamon	25
Cinnamon	4–5
Spearmint	85–90
Vanilla	15
Agrichemical	
Pyrethrins	20

the increasing consumer demand for natural products, rather than synthetic products, encourages the use of plant extracts. The market price of these products varies from a few U.S. dollars per kilogram to several thousand dollars for drugs such as digitoxin or shikonin, or fragrances such as jasmine oil, to several *hundred* thousand dollars per kilogram for anticancer drugs such as vincristine, vinblastine, and taxol. Table 14.1 shows the world market for some of the chemicals. Given the high price of these chemicals, it is little wonder that the biotechnology industry has tried to devise plant cell culture methods to produce them. Production in factories "at home" would make the developed countries independent of their supplies in Third World countries where many of the plants that produce these chemicals are found. The supply, cost, and quality of the raw materials—in this case, the plants—are affected by the climate, by diseases, and by political instability in the producing countries. All these problems are avoided by cost-effective plant cell culture production methods.

2   Establishing a plant cell culture for secondary metabolite production is a complex problem.

As noted earlier, when a piece of a plant organ is cultured on a solidified medium that contains only sucrose, minerals, a few vitamins, and two plant hormones, the cells divide and form a callus, a mass of relatively undifferentiated cells. When small pieces of such a callus are swirled in a liquid medium, cells break off and continue to divide, forming a cell suspension culture in which the cells divide every 36–48 hours.

In some cases, such cultures produce secondary metabolites. In the plant, secondary metabolites are often produced in only one organ or even one cell type—leaf hairs, for example—so the cell culture must be started with the right tissue. In genetic terms, the genes that encode the necessary enzymes are present in all cells but are not active everywhere. They are only active in

(a)   (b)

those cells that produce the secondary metabolite. Culturing the cells or tissue is often accompanied by an inhibition of the expression of these genes so that the production of the metabolite slows down dramatically. This dramatic drop in metabolite production usually occurs in most cells, but not necessarily in all cells. To find the cells that are still producing metabolite at a high rate, one must use a clonal selection procedure, in which small calluses (clones) are derived from single cells or protoplasts. From the clones one can then select those that are producing the metabolite at a high rate. One needs, of course, a good assay procedure for the metabolite. In the case of colored metabolites, visual inspection of the calluses is sufficient.

In addition, it may be necessary to start with different sources (such as varieties or cultivars) of the same plant species, because genetic differences between them may not be apparent at first. For example, when one specific cultivar of *Catharanthus roseus* was used for tissue culture to produce catharanthine, 62% of the clones produced this metabolite, compared to 0.3% of the clones when starting with another cultivar.

Genetic selection of a good maternal plant to start the clonal selection is an important first step. Having obtained a high-producing clone in tissue culture, one must find out whether or not the clone is stable. Repeated subculture of the clone should not reduce metabolite production.

Selecting a stable, high-producing clone must be followed by optimizing the conditions for cell culture (Figure 14.1). Unfortunately, no general culture medium is best for all cells. Rather, many variables must be adjusted for every different plant species and cell culture that produces a specific metabolite. Both chemical and physical conditions must be considered. Chemical variables include sugar or other energy source, minerals, vitamins, plant hormones, and oxygen. Physical variables include method of aeration, light and temperature regimes, and entrapment of the cells in a solid medium. Often two sets of

Table 14.2	Examples of secondary metabolites produced in cell culture and root culture		
Compounds	**Plant Species**	**Type of Culture**	**Yield**
Berberine	*Coptis japonica*	Cell	0.8 g/L
Rosmarinic acid	*Coleus blumei*	Cell	3.6 g/L
Shikonin	*Lithospermum erythrorhizon*	Cell	1.5–4.0 g/L
Atropine	*Atropa belladonna*	Root	3.7 g/kg dry weight
Scopolamine	*Atropa belladonna*	Root	0.24 g/kg dry weight
Scopolamine	*Duboisia leichhardtii*	Root	11.6 g/kg dry weight

conditions are devised. One set produces rapid cell division and growth of the culture. The conditions are then altered to elicit maximum production of the metabolite and its accumulation over a long period (20–100 days).

At the end of the culture period, the fermentor is dismantled and the cells are harvested so that the metabolite can be extracted. Often, valuable metabolites are not secreted into the culture medium, but remain intracellular, accumulating in the vacuoles. A major advantage of obtaining the metabolite from cell culture rather than plants is that the metabolite content and the daily production are both higher in the cultured cells. This is illustrated in the production of berberine by *Coptis japonica* (see Table 14.2). In roots of this plant, berberine accounts for 5% of the dry weight of the roots after five years of growth; this corresponds to a productivity of 0.17 mg per g per week. Selected cell lines of *Coptis japonica,* grown in tissue culture, contain 13.2% of the dry weight of the cells as berberine after only three weeks in culture, making their rate of berberine production 44 mg per g per week, or 250 times higher.

The biotechnological potential of plant cell culture can best be illustrated by describing the development of the industrial production of shikonin, a red pigment that is found in the purple roots of *Lithospermum erythrorhizon,* a perennial herb that is native to Japan, Korea, and China. Shikonin is used in dyes, ointments, and cosmetics.

The successful industrial production of shikonin, in cell culture, came about through the efforts of two Japanese scientists, M. Tabata and Y. Fujita, of Kyoto University. Tabata and Fujita first showed that it was necessary to start with clonal selection to obtain stable high-shikonin-producing lines. This clonal selection had to be initiated using single protoplasts, not small cell clumps, to obtain stable clones.

A second important observation was that changes in the nutrient medium could bring about a 10-fold increase in the yield of shikonin. They systematically investigated the effects of temperature, light quality, oxygen supply, growth hormones, and energy and nitrogen sources on shikonin production. They found, for example, that a 30-fold increase in calcium brought about a 3-fold increase in shikonin yield. Having realized that many factors affect shikonin yield without influencing cell growth, they devised a two-stage procedure in which the cells are allowed to multiply first in a 750-liter fermentor. The fresh medium is then added, and 1.2 kg of shikonin is produced in the subsequent two-week period. After much basic research, shikonin production was commercialized in Japan in 1985.

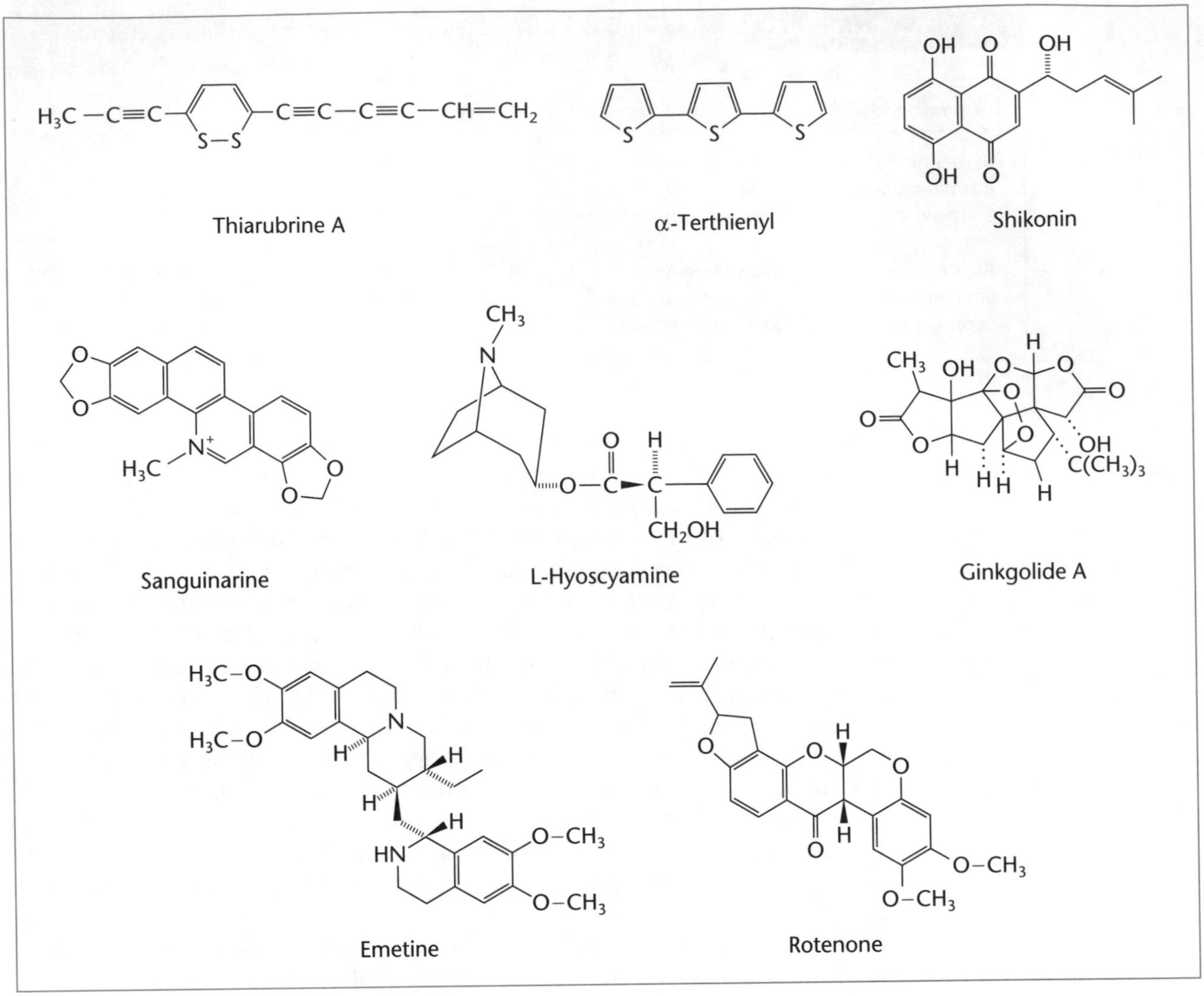

Figure 14.2

Structures of Biologically Active Compounds Found in the Underground Organs of Various Plant Species.

3 | Metabolites can be produced in root cultures.

The roots of plants produce a complex array of secondary metabolites, possibly because they live in an environment—the soil— that is permeated with pathogens and predators. For this reason, they have evolved capacities for synthesizing all manner of defense chemicals (Figure 14.2) that may accumulate in the roots, be secreted into the rhizosphere, or be transported to the shoot. For example, nicotine, found in the leaves of tobacco, is actually synthesized by the roots and transported to the shoot. In some cases, the secondary metabolites are produced by the roots at low levels until the roots are confronted with a fungal pathogen, at which time they may increase the synthesis and secretion of the metabolite to try to ward off the invader.

The production of a number of secondary metabolites is somehow inextricably linked with the existence of an organized root structure. For example, the alkaloid scopolamine is produced abundantly in roots, but not in cultured

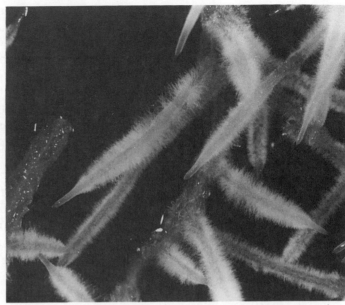

(a)   **Figure 14.3**   (b)

Infection by **Agrobacterium** **rhizogenes** *Causes the Formation of Hairy Roots.* **(a)** *Hairy roots are growing from the cut end of a tomato stem.* Source: **(a)** *Courtesy of S. Satoh.* **(b)** *Closeup photo of hairy roots.* Source: *Courtesy of H. E. Flores.*

cells, even after clonal selection. This problem can be solved, in some cases, by carrying out organ culture instead of cell culture, an approach that is especially applicable to roots. As noted in Chapter 3, roots can readily be cultured in sterile conditions and can be maintained indefinitely. However, one problem is that roots grow much more slowly than cultured cells, and for this reason more research has been done on cultured cells. Recently scientists have used the rapidly growing roots obtained by infecting plants with *Agrobacterium rhizogenes*, which causes hairy root disease. In Chapter 3, we mentioned that the soil bacterium *Agrobacterium tumefaciens* causes plant tumors by deregulating the hormonal balance of the cells. A portion of its DNA, encoding enzymes for auxin and cytokinin synthesis, is transferred to the plant cells and integrated into the plant chromosomes. This transformation of the plant cells results in excessive hormone synthesis and tumor growth.

Agrobacterium rhizogenes can also infect wound sites, and when it does so, roots with multiple side roots and covered with root hairs grow out of these wound sites (Figure 14.3). During the infection, the bacteria transfer some of their DNA to the plant cells at the site of infection, and these transformed cells are now much more sensitive to the hormones they produce. This increased sensitivity to the hormone auxin causes roots to form at the wound site, even if that wound site is on a leaf or stem. When these roots are transferred to a culture dish or flask, they continue to grow very rapidly, and they produce the same types of secondary metabolites that "normal" roots produce.

Because *Agrobacterium rhizogenes* can infect many different dicot species, this culturing system is applicable to many plants (Table 14.2). Hector Flores from the Biotechnology Institute at Pennsylvania State University has established such hairy root cultures from a number of plant species, including marigolds (*Tagetes*). Marigold roots synthesize potent nematocides, and their release into the soil provides the chemical basis for controlling nematode damage in a variety of garden crops.

Fungi often induce roots to start synthesizing secondary metabolites, and the effect of a fungal infection on the production of secondary compounds can easily be mimicked in root culture without introducing the fungus. It is sufficient to add to the root culture a fungal elicitor, a mixture of cell wall fragments of the fungus made by heat-killing live fungal cells. The roots react to this challenge as if they were being invaded by a fungus, and they start producing the defense chemical.

A major advantage of root cultures over cell cultures is that the secondary metabolites are often secreted by the roots into the culture medium, or can be released from the roots by a brief heat shock. When metabolites are present in the culture medium, they can be purified relatively easily. Moreover, adding adsorbants, such as activated charcoal, to the culture medium stimulates the production and secretion of metabolites that are continuously removed from the medium by adsorption on the charcoal.

An advantage of cell cultures, in contrast, is that they can be easily scaled up in large fermentors. No similar scaling-up has yet been devised for roots. When roots are grown in a shaking culture, they tend to grow as a root "ball" with young, rapidly growing roots at the periphery of the ball, and older, aging tissues on the inside.

4   Taxol is a unique anticancer drug from the bark of the Pacific yew.

Plants contain thousands of secondary metabolites whose effects on humans are unknown. Some 30 years ago, the National Cancer Institute, in collaboration with the U.S. Department of Agriculture, contracted to screen the extracts of 100,000 different plants and plant parts for their effects on mouse leukemia cells. The finding that a crude alcohol extract of the dried bark of the Pacific yew (*Taxus brevifolia*) killed the leukemia cells led to the identification of taxol, a new and potent anticancer drug. Once the activity of taxol against ovarian cancer, breast cancer, melanoma, and colon cancer became known, the Pacific yew became a highly prized commodity. And herein lies an important lesson about the value of biodiversity (see Box 14.2).

Taxol inhibits the correct functioning of microtubules, minute proteinaceous filaments in the cells of animals and plants that play many important roles in the cell, including the separation of the chromosomes during mitosis (see Chapter 9). In the presence of taxol, cell division stops. Furthermore, taxol also prevents the migration of animal cells and therefore may prevent the spread of metastatic cancer cells in the body.

Like many secondary metabolites, taxol is very potent, but not very abundant. Current estimates are that treatment of cancer patients in the United States alone would require about 250 kg of the pure drug each year. To obtain this we would need to strip the bark of 360,000 mature (60- to 75-year-old) trees each year. Because the Pacific yew was always considered a weed by foresters in the Pacific Northwest, no one really knows how many of these slow-growing trees there are, but it is clear that harvesting on this scale could not be sustained for very long. Stripping the bark kills the trees, so that repeated harvesting of the same trees is not possible. Taxol is equally abundant in the needles of the Pacific yew, and needles constitute a renewable resource (they will grow again after trees are shorn of their twigs). However, the U.S. government has not yet approved taxol extracted from needles for use in cancer patients. Because the supply of trees is limited, and the

Box 14.2

*Taxol and the
Need for
Biodiversity*

The discovery that the bark of the Pacific yew (*Taxus brevifolia*) contains a potent anticancer drug illustrates the need to maintain the biodiversity of our forests and the danger of replacing natural forests with stands of commercially productive species. Natural forests harbor many species of plants and animals; this richness is generally referred to as *biodiversity.*

Seedlings of the Pacific Yew. Source: *Courtesy of Phototake.*

When forests are managed by commercial loggers, the trees that contain valuable timber are first harvested and all the remaining trees are slashed and burned before the valuable species are replanted. As a result, the trees that have no apparent commercial value disappear from the forest. Their regrowth is not encouraged. Foresters generally lack a good inventory of the species in a forest they are "managing" and are interested primarily in encouraging fast-growing and/or valued tree species.

In this management philosophy, the Pacific yew is seen as a weed that takes up space and competes for resources with other trees. The yew is an extremely slow-growing tree that takes up to 50 years to reach maturity, and at that age the tree trunks may still only measure 10 cm in diameter. These moderate-size trees grow in a thin band along the Pacific Coast from the southeast coast of Alaska to San Francisco and in an area from eastern Washington to western Montana. The Pacific yew is an understory tree that does poorly in direct sunshine and prefers the shady, damp interior of old-growth forests.

It is somewhat ironic that the European yew (*Taxus baccata*) was highly prized a thousand years ago because its branches furnished the wood for the bows on which all European armies depended. The famous battle of Agincourt in 1415, in which a small English army defeated a superior French force of armored knights, was won with the aid of longbows made of yew. However, the Europeans managed this precious resource badly and by the end of the sixteenth century the European yew was close to extinction.

> The unwise forest practices of the past are forcing us to make a difficult choice. On the one hand, we would like to meet the needs of 100,000 cancer patients each year who need taxol. On the other hand, meeting the needs of all today's cancer patients would preclude harvesting yew trees on a sustainable basis and would result in yet more destruction of old-growth forests. The real reason to preserve not only old-growth forests, but all natural ecosystems, is that we do not know from which plant species the next drug or the next important gene will come.

harvesting of so many trees is severely opposed by many environmentalists, it is necessary to look for other sources of taxol.

Likely routes to a plentiful taxol supply are plant tissue culture, fungal culture, or partial synthesis starting from related natural compounds that are more abundant. Because the price of taxol is expected to be around $200,000 to $300,000 per kg, tissue culture is certainly a feasible alternative to extraction from the yew trees. Tissue culture research was started by screening the bark tissues of Pacific yews from different parts of the entire range of the tree in the Pacific Northwest to find the highest producers. Twenty-five-fold differences in taxol abundance were found among different trees. When cells of the inner bark were put into culture, taxol was secreted into the tissue culture medium—a decided advantage for the subsequent purification of the chemical.

With the presently available strains of cells, one liter of tissue culture medium contains about 1 to 3 mg taxol, a yield that can also be obtained by extracting 25 g of bark. The procedure is being refined at Phyton Catalytic in Ithaca, New York, and elsewhere, and scientists will have to apply all the selection procedures outlined elsewhere in this chapter to increase taxol production in tissue culture to commercially viable levels.

In a promising recent development, Gary Strobel from Montana State University demonstrated that a fungus that normally colonizes yew trees makes taxol when it is grown in sterile culture in the laboratory. This finding opens the way for a rapid scale-up of taxol production, because selection of fungal strains that produce high levels of a specific chemical is usually much faster than the selection of plant cells in tissue culture for the same purpose. Furthermore, the large-scale industrial production of chemicals by fungal cultures is much better worked out than that by plant cell cultures.

Because taxol is such a complex molecule (Figure 14.4), organic chemists are not pursuing the total synthesis in the laboratory. Rather, they are looking for taxol-like molecules or taxol precursors in other plants. One such chemical precursor, called *baccatin III*, was isolated from the needles of the European yew (*Taxus baccata*) by French scientists, who converted it in the laboratory to taxotere, a chemical analog of taxol. Chemists already know from experience that when a new drug is discovered, it is always possible to find analogs, molecules with small variations in their structure that are equally effective. The problem is that baccatin III is not any more abundant than taxol and its supply will depend on harvesting needles of European yew trees growing in plantations specifically for that purpose. Other chemists are trying to make taxol from simpler, but much more abundant precursors; that type of synthesis requires many more synthetic steps to be carried out in the laboratory.

Figure 14.4

*Structural Formula
of Taxol.*

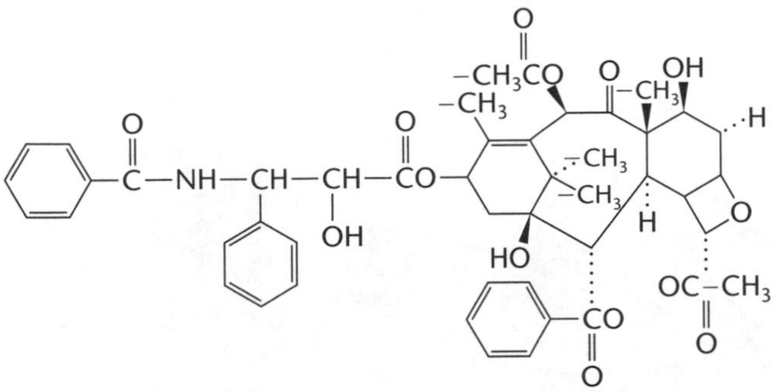

5 | The economics of large-scale plant cell culture favor only a few products at the present time.

Although much progress has been made in the past 10 years in the large-scale culture of plant cells for the production of secondary metabolites, only a few processes have been commercialized (Table 14.3). This is partly due to economic considerations. It has been estimated that research and development for commercial production of a new product require a 10-year investment. The cost of the physical plant must be recovered, and the required research effort over a 10-year period can only be achieved if the product has an annual market of U.S.$50 million, at a selling price of at least U.S.$400–500 per kg. Not many products fall into this category. For example, both jasmine and ajmalicine are quite expensive (jasmine costs U.S.$5,000 per kg and ajmalicine costs U.S.$1,500 per kg), but the total market for jasmine is estimated to be only U.S.$500,000 and for ajmalicine, U.S.$5 million. Spearmint oil, in contrast, has a huge market (U.S.$100 million), but the price is low (U.S.$30 per kg). In the future, research will be directed toward new products that will open up new markets and lead to the development of production processes that are cheaper, enabling more substances to be produced by cell culture.

An alternative approach would be to **isolate the genes** that encode the enzymes necessary for synthesizing some of these compounds and introduce the genes into cell lines or crops (such as potatoes or tobacco) that are already widely grown. This approach is only possible for compounds whose synthesis does not depend on a large number of enzymes not already present in these plants.

Table 14.3	Plant tissue cultures developed for industrial application		
Product	**Species**	**Company**	**Country**
Shikonin	*Lithospermum erythrorhizon*	Mitsui	Japan
Berberine	*Coptis japonica*	Mitsui	Japan
Biomass	*Panax ginseng*	Nitto Denki	Japan
Peroxidase	*Raphanus*	Toyobo	Japan
Geraniol	*Geranium*	Kanebo	Japan
Rosmarinic acid	*Coleus blumei*	Natterman	Germany
Digoxin	*Digitalis lanata*	Boehringer Mannheim	Germany

If researchers can establish which enzyme limits the production of a particular metabolite, then the gene that encodes this enzyme can be used to transform the cell line or plant. This can increase the level of the enzyme and the biosynthesis of the secondary metabolite. If the metabolite accumulates in the transgenic plants to higher levels, then the process becomes more cost effective.

Another way that the synthesis of complex secondary metabolites could be made cost effective is by using plant cells as **bioreactors.** In this procedure, the plant cells need not synthesize the metabolite in its entirety. Rather, the cells are fed a particular metabolite that is already quite complex but easily obtained, and the plant cells convert this metabolite into a high-value product. This method is usually chosen if the synthetic step(s) that must be accomplished are simply too difficult to do in the laboratory via organic synthesis and/or because the yield is too low. The function of the plant cells in this case is to synthesize the one or two crucial enzymes, to take up the precursor metabolite from the medium, and to carry out the required enzyme-catalyzed biosynthetic steps. The valuable metabolite continues to accumulate until the cells are ready for harvest. These are thoroughly domesticated plant cells, indeed! The ultimate goal, as always, will be to produce new chemicals or to bring down the cost of producing those whose production is not cost effective enough to compete with extraction from plants.

An example of this approach is the commercial production of pyrethrins. Pyrethrins are popular household pesticides that are extracted from flowers of a species of chrysanthemum. Chrysanthemum flower cultivation and pyrethrin extraction are important industries in Myanmar (Figure 14.5). Pyrethrin can readily be synthesized from chrysanthamic acid (by reaction with an alcohol), but chrysanthamic acid is difficult to synthesize in the laboratory. However, a single enzyme converts a metabolite found in many plants into chrysanthamic acid. This means that introducing a single gene into many plant species would convert them into good sources of chrysanthamic acid for the chemical manufacture of pyrethrin.

6 | **Producing secondary metabolites in tissue culture may have a negative impact on the economies of Third World countries.**

Experts such as the Kenyan economist Calestous Juma warn that the transfer of the production of secondary metabolites from farms in Third World countries to factories in developed countries will have a negative impact on the economies of the Third World countries. They will lose an important source of foreign exchange, and many farmers will lose their livelihoods. In his fascinating book *The Gene Hunters: Biotechnology and the Scramble for Seeds* (1989, Princeton, NJ: Princeton University Press, p. 143), Juma writes,

> The direction of this research is aimed at reducing dependence on imported raw materials. The impact on the countries exporting this product will be profound and irreversible. Over the years, large sections of the population have organized their lifestyles around the production of these crops. This is going to be changed by current developments in biotechnology. The impacts will not be only economic, but will have long-run political implications and the attempt by communities to reorganize themselves in response to the changed conditions. It is, therefore, in the interest of raw material exporters to closely monitor current trends in biotechnology and the use of genetic resources and modify their internal policies in anticipation of potential long-term effects.

A list of secondary metabolites, the plant species from which they are derived, and their present countries of production is shown in Table 14.4. Research for all these products is now going on in developed countries to try to produce them in tissue culture in a cost-effective way.

| Table 14.4 | Research in developed countries on high-value secondary metabolites of plants from Third World countries |

Plant	Product	Origin
Catharanthus	Vincristine	Mozambique, Israel, India, Sri Lanka, Thailand, Vietnam, Madagascar
Cinchona	Quinine	Indonesia, South America, Kenya, Tanzania
Digitalis	Digitoxin Digoxin Ditoxin	Eastern Europe, India
Dioscorea	Diosgenin	Mexico, India, China
Jasminum	Jasmine	—
Lithospermum	Shikonin	Korea, China
Papaver	Codeine, opium	Turkey, Thailand
Pyrethrum	Pyrethrins	Tanzania, India, Kenya, Ecuador
Rauwolfia	Reserpine, rescinnamine	India, Mozambique, Thailand, Zaire
Sapota	Chicle	Central America
Thaumatococcus	Thaumatin	Liberia, Ghana, Malaysia
Vanilla	Vanilla	Madagascar, Indonesia, Comoro Islands, Reunion

Source: C. Juma (1989), *The Gene Hunters: Biotechnology and the Scramble for Seeds* (Princeton, NJ: Princeton University Press), p. 137.

Figure 14.6

Vanilla Cultivation in Madagascar (Africa).
Vanilla is a group of climbing orchids that are native to America and Asia. The fruits are large, fleshy pods (called beans) that contain many tiny seeds. Vanilla is extracted from the pods after they have been allowed to ferment and dry out. This worker demonstrates artificial pollination of a vanilla flower. Source: Courtesy of the Food and Agriculture Organization, Rome.

This trend can best be illustrated by analyzing the present production of vanilla (Figure 14.6). The plant *Vanilla planifolia* is indigenous to Central and South America. Commercial production of vanilla occurs largely in Madagascar, the Comoro Islands, and Reunion. Madagascar alone accounts for 75% of the world market (about U.S.$70 million) and earns more than U.S.$50 million in foreign exchange. Vanilla beans account for 10% of Madagascar's export earnings. The production of vanilla involves over 70,000 small landholders (owners of small farms) whose incomes depend on vanilla exports. The current price of vanilla is about U.S.$70–75 per kg. By the two criteria mentioned in the previous section, price per kilogram and size of the market, vanilla is presently not threatened by plant tissue culture. Yet research is underway to try to change this. The price of tissue-cultured vanilla is still in excess of $2,000 per kg, but new techniques may well lower this by a factor of 40, bringing it within range of vanilla extracted from cultivated vanilla beans.

Another example of the relocation of production from Third World countries to developed countries is provided by the newly discovered sweet proteins such as thaumatin. Certain tropical fruits and leaves of tropical shrubs contain small proteins that are 1,500–3,000 times sweeter (on a weight basis) than sucrose. Some companies have established plantations in the tropics to produce these proteins. They can be used as sugar substitutes and advertised as "natural" sweeteners that are safer than synthetic products, such as saccharin or aspartame. Because they are proteins, each is encoded by a single gene, and such genes can easily be isolated and introduced into bacteria, yeast, or plants, which will then produce the sweet protein. Work on the genes of sweet proteins has attracted capital from some of the world's biggest companies, indicating the importance they attach to this emerging technology. Tate and Lyle, the British company that established plantations in Liberia, Ghana, and Malaysia to produce thaumatin, has also introduced the thaumatin gene into yeast by genetic engineering, allowing the company to manufacture the sweetener in large fermentors. The cheaper process will eventually displace the more expensive one.

It is important to note that the relocation of industries and production processes from one country to another is occurring on a large scale in the world today, and the biotechnology industry is not particularly to blame. Economic and political forces, rather than social considerations, are driving this relocation process.

Summary

Plants make a number of valuable chemicals, mostly secondary metabolites, that are also produced at low levels by cells in tissue culture. By carefully selecting clones and manipulating culture conditions, production can be raised to an economically rewarding level. Biotechnology companies have entered this field and are now beginning to produce these chemicals in tissue culture. In addition, research is under way to use genetically engineered cell lines, cultured roots, or whole plants to produce these chemicals more cheaply. With no accompanying changes in global, political, and economic structures, such developments will probably harm the economies of some Third World countries, where most of the plants that produce these valuable chemicals are now grown.

Further Reading

Balandrin, M. 1985. Natural plant chemicals: Sources of industrial and medicinal materials. *Science* 228:1154–1160.

Buitelaar, R. M., and J. Tramper. 1992. Strategies to improve production of secondary metabolites with plant cell cultures: A literature review. *Journal of Biotechnology* 23:111–141.

Chadwick, D. J., and J. Whelan, eds. 1992. *Secondary Metabolites: Their Function and Evolution.* New York: Wiley.

Charlwood, B., and Rhodes, M. J. C. 1990. *Secondary Products from Plant Tissue Culture.* New York: Oxford University Press.

Flores, E. H., M. W. Hoy, and J. J. Pickard. 1987. Secondary metabolites from root cultures. *Trends in Biotechnology* 5:64–69.

Graham, T. L., and M. Y. Graham. 1991. Cellular coordination of responses in plant defense. *Molecular Plant-Microbe Interactions* 4:415–422.

Hartzell, H. 1991. *The Yew Tree: A Thousand Whispers.* Eugene, OR: Hulogosi Press.

Kreis, W., and E. Reinhard. 1989. Production of secondary metabolites by plant cells cultivated in bioreactors. *Planta Medica* 55:409–416.

Sighs, M. W., and H. E. Flores. 1990. The biosynthetic potential of plant roots. *BioEssays* 12:7–13.

Willmitzer, L., and R. Toepfer. 1992. Manipulation of oil, starch and protein composition. *Current Opinion in Biotechnology* 3:176–180.

Chapter 15

Plant Genetic Engineering
New Genes in Old Crops

Genetic engineering involves the transfer of genes from one organism to another. Gene transfers resulting from sexual crosses have taken place throughout the course of plant evolution and are mostly restricted to organisms belonging to the same species. However, gene transfer between very different organisms also occurs in nature. Transfer between *Agrobacterium tumefaciens* (the cause of crown gall disease) and certain plants has provided scientists with the most powerful tool to genetically transform plants. Although other methods are also being used, *Agrobacterium*-mediated plant transformation remains the most convenient. By using this method, scientists merely piggyback the genes they want to transfer onto a naturally occurring gene transfer system. This lets them create transformed or transgenic plants.

Even before the advent of plant transformation, crop plants have been enriched by a large number of genes from wild species, either through natural hybridization or through the near-heroic efforts of plant breeders. Plant transformation procedures can break down the sexual barriers, so that any gene from any organism can be introduced into a plant. To be useful, the gene must be correctly expressed—at the right time and in the right organ, tissue, or cell—and the protein that the gene encodes must have the right function. Only our ability to identify useful genes and the difficulty of introducing genes into certain important crop plants are holding genetic engineers back at the moment. However, progress in plant transformation is so fast (see Table 15.1) that it is safe to predict that plant scientists will be able to transform all the major crop plants with relative ease within the next five years.

1

> ***Agrobacterium*-mediated gene transfer is the most convenient way to create transformed plants.**

In Chapter 3, we noted that when the plant-pathogenic *Agrobacterium tumefaciens* infects a wound site, it causes crown gall disease by transferring some of its own genes to plant cells. These genes, some of which encode

Table 15.1	The history of plant transformation

1983	Tobacco
1984	Carrot, *Lotus*
1985	Oilseed rape, petunia
1986	Alfalfa, *Arabidopsis*, cucumber, tomato
1987	Asparagus, cotton, flax, horseradish, lettuce, poplar, potato, rye, sunflower
1988	Cauliflower, celery, eggplant, corn, orchardgrass, rice, soybean, walnut
1989	Apple
1990	Buckwheat, birch, chrysanthemum, citrus, clover, grapevines, mustard, papaya, strawberry
1991	Carnation, cowpea, kiwi, melon, plum
1992	Sugarbeet, wheat
1993	Pea, barley

Source: Courtesy M. Van Montagu, University of Ghent, Belgium.

enzymes for biosynthesizing plant hormones, are stably integrated into the chromosomes of a single plant cell and subsequently inherited by all the progeny of this cell. This transfer of genes "transforms" the plant cell. The transformed cells synthesize excessive amounts of auxin and cytokinin, and a tumorlike growth or gall is formed. The genes transferred from the bacterium to the plant are carried on a bacterial **plasmid**, a circular DNA molecule. Like many other bacteria, *Agrobacteria* have most of their genes on a single circular chromosome, but have, in addition, some of their genes on a plasmid. The plasmid of *A. tumefaciens* is called the *Ti* or *tumor-inducing plasmid*, because some of the genes on the plasmid are required for tumor induction. The transfer of DNA occurs only after the bacteria have attached themselves to the cell walls of wounded plant cells. Substances that leak out of the wound site activate bacterial genes that encode enzymes necessary for cutting the DNA out of the bacterial plasmid and re-integrating it into the plant DNA.

So how does one go about using this natural gene transfer system to introduce an interesting gene into a plant? Small pieces of plant tissue (leaves, roots, or stems cut into pieces about 2–5 mm long) are cultivated with the *Agrobacteria* that carry the interesting gene in their *Ti* plasmid. This allows the bacteria to transfer their plasmid DNA to individual plant cells at the cut surface of the plant tissue. These transformed plant cells will eventually give rise to a transformed plant under the correct conditions of tissue culture. Thus, plant transformation has three phases:

1 Creating a strain of *Agrobacterium* that has a Ti plasmid with the gene scientists want to transfer to the plant

2 Culturing the bacteria with pieces of plant tissue so that DNA transfer can occur

3 Regenerating a plant from the transformed cells at the same time as the non-transformed cells are being killed by antibiotic

The first phase involves molecular manipulations of DNA. The genes in the Ti plasmid (about seven in all) that cause the tumorous growth of the plant must be removed and be replaced by the genes someone wants to transfer to the plant. These genes on the Ti plasmid occur in one continuous DNA segment that is flanked by a right border (RB) and a left border (LB) as shown in Figure

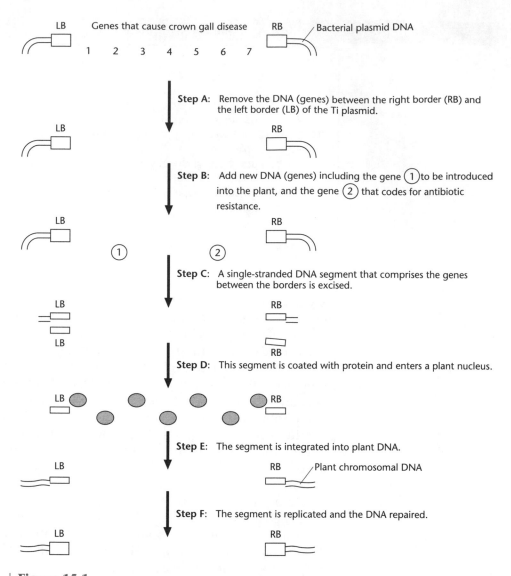

Figure 15.1

Manipulations of Ti Plasmid DNA in the Test Tube to Introduce New Genes (steps A and B), and Transfer of That DNA to the Plant DNA (steps C through F).

15.1. Restriction enzymes can be used to remove these genes, but care must be taken to leave the borders intact, as they are crucial for DNA transfer to the plant and integration into the plant DNA. Next one introduces the target gene, equipped with a suitable regulatory region, and an antibiotic resistance gene, also with its own regulatory region. The product of the antibiotic resistance gene—usually an enzyme that inactivates a specific antibiotic—will let the cells that have this gene (all the cells that have been transformed) survive in a culture medium that contains the antibiotic. The newly constructed Ti plasmid must then be reintroduced into a suitable strain of *Agrobacterium tumefaciens*.

In the second phase, the *A. tumefaciens* bacteria are co-cultivated with small pieces of leaf or root. Specific chemicals leak out of the wound sites where the tissue has been cut, making DNA transfer possible by activating certain genes in the bacteria. Transfer involves cutting one strand of the double-stranded DNA molecule, first at the right border and later at the left border, with the simultaneous copying of the DNA into small single-stranded molecules that contain only the genes contained between the two borders (Figure 15.1). These DNA molecules, which are coated with protein to protect them against degradation, enter the nucleus of the plant cell to which the *Agrobacterium* attached itself and are then integrated into the plant DNA. Integration is followed by synthesis of the second strand and repair of the breaks. The plant DNA is "as good as new," but carries with it some new genes that will be faithfully duplicated each time the DNA replicates and the cell divides.

The pieces of plant tissue are then transferred to a culture medium with two antibiotics: one to kill the bacteria (they are not needed anymore) and one to kill the untransformed plant cells. The transformed cells will now grow into small calluses if the necessary hormones are present in the culture medium to which the pieces of leaf or root were transferred. In the third phase, the small callus pieces can be transferred to a medium that induces shoots (has higher cytokinin levels). Finally, the shoots can be cut off and transplanted to a rooting medium (Figure 15.2).

In transformation, any gene placed between the two borders is transferred to the plant DNA. That includes not only the antibiotic resistance gene, but any other gene the researcher wants to transfer to the plant. All genes should be accompanied by a suitable control region, or promoter, that determines when the gene will be active, in which tissues, and as a result of which hormonal or environmental signals.

2 | **DNA can be introduced into cells by bombardment with DNA-coated particles or by electroporation.**

Although *Agrobacterium tumefaciens* has been tremendously useful in helping transform a large number of plants, many plant species, especially the grasses and cereal grains, are not susceptible to infection by *Agrobacterium*. In addition, a number of important legumes, such as soybean, are not easily regenerated in tissue culture from calluses obtained from *Agrobacterium*-transformed cells. This means that this transformation method cannot be applied to some of our most important crops, such as rice, corn, wheat, sorghum, sugarcane, beans and soybeans. It is quite easy to introduce DNA into the protoplasts of such species, but it is difficult to regenerate fertile plants from the protoplasts.

To overcome these barriers to plant transformation, J. C. Sanford and his colleagues at Cornell University developed a **microprojectile gun** to deliver DNA directly into plant cells by shooting it through the cell wall and the cell membrane. They coated small (4 μm in diameter) tungsten particles with DNA and constructed a gun that could accelerate the particles with enough speed that the particles would enter the first cell layer of a tissue. After lodging in the cells, the DNA was transcribed into RNA and the RNA was translated into protein. Thus the introduced DNA was genetically active. The device used to accelerate the microprojectiles was a modified standard bullet gun (Figure 15.3).

Within a year, scientists at Agracetus, a biotechnology company in Middleton, Wisconsin, refined the technique by using an electrical discharge to accelerate the particles. The spray of particles was aimed at meristematic tissues of young soybean embryos. For reasons not yet clear, DNA that enters the cells this way is also integrated in the chromosomes and passed on to the cells' progeny. The technique does not let all cells in an embryo be hit by

Figure 15.2

Schematic Representation of Two Different Methods to Create Transgenic Plants. In the Agrobacterium *method (left), DNA carrying desired genes is inserted into the tumor-inducing (Ti) plasmid of the bacterium, and when the bacterium infects a wounded tissue, this DNA is transferred to a cell nucleus and integrated into the chromosome. In the particle gun method, metal particles coated with DNA are fired into plant cells, and the DNA becomes integrated into the plant chromosome. When a new plant is regenerated from a single transformed cell, all the cells in the plant carry the new genes. Adapted from "Transgenic Crops" by C. S. Gasser and R. T. Fraley,* Scientific American, *June 1992. Copyright 1992 by* Scientific American, Inc. *All rights reserved.*

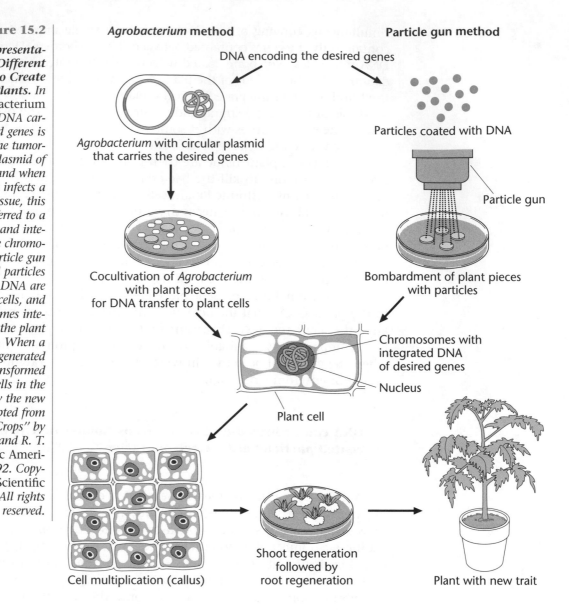

Agrobacterium **method**   Particle gun **method**

DNA encoding the desired genes

Agrobacterium with circular plasmid that carries the desired genes

Particles coated with DNA

Particle gun

Cocultivation of *Agrobacterium* with plant pieces for DNA transfer to plant cells

Bombardment of plant pieces with particles

Chromosomes with integrated DNA of desired genes

Nucleus

Plant cell

Cell multiplication (callus)

Shoot regeneration followed by root regeneration

Plant with new trait

particles, so untransformed cells usually coexist with transformed cells. However, when the embryos are cultured and grown into mature plants, a single cell in the embryo may give rise to an entire sector of a mature plant. Embryo bombardment results in plants with transformed and untransformed sectors. Some seeds of such plants may be completely transformed because reproductive tissues are derived from small groups of cells that arise relatively late in plant development. Thus an entire flower can be derived from a small cluster of transformed cells that grew out of a single transformed cell. The DNA originally present as a coating on the particles becomes integrated into the chromosomes of the plant cells. As these cells divide and pass on their DNA, the introduced DNA is present in all progeny of the cell that was originally transformed. The bombardment procedure has been used by scientists at Agracetus to transform soybeans and beans (Figure 15.2).

Electroporation is a totally different way of introducing DNA into cells. It is usually applied to protoplasts (cells whose cell walls have been removed by enzymes), but has more recently been proven successful for introducing DNA into the cells of apical meristems that are partially digested with

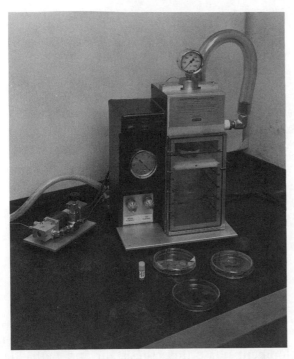
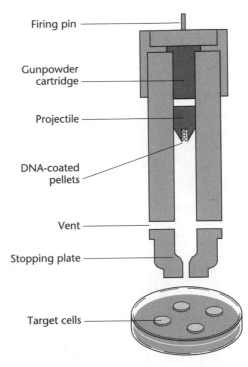

Figure 15.3

DNA Particle Gun. *The DNA particle gun, developed by John C. Sanford of Cornell University, fires tungsten pellets coated with DNA into plant cells. The pellets are held by a plastic microprojectile, which is accelerated by a gunpowder charge. The plate stops the microprojectile; momentum sends the DNA-coated pellets into the target. The instrument shown is the Biolistic® system from Bio-Rad, but other instruments using variations of this basic principle have been developed.*

enzymes that degrade cell walls. Electroporation involves a sudden electrical discharge in a very small container that holds plant protoplasts and DNA molecules in a liquid suspension. The discharge opens the plant plasma membrane simultaneously in many different places, allowing the DNA molecules to go into the cell. The next step is to culture the protoplasts and try to regenerate plants from them. This is the most difficult step, because many plant species cannot readily be regenerated from single protoplasts.

Scientists from Plant Genetic Systems in Belgium recently applied electroporation to an organized meristem that was partially digested by cell-wall-degrading enzymes, then rinsed away the enzymes and transferred these meristems to a tissue culture medium. Using normal selection procedures, they obtained transformed shoots that were grown into plants. In this way, they circumvented the problem of regenerating plants from single protoplasts. This method holds great promise for transforming important crop species, such as wheat and corn, that so far have been difficult to transform.

3 | **Genetically engineered crops are on the horizon.**

In previous chapters, we described numerous examples that show how genetic engineering of crop plants will let scientists change agricultural and food production practices. Table 15.2 lists approaches that are now on the

Table 15.2	Genetically engineered traits on the horizon

Improving Pest and Weed Management
    Herbicide tolerance
    Virus resistance
    Insect resistance
    Bacterial resistance
    Fungal resistance
	Improving Plant Breeding
	    Male sterility; production of hybrid
	    seeds

	Improving Nutritional Quality
	    High-methionine and high-lysine
	    seeds
	    Forage rich in sulfur amino acids

Improving Agronomic Properties
    Altering cold sensitivity
    Improving water stress tolerance
    Improving salt tolerance
	Molecular Farming
	    Oils
	    Starch
	    Plastic
	    Enzymes, pharmaceuticals

Improving Postharvest Qualities
    Delay of fruit ripening
    Delay of flower senescence
    High-solids tomatoes
    High-starch potatoes
    Sweeter vegetables
	Detoxifying Contaminated Soils

horizon. In some cases, the companies that are generating the transgenic crops are well into the development phase (insect-resistant and herbicide-tolerant cotton, and more slowly ripening tomatoes). In other cases, research has shown only that the approach is feasible. The various research efforts are listed under seven different headings to emphasize that genetically engineered plants are expected to impact all aspects of the farming, food production, and food processing industries. The production of more nutritious crops is not particularly high on the agenda of biotechnology companies. If more nutritious plants are produced, they will most likely impact the meat production industry first, as the primary goal appears to be feed grains that are now deficient in certain essential amino acids. Some effort is going into molecular farming. Crop plants could readily be used to produce both high-volume, low-value products, such as specialty starches and oils for industry; as well as low-volume, high-value products, such as vaccines and other pharmaceuticals.

In most cases, genetic engineering means adding one or more new genes to the plant's genetic material, so that a new enzyme will be made in the transgenic plant. However, genetic engineering can also suppress the appearance of a specific enzyme, by using "antisense" genes. An antisense genetic construct consists of a control region of a gene followed by the protein-coding region of a gene already present in the plant, in reverse orientation. After such a construct is introduced into a plant, antisense mRNA will be made, and this antisense messenger RNA will bind to the normal messenger RNA, thereby preventing synthesis of the protein (Figure 15.4). Thus, in the antisense plants the enzyme has been knocked out. Slowly ripening tomatoes in which an enzyme that catalyzes biosynthesis of the plant hormone ethylene has been knocked out are likely to be the first "antisense crops" to appear on the market.

Food surpluses are one of the major economic problems facing industrialized countries, such as Australia, the United States, France, and Germany. The low prices resulting from the surpluses create demands by farmers for a price support system in which the government becomes the major player in perpet-

Figure 15.4

*Schematic Represen-
tation of Sense and
Antisense Genetic
Constructs.*

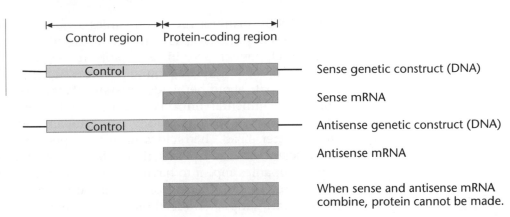

uating the system that makes the surpluses possible. Public funds collected as taxes from the general population are used to keep farmers and farming corporations producing crops for which there is no market.

One way out of this vicious cycle, without penalizing farmers, is to have alternate crops available that produce nonfood commodities needed by society. Advocates of molecular farming see a future in which farming corporations use the land to produce chemicals for industry in addition to food for people. This is not a new development, of course. Plant products, especially starch, cellulose, and oil, are already widely used for producing chemicals that have industrial, rather than food, uses. For example, more than half the starch produced in western Europe is not used as food. The conversion of sugarcane into alcohol as a gasoline substitute is a major industrial use of agricultural products in Brazil. Genetic engineering will make it possible to create crops that synthesize chemicals now produced only by exotic bacteria, such as polyhydroxybutyrate, a plastic material that can be used to make biodegradable packaging materials.

4 | Pest management can be improved by using herbicide-tolerant and pest-resistant plants.

Herbicide-Tolerant Plants. Many of the technical aspects of creating transgenic herbicide-tolerant plants are discussed in Chapter 13. The genes that make crops tolerant to herbicides have been derived from bacteria and encode enzymes that detoxify the herbicides or encode target enzymes that differ from the plant ones in that they are insensitive to the herbicide. This goal of the biotechnology industry has come in for much criticism. The arguments opposing it are summarized in a paper published in 1990 by the Biotechnology Working Group, and entitled *Biotechnology's Bitter Harvest: Herbicide Tolerant Plants and the Threat to Sustainable Agriculture.* The report points out that more than 27 private companies and 20 public research laboratories were doing or had engaged in research aimed at creating herbicide-tolerant plants. The reasons for this burst of research were that:

1 Such work was easy to do because the genes could readily be isolated from microorganisms and a single gene could make a plant herbicide tolerant.

2 Many agrichemical companies invest heavily in herbicide development and they want to get the most out of their investment.

In the 1980s, many of these companies acquired seed companies and their goal is to sell to the farmers both the transgenic seeds and the herbicide as a package. Opponents see this as a plot on the part of agribusiness, whereas proponents see it simply as sound business strategy.

The debate about herbicide-resistant plants revolves mostly around possible environmental impacts. "Good" herbicides are those that can be applied in low doses, are readily biodegradable, and do not seep into the groundwater. "Bad" herbicides have the opposite characteristics. Initially, plant scientists made plants that tolerated both types of herbicides, but most companies appear to have dropped research and development of crops that are tolerant of the environmentally unfriendly herbicides. If the herbicide-tolerant crops are indeed planted by the farmers, then the use of those herbicides to which they are tolerant will undoubtedly increase. One benefit could be that "good" herbicides will gradually replace "bad" ones. This would result in the use of less chemicals, and of more biodegradable chemicals, both desirable goals from an environmental point of view. Monsanto's glyphosate (Roundup®), Dupont's sulfonylureas and Hoechst's glufosinate are probably in the category of "good" herbicides. Some will argue, of course, that there are no "good" herbicides and that our goals should be a chemical-free agriculture. Some other herbicides are harder to categorize. For example, bromoxynil is readily biodegradable, but we do not know what happens to the breakdown products and what effects they have on the plants or the soil microorganisms. Furthermore, bromoxynil is a teratogen that may cause cancer.

The availability of herbicide-resistant plants may bring about a switch from pre-emergence to post-emergence herbicide treatment. Currently, farmers use many pre-emergence herbicides. They treat the fields with herbicides on the assumption that weeds will be a problem. This is the antithesis of integrated pest management. With widely applicable post-emergent sprays, a farmer will be able to use a different strategy: assess the weed population, estimate possible economic injury and make a decision about spraying based on the cost of the herbicide application. This could result in less herbicide use. However, one area where herbicide usage could grow enormously is in forestry. At the moment, only a small fraction of forest land is treated with herbicides, but genetic engineering of herbicide-tolerant tree species could greatly expand this use of herbicides. A number of tree species can readily be transformed, and herbicide-tolerant poplars are already available.

The bottom line for the farmer will be the yield of the herbicide-resistant plants that have been sprayed with herbicide. So far, the companies that make these plants have not yet released data to show that yields of transgenic herbicide-treated plants are comparable with those of non-transgenic plants. Economics will also enter into it. Will the price increase of the transgenic seeds that the farmer must buy, together with the cost of the herbicide, be less than the traditional herbicide treatment? And if the price of the former is greater, is it compensated for by an increase in yield caused by a reduction in weeds?

Insect-Resistant Plants. Many different insects damage crop plants, reducing the yield or quality of the harvested food. Some insects eat the roots and others the leaves, thereby diminishing plant growth and ultimately crop production. Others destroy the crop as it is growing, such as the cotton boll weevil and the corn earworm. Yet others, such as the bruchid beetles (Figure 15.5), infest stored seeds, which is particularly troublesome

Figure 15.5

Cowpea Weevil, Callosobruchus maculatus. Bruchid beetles lay eggs on cowpeas, and the young larvae burrow into the dry bean. These so-called storage pests cause serious food losses, especially in Third World countries. Source: Courtesy of L. Murdock, Purdue University.

for people who practice subsistence agriculture and must store their seeds without adequate facilities. We discussed in Chapter 13 several approaches to making transgenic plants that resist insects, including the use of enzyme inhibitors (protease inhibitors and amylase inhibitors), lectins, and Bt-toxins. The gene for Bt-toxin can be incorporated into the plant, or Bt-toxin can be delivered as a spray when it is present in other bacteria that have been heat-killed.

The plant genetic engineering approach to insect control is usually extremely specific. A gene will be introduced in a plant to kill just one type of insect, usually the major pest. The chemical insecticides that are used to kill insects are far less specific and affect a broad range of insects. This lack of specificity is both the strength and the weakness of chemical pesticides. It is a weakness because beneficial insects are killed as well, and the balance in the ecosystem is upset. It is a strength because more than one noxious insect can be killed at the same time. If farmers buy genetically engineered seeds that produce plants that are resistant to a single insect, they still must spray chemicals against the other insects, because crops have many insect pests. The expense of using chemical pesticides is causing many farmers to adopt integrated pest management, a system of pest control more in tune with sustainable agriculture. It is likely that transgenic insect-resistant crops will be a part of future integrated pest management programs.

A major drawback of chemical pesticides is that resistant pest strains arise relatively rapidly, because of the high selection pressure that results from killing the susceptible pest population. This leaves only resistant individuals, which can then multiply. This problem is not avoided by using transgenic plants. If the selection pressure on the insect population is high, then the genetic composition of the population changes rapidly and resistant strains arise. One way to avoid this problem may be to mix 1–2% seeds from nontransgenic plants with seeds from transgenic plants at planting time. In this way, a small population of susceptible plants is ravaged by the insects while the majority of plants are resistant. Another way—and according to population biologists, a better way—would be to set aside 1 to 2 percent of the land for planting susceptible lines of the same crop. These approaches reduce selection pressure on the insects and delay the emergence of insect strains that can attack the transgenic plants.

Virus-Resistant Plants.   Viruses cause substantial crop losses, especially in the tropics. Viruses multiply mostly in the leaves and cause the leaves to be yellow, smaller, and/or crinkled. The net result of virus infection is seriously inhibited photosynthesis and a reduced crop yield. No chemical sprays kill viruses, and to control them, two different approaches are used. Since most viruses are spread by aphids and other insects (generally referred to as *vectors*), chemical sprays can be used to kill the vectors. Some viruses are seed-transmitted, or transmitted with the planting stock in the case of vegetatively propagated plants such as potatoes. In that case, it is necessary to start with planting material or seeds that are free of the virus. Through micropropagation, biotechnology already has a major role in assuring the continuous supply of virus-free stocks of potatoes and other crops.

Genetic engineering of virus-resistant crop strains will be a major advantage for crop production. One of the methods that has already been used with several different crop species and for several different viruses was described in Chapter 13. This method depends on expressing the gene for the virus coat protein in the plant. When this protein is present in the cytoplasm of the plant cells, the virus cannot reproduce after infecting the cells. The first genetically engineered plants grown for commercial production anywhere in the world were tobacco plants resistant to tobacco mosaic virus that used the coat protein strategy. These plants were in the fields of commercial tobacco growers in China in 1993. This approach works for some but not for all viruses.

A second method that is being tried for killing viruses and other pathogens is to rapidly kill the cells that are being infected. If the cell in which the pathogen starts to multiply is killed rapidly by the infection, it will not support the multiplication of the pathogen and its spread to other cells. Killing only those cells that have been infected can be achieved by using a gene that encodes a cytotoxic enzyme, together with a control region that is only activated when viruses or pathogens are in the cell. Proteases that digest all cellular proteins, or ribonucleases that digest all RNA molecules, are good candidates for cytotoxic enzymes. This approach has already been shown to work for making male-sterile plants (see Chapter 10). Suitable gene control regions can be obtained from the pathogens themselves. Other approaches to genetically engineered plants resistant to bacterial and fungal pathogens were discussed in Chapter 13.

5 | **A major goal of plant biotechnology is to improve postharvest crop qualities.**

Regulating Fruit Ripening and Extending the Shelf Life of Vegetables and Flowers.   Senescence is a natural development process in plants that occurs at the end of an organ's life. On deciduous trees, leaves senesce and yellow in the autumn. In fruits, senescence is called *ripening,* and it is associated with desirable changes in the fruit such as softening, conversion of starch to sugar, loss of chlorophyll, synthesis of red pigments, and synthesis of aromatic compounds.

Like most other processes of plant development, senescence is controlled by hormones. In leaves, senescence does not occur as long as levels of auxins and cytokinins are high, but once these levels drop senescence can be greatly accelerated by the hormone, ethylene. Thus, when leafy vegetables are har-

Figure 15.6

Steps in the Biosynthesis of Ethylene. In 1, the enzyme ACC-synthase catalyzes the synthesis of aminocyclopropane carboxylic acid (ACC) from methionine. ACC is the key intermediate in ethylene biosynthesis; in 2, ACC-oxidase converts ACC into ethylene.

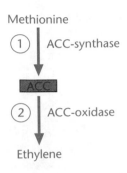

vested, the resulting drop in auxin and cytokinin and the synthesis of ethylene cause senescence. Vegetables senesce in the supermarket, and flowers start to senesce in the flower store. Refrigeration slows down the senescence processes, but cannot stop them.

Ripening fruits produce ethylene, and this ethylene production starts and accelerates the ripening process. The banana industry has taken advantage of scientists' understanding of the hormonal control of fruit ripening. Bananas are now picked green, transported half-way around the world and then provided with ethylene, the normal fruit-ripening hormone, to start the ripening process. What works reasonably well with bananas does not work nearly as well with other fruits, such as tomatoes. Tomatoes are also picked green, transported, and gassed with ethylene—but most consumers who know what vine-ripened tomatoes taste like consider the product for sale in markets only barely acceptable. To be tasty, tomatoes should be left on the plant much longer, but at that stage they are much harder to handle because of fruit softening. The regulation of fruit ripening and of senescence in flowers and leafy vegetables pose a challenge to the biotechnology industry, because the handling and storage result in considerable losses of fruits and vegetables.

Genetic engineering to alter ripening of tomatoes is now under way in several laboratories and the approach involves the suppression of ethylene synthesis by the tomato fruits themselves. The final two steps in the biosynthesis of ethylene, illustrated in Figure 15.6, involve first the synthesis of the key chemical aminocyclopropane carboxylic acid (ACC) by the enzyme ACC-synthase, and its subsequent conversion to ethylene by the enzyme ACC-oxidase. Ethylene synthesis can be suppressed with antisense technology using the genes that encode either one of these enzymes in the reverse orientation and hooked to a control region that directs gene expression in the entire plant. When such experiments were performed with tomato plants by research scientists in the United Kingdom and the United States, there was a dramatic reduction in ethylene synthesis, accompanied by a delay in fruit ripening. Ethylene synthesis in transformed fruits was down to 5% of control fruits, and ripening took four weeks instead of one week.

Scientists at Monsanto Company used a different approach to get the same result. They introduced into tomatoes a bacterial gene that encodes an enzyme that degrades ACC as soon as it is formed. The net result again is a suppression of ethylene synthesis and a delay in fruit ripening (Figure 15.7). The scientists from these different research groups hope that by making fruit ripening much slower, it will be possible to leave the fruits on the plants longer. The fruits can then be picked when they are turning red, can be transported, and can arrive in the grocery store a week before they become completely soft and red. This avoids picking the tomatoes green and having to

Figure 15.7

Slow-Ripening Tomatoes. The tomato on the left is a normal tomato, while the one on the right has a transgene that knocks out ethylene synthesis, reducing it to 10% of normal. The tomatoes were picked when pink and kept under sterile conditions at room temperature. By the time the normal tomato has completely shriveled up 137 days after harvest, the transgenic tomato is still firm and red (color not shown), because the entire ripening–senescence process has been slowed down. Source: Courtesy of H. Klee, Monsanto Company.

treat them with ethylene to initiate the ripening process. An added benefit is that the tomatoes will not need to be refrigerated during shipping. The normal ripening process will go on at a reduced rate during transport.

Genetic engineers are also hard at work to alter the softening of tomatoes that is associated with ripening. Softening occurs when the cell walls are degraded by cell-wall-degrading enzymes. At least two such enzymes have been identified in tomatoes: pectin methyl esterase and polygalacturonase. Scientists from Calgene in Davis, California, suppressed polygalacturonase using antisense genes. This produced tomatoes that can be left on the vine longer than normal tomatoes. The effect on softness is not spectacular, but the tomatoes do not spoil as easily and can therefore be harvested later. Calgene markets these tomatoes as FlavrSavr® tomatoes, but whether they are indeed as flavorful as ones picked from the vine is debatable. Any major advantage is likely to benefit the large tomato-processing industry in California, because it will make mechanical harvesting of tomatoes (Figure 15.8) easier and cheaper.

Many genetic engineering projects carried out by biotechnology firms are aimed at making harvesting and industrial food processing easier and less expensive, rather than at improving the nutritional value of the food. Previous classical plant-breeding objectives (for example, tomatoes with tough skins or compact cotton plants) had the same general goal. We do not necessarily need to criticize the companies for these priorities. More consumers are willing to pay extra for good looks and ease of preparation than for intrinsic food quality, and whatever its history, farming in the developed countries is a profit-oriented enterprise.

Engineering High-Starch Potatoes. If potatoes are already so starchy, why does industry want to make them starchier? In the United States, most potatoes are not consumed directly, but are processed into potato chips or frozen french fries. For these purposes, the potatoes should ideally contain about 25% starch by weight. Most potatoes processed contain only 21–22% starch, so more water needs to be evaporated, and more fat is absorbed, by the potatoes during cooking. Using a bacterial gene, scientists at Monsanto recently increased the starch content of potatoes from 22 to 25%, holding out hope that potato chips containing less cooking fat may be just around the corner.

The biochemistry behind this development involves the biosynthesis of starch itself. Starch, a polymer of glucose, is made by transferring glucose molecules one at a time from a glucose donor called *ADP-glucose* to a glucose

Worldwide, tomatoes are the most widely grown vegetable after potatoes and 50 million tons are marketed annually. In the United States, and especially in California, where 85% of U.S. tomatoes are grown, tomatoes form the basis of a multibillion-dollar industry that produces ketchup, soup, paste, sauce, canned tomatoes, and a variety of other products. These products are all made from so-called processing tomatoes, which come in many different varieties and flavors and differ from the fresh tomatoes sold in the supermarket. They are usually smaller in size and have tougher skins so that they can be harvested mechanically and transported to the factories in large, 25-ton gondolas without getting crushed.

Tomatoes are 95% water, and the remaining 5% consists of pulp, seeds, and soluble solids, mostly sugars, organic acids, and flavor compounds. Many tomato products (ketchup, paste, sauce) require that most of this water be removed, and much of the price of these tomato products is determined by the cost of removing water. Increasing the solids content of tomatoes from 5 to 6% would save the tomato industry in the United States about $75 million a year. It is not surprising, then, that plant breeders and genetic engineers are hard at work to achieve this goal.

One way to achieve this goal might be to increase the invertase content of tomatoes. Invertase is an enzyme that splits sucrose (cane or beet sugar) into the simple sugars glucose and fructose. When tomatoes ripen, they first store sugar as sucrose, but as ripening progresses the cells synthesize invertase. This enzyme converts the stored sucrose to glucose and fructose. It has been suggested that genetic engineering with invertase genes could increase the flux of sucrose from the plant into the fruits and then hasten its conversion to the simple sugars that make up most of the soluble solids. Such a gene could be expressed with a fruit-specific regulatory region that causes the gene to turn on at the right time during ripening.

Scientists at Purdue University unexpectedly stumbled on another approach. In trying to alter the ripening process of tomatoes, they suppressed the expression of pectin methyl esterase, an enzyme involved in cell wall degradation. To their surprise, they found an increase in soluble solids. The chemical nature of these soluble solids differs substantially from that of the soluble solids found normally, and whether these solids are desirable in good ketchup remains to be evaluated.

At an international meeting of plant scientists in Spain in 1993, genetic engineers from Monsanto company revealed yet another approach to the ketchup problem. By expressing the gene that encodes the enzyme ADP-glucose pyrophosphorylase in the tomato fruits, they were able to increase the starch content of the tomatoes—this did not make "starchy" tomatoes, as they still only contained a few percentage points of starch—and thereby caused a dramatic increase in the viscosity of the tomato juice. Ketchup manufacture may now require less water evaporation of the juice from such tomatoes. However, this would mean that the various flavor compounds would be less concentrated in the ketchup. We all want cheaper ketchup, but not at the price of changing its exquisite organoleptic properties.

acceptor, the growing starch molecule. The enzyme catalyzing this transfer seems to be plentiful in potatoes, but the enzyme that makes ADP-glucose is not as abundant. The gene for this enzyme was cloned and isolated from a bacterium by Jack Preiss and his collaborators at Michigan State University, who equipped it with a regulatory region specific to potato tubers before

Figure 15.8

Mechanical Harvesting of Tomatoes in California. Source: *U.S. Department of Agriculture.*

transforming it into potato plants. The net result was tubers on transgenic potato plants that synthesized ADP-glucose more rapidly and could therefore produce more starch.

6 | **Crop improvement is focused on producing more nutritious crops for people and animals.**

We noted in Chapter 4 that humans require eight essential amino acids that must be supplied by the foods they eat. When these are supplied by plant proteins, it is usually necessary to mix different foods because one or another essential amino acid may be below the optimal level. People generally prefer to eat a mixture of foods rather than a single food, and most human cultures have solved this essential amino acid deficiency problem by preparing meals based on a mixture of cereals and legumes: rice with tofu, tortillas with beans, or chapattis with dal. Other monogastric animals, such as poultry and pigs, have similar requirements for essential amino acids. They must also be fed a mixture of proteins, although the industry often prefers to supplement animal feed with chemically synthesized amino acids.

It has been argued that methionine-rich beans or lysine-rich corn could provide protein with a complete amino acid profile for the poorest and most poorly-fed people who rely on a single staple such as corn. Whereas this is true, efforts should probably be directed at providing people with a more diverse diet that is nutritionally adequate even if completely plant-based. This would enrich their lives more than a nutritionally-adequate single staple. To feed monogastric animals, such as pigs and poultry, the meat industry relies on supplementation with synthetic amino acids rather than mixing legumes and cereals, because the starch or oil content of such mixtures is too high to achieve maximal fattening rates. The animals grow faster on a high protein diet. To capture this market, Larry Beach and his colleagues at Pioneer, one of the largest seed companies in the United States, are transferring genes that encode methionine-rich proteins into soybeans. They reason that the feed industry will not have to purchase the synthetic amino acids if the transgenic soybeans are nutritionally adequate in methionine. Such genes have been found in a number of plants (sunflower, brazil nut), and the proteins they encode are present in small amounts in the seeds. Up to 20% or more of the amino acid content of these proteins is methionine. By using strong seed-specific gene control regions, it should be possible to increase the amount of this protein in the seeds of various crop plants and thus provide a more balanced amino acid profile in the seeds.

The same proteins rich in sulfur-containing amino acids are being expressed in the leaves of alfalfa and clover by T. J. V. Higgins and his colleagues in Australia. Their goal is to increase the sulfur amino acid level of feed consumed by sheep that produce wool. Wool is very high in sulfur amino acids (wool is a protein), and wool growth is limited by the sulfur amino acids in the diet of the sheep. By eating a diet richer in sulfur amino acids, the same number of sheep will produce more wool.

7 | Genetic engineering seeks to improve plant breeding and agronomic properties.

Most important agronomic properties of crops, such as drought resistance, cold-hardiness, or yield potential are multigene characteristics that are not easily dealt with using current biotechnological approaches. For example, winter rye is a particularly cold-tolerant cereal, but not nearly as valuable as wheat. Many genes are involved in the cold-hardiness of winter rye, and classical plant breeders have not yet succeeded in transferring the genes from rye to produce a wheat plant as hardy as rye. The recent successes in transforming wheat in at least three different laboratories open the way to introduce the cold-hardiness genes. However, scientists really do not have a good idea yet of which genes should be transferred to confer cold resistance to a crop. One successful approach has been to alter the nature of the phospholipids in the cellular membranes.

Cold-hardy plants have more unsaturated fatty acids in the phospholipids that form the cellular membrane than cold-sensitive plants. By introducing a gene that encodes an enzyme that preferentially uses unsaturated fatty acids to make phospholipids, scientists in Japan were able to make *Arabidopsis* plants more resistant to exposure at low temperatures (see Box 15.2). It is likely however, that other genes will need to be transferred to make cold-hardy plants.

A promising approach to making plants drought resistant was recently tried by Hans Bohnert from the University of Arizona. He reasoned that because many plants respond to drought stress by synthesizing a group of sugar derivatives called *polyols* (mannitol, sorbitol, and so on), plants that have high levels of polyols may be more resistant to stress. Using a bacterial gene that encodes an enzyme capable of synthesizing mannitol, he genetically engineered plants to accumulate rather high levels of mannitol (about 30 to 40 g of mannitol per kilogram of plant material). When these plants were subjected to a simulated drought, they performed better than the control plants. As in the case of cold hardiness, we will have to identify the many other genes that contribute to the drought-resistant phenotype.

To cope with the ever-changing environment, plants have evolved mechanisms that let them survive unusually low temperatures without injury. Many plants that grow in tropical or subtropical regions are sensitive to chilling. When such plants, which normally grow at 25–35°C, are suddenly cooled to 5–12°C, their leaves lose chlorophyll, develop chlorotic lesions, or appear unhealthy and water soaked. They are suffering from chilling injury. Similarly, when plants that grow in temperate regions at 12–22°C are suddenly cooled to –5°C, they may suffer cold or frost injury. However, if the temperature is lowered very gradually, about one or two degrees Centigrade each day, then these plants can easily withstand the lower temperature regime, because they have had time to become acclimated. Such acclimated plants are resistant to chilling or frost.

Plant breeders have long known that different varieties of the same crop—tomatoes, for example—show marked differences in their chilling sensitivity. Such observations indicate that properties such as chilling sensitivity and frost resistance have a genetic basis and that it may be possible to identify the genes responsible for this resistance. A plant's ability to become cold acclimated is also genetically determined.

The acclimation process, which makes plants cold or frost resistant, is accompanied by marked changes in cellular components: sugars and specific amino acids accumulate, "antifreeze" proteins are made, and there is a gradual change in the fatty acids of the phospholipids in the cellular membranes. When comparing the fatty acids in the membranes of chilling-resistant versus chilling-sensitive plants, or acclimated versus nonacclimated plants, biochemists found a greater abundance of unsaturated fatty acids with either two (linoleic) or three (linolenic) unsaturated bonds per fatty acids in the phospholipids of the cellular membranes of the acclimated plants. The unsaturated fatty acids make the membranes more fluid at the lower temperatures, preventing them from solidifying when the temperature drops below 10–12°C.

These observations prompted molecular biologists to isolate the genes for an enzyme involved in the biosynthesis of these phospholipids: a fatty acyl transferase that specifically attaches the unsaturated fatty acid to the glycerol backbone of the phospholipid (see Chapter 4). With the gene in hand, it became possible to make transgenic plants and assess the effect of introducing a single gene on the chilling sensitivity of plants.

The results obtained by a Japanese research group led by Norio Murata were startling. He transferred the gene for an acyl transferase enzyme from a chilling-sensitive plant (squash) and a chilling-resistant weed species (*Arabidopsis thaliana*) to tobacco plants. They used tobacco solely because it can be easily transformed with genes from other plants. They found that the introduction of the squash enzyme into the tobacco caused the membranes of the tobacco cells to have more saturated fatty acids, and this resulted in tobacco plants that were more chilling-sensitive. Conversely, introducing the *Arabidopsis* acyl transferase made tobacco more chilling-resistant with cellular membranes that have more unsaturated fatty acids in their phospholipids.

Chilling and frost resistance will undoubtedly turn out to be complex phenomena that involve many genes. Nevertheless, these promising experiments show that transfer of just one gene already altered the chilling sensitivity of a plant by a few degrees Centigrade.

These two examples show that we know very little about the complex adaptations of plants to environmental stresses, and that much basic research is needed in this field. We need to know how plants adapt themselves to the gradual imposition of an environmental stress, and which genes are turned on and what roles the proteins encoded by these genes play in the adaptation.

A major breakthrough in manipulating agronomic properties is likely to come from the recent development that makes it much easier to produce hybrid seeds. The introduction of hybrid corn in U.S. agriculture started a major period of increases in the productivity of this crop because of hybrid vigor, the increase in yield that occurs when one crosses two inbred lines. Because corn has separate male and female flowers, producing hybrid seeds turned out to be rather easy (see the discussion in Chapter 10). It is much more difficult to produce hybrid seeds if the male and female parts are in the same flower. However, in spite of the expense and difficulty, hybrid seeds for a number of vegetables are available. The increase in yield makes up for the high price of the seed ($5,000 per kg).

A genetically engineered male sterility system based on killing the pollen-producing cells with an enzyme that destroys all the RNA in the cell has recently been introduced by the Belgian biotechnology company Plant Genetic Systems (see Chapter 10 for a discussion). It is likely that this system will rapidly be engineered into a large number of vegetables, such as carrots, tomatoes, broccoli, cauliflower, and peppers, as well as into some major crops, such as wheat and rice. The system has already been introduced into corn and canola, and yield trials with hybrid canola are now underway in Canada. It remains to be determined whether these modified crops will be as productive as current varieties and whether their production will be economically feasible for producers.

8 | Molecular farming can produce oils, starches, and plastics.

Production of Valuable Industrial Oils. The oils produced by plants represent a vast renewable resource with a world production of 60 million tons. In the United States, about two-thirds of the plant oils are consumed by people, while the remainder is used by industry. We discussed both the molecular structure and the food value of oils and fats in Chapter 4. Oils (and fats) consist of a molecule of glycerol to which three fatty acids are attached. The properties of the fatty acids (their length and the number of unsaturated bonds) determine the physical and chemical properties of the oils and their uses in the industrial sector.

The edible oils—and many oils are not edible—all have rather similar fatty acids with 18 carbon atoms and either one, two, or three unsaturated bonds. Although some oils are more nutritionally desirable than others, the oils found in different crops are largely interchangeable. Exceptions are olive oil and canola oil, which are valued for their low levels of saturated fatty acids; and cocoa butter, which is valued for fatty acids that confer physical properties desirable for the confectionery and cosmetics industries.

A survey of tens of thousands of plant species shows that 95% of all plant oils are made from just six fatty acids. Yet more than a thousand different fatty acid structures have been identified, mostly in undomesticated plants.

Table 15.3		Some specialty uses of plant fatty acids and oils		
Lipid Type	Example	Major and Alternative Sources	Major Uses	Approximate U.S. Market Size (10^3t)
Medium chain (C8–C14)	Lauric acid	Palm kernel, coconut, *Cuphea*	Detergents	640
Long chain (C22)	Erucic acid	Rapeseed, *Crambe*	Lubricants, nylon, plasticizers	20
Epoxy	Vernolic acid	Epoxidized soybean oil, *Vernonia*	Plasticizers	64
Hydroxy	Ricinoleic acid	Castor bean, *Lesquerella*	Lubricants, coatings	45
Trienoic	Linolenic acid	Flax	Coatings, drying agents	30
Low-melting solid	Cocoa butter	Cocoa bean	Chocolate, cosmetics	100
Wax ester	Jojoba oil	Jojoba	Lubricants, cosmetics	0.35

Some of these fatty acids have found quite unique industrial uses. Some are obtained from domesticated plants, others from wild plants (Table 15.3). Some fatty acids might also find industrial uses if they could be produced in large-enough quantities. The approach of the biotechnology industry is not to domesticate the undomesticated plants that presently produce these fatty acids, but rather to transfer the necessary genes to oilseed crops that we already know how to grow and cultivate. Canola, soybean, flax, cotton, sunflower, and safflower all produce oil-rich seeds, and all have been transformed with foreign genes. The task ahead is to isolate the necessary genes, express them with seed-specific control regions, and transform the crop plants.

An example of what could be done is provided by the castor plant (*Ricinus communis*). Castor plants grow wild in many places and are cultivated on a small scale in India. Their seeds contain four potent plant defense chemicals: a toxic lectin, a potent allergen, a toxic alkaloid, and an oil that contains 90% ricinoleic acid, well known for its cathartic effects. Unlike most fatty acids, the hydrocarbon chain of ricinoleic acid has a hydroxyl group (as well as an unsaturated bond), and this makes it an extremely versatile natural oil. The industrial uses of castor oil include the synthesis of nylon, and the manufacture of lubricants, hydraulic fluids, plastics, cosmetics, and other materials.

Even if the plants were domesticated for higher oil production, we would still have to cope with the toxic chemicals in the seeds. Recent research shows that a single enzyme in castor seeds converts a normal fatty acid (oleic acid) into this unusual fatty acid (ricinoleic acid). Because oleic acid is synthesized by other oilseed crops, introducing this enzyme into the seeds of such a crop would result in the synthesis of ricinoleic acid by a crop that is already domesticated and that farmers know how to grow.

That such simple biotechnological goals can easily be achieved was demonstrated recently by a team of scientists at Calgene in California. Maelor Davies, Toni Voelker, and their collaborators transferred the capacity for short-chain fatty acid production from the California bay tree to oilseed rape. Short-chain fatty acids, such as lauric acid, are used very widely to make detergents and are now obtained primarily from tropical oils, such as coconut oil. Fatty acids are synthesized by a process of stepwise elongation, two carbon atoms at a time, while they are attached to a carrier protein (Figure 15.9). Cleaving the fatty acid from the carrier protein so that it can be used for synthesizing a fat molecule is

Figure 15.9

Scheme for Fatty Acid and Oil Synthesis. This scheme shows the importance of thioesterase (TE) in freeing a fatty acid of the correct length from the acyl carrier protein (ACP).

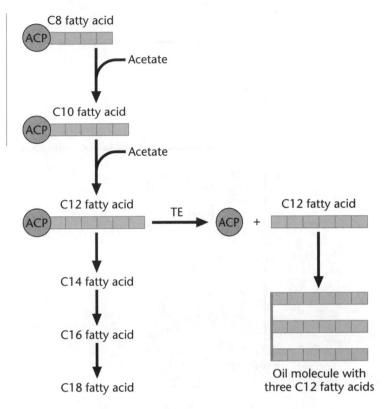

Figure 15.9

Scheme for Fatty Acid and Oil Synthesis. This scheme shows the importance of thioesterase (TE) in freeing a fatty acid of the correct length from the acyl carrier protein (ACP).

the function of a thioesterase enzyme, and there is a unique enzyme for each length of fatty acid. Thus, a plant that makes fats that have primarily C18 fatty acids has a very active C18 thioesterase. California bay, a small tree that grows in the foothills of northern California, has in its seeds fats with C12 fatty acids and a very active C12 thioesterase. When the gene for this C12 thioesterase was introduced in the oilseed rape plants with a seed-specific control region, the seeds made oils with C12 fatty acids (about 25%), as well as their normal C18 fatty acids (about 75%).

The next challenge will be to produce plants that have oils with 50–75% C12 fatty acids, the level needed for commercialization of the product. When this project is successful, it will transfer production of these specialty oils from Third World countries, such as the Philippines and Malaysia, that now heavily depend on these exports, to developed countries. Farmers in developed countries will reap the benefit of this biotechnology at the expense of farmers in underdeveloped countries.

Production of Specialty Starches for Industry. Industrialists do not think of starch as calories, but as raw material from which other foodstuffs (sweeteners, thickeners) or industrial products (adhesives, products for paper sizing) can be made. Fermentation processes can convert starch into other useful chemicals, and about 10% of the total starch production is used in this way.

Starch is a mixture of two types of large complex carbohydrates: amylose and amylopectin (see Figure 4.2). Amylose is a linear polymer consisting of 1,000 to 20,000 glucose molecules. The length of these glucose chains varies with the plant species from which the starch is obtained. Amylopectin is a branched polymer with from 300,000 to 3 million glucose molecules. The

Figure 15.10

Pathway of Polyhydroxybutyrate Synthesis in Bacteria and Plants. The enzymes and genes that are boxed represent the two enzymes whose genes had to be transferred from the bacterium Alcaligenes entrophus *to the plant to enable the plant to make PHB.*

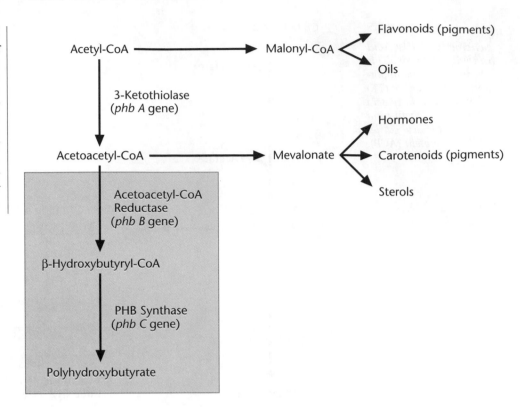

relative proportions of amylose and amylopectin determine the chemical and physical properties of starch, and therefore its usefulness to make certain industrial products. Starch is insoluble in water, but when a suspension of starch is heated, the starch grains swell and a gelatinous mass is formed. The temperature at which swelling occurs, the tendency to become insoluble again after it is solubilized, and the clarity, viscosity, and taste of the paste are all important characteristics of different starches, such as potato, corn, wheat, and tapioca starch. Every cook knows that it is easier to thicken a sauce with potato starch than with wheat starch.

The synthesis of starch depends on three different enzymes, and their relative levels determine the relative amounts of amylose and amylopectin. For example, a special variety of corn that lacks one of the enzymes cannot make amylose. Its starch is extremely rich in amylopectin, and is especially useful for certain industrial applications. Because starch synthesis depends on relatively few enzymes, and potatoes can be easily transformed, biotechnologists have started manipulating starch chemistry and properties.

Genetic engineering of a totally different type of starch was reported in 1991 by scientists from Calgene. They transferred to potatoes a bacterial gene that encodes an enzyme for synthesizing small cyclic starch molecules, rings that have seven or eight glucose molecules. These molecules, called *cyclodextrins*, can form complexes with other small molecules that are enclosed in the center of the ring. In this way, the "guest" molecules acquire new properties such as increased stability or greater solubility. Cyclodextrins are used in the pharmaceutical industry for drug delivery, and to remove unwanted compounds (such as caffeine) from foods. Cyclodextrins are expensive to produce, and their cost remains a problem

Figure 15.11

Small, Translucent Plastic Granules in a Leaf Cell of Arabidopsis thaliana. *The plant was transformed with two bacterial genes that encode enzymes necessary for polyhydroxybutyrate synthesis.* Source: Y. Poirier, D. E. Dennis, K. Klomparens, and C. Somerville (1992), Polyhydroxybutyrate, a biodegradable thermoplastic, produced in transgenic plants, Science 256:520–523. Copyright 1992 by the AAAS.

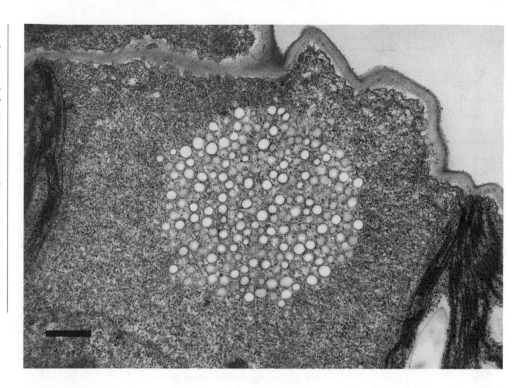

for large-scale use. The production of cyclodextrins in potatoes would undoubtedly decrease the cost of cyclodextrins so that they could now be used more widely by industry.

Production of Plastics by Plants. Polyhydroxybutyrate (PHB) is a polyester that accumulates as a natural storage material in many species of bacteria, in the same way that plants accumulate oil, starch, or sucrose as storage compounds. PHB produced by bacteria in large fermentors is presently used by industry as a renewable source for manufacturing biodegradable plastics. However, because producing biomaterials by fermentation is inherently much more expensive than their production by plants, Chris Somerville and his colleagues at Michigan State University investigated the possibility of transferring the necessary genes from bacteria to plants. Synthesizing PHB requires only three enzymes (see Figure 15.10), and the first enzyme in the pathway is already present in all plants. The research team introduced the genes encoding the two other enzymes into *Arabidopsis thaliana* plants and found that the transgenic plants synthesized PHB and that the plastic accumulated as small granules (0.2–0.5 μm) in the cytoplasm, nucleus, and vacuoles of the cells (Figure 15.11). Unfortunately, plant growth was severely stunted, possibly as a result of draining the important intermediate product acetoacetyl-CoA from the other pathways for which it is needed.

In these experiments, the two genes were expressed with general control regions rather than with organ-specific control regions. The next round of experiments will probably involve the expression of the genes with seed-specific promoters in oilseed rape or another oilseed crop, because acetyl-CoA, a starting product for PHB synthesis, is also the starting product for oil synthesis.

Table 15.4	The use of enzymes in the food industry

Enzyme(s)	Use(s)
Amylase and glucoamylase	Produced by *Aspergillus* fungus; completely hydrolyze starch to yield glucose
Glucose isomerase	Produced by bacteria or fungi; converts glucose to fructose, the main ingredient of corn syrup; corn syrup is the most widely used sweetener
Pectinase	Produced by fungi; breaks down complex carbohydrates of the cell wall; used to "clear" fruit extracts so that they can be filtered; complex carbohydrates would clog the filters
Glucanase	Produced by fungi; breaks down complex carbohydrates of the cell wall of cereals; used to make beer filtrable
Lipoxygenase	Produced by plants (legume seeds); used to whiten wheat flour; the same effect can be obtained by adding a small amount of soybean or pea flour to the wheat flour
Rennet extract	Produced in the abomasum of calf and cow; contains several enzymes of which chymosin is the most important; coagulates milk proteins so that whey and curds can be separated
Lactase	Produced by yeast and fungi; converts lactose (milk sugar) into galactose and glucose; used to prepare milk products such as Lactaid® for people who have lactase deficiency and cannot tolerate lactose
Proteases	Bacterial, fungal, and plant proteases find wide application in the cheese industry (substitution for rennet, and ripening of cheese) and as meat tenderizers
Lipases	Produced by calves and cows in rennet (see above), but also by micro-organisms; degrade butter fat in cheese to release fatty acids; involved in cheese ripening and creation of flavor compounds
Phytase	Produced by plants and micro-organisms; breaks down phytate in seeds to release phosphate; can be used to treat animal feed so that more of the phosphate can be absorbed by the animal, resulting in less phosphate in the environment

9 | Molecular farming can produce enzymes for food processing and other uses.

Enzymes are used extensively in the food industry (Table 15.4), and the commercial production of these enzymes is a $400-million industry. Enzyme suppliers may list as many as 20 different enzymes that can be purchased in bulk. These enzymes are produced by industrial fermentation, using yeasts and other fungi. Many of the enzymes are heat stable and are optimally active between 55°C and 90°C. They are produced by micro-organisms that have been carefully selected (domesticated) over the years. Such strains do not occur in nature. The genes for these enzymes can easily be introduced into plants and expressed with leaf-specific or seed-specific control regions that will cause the enzymes to accumulate in these organs. The enzymes would have to be purified and separated from the other plant proteins and substances, before they could be used in the food industry.

One advantage of using fermentation technology is that very little enzyme purification is needed, because the enzymes are secreted into the fungal culture medium, where they accumulate at very high levels. The use of plant material that contains an enzyme (as opposed to using a purified enzyme) for modifying animal feeds is quite promising. For example, phytase produced by fermentation is added to animal feed to help digest the phytic acid it contains (see Box 15.3). It would be simple to produce phytase in tobacco leaves or potato tubers, and simply mix these with phytic acid-containing animal feed.

Cellulase is another example of an enzyme that could be produced in transgenic plants and used as an animal feed additive. When added to plant material, cellulase helps digest cellulose, and feedstuffs could be partially pre-digested if they were mixed with cellulase-rich plant material. In this case, the plant parts that contain cellulase would have to be finely dispersed, as a powder or a slurry, to release the cellulase from the tissues.

Recently biotechnologists in San Diego, California, introduced the gene for bovine lysozyme into tobacco plants and showed that the leaves contained high levels of this enzyme. This enzyme is extremely useful for ridding seeds of the pathogenic bacteria that may adhere to them. Whether the enzyme will need to be purified from a tobacco leaf extract, or can be applied to the seeds as a powder made from dried pulverized leaves remains to be investigated.

10 | Can genetically engineered plants become weeds, or can genetically engineered genes be transferred to weeds?

As we noted earlier, weeds are plants that invade agricultural ecosystems and, by competing with crop plants, depress yields.

Weed biologists have identified 13 characteristics that make a plant a weed, and most serious crop weeds have 11 or 12 of these 13 characteristics (see Chapter 12). Crop plants, in contrast, have only 5 or 6 of these 13 characteristics. Since each characteristic is governed by at least one and possibly by more genes, it is unlikely that crop plants could suddenly acquire all the necessary characteristics to become weeds. The addition of a single additional trait—a gene that is generally totally unrelated to weediness—is unlikely to allow crops to become weeds. The exception to this general rule is the introduction of herbicide resistance genes. If several crop species are made tolerant to the same herbicide and these crop species are used sequentially in a rotation, then there is considerable danger that "volunteers" from the first crop will become weeds in the subsequent crop. This phenomenon already occurs on a small scale now, but it will be greatly aggravated by herbicide-tolerant crops. The solution to this problem is to use crops tolerant to different herbicides in crop rotation programs.

All crop plants have wild relatives somewhere on the earth, and a certain amount of gene flow between the crop and the wild relative can occur where the two populations grow side by side. Gene flow from wild relatives to crop plants may even be encouraged by many subsistence farmers as a means of maintaining the broad genetic base of their landraces. Such gene flow does of course not occur when farmers buy their seeds annually from seed producers. However, in that case, gene flow in the other direction is still possible, and a gene from the crop may end up in the wild relative, especially if there is selection pressure for the maintenance of the gene. This is likely to happen if the gene confers a selective advantage to the wild relative.

Experiences with corn and sorghum provide good examples of crop–weed hybridization in the Western Hemisphere. There are no wild relatives of corn in the United States, but in Mexico a relative of corn called *teosinte* grows around the corn fields. Corn–teosinte hybrids are formed quite readily, and in some areas the teosinte plants look very much like corn, indicating that gene transfer is occurring. Another example is found in sorghum (*Sorghum bicolor*)

Box 15.3

*Phytase to Reduce
Phosphate
Pollution from Pig
and Poultry
Manure*

Plants store phosphate in the form of phytate, a molecule that contains six phosphate groups attached to a glucose derivative. When seeds develop, they synthesize and accumulate phytate in their storage organs (cotyledons or endosperm). During germination, the seeds synthesize an enzyme called *phytase* that breaks down the phytate and releases phosphate for use by the growing seedling. When ruminants (such as cows and sheep) eat plants, phytate is broken down by enzymes produced by bacteria.

However, nonruminants such as chickens and pigs cannot break down the phytate in their diet very efficiently, and their feed must be supplemented with phosphate. In fact, these animals can use only one-third of the phosphate in the diet, and two-thirds passes through the body and ends up in streams and waterways. This problem is especially acute in The Netherlands, a small country with a serious manure problem. The Dutch fermentation biotechnology company Gist Brocades came to the rescue with industrially produced microbial phytase (Natuphos®) that can be added to the diet and increases phosphate use up to 50% of intake. A sample calculation goes as follows: when a pig is fattened from 20 to 60 kg, it needs to take up 175 g of phosphate to build its body. This requires 450 g of phosphate in the feed, and 270 g of this phosphate ends up in the manure. With phytase in the diet, the phosphate in the feed can be lowered to 350 g and only 170 g are excreted. Thus, phytase supplements reduce phosphate pollution of the environment.

Microbial phytase could be replaced by phytase produced in plants that have been genetically engineered to express the microbial phytase gene. The phytase need not be added as a pure enzyme to the feed; it could simply be present in plants that are part of the animal feed mixture. Genetic engineers at Mogen, another Dutch biotechnology company, took this approach, and their results were recently published in the international journal *Bio/Technology*. Tobacco plants were transformed with the microbial phytase gene hooked to a seed-specific control region, and the phytase-containing tobacco seeds were fed to broiler chickens over a four-week period. The chickens were kept on a diet in which two-thirds of the phosphate was present as phytate. The chickens grew equally well when this diet was supplemented with microbial phytase or with tobacco seeds expressing the enzyme. Without such supplementation, the chickens grew poorly. This experiment demonstrates that phytase in the milled tobacco seeds was available to break down the phytate in the other components of the diet, thereby mobilizing the phosphate.

Although tobacco seeds could be used in animal diets, Mogen scientists envision using seeds from crops that are already being used as animal feed, such as corn or cotton. Such phytase-containing seeds may have other useful applications. Phytase is considered to be an antinutritional factor because it binds essential minerals, such as calcium, iron and zinc, making them unavailable for uptake by the body. By mixing phytrate-rich feedstuffs with phytase-containing seeds, it may be possible to enhance the availability of these minerals and to decrease phosphate pollution at the same time.

a plant that was domesticated in Africa, but has wild relatives in the United States. Shattercanes (*Sorghum bicolor*) and Johnsongrass (*Sorghum halepense*) are two wild relatives of sorghum that are found in the United States and that cross quite readily with sorghum. The excessive vigor of North American populations of Johnsongrass is probably due to a constant flow of genes from cultivated sorghum varieties to this weed. Efforts to improve sorghum, there-fore, led to "improved" Johnsongrass. Similarly, gene flow from sorghum to shattercanes results in different strains of this weed that mimic the different cultivated sorghums. The more a weed resembles a crop, the more likely it is to be a serious weed in that crop and the more difficult it will be to control.

It seems likely that gene flow from crops to wild relatives will be espe-cially problematic with herbicide tolerance genes. Once the herbicide toler-ance gene has been transferred to the wild relative, there will be a strong selection pressure to maintain it there if the same herbicide is used, and as a result the weedy wild relatives will be more difficult to control. At present, tests with genetically engineered plants are conducted in such a way as to minimize the possibility of gene exchange with wild relatives. First of all, tests are not conducted in areas where wild relatives are present. Second, if the same crop plant is also cultivated in the area, then the test plot is surrounded by a wide border of control plants (not genetically engineered). The plants in this border are harvested and destroyed. The assumption is that if any gene transfer were to take place, it will be with the plants in the border zone. The rigor with which these and other precautions are observed depends very much on the regulations and their enforcement in individual countries.

Unfortunately, no one can say for sure whether gene transfer to wild relatives will be a real problem. Even for herbicide resistance, it can be argued that past practices have already resulted in many herbicide-resistant weeds. The only way to control these herbicide-resistant weeds is by using alternative cultural practices and/or alternative herbicides. Thus, the herbicide-resistant weeds arising from gene flow to wild relatives may not be any more trouble-some than the weeds we already have.

11 | Will genetically engineered plants be safe to eat?

There is at the moment no reason to suppose that genetically engineered plants will not be safe to eat. Genetically engineered plants will have an additional gene (or genes) that is present in all the cells. People now eat about 100,000 different genes daily, and this DNA is efficiently broken down in the human intestinal system. Experiments have been conducted to show that genes added to plants by gene transfer are equally efficiently digested.

In addition to the gene that confers the new property to the plant, the first generation of genetically engineered plants will contain a bacterial gene that makes the plant cells resistant to an antibiotic. This antibiotic resistance of the plant cells is an important step in the plant transformation process. Although the presence of such a gene could pose a hazard, new transforma-tion procedures developed at DuPont and at the U.S. Department of Agricul-ture allow for the removal of the bacterial gene after transformation has been achieved. The bacterial antibiotic resistance gene is simply cut out of the plant DNA, using a special enzyme produced within the cells themselves.

In addition to the foreign gene, the plant will, of course, contain the new protein that is encoded by the gene. In many cases, this protein may not be present in the part of the plant that we eat. The presence or absence of a protein in a particular organ depends on the gene control region used to make the transgenic plant. Thus, if a Bt-toxin gene is expressed with a root-specific control region, then the protein will only be present in the roots. Such a control region could be used to control an insect that feeds on the roots of tomato plants or corn plants. In other cases, the protein will be present in the part we eat because that was the objective of making the transgenic plant (for example, seeds with a new methionine-rich protein). Whether such proteins could be detrimental to humans must be evaluated case by case. In some cases, these proteins may already be consumed in large quantities. For example, if the gene for α-amylase inhibitor found in beans is transferred to other legumes to inhibit the development of bruchid beetles, we should be aware that this inhibitor inhibits human α-amylase as well as insect α-amylase. Nevertheless, this protein is eaten in large quantities by millions of people all over the world. Therefore, its transfer to other crops is probably quite safe as long as the foods are cooked before being eaten. The reason we need to cook beans well before eating them is precisely because they contain α-amylase inhibitor, as well as phytohemagglutinin, another plant defense protein.

Although there is no reason to believe that transgenic plants will be harmful when they are eaten, this does not mean that they will be readily accepted by consumers. Consumers want to eat "natural" products, and they think of gene transfer as an unnatural process because it involves laboratory procedures in addition to field testing. The fact that all our crop plants have greatly benefited from gene transfer for 10,000 years, especially in the past 100 years, appears irrelevant to the public at large. Some experts believe that the best way to overcome consumer resistance will be to clearly label food products, especially fruits and vegetables, derived from transgenic plants. Some people will want to know if they are eating transgenic tomatoes, but probably will not care if they are wearing shirts made from transgenic cotton. To allay the public's apprehension, Calgene intends to clearly label its transgenic FlavrSavr® tomatoes.

The Food and Drug Administration (FDA), the agency of the U.S. government that oversees food safety, made its policy public in May 1992. This policy states that food obtained from transgenic plants need not be labeled as such. The FDA considers that what is important to the consumer is the material content of the food (nutritional, allergenic, pesticidal, and so on), and not the process to generate the plants. At present, we do not label wheat flour based on the breeding processes used to generate common wheat lines, or the source of the genes in improved lines. Of importance to the consumers are the flour's protein content, presence of gluten and/or other allergens, bran content, and added fortified ingredients such as vitamins.

Of particular concern to consumers and the FDA are foods that cause allergic reactions. In some cases, scientists have identified the specific proteins that are allergenic, but often that information is unavailable. Therefore, when a gene is transferred from a plant known to have allergenic proteins, there is the possibility that the gene may encode such a protein. In that case, the genetically engineered food will have to be tested, and the company that wants to market this new food will have to demonstrate that it is not allergenic.

Safety assessment procedures of the FDA for all new food plants and foods are shown in Figure 15.12. The scheme shows the issues that concern the FDA (allergens, nutrients, and toxicants) and when biotechnologists and plant breeders should consult with the FDA. Expected and unexpected effects are separated. As with all such efforts, unexpected or unintended side effects may result from crop improvement. This is true whether one uses molecular or classical techniques of crop improvement. There are several known cases of high levels of toxic compounds in "improved" crop varieties produced by traditional plant-breeding techniques.

Figure 15.12

FDA Safety Assessment Flow Chart of Foods from New Plants Whether Produced by Traditional or Molecular Techniques. Source: *U.S. Food and Drug Administration.*

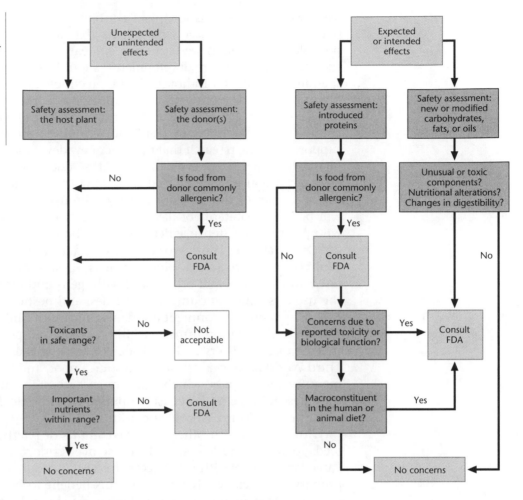

Requirements to label foods as the product of genetic engineering would have a major impact on introducing such foods, for two reasons. First, many consumers may resist such foods, for the reasons discussed. Second, it would necessitate keeping foods from transgenic plants separate from nontransgenic plants during food processing. Manufacturers would be very reluctant to do this. If Campbell and Hunt, two major food processors, need to separate processing operations for normal and transgenic tomatoes to produce normal and transgenic tomato soup and tomato ketchup, the introduction of transgenic crops will be greatly delayed. When milling flour to make bread, food manufacturers deliberately mix different strains of wheat (or other cereals) that may result from different breeding techniques. Following that logic, the FDA believes that there should be no separate processing facilities for genetically engineered plants. What counts is the end product, not the methods used to produce the elite plant varieties.

12 | Who will benefit from plant genetic engineering?

The survey of objectives and recent accomplishments of plant genetic engineering presented in this chapter makes it clear that the development and application of this revolutionary technology are closely tied to the needs of agriculture and the agricultural industries in the developed world. There are at least two major reasons for this. First, biotechnology originated in the developed world; all the basic experiments that make plant genetic engineering possible were done in western Europe, Japan, or North America. It is natural, therefore, that further developments should also occur there. Second, by spending liberally for in-house plant biotechnology research and by encouraging university scientists with small grants, biotechnology companies in developed countries can set the research agenda. Governmental institutions are relatively minor players in plant biotechnology, although they continue to support the basic research that underlies new developments. Herbicide-tolerant plants, high-starch potatoes, plants that produce specialty oils or plastics, and enzymes produced for the food processing industry are all desirable goals for the industries of the developed world. The beneficiaries of these new plants will be the inhabitants of developed countries, not the Third World farmers who practice subsistence agriculture and whose needs are very different.

One way the Third World countries (as well as developing countries) could benefit from genetic engineering is through introducing plants that are resistant to viruses and other plant pathogens and insects. Such plants allow the use of smaller amounts of pesticides, and pesticides are one of the most expensive and environmentally damaging inputs into modern agriculture. The Green Revolution introduced into underdeveloped countries new strains that require fertilizers and pesticides to achieve their high yields. However, Third World farmers are generally resource poor, and they cannot afford these manufactured inputs, which often must be bought in the developed countries. Less reliance on purchased inputs such as pesticides would therefore be a real plus for Third World agriculture.

However, as with other improved strains of plants, distributing the new seeds to small landholders and subsistence farmers will be a major obstacle. Large landholders with ready access to capital and markets are likely to be the primary beneficiaries. But for even this benefit to be realized, international agencies must finance the development of these disease-resistant strains of tropical crops. Biotechnology firms in the developed countries are focusing on major crops, such as soybeans and corn, and on minor high-value vegetable crops, such as tomatoes. They are not working hard on cassava, cowpeas, or yams, some of the major staples in Africa. However, in this respect they do not differ from commercial plant-breeding companies, who do not produce improved lines for use in Third World countries, where there are no markets for such lines.

One other major constraint in passing along the benefits of agricultural biotechnology to the Third World is the absence of a seed industry in developing countries. In developed countries, seed companies produce hybrid seed or other elite lines of seed, and an efficient distribution system ensures that the farmers have access to the latest genetic lines that produce the highest yields. In recent years, many seed companies have been taken over by agrichemical companies and these are excellently poised to sell and profit from genetically engineered seeds when they become available. But there is almost no seed industry in the Third World, and most farmers still save part of their harvest for planting the next year. That is an acceptable practice for the traditionally bred varieties introduced during the Green Revolution. However, the genetic uniformity of these strains resulted in a need to replace them every six to eight years with new varieties that were resistant against the ever-evolving strains of plant pathogens. Genetically engineered seeds, whether produced by biotech companies or by government research organi-

Box 15.4

*Which of These
Genetically
Engineered
Products Would
You Accept?*

1 Human insulin made by genetically engineered microorganisms using a process of fermentation.

2 Furniture made from chestnut wood produced by trees protected from chestnut blight (a fungal disease) by the use of a genetically engineered virus that kills the blight.

3 Cotton shirts produced by genetically engineered cotton plants that contain a bacterial Bt gene to protect them against the devastating boll weevil.

4 Milk produced by cows fed on alfalfa and corn that contain insect resistance genes derived from other plants and introduced into these crops by genetic engineering.

5 Cheese made with rennet (an enzyme used in the manufacture of cheese) produced by genetically engineered microorganisms (instead of extracting it from calves' stomachs).

6 Papayas from Mexico produced by trees that are resistant to papaya ringspot virus by the incorporation of a gene derived from the virus itself.

7 Mangoes from South America produced by trees that contain a genetically engineered gene that slows down ethylene production and delays ripening of the mangoes. This delayed ripening makes it possible to ship the mangoes to faraway markets.

8 Tomatoes produced by plants that have been sprayed with dead bacteria containing a genetically engineered protein that is toxic to various caterpillars (tomato hornworm) but harmless to animals.

9 Wine made from grapes produced by vines genetically engineered to resist phylloxera disease.

10 Tofu made from soybeans that are tolerant to the herbicide glyphosate because the plants have been engineered to contain a mutant plant gene.

11 FlavrSavr® tomatoes that contain a plant gene that slightly slows the softening of the tomatoes so that they can stay on the plant a few days longer.

zations and agricultural experiment stations, can also be propagated in the same way. Yet the absence of an effective seed distribution system in Third World countries makes it difficult to distribute the genetically engineered seeds in the first place.

The yield advantage that results from the use of hybrid seeds will probably mean that genetically engineered traits, such as disease resistance or herbicide tolerance, will be combined with genetically engineered hybrids. However, hybrids cannot be propagated on the farm. When a farmer uses hybrid seeds, she cannot keep a portion of this year's harvest and plant it out next year. All the advantages of hybrid vigor are lost if this is done. Developing a seed industry and a seed distribution system should be a high priority if Third World countries are to benefit from the advances of agricultural biotechnology. The broader question, whether this is the direction that less developed countries should take, is further explored in the next chapter.

Some of the big agrichemical companies anticipate their greatest potential profits in selling farmers integrated packages of seeds, fertilizers, and pesticides. Most seed companies are owned by large agrichemical companies—a relatively recent development in the developed world—and the aim is to produce transgenic seeds that are compatible with the chemicals produced by the parent company. This strategy is likely to increase rather than decrease the reliance on agricultural chemicals. Furthermore, this trend exacerbates the evolution of the dominant agricultural system in the direction of a high-input agriculture, rather than a sustainable agriculture.

Plant genetic engineering is not incompatible with sustainable agriculture. However, if the agrichemical companies set the research agenda, and if the big corporation farms allied with farming organizations representing their interests are allowed to set national production policies, then genetically engineered seeds will simply become another high input in the chemical–genetic agricultural production system. Given the right government policies, these new seeds could also accelerate the trend toward a sustainable agriculture.

13 | Opposition to genetic engineering and genetically engineered foods.

In a number of countries, groups and individuals oppose the use of genetically engineered organisms. They view genetic engineering as fundamentally different from all other genetic and tissue-culture manipulations that have been used heretofore to create new organisms (such as triticale, a wheat-rye hybrid that could not normally arise in nature). Their opposition is not primarily directed at the development of pharmaceuticals, such as interferon or insulin, but to plants and animals, as well as microorganisms that may have an indirect role in food production. In the 1980s, the efforts of these opponents were directed at the scientists and their research laboratories, and some activists even destroyed genetically engineered plants that were part of field trials (in The Netherlands and California, for example) or invaded research laboratories to dramatize their concerns and opposition to genetic engineering. In the United States, the cause of these small groups was taken up by certain environmental organizations, whereas in Europe certain political parties (the "Greens," for example) espoused opposition to biotechnology.

In the 1990s, opposition has taken the form of consumer boycotts. Thus, the Pure Foods Campaign in the United States has organized a number of chefs of well-known restaurants to declare that they will not serve foods of genetically engineered plants. Using a different approach, the opponents of genetic engineering are urging city councils in the United States to pass ordinances that require each store where such food is sold to post a warning sign. The reasons to oppose genetically engineered organisms appear to fall into five categories, which are described here with counter-arguments of proponents of biotechnology:

1   **Ethical Considerations.** Transferring genes between organisms that are not from the same species (and do not normally have sex) is unethical. Human beings should not alter other organisms in such profound ways. This argument ignores the fact that humanity has had a very profound impact on the evolution of many species, including gene transfer between species by traditional

breeding. Furthermore, gene transfers between unrelated organisms occur in nature.

2 **Safety Considerations.** First, releasing genetically engineered organisms may have unforeseen ecological consequences. This is true, say proponents, but not more so than the unforeseen consequences that may result and have resulted from moving organisms between continents. Some such movements have created pests, but biological control methods have brought them back to equilibrium. Second, opponents say that genetically engineered foods may not be safe to eat. On the other hand, since we know exactly which gene is being introduced, this is highly unlikely, and less of an issue than with conventional breeding, where whole segments of DNA with many genes are transferred between plants. The possibility exists that genetically engineered foods will contain novel allergens. However, this can readily be tested.

3 **Anti-Corporate Arguments.** Opponents say the purpose of corporations is to make money, not to look after human welfare. Corporations often twist the truth to serve their own needs (making money) and we cannot rely on them to give us correct information. While there is some truth in this argument, corporations often promote human welfare by making new products. They change the products they make in response to demand, which may be for better nutrition or health. Biotechnology corporations are not more "evil" then other corporations.

4 **Sustainability Considerations.** Biotechnology is driving us further towards high input agriculture and will not contribute to making agriculture more sustainable. Transgenic plants may increase the use of chemicals (herbicide tolerant plants) or decrease that use (pest resistant plants). But proponents say that it is the combination of technology, government regulations and tax laws that determines the direction of agriculture, not technology by itself.

5 **Philosophical Considerations.** Many opponents to biotechnology say that we must return to an ecologically-based stewardship of the earth and renounce the exploitative mode that now prevails. However, say proponents, in the meantime, we must feed the 5 billion people we have, and the 10 billion we can expect to have. Therefore, we must work towards a world in which all resources are used more wisely.

Genetic engineering is a technology that helps create new products. The difficulties that arise in accepting or rejecting the products that result from this technology become apparent if one considers individual products. In some of these, the connection between food and genetic engineering is direct, but in others it is more remote.

Summary

Transforming plants with DNA is both conceptually and practically a simple process. Although a number of crops cannot yet be transformed with ease, others are transformed routinely in many laboratories. Two procedures are used to obtain transformed plants. The most commonly used method relies on a natural gene transfer system of a plant-pathogenic bacterium. The second procedure involves bombarding cells with particles coated with DNA. In either case, tissue culture procedures are used for the transformed cells to grow into calluses and then into whole plants.

Crop improvement via genetic engineering is limited only by the ability to transform certain plants and to isolate genes that will give these plants new useful properties. Projects already underway include regulating fruit ripening in tomatoes, making plants resistant to insects and to viruses, creating herbicide-tolerant plants, and facilitating plant breeding with genetically engineered male-sterile lines. Genetic engineering will also greatly expand the possibilities for molecular farming. The goal of molecular farming is to use plants not only as a source of food, but also as a source of products that are needed by industry. Industry already makes many valuable products from plants. By incorporating one additional gene into an already established crop plant, it will be possible to create plants that make valuable oils, starches, and even plastics. Plants can also be used to produce enzymes for the food industry, and pharmaceuticals.

Are there only benefits, or are there also risks? It seems unlikely that genetically engineered plants will become weeds, but there is a possibility that making herbicide-tolerant crops and trees will result in a greater reliance on chemical herbicides. Another problem that will need to be watched carefully is the spread of genes from cultivated plants to wild relatives. The spread of herbicide tolerance to wild relatives will result in problems similar to the ones encountered now with certain herbicide-resistant weeds that have become difficult to control. Furthermore, molecular farming—the production of chemicals for industry—is likely to accentuate the trend toward high-input agriculture rather than reverse it. In addition, the lively debate between opponents and proponents of genetic engineering will probably continue for some time, at least in the developed countries.

Further Reading

Christou, P., D. McCabe, B. Martinell, and W. F. Swain. 1990. Soybean genetic engineering—commercial production of transgenic plants. *Trends in Biotechnology* 8:145–151.

Corbin, D. R., and H. J. Klee, 1991. *Agrobacterium tumefaciens* mediated plant transformation. *Current Opinion in Biotechnology* 2:147–152.

DaSilva, E. J., ed. 1992. *Biotechnology: Economic and Social Aspects.* Cambridge, UK: Cambridge University Press.

Gasser, C. S., and R. T. Fraley. 1992. Transgenic crops. *Scientific American* 266 (6):62–67.

Hiatt, A., ed. 1992. *Transgenic Plants.* New York: Dekker.

Khush, G. S., and G. H. Toenniessen. *Rice Biotechnology.* London: CAB International.

Klein, M. 1992. Transformation of microbes, plants and animals by particle bombardment. *Biotechnology* 10:286–291.

Persley, G. 1990. *Beyond Mendel's Garden.* London: CAB International.

Potrykus, I. 1989. Gene transfer to cereals: An assessment. *Trends in Biotechnology* 7:269–273.

Snape, J. W., C. N. Law, A. J. Worland, and B. B. Parker. 1990. *Targeting Genes for Genetic Manipulation of Crop Species.* Stadler Genetics Symposia. New York: Plenum Press.

Walgate, R. 1990. *Miracle or Menace? Biotechnology and the Third World.* Washington, DC: Panos Institute.

World Bank. *Agricultural Biotechnology: The Next Green Revolution.* Washington, DC: World Bank.

Chapter 16

Toward a Green Agriculture

Guthrie County, Iowa, is in the middle of the U.S. Corn Belt. By 1960, even with the adoption of hybrid corn, farmers in this county were hard pressed to produce enough of this important feed grain to keep up with the demands of the burgeoning cattle industry. The hybrids depleted the soil of its nitrogen to such an extent that cropping could only be done every other year.

Chemical fertilizers solved this nutritional problem. By the late 1960s, corn was being grown every year, with the result that the county's production shot up by over 50%. But this intensive monoculture had effects that were unforeseen at the time. Pests previously unable to build up their population to an economic injury level because the crop was not there every year, now thrived. A serious infestation of rootworms was controlled with heavy doses of insecticides. But by the 1980s these heavy doses of chemicals had led to worrisome amounts of pesticides and nitrates found in local drinking water. The state of Iowa began to tax fertilizers to encourage less usage, and by the late 1980s farmers were indeed lowering the amounts of chemicals they applied to the soil.

The increase in land use intensity also had a more long-term effect on the soil. Whereas previously the soil had a chance to replenish itself and build up organic matter in the alternate years when the corn was not grown, annual cultivation reduced the soil's ability to retain organic matter and water. It was now more susceptible to erosion, and signs of this degradation were becoming apparent by the 1980s.

In addition to the environmental effects, increased corn production in Guthrie County had another unpredicted effect: it was simply too successful. As the economy, both locally and globally, entered the recession of the late 1980s and early 1990s, demand for corn on the market fell significantly. To avoid surpluses, the U.S. government paid county farmers not to produce corn. Up to one-fourth of all corn acreage was idled in a typical crop year, while across the Atlantic in Africa, drought continued to cause crop failures and hunger.

These events in a small region of Iowa reflect similar events all over the world. For whatever reason—to feed people or to feed animals—intensifying agriculture may result in higher productivity in the short run, but may undermine the resource base in the long run and therefore may imperil food production in the future. Moreover, even in the short term the heavy use of inputs can have serious environmental costs in terms of pollution. Finally, while market-based "industrial agriculture" has certainly led to improved efficiencies and stimulated productivity, the policies of governments in both developed and developing regions sometimes blatantly ignore people in favor of economics.

The idea that agriculture must evolve to become *sustainable* and less dependent on market-purchased inputs is attracting increasing scientific and political attention. Agriculture must become green again. A number of studies show that high levels of agricultural production are quite compatible with a low-input agriculture that relies heavily on the best genetic lines, integrated pest management, and sound agricultural practices, and that depends less on purchased inputs of fertilizers and pesticides.

1 | **Many agricultural practices have adverse environmental effects on the farm.**

Although high-technology, high-input agriculture has certainly resulted in spectacular gains in productivity, it has degraded some of the natural bases on which this system rests. The ultimate costs of this degradation on the farm are borne by the farmers themselves and by society at large, be they in lowered productivity, environmental cleanup, or further technology to solve a technologically induced problem. Some of these on-farm impacts mentioned in earlier chapters are summarized here.

1   *Soil erosion.* As the farmers in Iowa found out, growing crops with shallow roots year after year, plowing deeply, and removing the crops all deplete the topsoil and the organic matter that holds it together. As the soil becomes more compacted, it also becomes more susceptible to erosion by raindrops and wind. During the late 1970s, the U.S. Department of Agriculture exhaustively inventoried U.S. agricultural soils and came up with an alarming result: up to half of the cropland showed serious erosion of its precious, nutrient-rich topsoil. The numbers are hard for a layperson to comprehend: over 15 tons per hectare of topsoil lost, on average. Global figures are just as alarming: the United Nations is using satellites to track soil erosion, and estimates that in northern India losses per hectare exceed those of the United States. As another example, these data indicate that in Chile 75% of the cropland is moderately to seriously eroded. The effect of these losses of topsoil on the farm is devastating. As wind and water typically remove the finer particles, the subsoil that is left is coarser, with less organic matter, and binds less water and minerals. More tillage is needed to break up the soil, and more fertilizers are needed to make it support plant growth. Ultimately, eroded soils can become useless for crop production.

2   *Irrigation and water availability.* Water is an essential component for plant growth, and in arid or semiarid regions, the availability of

Table 16.1	Irrigated acreage of surplus crops in areas of groundwater decline (1982)		

California	Cotton	240,000	hectares
Texas	Cotton	440,000	
	Sorghum	400,000	
	Small grains	400,000	
Nebraska	Corn	600,000	
Kansas	Corn	264,000	
	Sorghum	220,000	
	Small grains	270,000	

Source: U.S. Department of Agriculture (1987), *U.S. Irrigation—Extent and Economic Importance* (Washington, DC: USDA).

water for agriculture must be improved through irrigation. According to the FAO, 270 million ha in the world are irrigated, and this consumes 70% of the world's total human-induced water consumption. In the United States Southwest, irrigation consumes over 80% of all the water used in people's activities.

About half the water used in the United States for irrigation comes from underground aquifers (for example, the Ogalalla aquifer in the Southwest), from which it is pumped to the surface. The same is true in the Indo-Gangetic plain of India. These aquifers are slowly and constantly recharged through rain percolation in the soil. But the rate at which they are being depleted is much faster than this slow recharge, and the level of the water table is falling.

Paradoxically, in the United States this rate of decline is being primed by crops that are in surplus. Cotton, sorghum, and small grains, which are in surplus in most years, are being grown and are depleting the aquifer (Table 16.1). Different crop plants have different needs for water (Chapter 7). When this is expressed in economic terms, the cash value of crop per volume of water used is seen to vary greatly (Table 16.2). Yet governments often have "cheap water" policies, providing subsidized water to those farmers who grow crops that are water hungry in locations that are water poor. This promotes depletion of water resources.

3   *Salinization.* Irrigation of salt-rich arid lands has an added environmental cost. When water use is excessive, the underground water table rises, and brings with it dissolved salts from the lower regions of the soil. This initially lowers crop yield, and, if the water evaporates and leaves salts caked on the surface, ultimately renders the soil agronomically useless unless drastic and expensive measures are taken to remove the salts (Figure 16.1). The extent of salinization is more difficult to measure than simple soil erosion, because the effects of salinized soil are underground, at least initially. But estimates are that in Bangladesh river delta salinization is significantly reducing yields in about one-third of the cropland, while in the desert of the U.S. Southwest the figure approaches one-fourth.

Table 16.2	Economic efficiency of irrigation in California

Crop	Efficiency (water use in M² per $ value of crop)
Rice	19.8
Alfalfa	15.8
Almonds	13.2
Cotton	7.8
Apricots	3.9
Tomatoes	2.3
Broccoli	1.6
Onions	1.4
Lettuce	1.0
Strawberries	0.3

Source: Crop Budgets, University of California, Statistical Report, California Department of Agriculture.

4 *Fertilizer and pesticide contamination.* Applying nitrogen fertilizer in excess of the amount removed each year with the crop (Figure 16.2) can lead to loss of nitrogen from the soil environment. Because nitrate is negatively charged, it does not bind to negatively charged soil particles and so is poorly stored in the soil. When soil water exceeds plant needs (through rain or irrigation), nitrate leaches into soil groundwater. This can contaminate wells used by farm animals, and nitrates in drinking water pose a health problem to the animals and people who use the well.

The heavy use of pesticides poses a more direct problem for farm workers. Because these people are generally outside of the mainstream of health care, estimates of the extent of pesticide-induced health problems are fraught with difficulty. But incidents of acute poisoning certainly occur with alarming frequency, and epidemiological data show that workers who handle pesticides more than 20 days a year have an increased risk of developing certain types of cancer.

When a pesticide kills one pest, another may replace it. For example, in the early 1900s, the major pests of cotton were the boll weevil and the cotton leafworm, but since the extensive use of insecticides, the cotton bollworm and the tobacco budworm have become serious pests. The insecticides destroyed the natural enemies of the secondary pests of the cotton bollworm and the tobacco budworm, thus allowing their populations to rise. In 1978, it was estimated that in California 24 of the 25 top agricultural pests were secondary pests. That means that they became pests as a direct result of pesticide usage.

5 *Genetic erosion.* Modern agriculture has led to a narrowing of the genetic base of crop plants. The initial successes of newly bred and selected strains has repeatedly led to their rapid adoption by all farmers, and abandonment of the richly diverse landraces and wild relatives. For example, in the United States six cultivars (out

Figure 16.1

Figure 16.1

Extensive Salinity Damage in Abandoned Cropland in California's Coachella Valley. This was previously desert land, and was converted to agriculture by extensive irrigation. Salts already in the soil were dissolved in the water and rose to the surface.

of tens of thousands worldwide) account for a third of the wheat planted. In Greece, virtually all landraces of wheat have disappeared over the past 40 years. The dangers of such a narrow genetic base are obvious in terms of susceptibility to pests (for example, the tungro virus that devastated IR-8 rice in the Philippines during the 1970s).

Figure 16.2

Annual Average Addition of Fertilizer Nitrogen (straight line) and Removal of Crop Nitrogen (dotted line) for Corn in Central Illinois. Source: *National Research Council, National Academy of Sciences. Nitrates: An Environmental Assessment (1978), Washington, DC: NRC.*

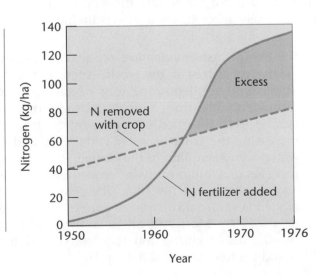

(a)   (b)

Figure 16.3

Hillside Farming Requires Careful Land Management to Prevent Soil Erosion.
(a) *Peruvian Indians harvesting barley in the* altiplano *of the Andes. Notice the completely eroded hillside above the plain. Areas that are so badly eroded cannot be reclaimed.* **(b)** *Hillside cultivation of coffee in Colombia. Because coffee is a perennial, it is particularly well-suited for hillside cultivation. Care must be taken to maintain a ground cover at all times, otherwise the soil will disappear. This farmer carries a basket of fresh coffee pulp to spread as fertilizer between the coffee trees.* Sources: *(a) Courtesy United Nations, ES/r (b) Courtesy Centro Internacional de Agricultura Tropical.*

2   Agricultural practices can have equally serious impacts on the environment away from the farm.

In addition to the serious effects of high-intensity farming in degrading the resource base for future farming, the environment away from the farm can be adversely affected. These effects include

1. *Soil erosion.* Astronauts have repeatedly commented on how the great river deltas of the world (such as the Nile, Amazon, and Mississippi) are depositing tons of silt into the oceans, silt that ultimately came from agricultural soil erosion. Eroded soil particles, especially carried by water, often end up filling reservoirs, increasing the costs of water extraction, and clogging navigable waterways. According to the United States Department of Agriculture, about 1 billion tons of eroded soils are deposited in waterways each year in the USA. The removal of 10% of the highly eroded lands from cultivation currently would reduce this deposition by 20%. Most difficult to cultivate without erosion are hillsides. Hillside cultivation requires careful management if soil erosion is to be avoided (Figure 16.3).

2   *Fertilizer and pesticide contamination.* Agriculture is a major source of water pollution by fertilizers and pesticides dissolved or suspended in soil water. Nutrient loading of lakes and estuaries stimulates the natural process of eutrophication. In many lakes and estuaries, algal growth is limited by the amount of phosphate available in the water. Pollution from phosphate fertilizers causes algal blooms. When the algae die and decay, the decay processes use up all the oxygen in the water, reducing fish and plant populations. The accelerated eutrophication of the Chesapeake Bay in the United States, caused by nutrients from agricultural and municipal sources, contributed significantly to the decline of the bay's fishing industry.

As noted, excess nitrogen fertilization can contaminate drinking water. It is estimated that 25% of the people living in the countries of the European Community are drinking water with nitrates above recommended levels (25 mg per liter). Pesticide contamination of groundwater, and then drinking water, is also a potentially serious problem. Although the effects of DDT on certain bird populations are well known, the effects of low concentrations of other pesticides on people have not been established. In the United States, detectable levels of 39 pesticides have been observed in groundwater in 34 states. Although in most cases, the levels are below those deemed unsafe, long-term effects of this exposure are not known.

3 | **Environmental degradation is associated with rural poverty in developing regions.**

Until recently in human history, there was enough land and associated resources to meet the needs of rural populations around the world. But now the easily converted land and easily extractable resources are already being used for agriculture, and an increasing number of poor people lack the economic resources to exploit the available land for optimal crop growth.

In fact, many people in the world have no choice but to exploit more and more marginal lands (Figure 16.3). As they begin to use the Amazon rainforest and Himalayan foothills, the potential for environmental damage rises. There is a double-edged sword here: to use the land productively, great inputs of agricultural technologies will be needed, and this can have the adverse effects noted earlier. In the far more likely case of these people being unable to afford these inputs, the yield of crops will be low and still more land will be put under the plow so that the people can eat. Unlike the situation in the developed regions, where market-driven productivity is the key, in the developing regions crop growth is a base for human survival. The pressure to produce more food for the expanding population has resulted in the use of marginal lands throughout the world (Figure 16.4).

The cycle of resource degradation, hunger, and poverty can best be illustrated by considering a rural family living in northern India. Conversion of forest lands to farming, and the needs of a growing population for wood for cooking and heating, have led to a serious decline in the availability of fuel

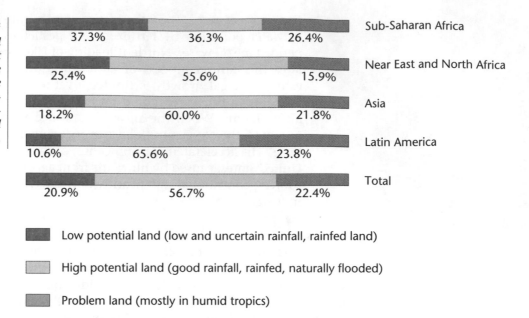

Figure 16.4

Shares of Harvested Land of Different Potentials in the Developing World. Note the large amount of marginal land under production. Source: *Food and Agriculture Organization.*

Sub-Saharan Africa: 37.3% | 36.3% | 26.4%

Near East and North Africa: 25.4% | 55.6% | 15.9%

Asia: 18.2% | 60.0% | 21.8%

Latin America: 10.6% | 65.6% | 23.8%

Total: 20.9% | 56.7% | 22.4%

 Low potential land (low and uncertain rainfall, rainfed land)

High potential land (good rainfall, rainfed, naturally flooded)

Problem land (mostly in humid tropics)

wood. Animal manure and crop residues may be used instead of wood for fuel; but this eliminates them from the soil as fertilizer, with concomitant reductions in crop yields and increases in erosion. If the local market is used to buy fuel wood, the price may be high and this results in less money available to purchase agricultural inputs; once again, yields fall and the land is not cared for.

Ultimately, the small farmer must make difficult decisions, and it is a no-win situation. In the struggle to survive, long-term considerations of land productivity are far from paramount. Many subsistence farmers in developing countries face a dilemma. Either they must eke out a living with dignity on marginal land, thereby contributing to deforestation and land degradation, or they must go to the rapidly growing cities where there are few jobs and where life itself is degrading for so many.

4   Economic and political considerations exacerbate the adverse effects of nonsustainable agriculture.

Two objectives have driven the agricultural policies of the *developed* countries:

1   To provide adequate food for the people
2   To provide adequate incomes for farmers

Left unstated but surely considered are such aspects as the preservation of the family farm, foreign trade balances, the incomes of ancillary industries that provide inputs, political voting blocs, and so on. And left almost unthought are such things as nutrition and nutritional diseases, the distribution of food to all the population, and the environmental costs of agricultural production.

In the United States, the policy that has the greatest impact on farming is the Farm Law, a commodity program under which the federal government guarantees the farmer a minimum price for crops produced on the farm. Two-thirds of all U.S. croplands are now enrolled in commodity programs, with

Table 16.3		Cost–benefit analysis of applying nitrogen (N) fertilizers to corn			
N Application (kg/ha)	Yield (kg/ha)	Added N cost (N @ 26¢/kg)	Added Corn Returns (Corn @ $10/100 kg)	Added N Cost (N @ 52¢/kg)	Added Corn Returns (Corn @ $13.50/100 kg)
67	5,785	—	—	—	—
90	6,602	$6.00	$81.70	$12.00	$110.30
113	7,293	6.00	69.10	12.00	93.30
136	7,860	6.00	56.70	12.00	76.50
159	8,237	6.00	37.70	12.00	50.80
182	8,361	6.00	12.40	12.00	16.74
205	8,425	6.00	6.40	12.00	8.64
228	8,487	6.00	6.20	12.00	8.37

Note: The yield data (columns 1 and 2) are for central Indiana 1967–1969. The calculations show two different prices for nitrogen fertilizer (26¢ and 52¢ per kg) and two different prices for corn. Each line gives the *increment* associated with adding another 23 kg per ha of nitrogen fertilizer.

Source: J. T. Pierce (1990), *The Food Resource* (New York: Wiley), p. 271.

80–95% of the acreage planted to rice, corn, wheat, sorghum, and cotton participating. Total payments are about $25 billion per year.

Payments are determined by three factors: acreage planted to the crop in a multiyear period, yield of the crop (also over several years), and a target price. If the actual market price falls below the target, the government guarantees to make up the difference to the farmer. To maximize payments from the government, therefore, farmers want to plant a lot of acres that yield a lot of the crop.

In this sense, farming in the United States has become a taxpayer-supported industry, where the government contracts for production, buys products, and indirectly pays workers. One result of the commodity program is that farmers cannot readily switch to other crops or other types of farming. Farmers cannot easily introduce crop rotation, because they have no "authorization" to grow the crops that might fit in a rotation schedule. The 1990 revision to the Farm Bill began to deal with this, and allows a supported farmer to plant up to 25% of the acreage with other crops without losing acreage-based supports.

The subsidy program also encourages the excessive use of fertilizers and pesticides, to keep yields high. If yields are high, the farmer safeguards future financial returns, even if it costs more to produce the crop than it is worth. Because of the law of diminishing returns, it takes much more fertilizer and other inputs to raise production from 6 to 8 tons per hectare, than from 4 to 6 tons per hectare. Pushing production up to reach such a high level may not be cost-effective, but the taxpayers foot the bill (Table 16.3). However, the large amounts of fertilizer and pesticide required to reach this production level put an unnecessarily heavy load on the environment.

In the European Community, the Common Agricultural Policy regulates prices and price supports with objectives (and effects) similar to those of the U.S. policy. Once again, subsidies are well above world prices, and environmental costs are not considered. In Europe, an additional strong consideration is the family farm. Many Europeans have close ancestors who were farmers, and still retain an affection for the romance of rural life. As taxpayers in urban areas, they seem more than willing to keep price supports high to

Box 16.1

Agricultural Change in the Yucatán

The Maya, who live in the northern part of the Yucatán peninsula, have practiced a shifting slash-and-burn agriculture for the past 3,000 years. This part of the Yucatán consists of a limestone plateau with a soil so thin that the sixteenth-century Franciscan bishop Diego de Landa commented, "Yucatán is the country with the least soil that I have seen." Shifting agriculture, although seemingly simple, requires quite a bit of experience on the farmer's part, who must prepare a field, or *milpa,* then coax a crop of corn, squash, beans, and peppers out of the thin, poor soil. Because the soil is nutrient poor, the useful life of a *milpa* is only two years, and the farmer must then let the field lie fallow for 15–20 years to allow the soil to regenerate. The Maya, then as now, practiced intercropping with bean vines growing up the corn stalks, and squash vines covering the soil below (see the figure, part a). In addition to these food crops, the Maya used a variety of wild plants, including an agave (*Agave fourcroydes*), locally known as *henequen,* to produce the sisal fiber they used to make ropes, hammocks, and mats.

(a)

A Milpa in Yucatán. Corn plants are growing in a thin and stony soil.

(b)

A Small Henequen Plantation.

The Spanish conquest and the establishment of the state of Mexico did not greatly affect agricultural life in this far corner of the Republic, except that the Spanish-speaking landowners established haciendas to produce sisal for export. Production of this cash crop was slow to take off because separating the fibers from the pulp of the thick agave leaves remained difficult and laborious until the invention of the mechanical decorticator by José Esteban Solís in 1861. This decorticator did for the sisal industry in Yucatán what the cotton gin in the United States did for the cotton industry. The advent of steam power and the increased demand for the sisal fibers in Europe and the United States caused henequen plantations to expand enormously. By 1890 a thousand steam-driven decorticators were in operation, processing the leaves produced on as many haciendas. The henequen plant (see figure, part b) has a useful life of 20–25 years, and mature leaves can be harvested every six months. Because the plants are perennials, and the leaves are not particularly rich in plant nutrients, the problems of soil exhaustion and soil degradation, normally encountered with the cultivation of annual food plants, can be avoided. The economic boom of this cash crop did not greatly benefit the local population, because the people were kept in bondage by the landowners. John Kenneth Turner, who traveled and studied in this region, wrote in 1908 that he found 8,000 Yaqui slaves, 3,000 indentured Asians (Chinese and Koreans), and more than 100,000 indentured Maya working on the haciendas. The breaking up of the haciendas by the movement for land reform in the late 1930s coincided with a gradual decline in the demand for sisal, and the decorticators slowly fell silent.

To reinvigorate the agricultural life of the region, the Mexican government now provides subsidies for a new cash crop industry: oranges for local consumption and processing into juice. The town of Oxkutscab has a weekly market where the small producers can sell their fruit. At the same time, several government research laboratories in Mérida are investigating the potential of plant biotechnology, especially *in vitro* propagation of plants and the production of valuable secondary products through tissue culture. Meanwhile, the cultivation of food crops in the villages of rural Yucatán continues as it has for hundreds of years, because intensive high-input agriculture is not possible on these thin, nutrient-poor soils.

keep production local. This results in total subsidies as high as the more productive United States (about $24 billion per year).

In Japan, the desire for local production of rice rests not only on tradition, but also on the historically strong tendency for self-sufficiency. With China as an unreliable trading partner and strong local political pressures to support rice farmers, the government of Japan has the highest price supports of the industrialized world (Figure 16.5).

Government pricing can profoundly affect decisions of farmers on what to grow; if a crop is unprofitable in the "doctored" market, a farmer will simply switch to another crop. For instance, in the European Community, the prices of corn and sugar beets have been kept high, while those of legumes and root crops are kept low. Not surprisingly, there has been a switch away from legumes and root crops toward corn. Farmers who previously planted corn and legumes now plant corn only, with predictably negative effects on declining soil fertility. High milk prices, combined with low prices for imported feed for dairy cows, have shifted milk production to the coastal

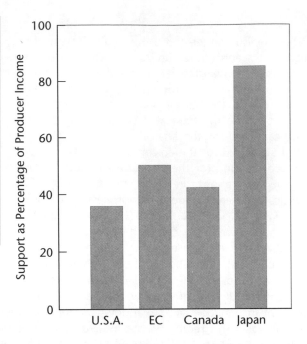

Figure 16.5

Average Producer Subsidy for Grains, Livestock, Dairy, Oilseeds, and Sugar in 1988. EC = European Economic Community. Although U.S. farmers derive a considerable proportion of their income from government subsidies, the percentages are even higher in other developed countries. Source: *U.S. Department of Agriculture.*

regions, near the ports where animal feed arrives. Concentrated livestock production is rapidly marginalizing the land base in this region.

In addition to subsidizing the price to the farmers, governments encourage high-input agriculture via tax policies. Although the U.S. Tax Reform Act of 1986 eliminated the financial incentives to convert wetlands and erodible fields into farmland, until that time, it was highly profitable to buy cheap rangeland, wetlands, forest land, and erodible fields, turn them into cropland, and sell them at a profit. Much of the profit came from the favorable tax treatment of the earnings. Thus, the enormous expansion of cropland into these ecologically sensitive areas was not driven by the need to produce more food, but by a favorable tax treatment for those who had access to the necessary capital to take part in this type of "development."

Low capital gains taxes help make speculation in farmland profitable, destabilize land prices, and generally accelerate the trends toward fewer and bigger farms. Accelerated depreciation and investment tax credits (two instruments of tax policy) also promoted the rapid growth of custom beef feedlots and reshaped this industry between 1960 and 1985. Beef feedlots produce mountains of manure that are often not properly recycled back to the land as fertilizers. Forcing the operators to do so would have diminished their financial gain and perhaps increased the price of beef, but could have been built in as part of the financial package.

The *trade* implications of these local policies have come sharply into focus as the governments of the industrialized world try to make sense out of the byzantine rules that govern agricultural sales from one country to another. Often a country will put high tariffs on an import to support local producers and discourage foreign competition. For instance, the United States has long protected its sugar producers by giving them subsidies far in excess of world prices (the recent figure was twice the world price) and putting prohibitive tariffs on imports (much of this from developing countries).

Another example is soybeans. Thirty years ago, the United States was granted "no tariff" status for its soybean sales to Europe. But over time the

Figure 16.6

European Farmers Demonstrate Against Their Own Government, Which Is Trying to Reduce Its Price Supports for Oilseeds. European governments are under pressure from the U.S. government, which wishes to sell soybeans produced by U.S. farmers (also with price supports) to the Europeans. Source: *Agence France Presse.*

Europeans have subsidized their farmers to produce other oilseeds, such as rape and sunflower, thereby obviating the need to buy U.S. soybeans. As more oilseeds are grown in Europe, prices of all oilseeds, including soybeans, have fallen. With the incentive of an artificially lower price, European consumers are buying the locally produced products. U.S. soybean producers are angry, because their market is shrinking; but the European governments, under pressure from their farmers (Figure 16.6), will not change their policies.

The General Agreement on Tariffs and Trade (GATT) governs trade between many countries. In the most recent negotiations, the Uruguay Round (1986–1992), it is not surprising that a major stumbling block was agriculture. The United States put forward a bold proposal to eliminate tariffs and subsidies on agricultural products altogether. But this was rejected by the Europeans, for domestic political reasons. Nevertheless, these negotiations were the hopeful beginning of the process of removing government subsidies and allowing the free market to govern trade in food and feed.

Although freeing the international agricultural markets would make the worldwide trade in these commodities more rational, it could have adverse effects on the developing countries. All over the developing world, except for China, a period of profound economic change is underway. Called **structural adjustment**, this is in its broadest sense a transition from socialism to free-market economics. It is having, and will continue to have, great implications for food production and consumption.

From the 1960s onward, many developing countries attempted to mimic demographic transitions that had occurred earlier in Europe (see Chapter 1).

Part of the development strategy was to encourage industrialization in the cities, and one incentive for this was keeping food prices low through subsidies. Low prices tended to discourage market farming, and as farmers left the land, they were supposed to find abundant work in the expanding cities. At the same time, education improved both in terms of literacy rates and in terms of quality. When urban jobs were slow in coming, and wage rates low, and when the rural poorest people got poorer because of the food policies aimed at the cities, attempts were made to create a social safety net to help them through the difficult period.

These subsidies and social programs required money, and assuming that the economy would expand, banks in the developed countries eagerly lent the funds. Then came the recession of the 1980s. Economic expansion in the developing countries slowed significantly, but they still had great debts to the rich countries. In conference after conference, at the World Bank and the International Monetary Fund, a solution was sought to the debt crisis.

The result was to formulate a new policy: the Third World countries would get out of debt by expanding their economies in the free-market model. The idea is to encourage modernization and growth by improving efficiency, devaluing currency to spur investment, and removing distortions in the market caused by government intervention. For many countries, this has meant three things:

1 Industries have been privatized and sold to foreign investors to gain foreign currency. In agriculture, this has meant a partial return to the situation that existed during colonial times, when the natural resources of the developing world were owned by the developed world. Participation in the world market economy means more emphasis on cash or export cash crops and less on food crops.

2 Those countries that have natural resources that are still locally owned are producing goods and selling them at the lowest possible price on the world market. This usually means that labor remains cheap (poverty), and environmental considerations are put into the background because of the urgency of the situation. Cheap labor means low purchasing power, the primary reason for hunger. The push by Mexico for a free-trade agreement with the United States and Canada is an example of a country greatly in need of selling its goods without tariffs.

3 With little money left for them, the social programs and safety net are receiving far less support from developing country governments. The rural and urban poor are getting poorer. Increasingly, this is leading to political instability, especially in Africa.

5 **Environmental accounting is a new way to formulate government policies.**

As economies around the world undergo these rapid changes, their impacts on the environment are not always considered. One reason is that the politicians, and the senior civil servants who advise them, are not aware of the real costs of soil erosion, depletion of nutrients, and pollution. Those who are aware of these factors do not know how to integrate them into a policy framework; or they may simply ignore them in responding to the urgent situation of the moment.

Economists typically estimate the cost of a commodity (such as corn) by its trading value on the market (such as price per ton). This price is determined both by its value (cost of production and profit) and by the laws of supply and demand on the market (if demand is high and supply is low, the price rises). But does the production cost really reflect the finite value of resources such as land and water?

Putting a dollar value on an environment such as a piece of land or a body of water is difficult. An approach to doing this is to survey the resource (such as land) and quantify its possible economic uses (such as forest, corn growth, soybean growth, fallow). Then some measure of ecological stress is developed (for example, nutrient depletion leads to a transition from crop growth to fallow) and an economic value is put on this change. This must include not only the reduced yields from the resource, but also the costs of cleanup (for pollution) and regulation. If this process is done for a region or country, a monetary value can be put on the rate of resource depletion leading to lowered economic value.

These kinds of calculations have been made by the United Nations for Indonesia for 1971–1984. During this period, this country had impressive economic growth, with net domestic production growing at the rate of over 7% per year. But when the depletion of nonrenewable timber and petroleum reserves, as well as soil erosion in Java, were figured in, the growth rate fell to 4%.

Some countries, notably Norway and France, have attempted to make annual environmental balance sheets, reporting "stocks" of renewable resources (such as farmland), nonrenewable resources (such as minerals) and cyclical resources (such as air and water). These are then used as a guide to government policies in using these resources. These approaches require not only trained people to conduct the inventories of the natural environment and analyze them, but policymakers educated in these subjects.

6 **Food security for people is often a missing dimension in agriculture policies.**

The Green Revolution and its aftermath have led to impressive increases in worldwide food production. When policymakers use the term "food security," they are usually referring to grain stocks in excess of demand, left in storage against bad crop years (see Chapter 11). This has been a concern of planners since agriculture began.

But there is another aspect of food security, at the level of the individual household. True food security means that each person has a reliable, available, and affordable food supply. The United Nations-sponsored World Commission on Environment and Development, set up to prepare for the U.N. conference on these subjects in Rio de Janeiro in 1992, used the term *equity* as a criterion for distributing the world's resources, including food, and defined the term in two ways:

1   Equity for people living now who do not have equal access to
    natural resources or to social and economic "goods"

2   Equity for the human generations to come, whose interests are not
    represented by current economic analyses or by market forces that
    do not take the future into account.

Environmental accounting (see earlier discussion) and conservation
strategies (see later) can deal with the second aspect, the future, however
imperfectly. But it is the present, the first aspect, that is troublesome. Both
within countries, and between countries, there is political and social ten-
sion as the need for the poorest to literally survive becomes starkly
apparent.

We have seen that the current structural readjustments to the economies
of many countries are leaving out the poor in the transition to a more efficient, market-driven system. For example, for two decades, Brazil has undergone what some call an "economic miracle," with the economy growing in
some years at the rate of 8% a year. (More recently, declines in oil price and
the worldwide recession have slowed this rate, but it is still well above the
Third World average.)

But the poorest Brazilians have not shared in the bounty; the income per
person overall is over $2,600; but for the poorest 40% (2 people in 5!), the
income is less than $350. Although one-fourth of the spending is done by the
government, in the rush to support economic development only 2% is spent
on social programs. This typically results in a bimodal development where
the gains of the few come at the expense of the many. Moreover, the environmental costs of development (for example, the Amazon rainforest) have been
high. Thus the "miracle" has failed to fulfill either of the two aspects of the
U.N.'s definition of equity: both present and future generations are not shar-
ing in the available resources. Policymakers have realized their mistake (certainly the presence of the World Environment conference brought it to their
attention) and more rational development, coupled with land reform, has
been promised.

At the Rio de Janeiro conference, the ownership of resources was a major
topic of discussion, especially involving the Third World countries and the
United States. At issue was whether a country owns not only its resources, but
their potential uses by its own citizens as well as by foreigners. For example,
the Brazilian rainforest may contain unique plants that have chemicals that
can be used to treat cancer. In the past, a European or U.S. pharmaceutical
country could come into Brazil, pay the landowners to explore the land,
remove a plant, and then isolate a compound that fights cancer. Once the
drug is marketed, the company reaps its research and development profits.
But what of Brazil, where the plant originated?

Currently, Brazil gets nothing. But at the 1992 conference, Third World
countries demanded that in the name of equity, their people should share
in the bounty from resources on their own land. This could be extended to
agricultural resources, such as genetic strains of crop plants. Because much
of the biological diversity is in the poorer countries, they could derive monetary benefit by "ownership." Although most countries agreed to this principle, the United States initially refused to sign the treaty, under pressure
from some industries that felt that exploration would be made less economically viable.

7 | **Alternative agriculture means organic farming, low-input agriculture, and/or sustainable agriculture.**

Throughout this book, we have stressed that plant growth and crop production are complex processes that depend on many interactions between organisms. The trend of modern agriculture has been to simplify:

- To substitute monoculture and continuous culture for crop rotation and diversified agriculture
- To use herbicides and pesticides to combat pests, rather than more complex biological control mechanisms
- To use genetically more uniform plants, that have a narrow genetic base
- To make bigger fields by eliminating all vegetation between them
- To use inorganic fertilizers, rather than the more difficult-to-use organic manures, in combination with green manures

Recently, there has been a resurgence of interest in "alternative agriculture," "organic farming," and "sustainable agriculture." In 1989, the Board on Agriculture of the National Research Council of the United States lent credence to this movement with the publication of a major study called "Alternative Agriculture." Alternative agriculture is not a single system of farming practices, but encompasses many different systems known as "organic," "low-input," "regenerative," or "sustainable farming." All these systems share an emphasis on management practices and on biological relationships between organisms.

Alternative agriculture recognizes that a piece of land on which crop plants are grown is first and foremost an ecosystem, and not a factory. An ecosystem has many interacting organisms that must remain in balance. Many natural processes occur in such an ecosystem, and farmers should take advantage of these natural processes, rather than try to circumvent them or destroy them with chemicals. Alternative agriculture rejects certain practices (such as heavy use of inorganic fertilizers), but most important are the practices it favors:

- Integrated pest management for pest control
- Tillage that minimizes soil erosion even if it is more expensive
- Reliance on animal manures and green manures with minimal input of inorganic fertilizers
- Management systems, such as crop rotations, that help control weeds and disease organisms

These techniques are not new of course, although some of them have been refined in recent years.

Organic farmers emphasize using only organic fertilizers for fertility maintenance and banning nearly all chemical pest control methods. However, gypsum, an inorganic calcium salt, and rock phosphate are allowed as fertilizer, and a number of "natural" sprays are also allowed. Among these are sulfur dust, extracts of toxic plants, insecticidal and herbicidal soaps, and

Figure 16.7

An Organic Produce Market. Such markets, which often occur once a week in an outdoor location, have proliferated in the United States, thanks in part to federal and local government support. These markets offer an outlet for producers of organically grown vegetables and fruits. Source: *Courtesy of D. Ott.*

virus sprays. Organic farmers use antibiotics derived from fermentation (such as streptomycin), as long as the fungi that produce them are not genetically engineered. Genetically engineered crops are not allowed, but crops improved by traditional breeding are permitted.

Until well into the twentieth century, organic farming was not alternative agriculture: it was the worldwide way of life. It still is, in many of the poorer regions of the Third World, where farmers cannot afford the technological inputs of modern agriculture. In North America, the Mennonites and Amish have long practiced organic farming out of tradition and choice. Studies of their farms show somewhat lower yields (5–15%) than conventional farming in nearby areas. But their net return on investment is usually higher, because they consume less inputs, and when environmental costs are taken into account (they seldom are), the organic alternative is clearly superior.

In many respects, organic farming is a way of life as much as it is a method of farming. The profitability of organic farms depends on the higher prices that their products command in the market place. To stimulate organic farming, some governments have passed laws that create a demand for organic foods. For example, in some U.S. states, poor people who receive food aid get coupons only redeemable at organic markets. Cities have created farmer's markets, where organic producers can sell their goods (Figure 16.7).

Sustainable agriculture, on the other hand, emphasizes the conservation of its own resources. For a farm to be sustainable, it must produce adequate amounts of high-quality foods, be environmentally safe, and, where appropriate, be profitable. Sustainable farms minimize their purchased inputs (fertilizers, energy, equipment) and rely, as much as possible on the renewable resources of the farm itself. This is especially important in the 90% of farms that exist in poorer parts of the world, where these inputs are often not available or affordable.

Sustainable farming uses some form of integrated pest management for pest control, and this can include the use of chemical pesticides that are not used by organic farmers. Thus, sustainable agriculture does not mean a

Figure 16.8

Profits from Sustainable Farms Can Exceed Those of Conventional Farms. According to calculations made by S. L. Kraten, the cash incomes per acre for the two types of farms were comparable over two years, but because the input costs of sustainable agriculture are lower, its net returns are 22.4% higher. Variable costs include those for fuel, machinery maintenance, seed, fertilizer, pesticide, and labor. Among the fixed costs are property taxes and interest on loans. Source: J. P. Reganold, R. I. Papendick, and J. F. Parr (1990), *Sustainable agriculture,* Scientific American 78:112–120. Copyright 1992 by Scientific American, Inc. *All rights reserved.*

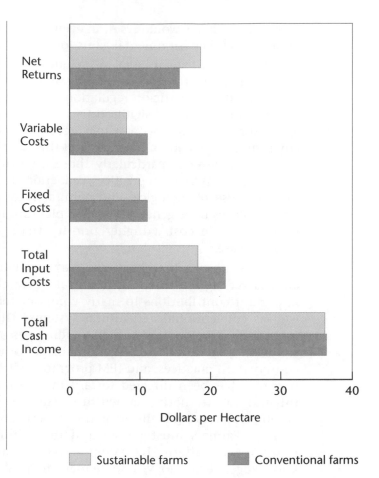

return to the farming methods of the late 1800s. Rather, it combines traditional techniques that stress conservation with modern technologies, such as certified seed, modern equipment for low-tillage practices, integrated pest management that relies heavily on biological control principles, and weed control that depends on crop rotations. Sustainable farms try to use wind or solar energy instead of purchased energy, and use organic animal manure and nitrogen-fixing legumes as green manure to maintain soil fertility, as much as possible, thereby minimizing the need to purchase inputs from outside the farm. The use of genetically engineered crop strains is certainly not excluded by sustainable farming. The emphasis is on maintaining the environment, not on rules about what can or cannot be done. Profits from sustainable farms can exceed those of conventional farms (Figure 16.8).

Although all pay some lip service to the concept, only a minority of farmers in the developed countries practice alternative agriculture to a significant degree. After studying such farmers and their practices in the United States, the National Research Council arrived at the following conclusions:

1   A small number of farmers in most sectors of U.S. agriculture currently use alternative farming systems, although components of alternative systems are used more widely. Farmers successfully adopting these systems generally derive significant sustained economic and environmental benefits. Wider adoption of proven

alternative systems would result in even greater economic benefits to farmers and environmental gains for the nation.

2 A wide range of federal policies, including commodity programs, trade policy, research, extension programs, food-grading and cosmetic standards, pesticide regulation, water quality and supply policies, and tax policy significantly influence farmers' choices of agricultural practices. As a whole, federal policies work against environmentally benign practices and the adoption of alternative agricultural systems, particularly those involving crop rotations, certain soil conservation practices, reductions in pesticide use, and increased use of biological and cultural means of pest control. These policies have generally made a plentiful food supply at the lowest possible cost a higher priority than protection of the resource base.

3 A systems approach to research is essential to the progress of alternative agriculture. Agricultural researchers have made important contributions to many components of alternative as well as conventional agricultural systems. These contributions include the development of high-yielding pest-resistant cultivars, soil-testing methods, conservation tillage, other soil and water conservation practices, and IPM programs. Little recent research, however, has been directed toward many on-farm interactions integral to alternative agriculture, such as the relationship among crop rotations, tillage methods, pest control, and nutrient cycling. Farmers must understand these interactions as they move toward alternative systems. As a result, the scientific knowledge, technology, and management skills necessary for widespread adoption of alternative agriculture are not widely available or well defined. Because of differences among regions and crops, research needs vary.

4 Innovative farmers have developed many alternative farming methods and systems. These systems consist of a wide variety of integrated practices and methods suited to the specific needs, limitations, resource bases, and economic conditions of different farms. To make wider adoption possible, however, farmers need to receive information and technical assistance in developing new management skills.

8 | Many farming techniques are used in sustainable agriculture.

Because the principle used (conservation of the land being a higher priority) differs from the production-based aims of technological agriculture, sustainable agriculture employs a variety of methods to achieve its goals. Some of the differences between the two approaches are listed in Table 16.4. The major methods used in sustainable agriculture are described as follows.

Crop rotation is the successive planting of different crops in the same field, and is the opposite of continuous cropping. Continuous cropping is planting the same crop on the same field year after year. A typical example of a crop rotation would be corn followed by soybeans, followed by oats, followed by alfalfa. Intercropping means growing two crops on the same field at the same time. The economic and environmental benefits of crop rotations

| Table 16.4 | Comparison between "Green Revolution" and sustainable agriculture |

Characteristic	Green Revolution	Sustainable
Crops used	Wheat, rice, corn	All
Land use	Flat, irrigated	All, including marginal and rainfed
System	Monoculture	Polyculture
Inputs	Fertilizers, pesticides, fossil fuels, nonrenewable	Organic fertilizer, nitrogen fixation, renewable
Environmental impacts	Medium to high: erosion, salinization, water pollution	Low to medium
Cash needs	High	Low; inputs local

Source: Modified from M. Altieri (1992), Sustainable agricultural development in Latin America, *Agriculture, Ecosystems and Environment* 39:1–21.

Figure 16.9

Intercropping: Soybean Plants Growing in a Wheat Field. The soybeans have nitrogen-fixing bacteria in their roots and remove little nitrogen from the soil. This allows the soil to replenish its nitrogen content for the wheat crop. Source: *U.S. Department of Agriculture.*

and intercropping (Figure 16.9) are well documented, and derive, in part, from the use of legumes, which are nitrogen fixers. In this rotation, soybeans and alfalfa are nitrogen fixers, and can fix 250–400 kg nitrogen per hectare per year. This allows the soil to replenish itself for the next cereal crop (Table 16.5).

Biological nitrogen fixation can supply all or much of the nitrogen required in some cropping systems. For example, soybeans and forage legumes can supply all their own nitrogen needs and forage legumes in a rotation as a legume cover crop can supply nitrogen to subsequent crops.

However, even if corn follows wheat, the corn yield will be greater for the same amount of fertilizer applied. The main benefit of this crop rotation comes from the control of weeds, insects, and disease organisms, particularly

Table 16.5	Average fixation of nitrogen by legumes

Legume	Nitrogen Fixed (kg/ha)
Alfalfa	217
Red clover	128
Cowpea	100
Pea	72
Soybean	65
Bean	45
Peanut	44

Modified from: Tisdale, S. L. and W. L. Wilson (1966). Soil fertility and fertilizers. The Macmillan Company, New York, NY.

the pests and diseases that attack root systems (see Chapter 8, "Life Together in the Underground"). Another benefit comes from the observation that deep-rooted crops bring mineral nutrients to the surface that are then available to more shallow-rooted crops. Crops that are closely sown (wheat, barley, oats, alfalfa) also provide much better erosion control than row crops such as corn and soybean. Economic tradeoffs in rotations include lower-value crops (such as alfalfa or other leguminous hay crops), the need to buy new equipment, and government disincentives.

Green manuring is a form of fertility maintenance that depends on the nitrogen-fixing capacities of forage legumes (alfalfa, clover, vetch) that are plowed under, rather than harvested. In this way, all the mineral nutrients, as well as the fixed nitrogen, stay in the ecosystem, rather than being exported with the harvest. Because these plants can produce new growth from their root systems after the field has been moved, it is not unusual for the farmer to collect one or more crops of hay, and to plow under the last regrowth. Whether this results in a net gain depends very much on how long the last growth is allowed to develop. After the shoots are cut off, new growth generally uses starch and protein reserves that are already in the root system. Only after a new canopy of leaves is established is there enough photosynthate to provide energy for additional nitrogen fixation. Thus, if the last cut is made too late in the season, there will be little net gain when the entire crop is plowed under.

A special strain of alfalfa has been developed in Minnesota that fixes about 250 kg/ha of nitrogen between the time of the last harvest (early September) and the time (mid October) when the plants are killed by frost. Such a level of nitrogen will sustain an abundant corn crop the next year. Thus, crops can be developed not just for greater yield, but to be particularly well suited for a specific management practice.

Tomatoes are often grown with the soil covered with polyethylene to retain moisture. If instead hairy vetch, a legume, is used, the same conservation of moisture ensues with two added benefits: there is less loss due to weeds, and the tomatoes' needs for nitrogen fertilizer decline. The result is higher yields (Figure 16.10).

Green manuring takes a totally different form in Asian wetland rice cultivation. Here the rice fields are "inoculated" with the floating fernlike plant

Figure 16.10

Tomato Crop Grown with (a) and Without (b) a Mulch of Vetch. The vetch contributes nitrogen to the soil, causing the plants to grow bigger and produce more tomatoes. Source: *U.S. Department of Agriculture.*

(a)

(b)

Azolla pinnata and with nitrogen-fixing bacteria. The symbiotic relationship between the two is similar to that found in the nodules of legume plants and indeed fixes as much nitrogen. When the Azolla plants die, they sink to the bottom and are slowly decomposed, adding nitrogen that is now available to the rice plants.

Animal manuring can make a significant contribution to soil fertility, as manure contains abundant amounts of the three nutrients that most limit crop growth—nitrogen, phosphorus, and potassium. Indeed, 1 ton of cow manure contains 5.6 kg nitrogen, 1.5 kg phosphorus, and 3 kg potassium. It would take about 25 tons (from three cows in a year) to supply the needs for these nutrients of a hectare of corn. But in the United States only about 10% of the nitrogen requirements of corn are met by manure. The problems of

Figure 16.11

Ridge Tillage Advantages in Alternative Production Systems.
The planter tills 5 to 10 cm of soil in a 15-cm band on top of the ridges. Seeds are planted on top of the ridges, and soil from the ridges is mixed with crop residue between the ridges. Soil on ridges is generally warmer than soil in flat fields or between ridges. Warm soil facilitates crop germination, which slows weed emergence. Crop residue between the ridges also reduces soil erosion and increases moisture retention. Mechanical cultivation during the growing season helps to control weeds, reduces the need for herbicides, and rebuilds the ridges for the next season.

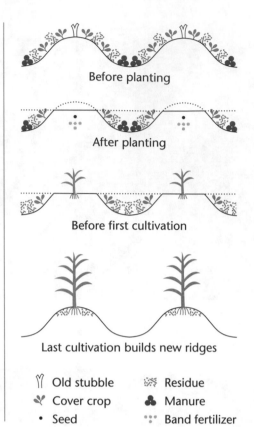

Before planting

After planting

Before first cultivation

Last cultivation builds new ridges

Ⓨ	Old stubble	⬚	Residue
✔	Cover crop	♣	Manure
•	Seed	⁖	Band fertilizer

storage and transport are the major drawbacks. In the developing countries, manure is far more important. China, especially, has been a leader in stimulating the use of this farm-produced resource instead of inorganic fertilizer.

Conservation tillage uses tilling and planting methods that leave residues (such as wheat stubble and straw) on the soil surface after a crop is harvested. These residues cover and protect the fields during the rainy months to protect topsoil from erosion, and retain moisture for the drier months.

Moldboard plowing is the conventional way to till the soil. The steel blades of the plow turn over furrows of soil and in this way, break up the soil, partially bury the weeds, and cut through patches of perennial weeds that form mats. However, it leaves the soil surface exposed to rain, wind, and sun, contributing to water loss and soil erosion.

The types of conservation tillage practices that reduce these losses include reduced tillage, no tillage, and ridge tillage (Figure 16.11). Although these practices have the beneficial effects of less erosion, more conservation of soil moisture, and less decay of soil organic matter, they may also increase the need for pesticides, especially herbicides. Conservation tillage also leaves a layer of crop residue on the soil, and that is where many pests may overwinter (Figure 16.12).

With a no-tillage regime, the weeds are killed with a herbicide and the seeds are then drilled into the soil together with the fertilizer. Ridge tillage is a tillage method that combines some of the advantages of conservation tillage without the disadvantages. The field is left in ridges after the harvest, and in the spring only the tops of the ridges are cultivated. This destroys the weeds in the immediate vicinity of where the seedlings will emerge. Cultivation destroys the ridges, but these are reformed later in the season.

Figure 16.12

Soybeans Growing in a Corn Stalk Residue in a No-Till System of Farming in Jackson Country, Iowa (United States). The residue helps retain water and reduces soil erosion. Source: *U.S. Department of Agriculture.*

Figure 16.12

Soybeans Growing in a Corn Stalk Residue in a No-Till System of Farming in Jackson Country, Iowa (United States). The residue helps retain water and reduces soil erosion. Source: *U.S. Department of Agriculture.*

Integrated pest management encompasses a variety of techniques, all aimed at reducing pesticide use to the minimal level necessary. Central to this concept is the notion of the economic injury level: the mere presence of pests in an agricultural field does not prove they will inflict economic damage by harvest time. IPM is an ecologically based pest control strategy that relies heavily on natural enemies, resistant plants, and crop management to limit the damage from pests. It was initially applied to cotton, where pesticide applications were particularly heavy. IPM depends heavily on scouting for pests (often using pheromone traps) and on predicting how pest populations will change given the age of the plants, the likely weather conditions, and so on. Computer programs can recommend the best strategy for pest control.

Although IPM was initially formulated for insects, its concepts can be extended to all plant pests. Most IPM programs result in a decreased pesticide use, by applying lower doses of pesticide at the time when pesticide application will be the most effective. It depends on the ability of "pest scouts" to accurately determine pest populations. Because this job requires considerable expertise, it is sometimes performed by small companies that specialize in IPM. Organic farmers use biological control methods exclusively, while traditional farmers may use chemical control, or a combination of both, depending on the exact situation (crop, insect pest, and so forth). In some crops, such as tomato and citrus, 75% of the acreage is already under IPM, while in soybeans and corn only 15–20% is under IPM. For many crops, more research is needed, especially in biological control methods, to extend the concept of IPM.

Calculations by David Pimentel and his colleagues show that total pesticide use in the United States could easily be reduced by 50% without reducing crop yields. It is interesting in this regard that several European countries (The Netherlands, Sweden, Denmark) have recently instituted national programs that have as their goal a 50% reduction in pesticide use over a 10– to 12–year period. These are not research goals, but implementation goals. Great

Britain has launched a research program to find out if one important crop, canola, can be grown without any pesticides.

| 9 | **Is there room for plant genetic engineering in a sustainable agriculture?** |

Although many organic farmers have rejected genetically engineered plants (and microbes) because they are "not natural," there is no *a priori* basic contradiction between plant genetic engineering and sustainable agriculture. Clearly, genetically improved plants have a place in sustainable agriculture, and it matters little whether they have been improved by traditional or molecular techniques. However, the goals of sustainable agriculture are to eliminate agronomic practices that lead to environmental degradation and to minimize inputs.

Most genetic engineering research is carried out by companies, and their goals appear to be to increase the use of inputs, not to decrease them. This is obvious from the major research projects they are pursuing, such as herbicide-tolerant plants. The purpose is to assure the continued use of herbicides and to market herbicides and seeds as one package. Hybrid seeds are another major goal. When planting hybrid seeds, farmers must purchase seeds each year from plant-breeding companies, and cannot plant next year the seeds they saved from this year's harvest. In a market economy, companies have no incentive to make farming less dependent on purchased inputs.

Those genetic engineering projects that are aimed at producing chemicals for industry (for example, plastics, or starch that can be converted into ethanol) could allow developed countries to find alternate uses for some farmland, and render unnecessary the government subsidies that now result in surplus food production in developed countries. However, such genetically engineered plants are more likely to be grown under high-input conditions and to perpetuate the present agricultural system rather than to help convert it to a sustainable system.

The major reason to doubt that plant genetic engineering will benefit sustainable agriculture has to do with the research in both fields. Powerful economic and social forces drive research in biotechnology and genetic engineering, not only in companies but also at major universities and research institutions. Genetic engineering is generally regarded as a major scientific advance, and Western society holds such advances in high esteem. Relatively little research is being done in sustainable agriculture, and most of it is on-farm research. Researchers are often driven to solve problems that society perceives as being important: curing cancer, finding a clean new energy source, or eliminating air pollution. No such aura of esteem is currently attached to making agriculture more sustainable. "Solving the problem of hunger" is seen as a major societal objective that is worthwhile pursuing through research, but our technologically oriented society favors technological solutions over low-input solutions. Similarly, a technological breakthrough to produce cheap energy is seen as much more desirable than taking hundreds of small measures to save energy. As a result of this infatuation with technology, far more research funds are expended for biotechnology as opposed to sustainable agriculture.

Should society decide that sustainable agriculture really is important and eliminate the financial incentives that maintain present agricultural practices, then plant genetic engineering could indeed make a major contribution to a green agriculture.

10 | Farming in nature's image is a new way of looking at food production.

A different type of sustainable agriculture is envisaged by Judith Soule and Jon Piper in their book, *Farming in Nature's Image* (1992, Washington, DC: Island Press). They urge us to think of a farm as an ecosystem that should reflect the natural ecosystem that existed before the land was put to the plow and native vegetation eliminated. At one time, deciduous forests covered much of the eastern United States and western Europe. Such forests have not only large trees but also understories of shrubs and herbaceous plants. Furthermore, the tree cover is not continuous, but there are clearings in as much as 10% of the area. Understory plants prosper whenever clearings are created by fallen trees.

Native Americans and early European settlers practiced agriculture in these deciduous forests by clearing the land in small parcels (8–80 hectares) and farming them for a few years until the soils were exhausted of their nutrients. This type of shifting agriculture is still practiced in some tropical regions, and in these cases it takes about 8–10 years until a site is abandoned and the forest invades what was a field.

Shifting agriculture mimics on a somewhat larger scale what happens in the natural clearings in the forest, where growth of smaller plants is very vigorous until the canopy closes again and the shade from the big trees reduces productivity of the understory. But what kind of modern agricultural system would imitate this ecological succession? Watching the devastation caused by the practices of Alabama farmers that resulted in rapid hillside erosion, J. Russell Smith proposed 40 years ago that we should "farm in nature's image." For those eroding hillsides, he proposed and developed an agricultural system similar to the one that existed before European immigrants arrived. It consisted of an **overstory** of legume trees (honey locust) and an **understory** of Chinese bush clover that can be cut as hay or grazed directly.

Smith's system was both ecologically sustainable and cost effective. The clover provided a permanent cover, so there was no erosion, just as in nature. The use of two legumes eliminated the need for nitrogen fertilizers, and grazing ensured that the nutrients were recycled, again as in nature. Purchased inputs, including labor, were low, and the resulting productivity (5 tons per hectare of hay and 3 tons per hectare honey locust nuts) was high.

A more challenging task is to devise an agricultural system that mimics the U.S. prairie. A typical Kansas tallgrass prairie is dominated by C3 and C4 grasses, but also contains legumes and Compositae, growing very densely. The mix of species varies with climate: C4 grasses thrive in the hot summers, while cool season grasses complete their life cycles before the heat of the summer arrives. The plants have different root systems and differ in the way they obtain nutrients. Some have shallow, fibrous root systems, while others have deep taproots (Figure 16.13). Some depend completely on mycorrhizal fungi for phosphate use, while others do not.

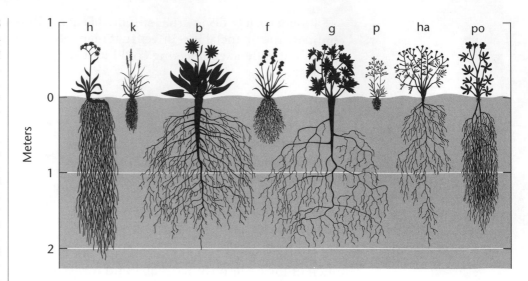

Figure 16.13

Schematic Bisect Showing Root and Stem Relations of Several Prairie Plants. *Shown are (h)* Hieracium scouleri *(hawkweed), (k)* Koeleria cristata *(Junegrass), (b)* Balsamorhiza sagittata *(balsam root), (f)* Festuca idahoensis *(fescue), (g)* Geranium viscosissimum *(geranium), (p)* Poa sandbergii *(bluegrass), (ha)* Haplopappus racemosus, *(po)* Potentilla gracilis *(cinquefoil).* Source: *Weaver (1919), in* The Ecological Relations of Roots *(Washington, DC: Carnegie Institution of Washington).*

The idea that **perennial polyculture** of grain crops could mimic this prairie was proposed by Wes Jackson in his book, *New Roots for Agriculture* (1985, Lincoln: University of Nebraska Press). Promising candidates for such a perennial, "domesticated prairie" include a composite (the Maximilian sunflower), a legume (the Illinois bundleflower), and several C3 and C4 grasses (Figures 16.14 and 16.15). Such a system would have great advantages over the present agriculture, which is based on annual plants:

1   Greater species diversity and genetic diversity within species would result in lower susceptibility to pests.

2   Perennials would obviate the need to prepare the field for planting every year, with reductions in erosion and energy costs.

3   The inclusion of a legume would reduce the need to add fertilizer to replace organic matter that is produced by the annual growth and decay of crop roots.

4   Continuous ground cover would reduce erosion and groundwater pollution.

5   The use of native species would ensure that water use would be related to rainfall patterns; irrigation needs would be reduced.

6   Since different species often send roots to different soil depths, more of the soil would be evenly exploited for nutrients.

Although the prospect is very attractive, at the moment, we are still far from establishing such a perennial agricultural system. It is likely that crop plants would have to be bred specifically for it, but most of the breeding efforts at present are directed to annuals. Because of their short life cycle, annual plants can be improved faster through traditional breeding. In addition, it is not clear how fertilizers would be handled in such a permanent system. If the harvest removes seeds with their abundant nutrients, these nutrients would have to be replaced in some way. A legume would help with nitrogen, but phosphorus, potassium, and sulfur would still have to be replenished. Even with these two potential problems, perennial polyculture deserves intensive investigation as a method to sustain productivity.

Figure 16.14

Plants of the Prairie.
The sunflower, a plant
native to America, could
become an important
part of a perennial eco-
system in which farming
is practiced "in nature's
image." Source: *Cour-*
tesy of David Monk.

11 | Sustainability in the Third World requires sensitivity to the social environment as well as to the biological and physical environments.

During the 1970s, the Lilongwe Land Development project was under-taken in Malawi to prevent erosion by controlling runoff from farmers' fields. Roads, soil breaks, and waterways were built at considerable cost, but the project did not succeed. Farmers had not been consulted and were not willing to commit labor to maintain these structures that they had not wanted in the first place. Because of the lack of maintenance, the structures ended up caus-ing erosion rather than preventing it.

Meanwhile, another project was undertaken in Malawi, this time to pre-vent an impending shortage of fuel wood. Instead of the government plant-ing trees and selling the wood at prices beyond the reach of the people who needed it, the farmers themselves were consulted. An initial idea was to have the farmers grow the trees on some of their land; but they said that the costs of taking land growing valuable corn and planting trees instead was prohibi-tive. Consultation with the people involved saved the Malawi government from another costly mistake. The solution was government-subsidized cost of trees.

The lessons learned from these and countless other examples have not been lost on planners in the Third World, as they consider the potential benefits of sustainable agriculture. Bonita Brindley, writing in the U.N. publication, *Ceres*, lists 10 principles of a people-oriented agricultural revolution:

Figure 16.15

Plants of the Prairie. (a) *Three perennials, from left to right, mammoth wildrye (a cool-season grass), Illinois bundleflower (a legume) and Eastern gamma grass (a warm season grass).* (b) *Close-up of mammoth wildrye (Leymus racemosus), a relative to wheat, rye and barley. These plants could be part of a perennial sustainable agriculture ecosystem. Source: Courtesy P. Kulakow and J. Shields, Salina, Kansas: The Land Institute.*

(a)

(b)

1 Consult with villagers, farmers, and all other participants.

2 Plan small-scale flexible projects.

3 Let the people benefiting from the project make the decisions.

4 Look for solutions that are widely applicable.

5 Provide education and training.

6 Keep external inputs to a minimum.

7 Build on what people are already doing well.

8 Assess the impacts of proposed changes.

9 Consider the balance between input and outcome.

10 Maintain and improve the participants' standard of living.

The Green Revolution was, and continues to be, a somewhat "top–down" approach to improving agricultural productivity. Scientists at research stations, such as the CGIAR institutes, develop new technologies for application in the field. Over the years they have come to realize that these technologies must be integrated into existing farming systems, and have been going out into the fields to try to understand the problems and to see how the solutions are working.

Although this may seem like an excellent model to most scientists, Paul Richards, after making an extensive study of farming in West Africa, argues in his book *Indigenous Agricultural Revolution* (1985, London: Unwin Hyman) that it is not sufficient to consult the people who are affected (as discussed earlier) but that in addition, farmers have to be involved directly in formulating the research agenda and in the research process. He maintains (p. 14) that "a thorough ecological understanding of the aim and methods of small-scale producers is necessary if inputs from scientific research and the development agencies are to complement and augment local trends and interest." The last 10–15 years have seen a renewed focus of interest on small landholder systems, and a reevaluation of the skills and knowledge of peasant farmers. It is generally true that practicing farmers have the most intimate and detailed knowledge of the farming system and its ecological requirements.

Agricultural researchers in Europe around the middle of the nineteenth century drew heavily on the experience of local farmers. This was certainly the case when John Lawes and Henry Gilbert, who set up the world's first agricultural experiment station outside London, first began to apply scientific principles to the study of crop productivity. Farmers everywhere are lifelong experimenters, whether they have a formal education and keep written records, or whether they are subsistence farmers who record their life's experiments in their heads. Local subsistence farmers are often the only experts on local ecological conditions. For example, in Veracruz, Mexico, over 200 wild plants and animals are used as food.

Unfortunately, the colonialist powers and their agricultural researchers labeled most indigenous cultivation as "primitive" and this stigma has not yet disappeared. In many quarters, "technology transfer" is still seen as the salvation of agriculture. The underlying assumption is that we do not need to spend too much time understanding the particulars of all these different agricultural systems, because new technology will solve the problem in any case.

There are two major problems with this approach. First, the large-scale adoption of technology-based agriculture can lead to a loss of knowledge of traditional farming methods, which were developed over many centuries and are adapted to the physical, biological, and social environments of the people. This knowledge may be useful in devising strategies for sustaining agriculture in the developing regions.

The environment in the foothills of the Andes Mountains in South America is a harsh one, with floods, droughts, and frosts. Yet, 3,000 years ago, the Incas devised a system of elevated fields, the remains of which can still be

(a)   (b)

Figure 16.16

Agroforestry and Alley Cropping. In agroforestry, tree crops and field crops are
intercropped, whereas alley cropping refers to a system in which rows of perennials
alternate with rows of annuals. The two systems are often combined. **(a)** A farmer in
Burundi (Africa) is weeding an agroforestry demonstration plot in which banana trees are
intercropped with sweet potatoes. **(b)** Alley cropping of leucania trees (shown on the right
and left) alternating with rows of mungbeans in tropical Africa. Source: (a) Courtesy of the
Food and Agriculture Organization, Rome, (b) Courtesy of the International Institute of
Tropical Agriculture. Photo by B. Fadare.

seen near Lake Titicaca, that circumvented these problems and produced high
yields of potatoes and quinoa, a grain. Recently, Peruvian farmers have begun
recreating this system of soil platforms surrounded by water-filled canals.
During drought periods, this water slowly moves up to the crop roots by
capillary action, and during floods, the ditches remove excess rain, preventing erosion. What is more, in the ditches plant remains and animal wastes
accumulate as a slurry rich in organic matter; this is removed periodically and
spread onto the crop soil to keep it fertile. As a result, current yields of potatoes and quinoa in these fields, cultivated by an ecologically sound yet
ancient method, are two to three times the average of other local fields.

The second danger in ignoring traditional agriculture is that these methods may be applicable for designing sustainable systems in the developed
countries. In other words, the rich and "sophisticated" can learn a lot from
the poor and "unsophisticated."

Intercropping is an excellent example where scientific research has confirmed the validity of what was once thought to be a primitive practice. Intercropping involves growing two or more crops in the same field at the same
time, and has been practiced for centuries in Africa. Because different crops
have roots that grow to different depths, intercropping results in better use of
available moisture and mineral nutrients. Some reports indicate an 80% yield
advantage when fast-maturing millet (85 days to harvest) is intercropped
with slow-maturing sorghum (150 days). Both are cereals, but they use the
soil's resources at different times because of their different maturation times.
Slow-maturing crops usually benefit from fast-maturing crops because the
root system of the fast-maturing crop dies in the soil after harvest, and when

Table 16.6	The effect of alley cropping with *Leucaena* on rain runoff, erosion, and corn yield		
	Rain Runoff	**Erosion**	**Corn Yield**
Plowed	232 mm	14.9 t/ha	3.6 t/ha
Leucaena at 4-m intervals	10 mm	0.2 t/ha	3.7 t/ha

Source: Data from R. Lal (1986), Soil surface management in the tropics for intensive land use and high and sustained production, *Advances in Soil Science* 51:1–108.

this organic matter decays, it contributes minerals for the growth of slow-maturing crops. This is especially true if the fast-maturing crop is a nitrogen-fixing legume.

Furthermore, intercropping tends to minimize weed problems; the two competing crops provide a more adequate ground cover and exclude weeds. The greater diversity of the agroecosystem also helps to keep down the explosive population increase of insect pests and the epidemic spread of plant diseases. Pests and diseases are present but remain at low levels. Intercropping is best suited to soils of low or medium fertility and is in this sense again an adaptation to the highly weathered soils found in many tropical areas. Another advantage of intercropping is that it minimizes the risk of crop failure, and many subsistence farmers prefer to minimize this risk, rather than maximize the potential crop output. Yield fluctuations are much smaller with intercropping than they are with single crops (monoculture). Intercropping is just one of the highly sophisticated agricultural adaptations to ecological conditions in specific areas of the tropics.

Alley cropping is an example of an agricultural system that mimics a natural ecosystem somewhat. It is a special type of agroforestry (Figure 16.16) in which crops are planted between rows of deep-rooted leguminous woody shrubs or trees. With this system, about 15% of the land is sacrificed to the shrubs that need to be pruned regularly to keep them from shading the crops. The prunings provide mulch as well as nutrients. This cropping system is well suited for the humid tropics, where it greatly reduces water runoff and soil erosion (Table 16.6).

12 On-farm research will help create sustainable livelihoods backed by sustainable farm productivity.

M. S. Swaminathan, the "father" of the Green Revolution in India and a former director-general of IRRI, has set up a new research institute in Madras in south India. His goal is not to duplicate what is already being done at the CGIAR institutes. According to Swaminathan, "The greening of agriculture requires the greening of both technology and public policy. Producing more food and agricultural commodities from less land, water and energy is a task that will call for the integration of the best in modern technology, with the ecological strengths of traditional farming practices." Thus, scientists like Swaminathan, who once championed the Green Revolution approach, realize that a new approach is needed in which agricultural development is guided by new principles.

Indeed, within the last ten years there have been important changes in the thinking of agricultural scientists and policy makers concerned with agricultural research in developing countries. It is clear that the type of research that led to the Green Revolution has benefited the resource-rich farmers, whereas resource-poor families have been left behind (see Chapter 11). It is increasingly recognized that who produces food and where it is produced may actually be as or more important than how much is produced. If resource-rich farmers produce food that subsistence farmers cannot purchase, there is no improvement in the food situation. Population projections show that many rural areas with fragile ecologies will have to continue supporting large populations. In the words of the editors of *Farmer First: Farmer Innovation and Agricultural Research,* "The priority has become not just sustainable agriculture, but sustainable livelihoods based on agriculture, not only for present populations but for hundreds of millions more people" (*Farmer First: Innovation and Agricultural Research.* Eds. R. Chambers, A. Pacey and L. A. Thrupp, 1989. Intermediate Technology Publications, London, U.K. p. xvii). Some 1.5 billion people are thought to live in areas in which traditional high-input agriculture or Green Revolution agriculture are not possible. These farm lands include a great variety of soil types, farming systems, climatic conditions, and terrains. The challenge for agricultural researchers is to improve and then sustain the productivity of the resource-poor subsistence farmers who cultivate these difficult regions.

Bringing this about will require an entirely new research approach in which researchers try to understand the local farming systems and ask the farmers to help them identify the productivity constraints. Small teams of researchers will have to meet frequently with individual farmers and groups of farmers to establish the research agenda. Since women are an important source of information about farming practices, the research teams that interview farmers and study local farming practices must also include a substantial number of women scientists. Local trials must be undertaken by the farmers themselves and they choose the varieties to be tested. Their choices are not always guided by maximal productivity of the varieties under consideration, but by the way the plant looks in the field, or how the crop can be expected to taste when it is cooked. For example, farmers in Colombia expressed a clear preference for small beans because small beans usually taste better. This type of research is known as "on-farm" or "farmer first" research and many farm experiments conducted around the world are summarized in the above-mentioned volume. To conduct research in this manner marks a major new direction in agricultural research. Agricultural scientists are slowly beginning to understand that subsistence farmers have a tremendous store of knowledge that must be tapped if productivity in these complex and fragile ecosystems is to be increased. On-farm research has shown again and again that practices which at first sight seemed irrational and wrong to outside professionals were in fact correct and rational. To be successful in such situations, scientists must develop a completely new attitude with respect to the contributions that subsistence farmers can make to the solution of the problems. In addition, new methodologies are needed, for the old methodologies (randomized plots, statistics) may not be sufficient or appropriate. Entirely new methodological questions crop up: for example, how to choose participants for on-farm research plots; how to conduct a cost-benefit analysis with costs and benefits that are not traded in the market place; how to evaluate the worth of the village commons; how to tap the knowledge that farmers have and that has accumulated over the centuries. It is hoped that on-farm-research will help raise productivity in those regions of the world which up to now have been bypassed by the research achievements of the 20th century.

Summary

The emergence of industrial agriculture, coupled with a push to put marginal lands to the plow as a result of financial incentives in developed countries and population pressures in Third World countries, is undermining the resource base of agriculture. Soil degradation, pesticide and fertilizer contamination, genetic erosion, and exhaustion of fossil water are a few examples of this erosion.

Food can be produced in a more sustainable way that protects the resource base. The techniques are known and available and used by some farmers. Although more research is needed, these techniques could find more widespread use given the proper incentives. Currently, the incentives (price supports, tax policies, and export subsidies) favor industrial agriculture. Society needs to replace them with incentives that favor sustainable agriculture. Because sustainable agriculture uses less inputs, this may cause some dislocation in the industrial sector and will be strongly resisted. In the Third World, much research on indigenous farming methods is needed, and the top–down approach to research needs to be replaced and/or supplemented by on-farm research.

Biotechnology and plant genetic engineering are not incompatible with sustainable agriculture. The economic forces presently at work make it unlikely that these scientific advances will help direct agriculture toward a more sustainable future. If society decides to reorder its priorities with respect to the type of agriculture that is best suited to our civilization, then products of technology and genetically engineered plants could make a contribution to a green agriculture.

Further Reading

Abelman, M. 1993. *From the Good Earth*. New York: Abrams.

Altieri, M. A. 1987. *Agroecology: The Scientific Basis of Alternative Agriculture*. Boulder, CO: Westview Press.

Altieri, M. A. 1992. Where the rhetoric of sustainability ends, agroecology begins. *Ceres* 134:33–39.

Bezdicek, D. F., ed. 1984. *Organic Farming*. Madison, WI: American Society of Agronomy.

Bunders, J. F. G., and J. W. E. Broerse, eds. 1991. *Appropriate Biotechnology for Small Scale Agriculture*. London: CAB International.

Chambers, R., A. Pacey, and L. A. Thrupp, eds. 1989. Farmer First: Farmer innovation and agricultural research. *Intermediate Technology Publications*, London, U.K.

Coway, G., and E. Barbier. 1990. *After the Green Revolution: Sustainable Agriculture for Development*. London: InBook.

Jackson, W. 1987. *Altars of Unhewn Stone: Science and the Earth*. San Francisco: North Point Press.

Morris, P., M. Bellinger, and A. Rosenfeld. 1992. *Trying to Take Root: Sustainable Agriculture in the US Heartland*. Public Voice Press.

Pearce, D. 1991. *Sustainable Development*. London: Earthscan.

Poincelot, P. R. 1986. *Sustainable Agriculture*. Westport, CT: AVI.

Reijntjes, C. 1992. *Farming for the Future*. New York: Macmillan.

Richards, P. 1985. *Indigenous Agricultural Revolution*. London: Unwin Hyman.

Soule, J. D., and J. K. Piper. 1992. *Farming in nature's image*. Island Press, Washington, D.C.

Swegle, W., ed. 1991. *Globalization of Agriculture*. Arlington, VA: Winrock.

United Nations, Food and Agriculture Organization. 1992. *Sustainable Development and the Environment*. Rome: FAO.

U.S. National Research Council. 1989. *Alternative Agriculture*. Washington, DC: NRC.

Wen, D., Y. Tang, X. Zheng, and Y. He. 1992. Sustainable and productive agricultural development in China. *Agriculture, Ecosystems Environment* 39:55–70.

Index

banana, 160
Bangladesh,
    agricultural development of, 54–56
    future food needs, 56
bean. *See* common bean
bean rust fungus, 372
berberine, 389
bio-active peptides, 75
biodegradable plastic, 74
biological, nitrogen fixation, 143
biological control, 364–369
biological nitrogen fixation, 211–213
    and the nitrogen cycle, 214
biological warfare, in soils, 232
biopesticides, 78
bioreactors, plant cells as, 396
biotechnology,
    comparison of medical and agricultural, 71
    impact on Third World, 397–399
    origin of, 58
birth rate, 2, 14
Borlaug, Norman, 298, 312
bran, 141
Brazil,
    development of, 20
    unequal development, 448
bread, 141
    wheat, 136
breeding,
    goals for crop improvement, 295
    methods of, 280–292
    use of polyploids, 292
bruchids, 349–350
Bt protein, 80
Bt toxin, 378

C3 photosynthesis, 168, 176
C4 photosynthesis, and water requirements, 189
C4 plants, 174, 176
cabbage family, as a food source, 155
calcium, role in bone formation, 100
callus, 65, 67
calorie, definition, 90
canavanine, 347, 350
cancer,
    anticancer drugs, 387
    of the colon, 88
    treatment by taxol, 392
cancer-causing chemicals, naturally occuring in plants, 345
capillaries, in the soil, 197
carbohydrates,
    content in US diet, 85
    as a source of energy for humans, 84
carbohydrates, simple, definition of, 84, 86
carbohydrates, complex, definition of, 84, 86
carbon dioxide, 82
    and global warming, 39
    concentrations in the air, 40
carbon fixation, net, 179
carcinogens,
    in our food, 117
    natural, 117
cardenolides, 346, 351
cardiac glycoside, 385

cardiac glycosides. *See* cardenolides
carpels, 132
carryover stocks, 324–325
Carson, Rachel, 356
cassava, 154
    as a source of animal feed, 28
    improvement of, 312
    yields on eroded soil, 216
    and cyanogenesis, 119
cassava mealybug, biological control, 366
cations, in the soil, 201
cell, structure of, 124–127
cell culture, 61, 65
    for production of metabolites, 387–389
cell differentiation, 151–152
cell division, 151–152
cell elongation, 151–152
cell suspension culture, 65
cell wall, 87, 125
cellulose,
    in the diet, 87
    structure of, 86, 87
centers of origin, of crops, 265
cereal, breakfast, 141, 142
cereal grains, 136–141
cereals, and nitrogen fertilizers, 210
CGIAR institutes, 308–312
chemical defenses,
    and insect adaptations, 332
    constitutive, 344
    inducible, 344
    of plants, 118
chilling injury, 416
China, site of rice domestication, 272
chitinase,
    for plant defense, 353
    for resistance to fungi, 382
chlorinated hydrocarbons, 361
chlorophyll, 165, 166
chloroplast, 125, 165, 167
cholesterol, 88
    in the blood, 91
chromosome mapping, 293–295
chromosome movement, 293
chromosomes, 246
CIAT, 309
CIMMYT, 308, 372
citrus canker, 340
climate, 30–36
climate change, 36
climate cycles, 36
climate modification, 37
clonal propagation, 61
clonal propagation of oil palms, 62
clonal selection, of plant cells, 388
clone, for metabolite production, 388
cloning of cells, 61
CO2 concentration, and photosynthesis, 176
CO2 enriched air, 177
CO2 in the soil, 197
coconut, 159
cocultivation, of bacteria with plant cells, 404
codon, 252
coenzymes, 98
coffee, hillside cultivation in Colombia, 438